MARINE FUNGI

Biotechnology and Bioprospects

Series

Progress in Mycological Research

- Fungi from Different Environments (2009)
- Systematics and Evolution of fungi (2011)
- Fungi from Different Substrates (2015)
- Fungi: Applications and Management Strategies (2016)
- Advances in Macrofungi: Diversity, Ecology and Biotechnology (2019)
- Advances in Macrofungi: Industrial Avenues and Prospects (2021)
- Advances in Macrofungi: Pharmaceuticals and Cosmeceuticals (2021)
- Fungal Biotechnology: Prospects and Avenues (2022)
- Bioprospects of Macrofungi: Recent Developments (2023)
- Ecology of Macrofungi: An Overview (2024)
- Advances in Cordyceps Research: Prospects and Avenues (2025)
- The Role of Entomopathogenic Fungi in Agriculture (2025)
- Truffles: Diversity, Ecology and Biotechnology (2025)
- Marine Fungi: Biodiversity, Genomics and Applications (2027)
- Marine Fungi: Biotechnology and Bioprospects (2027)

Series: Progress in Mycological Research

MARINE FUNGI

Biotechnology and Bioprospects

Editors

Kandikere R. Sridhar

Central Research Laboratory, NITTE (Deemed to be University)
KS Hegde Medical Academy (KSHEMA), Mangalore, India

Sunil K. Deshmukh

Scientific Advisor
Greenvention Biotech Pvt. Ltd. Pune, India

CRC Press
Taylor & Francis Group
Boca Raton London New York

CRC Press is an imprint of the
Taylor & Francis Group, an **informa** business
A SCIENCE PUBLISHERS BOOK

First edition published 2027
by CRC Press
2385 NW Executive Center Drive, Suite 320, Boca Raton FL 33431

and by CRC Press
4 Park Square, Milton Park, Abingdon, Oxon, OX14 4RN

CRC Press is an imprint of Taylor & Francis Group, LLC

Library of Congress Cataloging-in-Publication Data (applied for)

ISBN: 978-1-041-34915-0 (hbk)
ISBN: 978-1-041-34916-7 (pbk)
ISBN: 978-1-003-79510-0 (ebk)

DOI: 10.1201/9781003795100

Typeset in Times New Roman
by Prime Publishing Services

Preface

Mycological research has advanced from basics to applications as an important area of contemporary biological research. It has widened its arena into several biotechnological innovations in agriculture, biofuels, biomaterials, biotherapeutics, the environment, human health, industries, and medicine. Fungi are the major source of bioproducts like antibiotics, beverages, biofuels, biopesticides, drugs, enzymes, natural products, nutraceuticals, metabolites, organic acids, prebiotics, pigments, polysaccharides, and proteins (Niego et al., 2023; Hyde et al., 2024). The biodiversity of the fungal kingdom spanned from tiny unicellular chytrids to large multicellular mushrooms with varied morphological, ecological, and biotechnological dimensions, having different lifestyle strategies. However, despite the enormous diversity, exploration has achieved only up to 5% of the global conventional estimate (2.2–3.8 million) (Hawksworth and Lucking, 2017).

The marine habitats spread across about 67% of the biosphere and consist of up to 50% of the biodiversity, providing ample opportunities to explore fungal bioactive substances with diverse applications in health, nutrition and industries (Puri et al., 2023). Marine and marine-influenced ecosystems are the major biodiversity hotspots for various fungi, which serve as a rich source for basic study and biotechnological or bioprospecting applications. Obligate marine fungi live and sporulate exclusively in the marine environment; the facultative marine fungi have input into marine habitats from freshwater and terrestrial environments and are capable of being active and sporulating in the marine environment (Kohlmeyer and Kohlmeyer, 1979). Marine fungi have been isolated and studied from a wide range of substrates (algae, dead animal remains, driftwood, mangrove leaves, pneumatophores, roots, seagrasses, seaweeds, sediments, seedlings, invertebrates, and wood) (Jones et al., 2019). The current global estimate of fungi in marine habitats is 120,000 to 140,000 species, of which about 1,900 species have been described (Cunliffe, 2023). At present, 2,254 fungi are considered marine (882 genera, 12 phyla, and 36 classes) (marinefungi.org: 7th December, 2025). The first marine fungus was described from the shores of North Africa two centuries ago (Amend et al., 2025). However, the occurrence, lifestyle, functions and applications of marine fungi remain underexplored in marine habitats.

The book *Marine Fungi: Biotechnology and Bioprospects* emphasizes several applied aspects (abatement of pollution, applications in agriculture, bioactive metabolites, bioremediation, biosurfactants, health-promoting bioactive compounds, human nutrition, hydrophobins, and pigments). This volume comprises 16 chapters contributed by 58 authors from 12 countries divided into four subdivisions: (1) Agricultural and Nutraceutical Applications; (2) Industrial Applications; (3) Bioremediation; (4) Future Directions. The first subdivision, in three chapters, emphasizes the application of marine fungi in agriculture, aquaculture and human nutrition. In five chapters, the second subdivision discusses industrial-scale production of pigments and their applications; production and applications of hydrophobins and biosurfactants; lichens and their applications; and the significance of fungi from the Arctic and Antarctic. The third subdivision deals with advances in bioremediation for oil spills, the remediation of hydrocarbons, heavy metals, and textile dye reduction in five chapters. The last subdivision, as future directions, discusses the significance of deepsea fungi, geomycology, and synthetic biology in three chapters.

We anticipate that the expectations of the authors of chapters and editors in following different aspects of marine mycological research and applications will be valuable for readers with a wide interest in applications in marine mycology (agriculture, applied biology, biochemistry, bioremediation, biotechnology, food technology, health, medicine, nutrition, metabolites, and pollution control).

The editors appreciate the contributors and reviewers who delivered the chapters and evaluation, respectively. We are indebted to the CRC Press, Boca Raton, USA, for its keen interest in bringing out this book.

Mangalore, Karnataka, India — Kandikere R. Sridhar
Pune, Maharashtra, India — Sunil K. Deshmukh

References

Amend, A.S., Gunde-Cimerman, N., Coelho, M.A., Durkin, C.A. et al. (2025). Ten pressing questions (and answers) about marine fungi and opportunities for collaborations in the ocean sciences. Oceanography, 38(4): 40–49. https://doi.org/10.5670/oceanog.2025.e402.

Cunliffe, M. (2023). Who are the marine fungi? Environ. Microbiol., 25(1): 131–134. https://doi.org/10.1111/1462-2920.16240.

Hawksworth, D.L. and Lücking, R. (2017). Fungal diversity revisited: 2.2 to 3.8 million species. Microbiology Spectrum, 5: FUNK-0052-2016. https://doi.org/10.1128/microbiolspec.funk-0052-2016.

Hyde, K.D., Baldrin, P., Chen, Y., Chethana, K.W.T., De Hong, S. et al. (2024). Current trends, limitations, and future research in the fungi? Fungal Divers., 125: 1–71. https://doi.org/10.1007/s13225-023-00532-5.

Jones, E.B.G., Pang, K.L., Abdel-Wahab, M.A., Scholz, B., Hyde, K.D. et al. (2019). An online resource for marine fungi. Fungal Diver., 96(1): 347–433. https://doi.org/10.1007/s13225-019-00426-5.

Kohlmeyer, J. and Kohlmeyer, E. (1979). Marine Mycology: The Higher Fungi. Elsevier-Academic Press, New York and London, 705 pp.

Neigo, A.G.T., Lambert, C., Mortimer, P., Thongklang, N., Rapior, S. et al. (2023). The contribution of fungi to the global economy. Fungal Divers., 121: 95–137. https://doi.org/10.1007/s13225-023-00520-9.

Puri, S., Mahar, R. and Goswami, G. (2023). Bioactive compounds from aquatic ecosystem. pp. 387–407. *In*: Soni, R., Suyal, D.C., Morales-Oyervides, L. and Sungh Chauhan, J. (eds.). Current Status of Fresh Water Microbiology. Springer Nature Singapore. https://doi.org/10.1007/978-981-99-5018-8_18.

Contents

List of Contributors

Barreto, Débora L.C.
Departamento de Microbiologia, Universidade Federal de Minas Gerais, Belo Horizonte, MG, Brazil.

Barthelemy-Ake, S.
Instituto de Biología, Universidad Nacional Autónoma de México, Mexico City, Mexico; Posgrado en Ciencias Biológicas, Universidad Nacional Autónoma de México, Mexico City, Mexico.

Benegas, Gabriela Sosa
Laboratório de Biotecnologia Ambiental, Universidade Federal da Integração Latino-Americana (UNILA) (UNILA), Foz do Iguaçu, Paraná, Brazil.
Laboratorio de Agua y Sedimentos (LASED), Itaipu Binacional, Hernandarias, Paraguay.

Bernardo, Monelly da Silva
Universidade Federal de Alagoas (UFAL), Campus Arapiraca, Alagoas, Brazil.

Bora, Bhagyashree
Biological Sciences Division, CSIR-North East Institute of Science and Technology (CSIR-NEIST), Branch Laboratory-Itanagar, Naharlagun 791 110, Arunachal Pradesh, India.

Cantrell, Charles L.
Natural Products Utilization Research Unit, Agricultural Research Service, United States Department of Agriculture, Oxford, MS, USA.

Chikkaputtaiah, Channakeshavaiah
Academy of Scientific and Innovative Research (AcSIR), Ghaziabad 201 002, India; Biological Sciences and Technology Division, CSIR-North East Institute of Science and Technology (CSIR-NEIST), Jorhat 785 006, Assam, India.

Dabolkar, Sujata
Department of Botany, Government College of Arts, Science and Commerce, Quepem, Goa, India.

De Menezes, Graciéle Cunha Alves
Departamento de Microbiologia, Instituto de Ciências Biológicas, Universidade Federal de Minas Gerais, Belo Horizonte, Minas Gerais, Brazil.

De Oliveira, Miguel Lucas Ribeiro
Departamento de Microbiologia, Instituto de Ciências Biológicas, Universidade Federal de Minas Gerais, Belo Horizonte, Minas Gerais, Brazil.

De Souza, Sara Lemes
Departamento de Microbiologia, Universidade Federal de Minas Gerais, Belo Horizonte, MG, Brazil.

Dhara, Shubhajit
Division of Sustainable Agriculture, The Energy and Resources Institute, Gurugram, Haryana 122001, India.

Duarte, Alysson Wagner Fernandes
Universidade Federal de Alagoas (UFAL), Campus Arapiraca, Alagoas, Brazil.

Duke, Stephen O.
National Center for Natural Products Research, School of Pharmacy, University of Mississippi, Oxford, MS, USA.

Elfaky, Mahmoud A.
Department of Natural Products and Alternative Medicine, Faculty of Pharmacy, King Abdulaziz University, Jeddah 21589, Saudi Arabia; Centre for Artificial Intelligence in Precision Medicines, King Abdulaziz University, Jeddah 21589, Saudi Arabia.

Gershom, A.P. Roshan
Department of Biotechnology, Prathyusha Engineering College, Poonamallee Tiruvallur Road, Aranvoyal Kuppam, Tiruvallur 602025, Tamil Nadu, India.

Goel, Mayurika
Division of Sustainable Agriculture, The Energy and Resources Institute, Gurugram, Haryana 122001, India.

Gonçalves, Vívian N.
Departamento de Microbiologia, Universidade Federal de Minas Gerais, Belo Horizonte, MG, Brazil.

Guo, Sheng-Yu
Institute of Marine Biology and Center of Excellence for the Oceans, National Taiwan Ocean University, 2 Pei-Ning Road, Keelung 202301, Taiwan.

Hamzah, Tuan Noraida Tuan
Civil Engineering Department, Faculty of Engineering and Built Environment, Universiti Kebangsaan Malaysia, UKM Bangi 43600 Bangi, Selangor, Malaysia; Environmental Management Centre, Institute of Climate Change, Universiti Kebangsaan Malaysia, UKM Bangi 43600 Bangi, Selangor, Malaysia.

Hanafiah, Zarimah Mohd
Chemical Engineering Department, Faculty of Engineering, Universiti Malaya, Kuala Lumpur 50603, Malaysia.

Ibrahim, Sabrin Ragab Mohamed
Preparatory Year Program, Department of Chemistry, Batterjee Medical College, Jeddah 21442, Saudi Arabia; Department of Pharmacognosy, Faculty of Pharmacy, Assiut University, Assiut 71526, Egypt.

Iñiguez-Moreno, Maricarmen
Department of Physics and Mathematics, Universidad de Monterrey, Avenida Morones Prieto 4500, San Pedro Garza García 66238, NL, Mexico.

Jayalal, Udeni
Department of Natural Resources, Faculty of Applied Sciences, Sabaragamuwa University of Sri Lanka, Belihuloya 70140, Sri Lanka.

Karthik, Loganathan
ARQGENE, Vellore, Tamil Nadu, India.

Kaur, Mehak
Department of Botany, University of Delhi, Delhi, 110007, India; Division of Sustainable Agriculture, The Energy and Resources Institute, Gurugram, Haryana 122001, India.

Kong, Lam
Institute of Marine Biology and Center of Excellence for the Oceans, National Taiwan Ocean University, 2 Pei-Ning Road, Keelung 202301, Taiwan.

Kumar, T. Naveen
Department of Biotechnology, Vel Tech High Tech Dr. Rangarajan Dr. Sakunthala Engineering College, Avadi, Chennai 600062, Tamil Nadu, India.

Landeta-Salgado, Catalina
Departamento de Ciencia de los Alimentos y Tecnología Química, Facultad de Ciencias Químicas y Farmacéuticas, Universidad de Chile, St. Dr. Carlos Lorca Tobar 964, Independencia, Santiago 8380000, RM, Chile; Center for Biotechnology and Bioengineering (CeBiB), Department of Chemical Engineering, Biotechnology, and Materials, Faculty of Physical and Mathematical Sciences, University of Chile, Santiago, Beauchef 851 - 8370456 Chile.

Lienqueo, María Elena
Center for Biotechnology and Bioengineering (CeBiB), Department of Chemical Engineering, Biotechnology, and Materials, Faculty of Physical and Mathematical Sciences, University of Chile, Santiago, Beauchef 851 - 8370456 Chile.

Lopez-A.
Instituto de Biología, Universidad Nacional Autónoma de México, Mexico City, Mexico; Posgrado en Ciencias Biológicas, Universidad Nacional Autónoma de México, Mexico City, Mexico.

Manohar, Catherine Sumathi
Biological Oceanography Division, CSIR-National Institute of Oceanography, Dona Paula, Goa, India; Academy of Scientific and Innovative Research, Ghaziabad, India.

Mohamed, Gamal A.
Department of Natural Products and Alternative Medicine, Faculty of Pharmacy, King Abdulaziz University, Jeddah 21589, Saudi Arabia.

Mohamed, Hagar M.
Department of Medical Laboratory Analysis, College of Medical and Health Sciences, Liwa University, Abu Dhabi 41009, United Arab of Emirates; Department of Applied Medical Chemistry, Medical Research Institute, Alexandria University, Alexandria, Egypt.

Mohtar, Wan Hanna Melini Wan
Civil Engineering Department, Faculty of Engineering and Built Environment, Universiti Kebangsaan Malaysia, UKM Bangi 43600 Bangi, Selangor, Malaysia.

Olguin, J.
Posgrado en Ciencias del Mar y Limnología, Universidad Nacional Autónoma de México, Mexico City, Mexico; Instituto de Biología, Universidad Nacional Autónoma de México, Mexico City, Mexico.

Pan, Yen-Ju
Department of Aquaculture, National Taiwan Ocean University, 2 Pei-Ning Road, Keelung 202301, Taiwan.

Pang, Ka-Lai
Institute of Marine Biology and Center of Excellence for the Oceans, National Taiwan Ocean University, 2 Pei-Ning Road, Keelung 202301, Taiwan.

Passarini, Michel Rodrigo Zambrano
Laboratório de Biotecnologia Ambiental, Universidade Federal da Integração Latino-Americana (UNILA) (UNILA), Foz do Iguaçu, Paraná, Brazil.

Phatthongkleang, Thossaporn
Institute of Marine Biology and Center of Excellence for the Oceans, National Taiwan Ocean University, 2 Pei-Ning Road, Keelung 202301, Taiwan.

Rabelo, Natana Gontijo
Departamento de Microbiologia, Universidade Federal de Minas Gerais, Belo Horizonte, MG, Brazil.

Ramesh, A. Sai
Department of Bio-Engineering, Vels Institute of Science, Technology and Advanced Studies, Chennai 600117, Tamil Nadu, India.

Ranasinghe, Malshadi
Department of Natural Resources, Faculty of Applied Sciences, Sabaragamuwa University of Sri Lanka, Belihuloya 70140, Sri Lanka.

Redkar, Neha
Biological Oceanography Division, CSIR-National Institute of Oceanography, Dona Paula, Goa, India.

Ribeiro, Victor Pena
Natural Products Utilization Research Unit, Agricultural Research Service, United States Department of Agriculture, Oxford, MS, USA.

Rodríguez-Ramírez, Ivonne
Escuela de Biología, Universidad de Costa Rica, San Pedro de Montes de Oca, San José, Costa Rica.

Romero-Hernandez, L.
Instituto de Biología, Universidad Nacional Autónoma de México, México City, México.

Rosa, Luiz Henrique
Departamento de Microbiologia, Universidade Federal de Minas Gerais, Belo Horizonte, MG, Brazil.

Siqueira, Natana Gontijo
Departamento de Microbiologia, Universidade Federal de Minas Gerais, Belo Horizonte, MG, Brazil.

Takam, Akash
Biological Sciences Division, CSIR-North East Institute of Science and Technology (CSIR-NEIST), Branch Laboratory-Itanagar, Naharlagun 791 110, Arunachal Pradesh, India.

Tezocotitla, G.
Instituto de Biología, Universidad Nacional Autónoma de México, Mexico City, Mexico; Facultad de Ciencias, Universidad Autónoma del Estado de México, Toluca de Lerdo, Estado de México, Mexico.

Valdes, J.
Facultad de Estudios Superiores Zaragoza, Universidad Nacional Autónoma de México, Mexico City, Mexico.

Vallejo-Arróliga, Mónica
Escuela de Biología, Universidad de Costa Rica, San Pedro de Montes de Oca, San José, Costa Rica.

Velez, Patricia
Instituto de Biología, Universidad Nacional Autónoma de México, México City, México.

Velmurugan, Natarajan
Biological Sciences Division, CSIR-North East Institute of Science and Technology (CSIR-NEIST), Branch Laboratory-Itanagar, Naharlagun 791 110, Arunachal Pradesh, India; Academy of Scientific and Innovative Research (AcSIR), Ghaziabad 201 002, India.

Wan-Mohtar, Wan Abd Al Qadr Imad
Functional Omics and Bioprocess Development Laboratory, Institute of Biological Sciences, Faculty of Science, Universiti Malaya, Kuala Lumpur 50603, Malaysia.

Wong-Villarreal, A.
División Agroalimentaria, Universidad Tecnológica de la Selva, Ocosingo, Chiapas, México.

Yáñez-Ocampo, G.
Facultad de Ciencias, Universidad Autónoma del Estado de México, Toluca de Lerdo, Estado de México, México.

Agricultural and Nutritional Applications

1

Marine Fungi as a Source of Bioactive Compounds for Use in Agriculture

Vívian N. Gonçalves,[1] *Débora L.C. Barreto,*[1] *Natana Gontijo Rabelo,*[1] *Nicole Carraretto Siqueira,*[1] *Sara Lemes de Souza,*[1] *Victor Pena Ribeiro,*[2] *Charles L. Cantrell,*[2] *Stephen O. Duke*[3] **and** *Luiz Henrique Rosa*[1,*]

Marine environments are responsible for over 50% of the planet's biodiversity and harbor a community composed primarily of microorganisms. In these substrates, fungi are known to withstand low temperatures, lack of light, high salinity pressure, and low nutrient availability. Due to the harsh conditions, unique metabolic pathways can be selected, and the investigation of secondary metabolites by marine fungi has been the subject of numerous studies. One study shows that in 2021–2022, approximately 1,400 secondary metabolites obtained from marine fungi were detected. However, few of these compounds are studied as prototypes for new pharmaceuticals to be used as pesticides. Therefore, this chapter aimed to present the work carried

[1] Departamento de Microbiologia, Universidade Federal de Minas Gerais, Belo Horizonte, MG, Brazil.

[2] Natural Products Utilization Research Unit, Agricultural Research Service, United States Department of Agriculture, Oxford, MS, USA.

[3] National Center for Natural Products Research, School of Pharmacy, University of Mississippi, Oxford, MS, USA.

* Corresponding author: lhrosa@icb.ufmg.br

DOI: 10.1201/9781003795100-2

out with the marine fungal community that produces active compounds for agricultural applications. Secondary metabolites with phytotoxic activity include benzopyranones and aromatic derivatives, polycyclic perylenes, isoquinolones, diterpenes, sesquiterpenes and lactone derivatives, pyrones, macrocyclic polyketides, azaphilones, phenols, solanapyrones, xanthones, terphenyls, and anthraquinones, which were able to inhibit plant growth. Species of the genera *Alternaria*, *Aspergillus*, *Eurotium*, *Trichoderma*, and *Verruculina* were investigated for the inhibition of phytopathogenic bacteria and fungi, demonstrating potential use in crop disease management. Compounds produced by *Aspergillus luchuensis*, *A. sydowii*, and *T. longibrachiatum* exhibited toxic effects on plant-parasitic nematodes, such as *Meloidogyne incognita*. Metabolites identified as sesquiterpenes, naphthoic acids, aromatic alkaloids, and chloramphenicol derivatives exhibited regulatory or repellent activity against insects and mites. This research demonstrates the importance of studying marine fungal communities as a source of metabolites that can serve as prototypes for new pesticides. However, the difficulty in accessing different marine substrates and cultivating a unique fungal community, combined with the fact that many studies fail to investigate the compounds' mechanisms of action or their toxicological implications, hinders the transfer of knowledge to innovative products that can be used as pesticides in agriculture.

1. Introduction

The marine environment occupies approximately 67% of the biosphere and contains 50% of the planet's biodiversity, offering researchers promising sources of bioactive substances with diverse applications in the pharmaceutical and nutraceutical industries (Puri et al., 2023). Among the organisms in marine ecosystems, microorganisms exhibit adaptability to extreme conditions, such as low temperatures, absence of light, high salt concentrations, high pressures, and nutritional depletion (Puri et al., 2023). To adapt to these polyextreme environments, these microorganisms have been characterized as potential producers of unique and valuable bioactive metabolites with applications in the pharmaceutical and food industries, agriculture, and medicine (Namikoshi, 2006; Giddings and Newman, 2015; Karthikeyan et al., 2022; Puri et al., 2023).

Currently, several unique metabolites have been identified from the metabolism of fungi originating from marine environments. According to MarinLit, 42,843 natural products (secondary metabolites) have been described from these sources, and from 2021–2022, approximately 1,417 were derived from marine sources (Carroll et al., 2024). Few of these compounds have been evaluated for potential use as pesticides. Very few of those that have been evaluated for this type of biological activity have been adequately evaluated for activity against all classes of agricultural pests (plant pathogens, arthropods, and weeds). This review summarizes the papers that have reported effects of marine fungi and their metabolites on agricultural pests or organisms that are used for pesticide screens. Table 1 shows the overall summary.

Table 1. Bioactive compounds produced by marine fungi with reported activity against agricultural pests.

Fungi	**Origin**	**Compound**	**Activity (Target Species)**	**References**
Herbicidal Activity				
Alternaria sp.	Marine habitats (seawater, sediments, animals, and algae)	Benzopyranone, (+)-(2S,3R,4aR)-altenuene (**1**) (+)-Isoaltenuene (**2**) Altertoxin I (**3**) Stemphyperylenol (**4**) and Alterperylenol (**5**)	*Amaranthus retroflexus* and *Lactuca sativa*	Huang et al. (2018)
Fusarium equiseti	Unidentified sea plant	Fusarisetin C (**6**) Fusarisetin D (**7**) Fusarisetin B (**8**) Fusarisetin A (**9**) Equisetin (**10**) and Epi-equisetin (**11**)	*Amaranthus retroflexus* and *Lactuca sativa*	Zhao et al. (2019)
Trichoderma harzianum	Unidentified coral	Harzianelactones A (**12**) Harzianelactones B (**13**) Harzianones A (**14**) Harzianones B (**15**) Harzianones C (**16**) Harzianones D (**17**) and Harziane (**18**)	*Amaranthus retroflexus* and *Lactuca sativa*	Zhao et al. (2019)
Penicillium vancouverense	*Monodonta neritoides* (gastric flora)	Vancouverone B (**19**) α-pyrone, 10,11-dihydroxy-citreoviridin (**20**) Neocitreoviridinol (**21**) and PF1140 (**22**)	*Trifolium repens*, *Raphanus sativus* var. *longipinnatus* and *Brassica rapa* var. *perviridis*	Yokoyama et al. (2020)
Aspergillus alabamensis	*Enhalus acoroides*	Asperalacid D (**23**)	*Triticum aestivum*	Hu et al. (2022)
Penicillium sclerotiorum	Unidentified plant	Sclerotiorin B (**24**) Ochlephilone (**25**) and Isochromophilone I (**26**)	*Avena fatua, Lolium perenne, Amaranthus retroflexus* and *Abutilon theophrasti*	Wang et al. (2022)

Table 1 contd. ...

...Table 1 contd.

Fungi	**Origin**	**Compound**	**Activity (Target Species)**	**References**
Aspergillus terreus	Unidentified mangrove leaves	(+)-Terrein (**27**)	*Amaranthus retroflexus, Amaranthus hybridus, Celosia argentea, Portulaca oleracea* and *Trifolium repens*	Wang et al. (2023)
Alternaria iridiaustralis	*Suaeda glauca*	16-methoxy solanapyrone B (**28**) Betaenone I (**29**) Alternanone B (**30**) and Tenuazonic acid (**31**)	*Echinochloa crus-galli, Digitaria sanguinalis, Portulaca oleracea* and *Descurainia sophia*	Fan et al. (2023)
Penicillium palitans	Marine sediment	(−)-Penienone (**32**) and (−)-Palitantin (**33**)	*Lactuca sativa, Agrostis stolonifera* and *Lemna paucicostata*	Barreto et al. (2024)
Aspergillus sp.	Coral reef	Terphenyllin (**34**) 3-hydroxyterphenyllin (**35**) and Candidusin C (**36**)	*Triticum aestivum, Oryza sativa* and *Amaranthus retroflexus*	Abulaizi et al. (2024)
Aspergillus sp.	*Ianthella reticulata* (sponge)	Unguinol (**37**)	*Digitaria ciliaris*	Motti et al. (2007)
Antimicrobial Activity				
Trichoderma longibrachiatum	Roots of *Suaeda glauca*	Norsesquiterpene (**38**) Sesquiterpene (**39**)	*Colletotrichum lagenarium, C. fragariae, Botrytus cinérea, Fusarium oxysporum* f. sp. *cucumerinum* and *F. oxysporum* f. sp. *lycopersici*	Du et al. (2020)
Eurotium cristatum	Brown seaweed *Sargassum thunbergii*	Rubrumazine B (**40**)	*Magnaporthe grisea*	Du et al. (2017)
Aspergillus sp. D40	Red seaweed *Grateloupia filicina*	Penicillic acid (**41**)	*Acidovorax avenae, Clavibacter michiganensis, Erythema carafavora, Xanthomonas campestris,* and *X. citri*	Huang et al. (2020)

Pleosporales sp.	Marine sediment	Pleosporalone B (**42**) Pleosporalone C (**43**) and Pleosporalone D (**44**)	*Alternaria brassicicola, Botryosphaeria dothidea,* and *Fusarium oxysporum*	Cao et al. (2019)
Verruculina enalia	Driftwood	Cercosporamide (**45**)	*Colletotrichum acutatum* and *Magnaporthe grisea*	Bunyapaiboonsri et al. (2020)
Alternaria sp. (P8)	Unidentified plant	Stemphyperylenol (**46**) Alterperylenol (**47**)	*Pestallozzia theae, Alternaria brassicicola*, and *Clavibacter michiganensis*	Zhao et al. (2018)
Nematicidal Activity				
Lachnum papyraceum	Not reported	Lachnumon (**48**) Lachnumol A (**49**) and (+)-mycorrizina A (**50**) (+)-chloromycorrizina A (**51**) and (+)-dechloromycorrizina A (**52**)	*Caenorhabditis elegans*	Stadler and Anke (1993)
Trichoderma longibrachiatum	Root of *Suaeda glauca*	10(*E*)-cyclonerotriol (**53**) Homodestcardin (**54**) Trichomide B (**55**) and Homodestruxin B (**56**)	*Meloidogyne incógnita*	Du et al. (2020)
Aspergillus luchuensis	Seaweed	3-furoic acid (**57**)	*Meloidogyne incógnita*	Wang et al. (2023, 2025)
Aspergillus sydowii LW09	Not reported	Penicibisabolane (**58**) Expansol G (**59**) and (S)-sidonic acid (**60**)	*Meloidogyne incógnita*	Yang et al. (2023)
Eurotium cristatum EN-220	Seaweed	Variecolorin I (**61**) Alkaloid E-7 (**62**) and Didehydroechinulin (**63**)	*Panagrellus redivivus*	Du et al. (2017)
Purpureocillium lilacinum FDZ8Y1	Mariana Trench	Leucinostatin A (**64**) and Leucinostatin B (**65**)	*Caenorhabditis elegans*	Li et al. (2024)

Table 1 contd. ...

...Table 1 contd.

Fungi	**Origin**	**Compound**	**Activity (Target Species)**	**References**
Insecticidal Activity				
Neosartorya pseudofischeri	*Acanthaster planci*	Deacetylsesquiterpene (**66**) and 5-formyl-6-hydroxy-8-isopropyl 2-naphthoic acid (**67**)	*Spodoptera frugiperda*	Lan et al. (2016)
Aspergillus fumigatus JRJ111048	*Acrostichum speciosum*	Aspergide (**68**)	*Spodoptera litura*	Guo et al. (2017)
Acremonium vitellinum	Seaweed	Cloranfenicol (**69**) Oxazolidin-2-one cloranfenicol (**70**) and Acetamida cloranfenicol (**71**)	*Helicoverpa armigera*	Chen et al. (2018)

2. Bioactive Compounds with Phytotoxic Activity

In 2022, the total area of agricultural land worldwide was 4,781 million hectares, representing one-third of the global land area (FAO, 2024). Within this agricultural land, farmers face several challenges, including the presence of weeds, which significantly hinder global agricultural productivity (Curran, 2016; Kaur et al., 2019). Weeds compete with agricultural crops, leading to production reductions ranging from 20% to 50%, depending on the intensity and duration of infestation in the field (Curran, 2016; Kaur et al., 2019). Furthermore, weeds are a major problem in land used for farm animals (Martins et al., 2022). Herbicides are used to eliminate or control these unwanted plants in agriculture. These products can be applied to the soil before or immediately after planting (pre-emergent) or directly to plant leaves (post-emergent) (Casida and Bryant, 2017).

Herbicides can be classified as non-selective or selective (Duke et al., 2018). Non-selective herbicides such as glyphosate kill almost all plant species. They are also sometimes used to kill and dehydrate crops in order to facilitate harvesting (Duke, 2021). In contrast, most herbicides are classified as selective, targeting certain weeds without significantly harming the crops with which they are used (Duke et al., 2018). Another categorization of pesticides is based on their movement within the plant: (i) systemic herbicides move (translocate) internally after being applied to a specific plant part, while (ii) non-systemic or contact herbicides act only at the application site, without being translocated after being absorbed. Another way to classify herbicides is by their mechanism of action, which corresponds to the molecular target of the compound within the plant cell. This generally involves the inhibition of an essential enzyme or protein, triggering a series of metabolic reactions that alter the growth and development of the plant, which can lead to its death (Dayan, 2019). Herbicides that share the same mechanism of action exhibit similar activities and produce comparable injury symptoms. Currently, there are 261 commercial herbicides distributed across 23 mechanisms of action (HRAC, 2025).

The declining efficacy of commercial herbicides due to the evolution and spread of herbicide resistance, along with growing environmental and human health concerns related to herbicides, underscores the urgent need for new herbicides. This challenge is compounded by the fact that only one herbicide with a new mechanism of action has been introduced to the market in the past 40 years, making this search even more pressing (Duke and Dayan, 2022). In this context, secondary metabolites produced by fungi, which exhibit significant structural diversity and broad bioactivity, emerge as a promising approach for the discovery of new herbicidal compounds (Perotti et al., 2020). Numerous metabolites of terrestrial fungi, especially plant pathogens, are highly phytotoxic via mechanisms of action not utilized by commercial herbicides (Dayan and Duke, 2014). In recent decades, marine fungi have attracted increasing interest due to their remarkable diversity and ability to produce secondary metabolites with diverse biological activities (Blunt et al., 2018). Among these activities, phytotoxicity stands out as promising for the development of sustainable herbicides.

From the fungus *Penicillium vancouverense*, derived from the gastric microbiota of the marine snail *Monodonta neritoides*, collected off the coast of Shonai, Japan, three new metabolites were identified: two alkaloids, vancouverone A and vancouverone B (**19**), and an α-pyrone, 10,11-dihydroxy-citreoviridin (**20**) (Yokoyama et al., 2020). Additionally, two known compounds, neocitreoviridinol (**21**) and PF1140 (**22**), were identified. These compounds produced by *P. vancouverense* were evaluated for their phytotoxicity on the germination of lettuce seeds, with 2,4-dichlorophenoxyacetic acid (2,4-D) used as a positive control. In this assay, compounds **19–22** exhibited inhibitory effects. Compound **19** demonstrated the strongest inhibitory effect on root growth, leading to additional phytotoxicity tests on other plant species (*Trifolium repens* L., *Raphanus sativus* var. *longipinnatus*, and *Brassica rapa* var. *perviridis*). Vancouverone B (**19**) was phytotoxic to all plant species tested at concentrations ranging from 10 to 200 μg mL^{-1}. However, at lower concentrations, its effects on these species were less pronounced, with inhibition mainly on root growth. Although the phytotoxic activity of compound **19** was lower than that of available commercial phytotoxic compounds, these findings strongly suggest that vancouverone B may serve as a lead or prototype compound for developing more potent analogues with potential as herbicides (Yokoyama et al., 2020).

The fungus *Aspergillus alabamensis* was isolated from necrotic leaves of the seagrass *Enhalus acoroides* (Linnaeus f.) Royle, collected in the subtidal zone of Dongzhai Port, China, resulting in the discovery of five new carotane-type sesquiterpenoids, one new tricyclic sesquiterpenoid, and two known compounds (Hu et al., 2022). The effects of these compounds on root and shoot elongation of wheat (*Triticum aestivum* L.) were determined using a concentration range of 62.5– 2,000 ppm. Asperal acid D (**23**) strongly inhibited both shoot and root growth. At a concentration of 500 ppm, it inhibited shoot length by approximately 90% and root length by 80%, surpassing the herbicide terbutryn (87.0% and 44.8%, respectively). At 250 ppm, the inhibitory effects were 83.5% in the shoot and 72.3% in the root, compared to terbutryn (85.2% and 68.9%, respectively). Even at a concentration of 125 ppm, the compound was active, inhibiting shoot and root growth by 73.3% and 55.7%, respectively.

Seven pairs of azaphilone E/Z isomers, including eight novel compounds, were isolated from the mangrove fungus *Penicillium sclerotiorum* in coastal marine habitats of Haikou, China (Wang et al., 2022). Seed germination and growth bioassays were conducted with representative weeds from agricultural areas, including two grass species (*Avena fatua* L. and *Lolium perenne* L.) and two broadleaf species (redroot pigweed and *Abutilon theophrasti* Medik.). Sclerothiorin B (**24**), ochlefilone (**25**), and isochromophyllone I (**26**) was significantly phytotoxic against the growth of the radicle and shoot of redroot pigweed, with EC_{50} values ranging 235–321 μM, compared to the positive control (glufosinate ammonium), which had EC_{50} values of 555 μM for the radicle and 656 μM for the shoot. Ochlefilone (**25**) and isochromophyllone I (**26**) also inhibited the growth of velvetleaf (*Abutilon theophrasti*), with EC_{50} values ranging 769–1,202 μM.

From the endophytic fungus *Aspergillus terreus*, isolated from mangrove plant leaves in coastal habitats of the South China Sea, two novel terrein derivatives were identified, along with two known analogues and three previously reported butenolides (Wang et al., 2023). The phytotoxicity of the *A. terreus* compounds was evaluated using the foliar spray method on redroot pigweed and smooth pigweed (*Amaranthus hybridus*). Only (+)-terrein (**27**) showed significant phytotoxicity against redroot pigweed, causing leaf necrosis and wilting at both early and late stages. Compound **25** significantly reduced plant height and root length at 250 ppm. Furthermore, it demonstrated considerable phytotoxicity against redroot pigweed, leading to wilting and leaf chlorosis. At 500 ppm, compound **25** reduced root length, and at 1,000 ppm, it caused visible mass loss. Due to its strong phytotoxicity, compound **27** was tested against three other broadleaf weed species: *Celosia argentea* L., *Portulaca oleracea* L., and *Trifolium repens* L. The results showed clear phytotoxicity at 1,000 ppm, with leaf necrosis and wilting, suggesting its potential as a lead compound for natural herbicides (Wang et al., 2023).

Investigation of bioactive metabolites from the marine fungus *Alternaria iridiaustralis*, isolated from the root of *Suaeda glauca* (Bunge) Bunge, collected in the intertidal zone of the Yellow River Delta, China, led to the identification of nine compounds, five of which were novel (Fan et al., 2023). All compounds from *A. iridiaustralis* were tested for herbicidal activity against the weed species barnyardgrass (*Echinochloa crus-galli* (L.) P. Beauv). The compounds 16-methoxy solanapyrone B (**28**), betaenone I (**29**), alternanone B (**30**), and tenuazonic acid (**31**) demonstrated herbicidal potential against the root growth of *E. crus-galli*. Notably, compound **31** inhibited growth greater than 90% at concentrations of 20 and 40 ppm, outperforming the commercial herbicide acetochlor. Due to its significant herbicidal activity, tenuazonic acid was tested against other weed species such as *Digitaria sanguinalis* L., *Portulaca oleracea* L., and *Descurainia sophia* (L.) Webb ex Prantl, which are common in northern China and cause substantial economic losses. Tenuazonic acid had a broad herbicidal spectrum, with significant activity against *D. sanguinalis* (comparable to acetochlor) and moderate activity against *P. oleracea* and *D. sophia*. Tenuazonic acid has been shown to be phytotoxic before and has been studied as a potential herbicide in cotton (Zhou et al., 2019). Alternanone B (**30**) at 20 and 40 ppm exhibited moderate inhibition rates of 60.3% and 72.6%, respectively (Fan et al., 2023).

Metabolites derived from *Penicillium palitans*, isolated from deep-sea sediments in maritime Antarctica, were evaluated for their phytotoxic properties (Barreto et al., 2024). Chromatographic fractionation of the crude extract in dichloromethane resulted in the production of two structurally related compounds: (–)-penienone (**32**) and (–)-palitanthine (**33**). The phytotoxic effects of the compounds from *P. palitans* were tested on lettuce and *Agrostis stolonifera* seeds, and both compounds demonstrated selective activity against the monocotyledon *A. stolonifera*. At a concentration of 1,000 ppm, (–)-penienone completely inhibited seed germination, while (–)-palitanthine showed only moderate effects after 7 days of exposure. In assays with *Lemna paucicostata*, (–)-penienone showed an IC_{50} of 57 μM, significantly reducing growth, while (–)-palitanthine showed considerably lower activity. These findings highlight *P. palitans* as a promising source of phytotoxic metabolites, reinforcing the potential of extremophilic fungi, such as Antarctic *Penicillium* strains, in the discovery of new bioactive molecules with applications in pesticide development (Barreto et al., 2024).

Aspergillus sp. was obtained from coral collected in the South China Sea, near the Hainan Province of China. A detailed chemical investigation of the strain extract led to the identification of four novel secondary metabolites, in addition to four known compounds (Abulaizi et al., 2024). A phytotoxicity assay revealed that the *Aspergillus* sp. compounds terphenylin (**34**), 3-hydroxyterphenylin (**35**), and candidusin C (**36**) showed significant phytotoxic activity against wheat and rice (*Oryza sativa*) seed germination in a dose-dependent manner but had no effect on redroot pigweed. Thus, compounds **34**–**36** may exhibit specific phytotoxicity against monocotyledonous species, providing potential weed control with dicotyledonous crops. According to Abulaizi et al. (2024), the selective toxicity of compounds **34**–**36** suggests that they can be integrated into fertilizers to suppress the germination of monocotyledonous weeds in soil.

Another approach to mining marine fungi for potential herbicides is to bioassay extracts against a desired molecular target. Motti et al. (2007) screened 2,245 extracts of 449 marine fungi against the C_4 plant enzyme pyruvate phosphate dikinase (PPDK), a target that would theoretically be selective for grass weeds. Bioassay-guided isolation of unguinol (**37**) (also known as yasimin) from one of the extracts led to the discovery that it is a PPDK inhibitor. Unguinol was then shown to be toxic to a C_4, but not a C_3 plant. To date, we have seen no further development of this compound as an herbicide.

3. Bioactive Compounds with Antimicrobial Activity

Several studies evaluating crude extracts of marine fungi have demonstrated that the marine fungal communities hold promise in antimicrobial activity research. Among the various effects of marine fungal metabolites, activity against phytopathogens has been extensively studied (Li et al., 2021). According to the World Food Program (2025), 343 million people in 75 countries are currently facing acute hunger worldwide. To increase the yield of world crops, it is necessary to overcome the problem of phytopathogens, as they are responsible for 20–40% of global crop losses (Ayaz et al., 2023). The antimicrobials currently used have several negative effects on both the environment and humans, and many microorganisms have acquired resistance to them (Alengebawy et al., 2021).

Fungi of the genera *Penicillium* and *Aspergillus* are known to produce metabolites with antifungal activity, such as penicillic acid (Kearney et al., 1965; Li et al., 2022; Hu et al., 2005; Abad et al., 2023; Fan et al., 2024). Penicillic acid has proven effective against Gram-positive and Gram-negative bacteria. However, it is extremely toxic to rats and rabbits and is also cytotoxic, hepatotoxic, carcinogenic, and cardiotoxic. The effect of penicillic acid becomes synergistic with ochratoxin A, a mycotoxin commonly produced by *Aspergillus* species, and is toxic to bovine macrophage cell lines, increasing degeneration of nearby tubules in the kidney (Frisvad, 2018).

The fungal species *Trichoderma longibrachiatum* was isolated from the root of *Suaeda glauca* in the study by Du et al. (2020) and was able to produce two compounds, a rare norsesquiterpene (**38**) and a sesquiterpene (**39**) (Fig. 2) with antimicrobial activities. These compounds were antifungal to *Colletotrichum lagenarium*, *C. fragariae*, *Botrytis cinerea* (carbendazim-resistant), *Fusarium oxysporum* f. sp. *cucumerinum*, and *F. oxysporum* f. sp. *lycopersici*. The activity of these molecules was better than that of the fungicide carbendazim commonly used for infections caused by these species.

Du et al. (2017) isolated the fungus *Eurotium cristatum* from the algal species *Sargassum thunbergii*, which produced rubrumazine B in liquid fermentation (**40**). This molecule demonstrated moderate activity against *Magnaporthe grisea*, with an MIC of 64 ppm. The fungus *Aspergillus* sp. D40, isolated from the red algae *Grateloupia filicina*, produced 11 molecules, among which penicillic acid (**41**) showed antibacterial activity against the bacteria *Acidovorax avenae*, *Clavibacter michiganensis*, *Erwinia carotovora*, *Xanthomonas campestris*, *X. citri*, and *Ralstonia solanacearum*. Against the bacteria *A. avenae*, penicillic acid demonstrated a MIC of 51.3 ppm, slightly lower than that of streptomycin sulfate (56.9 ppm).

Pleosporales sp. CF09-1, obtained from marine sediment from the Bohai sea (Cao et al., 2016), produced pleosporalones B, C, and D (**42**–**44**), all of which exhibited MIC values lower than ketoconazole against the phytopathogenic fungi *Alternaria brassicicola*, *Botryosphaeria dothidea*, and *F. oxysporum* (Cao et al., 2019). Pleosporalone B demonstrated an MIC value of 1.6 μg mL^{-1} for two of the species tested, while ketoconazole presented values of 3.1 and 6.3 ppm.

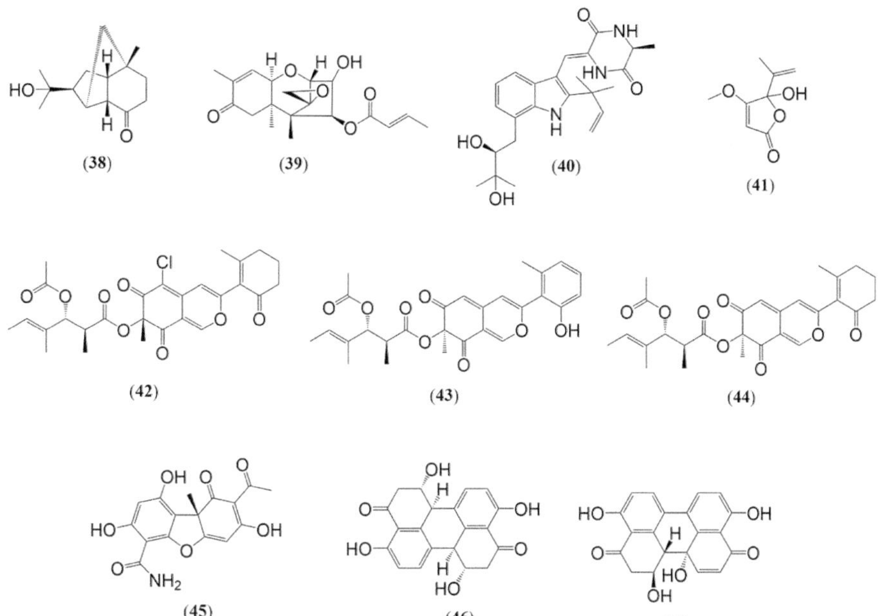

Fig. 2. Bioactive compounds with antimicrobial activity from marine fungi.

Bunyapaiboonsri et al. (2020) characterized nine compounds from the extract of the fungus *Verruculina enalia*, an obligate marine fungus. Of these substances, only cercosporamide (**45**) showed activity against *Colletotrichum acutatum* and *Magnaporthe grisea*. Cercosporamide (**45**) had IC_{50} values for these two fungi of 6.25 and 25 ppm, respectively.

Bioprospecting for marine fungi revealed that taxa such as *Fusarium equiseti*, *Alternaria* sp., and *Penicillium* sp. are capable of synthesizing metabolites with high antimicrobial potential (Zhao et al., 2018). The authors isolated about 141 fungi from marine plants and selected 31 for identification and evaluation of activity against phytopathogenic bacteria and fungi. Extracts of *F. equiseti*, *P. oxalicum*, and *P. chrysogenum* exhibited strong inhibitory activity against the bacterium *Pseudomonas syringae* and the fungus *Alternaria avenae*. Isolates of *Penicillium* sp. demonstrated the best activity against bacterial phytopathogens. Additionally, the 31 extracts showed selective inhibitory activity against *P. syringae*, *A. avenae*, *Erwinia carotovora*, and *Clavibacter michiganensis*. The *Alternaria* sp. extract was the only one capable of inhibiting the growth of all species, demonstrating a broad spectrum of action. Based on these results, this *Alternaria* sp. was selected, and four compounds were characterized, with stemphyperylenol (**46**) and alterperylenol (**47**) showing antimicrobial activity. Compound **46** exhibited potent antifungal activity against *Pestallozzia theae* and *Alternaria brassicicola*, with MIC values comparable to the commercial fungicide carbendazim (7.81 and 125 ppm, respectively). Compound **47** demonstrated remarkable antibacterial activity against *Clavibacter michiganensis*, with an MIC of 1.95 ppm, being twice as effective as streptomycin (MIC of 3.90 ppm). Furthermore, studies on the mechanism of action indicated that compound **47** causes hyperpolarization of the bacterial cell membrane without compromising its integrity, suggesting a form of non-destructive inhibition. These results highlight the compounds as potential candidates for developing new antimicrobial agents of marine origin.

Purić (2018) demonstrated that extracts of fungi isolated from Antarctic marine sediments showed activity against phytopathogens of agricultural interest. Extracts from the genera *Pseudogymnoascus*, *Penicillium*, *Paraconiothyrium*, *Cadophora*, and *Toxicocladosporium* were tested against *Xanthomonas citri*, *Xanthomonas euvesicatoria*, and *Xanthomonas axonopodis*. The extract from the genus *Pseudogymnoascus* was able to inhibit the growth of nearly 100% of *X. citri*, *X. euvesicatoria*, and *X. axonopodis*.

4. Bioactive Compounds with Nematicidal Activities

The annual loss caused to grain crops worldwide by plant parasitic nematodes (PPN) is estimated at more than US$100 billion (Burns et al., 2023). These PPNs comprise approximately 4,100 species of nematodes that feed on vascular, cortical, and pericycle tissues in plant roots, reducing nutrient absorption and resulting in stunted growth (Phani et al., 2021). PPNs develop a close pathogen-host relationship with the plants they parasitize, secreting substances that reprogram host cells to favor the parasite. This process opens space in the host roots for fungi and bacteria, potentially causing secondary infections in plant tissues, and also acts as viral vectors (Berg et al., 2009; Jones et al., 2013; Phani et al., 2021).

However, due to their microscopic size and symptoms that are barely visible on the shoots of infected plants, there is relatively little investment in controlling PPNs, leading to significant agricultural losses that could be avoided (Phani et al., 2021). The lower priority of research on controlling these nematodes is reflected in the lack of strategies and mechanisms to combat these parasites effectively, sustainably, and without harming the host plant (Phani et al., 2021).

The use of synthetic nematicidal substances has been a solution in this context and has contributed to the control of PPNs for decades. However, synthetic nematicides have raised concerns about their safety for humans and possible environmental damage (Burns et al., 2023). Bioactive nematicidal compounds of fungal origin could be a safer alternative (Wang et al., 2025).

Marine ecosystems harbor a wide variety of fungal species with the capacity to synthesize unique metabolites, including those with nematicidal potential (Wang et al., 2025). One of the pioneering works in identifying bioactive compounds with nematicidal activity from marine fungi was conducted by Stadler and Anke (1993), who identified five compounds (Fig. 3), lachnumon (**48**), lachnumol A (**49**), (+)-mycorrizin A (**50**), (+)-chloromycorrizin A (**51**), and (+)-dechloromycorrizin A (**52**) produced by the marine fungus *Lachnum papyraceum*. These compounds were evaluated for their nematicidal activity against *Caenorhabditis elegans*, and all showed activity at some of the concentrations tested. Notably, mycorrhizin A demonstrated the strongest nematicidal potential (Stadler and Anke, 1993).

Trichoderma longibrachiatum, isolated from seaweed, was studied for its biocidal action against plant pathogens (Du et al., 2020). This fungus produced a series of metabolites that were initially tested for their antifungal action and nematicidal action against the root-knot nematode (*Meloidogyne incognita*), a plant parasite that causes significant damage to agriculture. The nematicidal substances produced by *T. longibrachiatum* were isolated and identified as 10(E)-cyclonerotriol (**53**), homodestcardin (**54**), trichomide B (**55**), and homodestruxin B (**56**), all of which demonstrated the ability to kill juvenile *M. incognita*.

Another fungus isolated from seaweed that demonstrated potential nematicidal action against *M. incognita* was *Aspergillus luchuensis* (Wang et al., 2023; 2025). The secondary metabolite synthesized by *A. luchuensis* was identified as 3-furoic acid (**57**), which showed high stability in aqueous solution and a mortality rate of 50% at concentrations of 34.85–39.09 ppm after 24 h, in addition to causing mortality rates of 77% and 83% at concentrations of 50 and 75 ppm, respectively. The application of 3-furoic acid had detrimental effects on *M. incognita* by reducing motility, shortening lifespan, causing changes in morphology and embryonic development, generating the accumulation of reactive oxygen species, and impairing the nematode's chemostatic ability to recognize and infect cucumber cells.

The marine fungus *Aspergillus sydowii* LW09 synthesizes new sulfoxide-containing sesquiterpenoids in nine different analogous forms (Yang et al., 2023). Among these compounds, three, i.e., penicibisabolane (**58**), expansol G (**59**), and (S)-sidonic acid (**60**) - exhibited nematicidal activity against *M. incognita*, with a

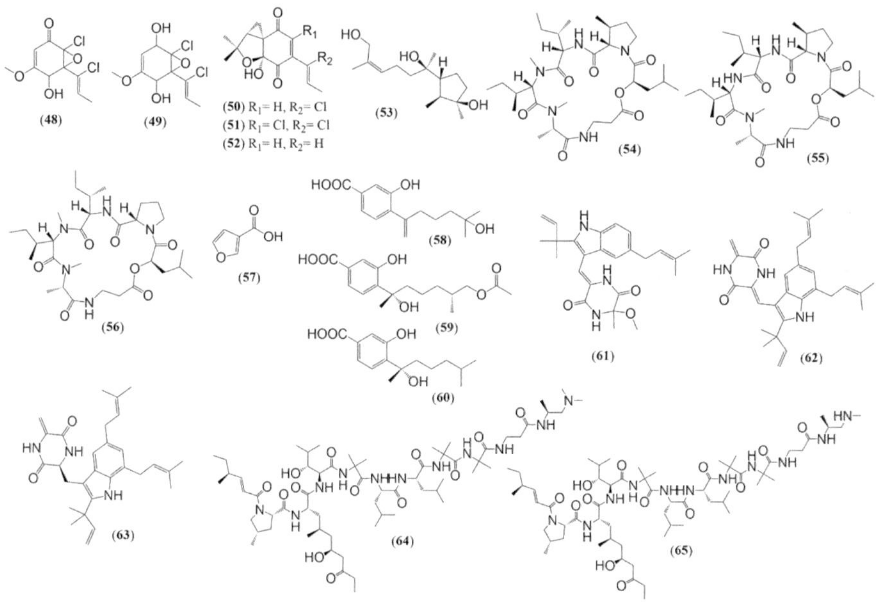

Fig. 3. Bioactive compounds with nematicidal activity from marine fungi.

lethality of almost 80% of juveniles at a concentration of 192 ppm. Additionally, they had a paralytic effect on these nematodes at lower concentrations, hindering their parasitic activity (Yang et al., 2023).

The fungus *Eurotium cristatum* EN-220, isolated from seaweed, produced 13 indole-diketopiperazine-derived compounds, including variecolorin I (**61**), alkaloid E-7 (**62**), and didehydroechinulin (**63**), which demonstrated weak lethality against the nematode *Panagrellus redivivus* and strong lethality against the brine shrimp (*Artemia* spp.) (Du et al., 2017). *Purpureocillium lilacinum* FDZ8Y1, isolated from deep sediments of the Mariana Trench, produced metabolites with 94.7% lethality against juvenile *C. elegans* (Li et al., 2024). The nematicide assay demonstrated that *P. lilacinum* hyphae exhibited growth and adhesion activity directed toward *C. elegans* eggs, as well as the synthesis and release of chitinase, an enzyme that digests the shells of these eggs (Li et al., 2024). Chromatography analyses of the compounds synthesized by *P. lilacinum* FDZ8Y1 indicated the production of leucinostatin A (**64**) and B (**65**), which are typically associated with predation mechanisms in fungi.

Some studies have evaluated the nematicidal activity of extracts produced by marine fungi. Meng et al. (2012) evaluated 52 extracts from marine fungi against the nematode *Bursaphelenchus xylophilus*. Among the fungi evaluated, *Acremonium* sp. BH0531 caused a lethality rate of 93% against *B. xylophilus*. The authors also assessed the optimal conditions for synthesizing nematicidal metabolites by *Acremonium* sp. BH0531 and demonstrated that neutral pH, mild temperature (24–26°C), and a culture medium containing 600 mL of seawater per liter produced the best conditions for obtaining compounds with nematicidal activity from this marine fungus. Ogaki et al. (2019) tested the nematicidal activity of 31 fungi isolated from marine sediment from islands in Antarctica against *C. elegans*. The extracts from all the fungi studied demonstrated nematicidal potential, causing the death of all the nematodes tested within 10 days. The bioactive fungi were identified as belonging to the genera *Acremonium*, *Penicillium*, and *Pseudogymnoascus*.

5. Bioactive Compounds with Activity Against Insect Agricultural Pests

Arthropods (primarily insects, ticks, and mites) are major pests in crop, poultry, and livestock production. Insecticides often have acaricidal (anti-mite and anti-tick) activity, as these pests are all arthropods with similar molecular targets.

Mites and ticks transmit several diseases in humans and animals and controlling them often relies on the repeated use of synthetic acaricides/insecticides. Natural products from marine organisms may have unique structures, reducing the chances of resistance in different mite and tick species and offering low side effects, making them advantageous compared to commercially available compounds (Benelli et al., 2017; Liao et al., 2023). In this context, some marine fungi can produce secondary metabolites with acaricidal activity (Ebani and Mancianti, 2021).

According to Chandler et al. (2010), acaripathogenic fungi are classified as: (i) mite-specific pathogens, (ii) non-specific fungal species, and (iii) fungal species that are rarely found as pathogens and are not normally studied for biological control purposes:

(i) The best-known genus in this group is *Hirsutella*, which includes species recognized as important regulators of phytopathogenic mites. Several species of this genus have been reported to infect mites. Another example is entomopathogenic fungi belonging to the genera *Neozygites* and *Conidiobolus* and the order *Entomophthorales*.

(ii) This group is characterized by infecting mites as well as insects. The main genera are *Beauveria*, *Metarhizium*, *Paecilomyces*, and *Verticillium*. The acaripathogenic species reported in the literature include *M. anisopliae*, *B. bassiana*, *B. brongniartii*, *V. lecanii*, *P. eriophyes*, *P. farinosus*, *P. fumosoroseus*, and *P. terricola*.

(iii) The third group consists of fungi occasionally isolated from carcasses of ticks and mites, such as *Aspergillus fumigatus*, *Penicillium insectivorum*, and *Trichothecium roseum*.

However, none of these are marine fungi. The only published report to date evaluating the production of bioactive substances with acaricidal activity from marine fungi is the study on destruxins. These are depsipeptides produced by *Beauveria felina*, capable of controlling the tick *Rhipicephalus microplus* (Morais-Urano et al., 2012). Further studies should be made of insecticidal compounds from marine fungi (see below) on mites and ticks to expand their potential utility.

The significant problems caused by insects in agriculture highlight the increasing importance of researching bioactive metabolites produced by various biological sources, including marine fungi. These compounds may offer an alternative to synthetic insecticides in agriculture and vector control, particularly given the rise in insecticide resistance and the environmental impacts of conventional insecticides. Marine fungi, adapted to extreme environments, produce various secondary metabolites, including those with insecticidal effects against pests such as *Spodoptera litura*, *Helicoverpa armigera*, and *Myzus persicae* (Blunt et al., 2018; Shin, 2020). Thus, using these marine microorganisms represents an innovative and sustainable strategy for developing bioinsecticides with potentially lower environmental toxicity and the potential for use in integrated pest management systems.

The crude extract of the marine fungus *Neosartorya pseudofischeri*, isolated from the starfish *Acanthaster planci*, significantly inhibited the growth of Sf9 cells of the caterpillar of the fall armyworm (*Spodoptera frugiperda*), a major insect pest that damages corn, sorghum, rice, sugarcane, cotton, and many other crops (Lan et al., 2016). *Neosartorya pseudofischeri* produced five new compounds and 11 known compounds. Twelve of the 15 identified compounds exhibited insecticidal activity. The new substances (Fig. 4), deacetylsesquiterpene (**66**) and 5-formyl-6-hydroxy-8-isopropyl 2-naphthoic acid (**67**), demonstrated a kill rate of over 90% against Sf9 cells derived from *S. frugiperda* at the concentration of 50 mg mL^{-1}. These results suggest that the metabolites of *N. pseudofischeri* can be further explored for developing natural insecticides.

Guo et al. (2017) investigated the metabolites with insecticidal activity of the marine endophytic fungus *Aspergillus fumigatus* JRJ111048, isolated from the mangrove plant *Acrostichum speciosum*, endemic to Hainan Island, China. Various bioactive compounds, including fumigaclavin A, fumitremorgin C, and verruculogen, were identified and isolated from the *A. fumigatus* culture, demonstrating significant insecticidal activity against tobacco cutworm (*Spodoptera litura*) larvae. The tested compounds were incorporated into the artificial diet of the larvae at a concentration of 20 ppm. A compound identified as aspergide (**68**) demonstrated significant insecticidal activity at 1,000 ppm, causing larval mortality of 40–60% on the 7th and 14th days after exposure, respectively. In comparison, the positive control with azadirachtin resulted in 86.6–100% mortality on days 7 and 14, respectively. Because azadirachtin has more than double the molecular weight of aspergide, the results would have been more similar if the bioassay had been done at the same molarity. These results suggest that although the efficacy of aspergide is inferior to the positive control under the conditions tested, it is still promising and can be used in future studies (Guo et al., 2017).

Chen et al. (2018) evaluated the insecticidal activity of three compounds derived from chloramphenicol: chloramphenicol (**69**), oxazolidin-2-one chloramphenicol (**70**), and acetamide chloramphenicol (**71**). These compounds were isolated from the fungus *Acremonium vitellinum*, obtained from red seaweed, and tested against the larvae of cotton bollworm (*Helicoverpa armigera*). Compound **70** exhibited the highest toxicity (LC_{50} = 560 ppm) and affected caterpillar defense enzymes, such as glutathione S-transferase (GST) and catalase (CAT), resulting in increased oxidative stress, which hindered the insect's ability to metabolize the compound. In addition, **70** promoted an increase in acetylcholinesterase (AChE) activity,

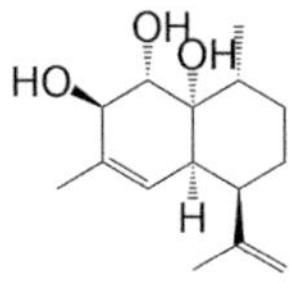

(66)

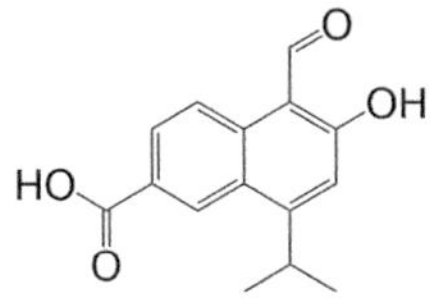

(67)

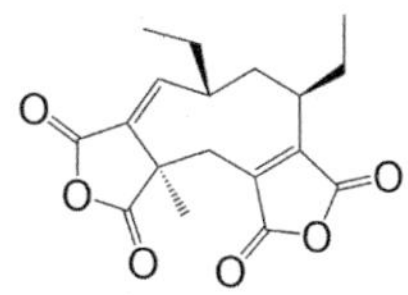

(68)

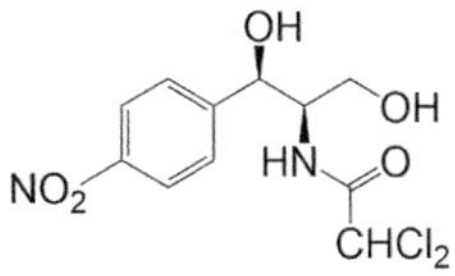

(69)

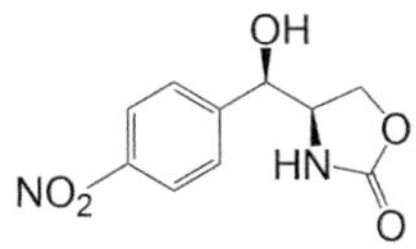

(70)

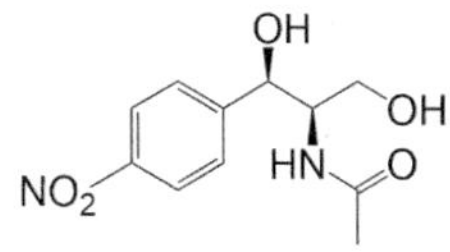

(71)

Fig. 4. Bioactive compounds with insecticidal activity from marine fungi.

indicating a possible neurological action in the larvae. Compounds **69** and **71** were less toxic, with LC_{50} values of 930 and 910 ppm, respectively, while the commercial insecticide used (matrine) had an LC_{50} of 240 ppm. The authors also performed transcriptomic analysis that revealed changes in the expression of genes involved in xenobiotic metabolism and cellular respiration, suggesting that compound **70** may be a potential sustainable bioinsecticide for the control of *H. armigera*.

A comprehensive study evaluated the activity of 181 marine fungal strains isolated from macroalgae and sediments in the intertidal zones of Qingdao, China, through assays targeting microcrustaceans, green peach aphids (*Myzus persicae*), and plant pathogens (Wang et al., 2022). Of the fungi evaluated, 85 strains (47%) produced extracts with high toxicity (> 70% mortality) against the microcrustacean *Artemia salina* (a brine shrimp) after 24 h. Among these, extracts from 14 fungal strains showed acute toxicity, with more than 75% mortality in just 4 h. Notably, extracts from seven strains, including *Aspergillus niger* (AS13), *Trichoderma scalesiae* (AS65), *Fusarium verticillioides* (AS99), *Emericella nidulans* (AS121), *Geotrichum* sp. (AS113), *Fusarium avenaceum* (AS133), and *Penicillium simplicissimum* (AS129), resulted in 100% mortality. In a second screen with *M. persicae*, 34 crude extracts at 10,000 ppm caused mortality greater than 70% after 48 h. These findings highlight the potential of marine fungi as sources of natural insecticidal compounds.

6. Conclusions and Outlook

The importance of expanding knowledge about marine fungi, their secondary metabolites, and their potential activities for use in agriculture is clear. These fungi, adapted to extreme environments, represent a valuable, virtually unexplored source of bioactive and effective compounds for potential use in agriculture. A better understanding of their mechanisms of secondary metabolite production may not only open new frontiers for biological pest control but also offer sustainable and ecological alternatives to traditional pesticides, creating a greener agricultural economy. Given the rich marine biodiversity and the wide variety of fungi yet to be discovered, investing in this area of research represents a strategic opportunity for advancing science and developing more sustainable solutions, with positive impacts on health and the environment.

However, the search for bioactive compounds from marine fungi with agricultural applications faces significant challenges, such as the difficulty of accessing and sampling in extreme marine environments, the limited taxonomy of many isolates, and the challenge of cultivating some species under laboratory conditions. Additionally, the limited production of metabolites in conventional cultures makes it difficult to obtain enough for biological tests, structural analysis, and toxicological studies. There are obstacles in transferring compounds from the laboratory to the field due to a lack of information on stability, mode of action, selectivity, and environmental impact. Imhoff (2016) discusses these obstacles, highlighting the underexploited potential of marine fungi as an emerging and strategic theme in discovering sustainable bioinputs for modern agriculture.

References

Abad, A.V., Manzanares, P., Marcos, J.F. and Martínez-Culebras, P.V. (2023). The *Penicillium digitatum* antifungal protein PdAfpB shows high activity against mycobiota involved in sliced bread spoilage. Food Microbiol., 109: 104142. https://doi.org/10.1016/j.fm.2022.104142.

Abulaizi, A., Wang, R., Xiong, Z., Zhang, S., Li, Y. et al. (2024). Secondary metabolites with agricultural antagonistic potential from *Aspergillus* sp. ITBBc1, a coral-associated marine fungus. Mar. Drugs, 22: 270. https://doi.org/10.3390/md22060270.

Alengebawy, A., Abdelkhalek, S.T., Qureshi, S.R. and Wang, M.Q. (2021). Heavy metals and pesticides toxicity in agricultural soil and plants: Ecological risks and human health implications. Toxics, 9(3): 42. https://doi.org/10.3390/toxics9030042.

Ayaz, M., Li, C.H., Ali, Q., Zhao, W., Chi, Y.K. et al. (2023). Bacterial and fungal biocontrol agents for plant disease protection: Journey from lab to field, current status, challenges, and global perspectives. Molecules, 28: 6735. https://doi.org/10.3390/molecules28186735.

Barreto, D.L.C., Cantrell, C.L., da Silva, K.M., de Carvalho, C.R., de Queiroz, S.C.N. et al. (2024). Phytotoxic and antifungal activity of (–)-penienone produced by *Penicillium palitans* (Ascomycota) isolated from deep sea sediments in the Southern Ocean, Maritime Antarctica. Chem. Biodivers., 22: e202401603. https://doi.org/10.1002/cbdv.202401603.

Benelli, G., Maggi, F., Romano, D., Stefanini, C., Vaseeharan, B. et al. (2017). Nanoparticles as effective acaricides against ticks: A review. Ticks Tick-borne Dis., 8: 821–826. https://doi.org/10.1016/j.ttbdis.2017.08.004.

Berg, R.H., Fester, T. and Taylor, C.G. (2009). Development of the root-knot nematode feeding cell. *In*: Berg, R.H. and Taylor, C.G. (eds.). Cell Biology of Plant Nematode Parasitism. Plant Cell Monographs (Vol. 15). Springer, Berlin, Heidelberg. https://doi.org/10.1007/7089_2008_30.

Blunt, J.W., Carroll, A.R., Copp, B.R., Davis, R.A., Keyzers, R.A. et al. (2018). Marine natural products. Nat. Prod. Rep., 35: 8–53. https://doi.org/10.1039/C7NP00052A.

Blunt, J.W., Copp, B.R., Keyzers, R.A., Munro, M.H.G. and Prinsep, M.R. (2018). Marine natural products. Nat. Prod. Rep., 35: 8–53. https://doi.org/10.1039/B415080P.

Bunyapaiboonsri, T., Yoiprommarat, S., Suntivich, R., Preedanon, S., Komwijit, S. et al. (2020). A cyclic lipodepsipeptide, a spirolactone, and a chromanone from the marine fungus *Verruculina enalia* (Kohlm) Kohlm & Volkm-Kohlm BCC 22226. Tetrahedron, 76: 131497. https://doi.org/10.1016/j.tet.2020.131497.

Burns, A.R., Baker, R.J., Kitner, M., Knox, J., Cooke, B. et al. (2023). Selective control of parasitic nematodes using bioactivated nematicides. Nature, 618: 102–109. https://doi.org/10.1038/s41586-023-06105-5.

Cao, F., Meng, Z.H., Mu, X., Yue, Y.F. and Zhu, H.J. (2019). Absolute configuration of bioactive azaphilones from the marine-derived fungus *Pleosporales* sp CF09-1. J. Nat. Prod., 82: 386–392. https://doi.org/10.1039/C7NP00052A.

Cao, F., Yang, J.K., Liu, Y.F., Zhu, H.J. and Wang, C.Y. (2016). Pleosporalone A, the first azaphilone characterized with aromatic A-ring from a marine-derived *Pleosporales* sp fungus. Nat. Prod. Rep., 30: 2448–2452. https://doi.org/10.1080/14786419.2016.1198352.

Carroll, A.R., Copp, B.R., Grkovic, T., Keyzers, R.A. and Prinsep, M.R. (2024). Marine natural products. Nat. Prod. Rep., 41: 162–207. https://doi.org/10.1039/D3NP00061C.

Casida, J.E. and Bryant, R.J. (2017). The ABCs of pesticide toxicology: Amounts, biology, and chemistry. Toxicol. Res., 6: 755–763. https://doi.org/10.1039/c7tx00198c.

Chandler, D., Davidson, G., Pell, J.K., Ball, B.V., Shaw, K. et al. (2000). Fungal biocontrol of Acari. Biocontrol. Sci. Technol., 10: 357–384. https://doi.org/10.1080/09583150050114972.

Chen, D., Zhang, P., Liu, T., Wang, X.F., Li, Z.X. et al. (2018). Insecticidal activities of chloramphenicol derivatives isolated from a marine alga-derived endophytic fungus, *Acremonium vitellinum*, against the Cotton Bollworm, *Helicoverpa armigera* (Hübner) (Lepidoptera: Noctuidae). Molecules, 23: 2995. https://doi.org/10.3390/molecules23112995.

Curran, W.S. (2016). Persistence of herbicides in soil. Crops Soils, 49: 16–21. https://doi.org/10.2134/cs2016-49-0504.

Dayan, F.E. (2019). Current status and future prospects in herbicide discovery. Plants, 8: 341. https://doi.org/10.3390/plants8090341.

Dayan, F.E. and Duke, S.O. (2014). Natural compounds as next generation herbicides. Plant Physiol., 166: 1090–1105. https://doi.org/10.1104/pp.114.239061.

Du, F.Y., Ju, G.L., Xiao, L., Zhou, Y.M. and Wu, X. (2020). Sesquiterpenes and cyclodepsipeptides from marine-derived fungus *Trichoderma longibrachiatum* and their antagonistic activities against soil-borne pathogens. Mar. Drugs, 18: 165. https://doi.org/10.3390/md18030165.

Du, F.Y., Li, X., Li, X.M., Zhu, L.W. and Wang, B.G. (2017). Indolediketopiperazine alkaloids from *Eurotium cristatum* EN-220, an endophytic fungus isolated from the marine alga *Sargassum thunbergii*. Mar. Drugs, 15: 24. https://doi.org/10.3390/md15020024.

Duke, S.O. (2021). Glyphosate: Uses other than in glyphosate-resistant crops, mode of action, degradation in plants, and effects on non-target plants and agricultural microbes. Rev. Environ. Contam. Toxicol., 255: 1–65. https://doi.org/10.1007/398_2020_53.

Duke, S.O. and Dayan, F.E. (2022). The search for new herbicide mechanisms of action: Is there a 'holy grail'? Pest Manag. Sci., 78: 1303–1313. https://doi.org/10.1002/ps.6726.

Duke, S.O., Owens, D.K. and Dayan, F.E. (2018). Natural product-based chemical herbicides. *In*: Weed Control. CRC Press, Boca Raton, pp. 153–165.

Ebani, V.V. and Manciant, F. (2021). Entomopathogenic fungi and bacteria in a veterinary perspective. Biology, 10: 479. https://doi.org/10.3390/biology10060479.

Fan, J., Guo, F., Zhao, C., Li, H., Qu, T. et al. (2023). Secondary metabolites with herbicidal and antifungal activities from marine-derived fungus *Alternaria iridiaustralis*. J. Fungi, 9: 716. https://doi.org/10.3390/jof9070716.

Fan, H., Shao, X.H., Wu, P.P., Hao, A.L., Luo, Z.W. et al. (2024). Exploring brominated aromatic butenolides from *Aspergillus terreus* EGF7-0-1 with their antifungal activities. J. Agric. Food Chem., 72(36): 19869–19882. https://doi.org/10.1021/acs.jafc.4c04728.

FAO (Food and Agriculture Organization) (2024). Land Statistics 2001–2022. Global, Regional, and Country trends. FAOSTAT Analytical Briefs No 88. Available at: https://www.fao.org/statistics/highlights-archive/highlights-detail/land-statistics-2001-2022.-global--regional-and-country-trends/en (Accessed 4 April 2025).

Frisvad, J.C. (2018). A critical review of producers of small lactone mycotoxins: Patulin, penicillic acid, and moniliformin. World Mycotoxin J., 11: 73–100. https://doi.org/10.3920/WMJ2017.2294.

Giddings, L.A. and Newman, D.J. (2015). Bioactive compounds from marine extremophiles. Springer Nature, 1–124. https://doi.org/10.1007/978-3-319-14361-3_1.

Zhou, B., Wang, H., Meng, B., Wei, R., Wang, L. et al. (2019). An evaluation of tenazonic acid, a potential biobased herbicide in cotton. Pest Manag. Sci., 75: 2482–2489. https://doi.org/10.1002/ps.5402.

Zhao, D., Han, X., Wang, D., Liu, M., Gou, J. et al. (2019a). Bioactive 3-decalinoyltetramic acids' derivatives from a marine-derived strain of the fungus *Fusarium equiseti* D39. Front. Microbiol., 10: 1285. https://doi.org/10.3389/fmicb.2019.01285.

Zhao, D.L., Wang, D., Tian, X.Y., Cao, F., Li, Y.Q. et al. (2018). Anti-phytopathogenic and cytotoxic activities of crude extracts and secondary metabolites of marine-derived fungi. Mar. Drugs, 16: 36. https://doi.org/10.3390/md16010036.

Zhao, D.L., Yang, L.J., Shi, T., Wang, C.Y., Shao, C.L. et al. (2019b). Potent phytotoxic harziane diterpenes from a soft coral-derived strain of the fungus *Trichoderma harzianum* XS-20090075. Sci. Rep., 9: 13345. https://doi.org/10.1038/s41598-019-49778-7.

2

Using Zoosporic Fungi and Fungus-Like Organisms in Aquaculture

Natarajan Velmurugan,[1] *Yen-Ju Pan,*[2] *Thossaporn Phatthongkleang,*[3] *Lam Kong,*[3] *Sheng-Yu Guo*[3] and *Ka-Lai Pang*[3,*]

Overfishing and climate change have impacted human intake of essential fatty acids, i.e., docosahexaenoic acid [DHA; C22:6(n-3)] and eicosapentaenoic acid [EPA; 20:5(n-3)], through consumption of marine deep-sea fish. Zoosporic fungi and fungus-like organisms (thraustochytrids, *Halophytophthora*, and related taxa) are lipid-producing eukaryotic microorganisms, which have great potential in industrial-scale production of DHA and EPA and aquaculture as a feed supplement. Cells of thraustochytrids are rich in DHA, while EPA was detected in mycelia of *Halophytophthora* and related taxa and cells of zoosporic fungi. The pollutant-free fatty acids from these microorganisms can directly supplement aquaculture feed or be used for the enrichment of agricultural waste products. Copepods fed with thraustochytrids, *Halophytophthora*, or zoosporic fungi were found to increase in lipid content and may serve as

[1] Biological Sciences Division, CSIR-North East Institute of Science and Technology (CSIR-NEIST), Branch Laboratory-Itanagar, Naharlagun 791 110, Arunachal Pradesh, India.

[2] Department of Aquaculture, National Taiwan Ocean University, 2 Pei-Ning Road, Keelung 202301, Taiwan.

[3] Institute of Marine Biology and Center of Excellence for the Oceans, National Taiwan Ocean University, 2 Pei-Ning Road, Keelung 202301, Taiwan.

* Corresponding author: klpang@ntou.edu.tw

DOI: 10.1201/9781003795100-3

a live feed for fish larvae. The aquaculture industry is predicted to expand further in the years to come to meet with the increase in human population, and so does the demand for efficient, toxin-free, economical, and nutritious aquaculture feed. DHA and EPA from the heterotrophic zoosporic fungi and fungus-like organisms may provide alternative sources of essential fatty acids in the aquaculture feed industry.

1. Introduction

The aquaculture industry is the fastest-growing food production industry worldwide. A very recent report by the Food and Agriculture Organization stated that aquaculture production of aquatic animals has now surpassed capture fisheries, highlighting its significant role in global food security (FAO Report, 2024; https://www.fao.org/newsroom/detail/fao-report-global-fisheries-and-aquaculture-production-reaches-a-new-record-high/en). Approximately 89% of aquatic animal production is used for human consumption, while the remaining portion primarily supports the production of fish oil and fishmeal. By 2050, as the human population will likely reach over 9 billion, the demand for food products, especially aquatic food material, will also rise (Galanakis, 2022). The aquaculture animals are the cheapest protein source, and so they represent one of the potential food sources to meet the demand. However, the aquaculture industry faces many challenges; among them, developing nutritious and economical feeds is one of the bottlenecks yet to be addressed.

Marine fish are well-known for the omega-3 fatty acid contents in their tissues, but they cannot synthesize them *in vivo* and obtain them from organisms in the lower trophic levels of the aquatic food web, such as phytoplankton. Overfishing depletes fish stocks, disrupts marine ecosystems, and negatively impacts a steady supply of omega-3 fatty acids from fish for human consumption (Naylor et al., 2000). Also, increasing global temperature affects the production of cold-protective omega-3 fatty acids in phytoplankton, resulting in reduced amounts of docosahexaenoic acid (DHA) and eicosapentaenoic acid (EPA) in fish. In aquaculture farms, omega-3 fatty acids in fish are fulfilled by fishmeal and fish oil in fish feed. The feed material offers essential amino acids, energy, and palatability that enhance growth and improve omega-3 fatty acid content in fish (Jackson, 2006). Because of these beneficial properties, fishmeal and fish oil have emerged as staples, not only in aquaculture farms but also in the poultry and swine industries.

However, the heavy reliance on these marine-derived resources raises significant concerns related to sustainability. Furthermore, fish oil can contain contaminants due to the bioaccumulation of pollutants in marine ecosystems. As fish occupy higher trophic levels in the food chain, they accumulate substances from their diet and environment, which may ultimately contaminate fish oil (Mao and Chen, 2018). The use of such dietary fish oil in aquafeeds can transfer these marine pollutants to artificially cultivated fish and, consequently, to humans through fish consumption (Llobet et al., 2003; Sun et al., 2018). The deficit between supply and demand for fish oil is substantial and was estimated to be over one million tons (Colombo et al., 2020).

To mitigate these challenges, vegetable oils are increasingly used in aquafeeds, but this replacement reduces the levels of EPA and DHA in farmed fish, which are critical for their growth and health (Cahu et al., 2004; Ventura et al., 2024). DHA is particularly vital in aquaculture feed formulations, as its supplementation significantly enhances growth rate and reduces mortality in fish larvae (Yamasaki et al., 2007). A higher percentage of DHA over EPA also induced a higher lipid accumulation in the flesh of the marine flatfish, tongue sole (*Cynoglossus semilaevis*) (Xu et al., 2018). The aquaculture sector has been looking for alternate dietary lipid sources due to the declining worldwide supply of fish oil and its extremely fluctuating prices.

Marine zoosporic fungi and fungal-like organisms (thraustochytrids, *Halophytophthora*) are oil-producing and oil-accumulating (including DHA and EPA) eukaryotic organisms, which produce oil in their intracellular lipid bodies (Jakobsen et al., 2008; Jiang et al., 2004). These heterotrophic organisms are emerging as a sustainable alternative to marine fish and algal oils due to their high DHA content (in thraustochytrids) and ethical production advantages, i.e., in fermenters (Raghukumar, 2002). Thraustochytrids produce 10–50% of their biomass as total lipids, with DHA accounting for 30–70% of the lipid content (Aasen et al., 2016; Parrish, 2024). The DHA production levels can be manipulated in fermentation by adjusting growth conditions (Jaseera and Kaladharan, 2019; Burja et al., 2006). Their rapid growth, omega-3 fatty acid profile, and ability to thrive under variable fermentation conditions make them commercially feasible for DHA production (Armenta and Valentine, 2013).

Unlike photoautotrophic microorganisms (those requiring light for growth), marine zoosporic fungi and fungal-like microorganisms can achieve high biomass and DHA production through heterotrophic growth, which in fact is much faster and more efficient (Meng et al., 2009; Morabito et al., 2019). Their simpler omega-3 profiles reduce production costs compared to photosynthetic microalgal oils, which often have complex fatty acid compositions (Lewis et al., 1999).

Marine fungi are an ecological group of predominantly saprobic eukaryotes, which include true fungi mainly belonging to the Ascomycota and the Basidiomycota, and fungus-like organisms such as the Oomycota and the Labyrinthulomycetes (Fig. 1). Early research on marine fungi was mainly taxonomic, while later reports revealed their capability to produce various molecules that are of biotechnological potential, such as bioactive compounds (Rateb and Ebel, 2011) and some essential polyunsaturated fatty acids (PUFAs) (Pang et al., 2015b; Wang et al., 2024). Marine *Halophytophthora* (Oomycota) and allied taxa are known to produce high quantities of EPA in mycelia (Pang et al., 2015b), while thraustochytrids (Labyrinthulomycetes) are prolific producers of DHA in cells (Ou et al., 2016). The Chytridiomycota also produced EPA, among other saturated and unsaturated fatty acids (Gerphagnon et al., 2019). This chapter provides a brief overview of the fatty acids of biotechnological potential produced by thraustochytrids, marine Oomycota, and true chytrids, and their possible application in the aquaculture industry.

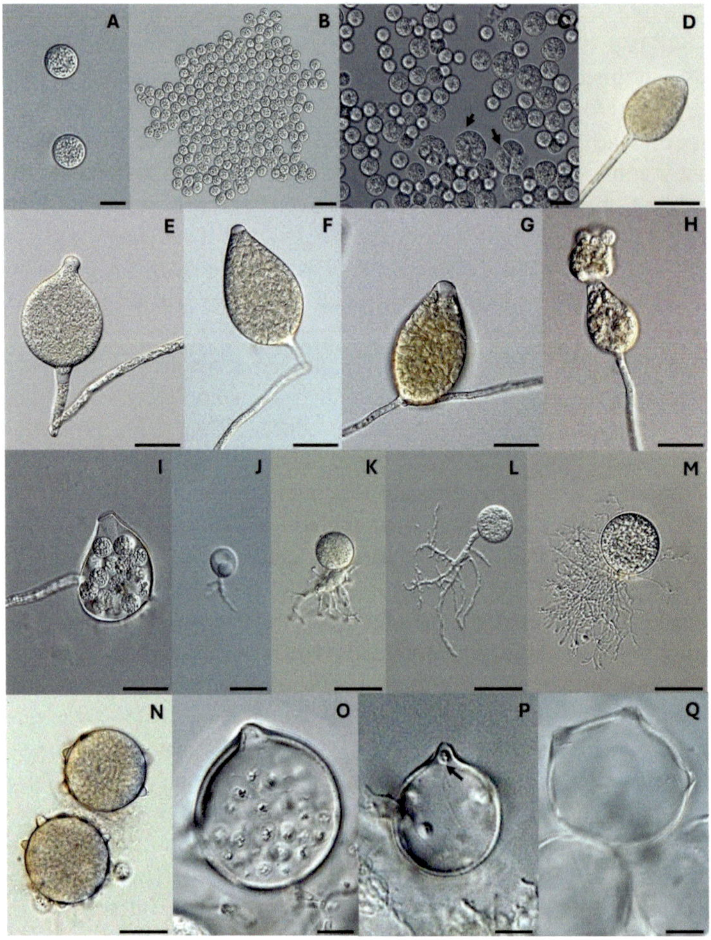

Fig. 1. Differential life stages and morphological features via light microscopy of thraustochytrids, *Halophytophthora* spp., and the Chytridiomycota. (A–C) Thraustochytrids, (A) spherical thalli, (B) cluster of developing sporangia, (C) sporogenous cells undergoing rapid cell division (three to four visible cells; arrows); (D–I) *Halophytophthora* spp., (D–E) immature sporangia, (F–G) mature sporangia with zoospores, (H–I) zoospore discharge through a vesicle; (J–Q) The Chytridiomycota, (J) encysted zoospore, (K–L) young sporangia with rhizoidal structures, (M) mature sporangium with extensive rhizoid network, (N–O) zoospore-filled operculate sporangia; (P–Q) empty sporangial wall after zoospore discharge (arrow, zoospores) (***Scale bars***: A-C, J, O, Q = 10 μm; D-I, K-N, P = 30 μm).

2. Zoosporic Fungi and Fungus-Like Organisms

In recent years, there has been a dramatic change in the classification of fungi. Alexopoulos et al. (1996) outlined four phyla (Chytridiomycota, Zygomycota, Ascomycota, and Basidiomycota) in the kingdom Fungi. The recent phylogenetic analysis suggested that both Zygomycota and Chytridiomycota were polyphyletic, and subsequently each of them was split into multiple phyla (Tedersoo et al., 2018; Wijayawardene et al., 2018; Voigt et al., 2021). Aphelidiomycota,

Blastocladiomycota, Caulochytriomycota, Chytridiomycota, Monoblepharomycota, Neocallimastigomycota, Olpidiomycota, Rozellomycota, and Sanchytriomycota are phyla of fungi producing zoospores (zoosporic fungi).

The life cycle of zoosporic fungi is complex and is taxon-dependent (Fig. 1). For example, in the Chytridiomycota, motile zoospores colonize new substrata with retraction of the cilium and production of thallus and rhizoids; cell growth occurs, and multiple nuclei are produced within the thallus, which eventually develop into a sporangium; each nucleus and new cilium are encapsulated to become a zoospore; and zoospores are released to the aquatic environment to colonize new substrata (Medina and Buchler, 2020).

Traditionally, marine fungi are an ecological group comprising true fungi and fungus-like organisms (Jones et al., 2009). These marine zoosporic fungus-like organisms include Mesomycetozoea, Hyphochytriales, Phytomyxea, Oomycota, and Labyrinthulomycetes (thraustochytrids) (Jones and Pang, 2012; Neuhauser et al., 2012). These lineages share many features with zoosporic fungi, including heterotrophy, reproduction through flagellated spores, chitin in their cell walls, and the presence of fungus-like structures such as hyphae, resistant spores, and similar life cycles (Neuhauser et al., 2012; Hassett et al., 2020).

Thraustochytrids (Fig. 1) are included in the Labyrinthulomycetes, belonging to the Stramenipila, Alveolate, Rhizaria (SAR) group (Adl et al., 2012). As single-celled organisms, Thraustochytrids reproduce asexually by releasing zoospores from zoosporangia; when zoospores settle, the flagella are lost and the cells subsequently develop and enlarge into new thalli (Bongiorni et al., 2005; Pang et al., 2015a). Sexual reproduction was observed recently in *Aurantiochytrium acetophilum* through fusion of amoeboid cells (Ganuza et al., 2019). The study by Sparrow (1936) reported thraustochytrids from diverse marine environments and showed that thraustochytrids are crucial to the ecosystem and contribute significantly to biomass accumulation.

Thraustochytrids are associated with living marine animals and plants, are degraders, mineralize herbaceous materials, and may act as secondary producers (Fan et al., 2002; Leaño et al., 2000; Raghukumar, 2008; Neuhauser et al., 2012). Thraustochytrids are abundant in coastal waters, e.g., up to 5.6×10^4 cells l^{-1} (Naganuma et al., 1998) and up to 17.57×10^5 cells l^{-1} (Liu et al., 2017), in water columns, in sediments, and on a variety of marine organisms such as sponges, corals, hydroids, bivalves, octopus, squid, nudibranchs, echinoids, diatoms, seagrass, and more (Porter, 1990). Currently, there are 12 genera of thraustochytrids, including *Althornia*, *Aurantiochytrium*, *Botryochytrium*, *Diplophrys*, *Elina*, *Japonochytrium*, *Oblongichytrium*, *Parietichytrium*, *Schizochytrium*, *Sicyoidochytrium*, *Thraustochytrium*, and *Ulkenia*, with approximately 40 species (Liu et al., 2014).

Oomycota also belong to the SAR group (Marano et al., 2012; Beakes and Thines, 2017). The life cycle of marine Oomycota is characterized by a vegetative phase with coenocytic hyphae, an asexual reproductive phase with sporangiophores, papillate/nonpapillate zoosporangia with smooth or ornamented walls (Fig. 1), which are equipped with an operculum, vesicle, or plug and release motile animal feeds (Kuratko and Salem, 2013; Marchan et al., 2018; Nazir et al., 2020). EPA from marine Oomycota and zoosporic fungi has not been explored for large-scale production.

As the majority of the thraustochytrids have been cultured from high lignocellulose content substrates such as mangrove leaves and debris, they possess characteristics to grow in agricultural wastes such as fruit peels, enriching them with PUFA, which can be used as aquaculture feed. These low-cost substrates contain a significant amount of organic carbon compounds, which are usable for growth and fatty acid production of thraustochytrids. Utilization of agricultural waste substrates for growth and accumulation of fatty acids significantly reduces production cost in the different industries. In a recent study, Saikia et al. (2022) enriched fruit peels with PUFA produced by thraustochytrids during solid-state fermentation (SSF); glucose (6.99%) and maltose (6.22%) in the fruit peel waste were efficiently utilized by thraustochytrid cells for growth and PUFA production. Also, the optimized SSF composite was enriched with 8.37% of thraustochytrid-derived PUFA. The work opens new doors to combine biotechnology of thraustochytrid fatty acid production technology and waste management practices. Abdel-Wahab et al. (2022) also successfully utilized date fruits, molasses, and glycerol to grow thraustochytrids.

Optimization of efficient fermentation conditions/parameters with agricultural wastes is a complex task (Jakobsen et al., 2008), considering that carbon sources affect fatty acid composition in thraustochytrids. Taguchi methodology and response surface methodology (RSM) can be potentially used for optimizing the multi-parameter-based SSF fermentation conditions suitable for thraustochytrids. In addition to fatty acid biosynthesis, fermentation of thraustochytrids may also yield other byproducts such as pigments, exopolysaccharides, and enzymes. Patel et al. (2020) simultaneously yielded squalene and DHA from thraustochytrids when cells were cultivated on pretreated spruce hydrolysate. The highly prized nutraceutical compound, astaxanthin, was produced by thraustochytrid isolates (Yokoyama and Honda, 2007). For economical sustainability, it is essential to explore large-scale upstream and downstream process optimization of low-cost, hazardous-free substrates to produce PUFAs by thraustochytrids.

Palmitic acid, a saturated fatty acid, is dominant in both thraustochytrids and marine Oomycota (Ou et al., 2016; Su et al., 2021). High levels of saturated fatty acids implicate their potential application in biofuel biotechnology, acting as alternative sources of energy to fossil fuels (Sohedein et al., 2020).

5. Applications in Aquaculture

Due to their high PUFA content, zoosporic fungi and fungus-like organisms have great potential in marine aquaculture. Thraustochytrids, in particular, have been used to enrich live feed for aquacultured fish larvae. The DHA and EPA contents in *Artemia franciscana* nauplii and rotifer *Brachionus plicatilis* increased when they were fed with spray-dried cells of *Schizochytrium* sp. (Barclay and Zeller, 1996). Moreover, a diet based on *Schizochytrium*-enriched *Artemia* nauplii has been shown to enhance the DHA levels in pompano (*Trachinotus blochii*) larvae, resulting in a higher survival rate, improved growth performances, and increased stress tolerance (Pedroso and Moquera, 2023). These studies highlight the feasibility of using this emerging biomass source as a nutritional booster for live feed enrichment, offering a sustainable alternative to fish oil.

Copepods are a group of microscopic crustaceans that play an important trophic linkage between the primary producer (phytoplankton) and the consumers in the aquatic ecosystems (Schminke, 2007). In aquaculture, the supplements of copepod-based diets for fish and shrimp larvae have been shown to be beneficial due to their superior nutritional value and high palatability (Sargent et al., 1997; Shields et al., 1999; Rayner et al., 2015). Notably, copepod-based diets have been shown to improve larval survival (Kraul et al., 1992; Payne et al., 2001), growth rates (Engell et al., 2006), and skin pigmentation (Sargent et al., 1997; Ananthi et al., 2011).

Copepods, like other filter feeders, rely extensively on microalgae as diets in aquaculture environments. The evaluations of suitable microalgal diets, as copepods' natural food items, have been carried out for the copepods belonging to the three orders, including Calanoida (Pan et al., 2014, Siqwepu et al., 2017), Harpacticoida (Arndt and Sommer, 2014), and Cyclopoida (Pan et al., 2018; Rasdi and Qin, 2018) (Fig. 2). While microalgae are well-suited for many filter-feeding organisms, the maintenance of live microalgae production systems remains costly and labor-consuming. Additionally, the risk of unexpected microalgae culture crashes can hinder the utilization of microalgae in aquaculture.

A recent study demonstrated that *Mytilus galloprovincialis* spat achieved similar growth when fed microcapsules made from *Schizochytrium* compared to those fed with commercial diets, highlighting the potential of *Schizochytrium* as a viable and cost-effective feed alternative for filter feeders (Campanati et al., 2023). In our experiment, individuals of the copepod *Apocyclops royi* accumulated numerous large lipid droplets in their bodies (Fig. 2) and accomplished a higher population growth rate when fed with freeze-dried mycelia of *Salispina* sp. (marine Oomycota), compared to those fed with microalgae. This finding suggests that marine oleaginous microorganisms may serve as a solution that replaces live microalgae as feed for copepods, potentially reducing the challenges associated with mass production and enhancing the economic and industrial feasibility of copepod cultivation.

On the other hand, the biomass of zoosporic fungi and fungus-like organisms could serve as an alternative lipid source in the formulation of dry fish pellets. Dietary supplementation with dried cells of *Thraustochytrium* sp. SB04 and *Schizochytrium* sp. SB11 led to significant improvements in body length and weight in Nile tilapia (*Oreochromis niloticus*) compared to those fed a conventional fish meal-based diet (Atienza et al., 2012). The addition of dried cells of *Schizochytrium* sp. to the diet of channel catfish (*Ictalurus punctatus*) resulted in increased body weight as well as n–3 long-chain polyunsaturated fatty acids (Li et al., 2009). In Atlantic salmon juveniles (*Salmo salar*), no significant differences in growth or biochemical composition were observed between fish oil-based and thraustochytrid-based diets (Carter et al., 2003). In another study, thraustochytrid-derived fatty acids were indeed detected in cultured Atlantic salmon fingerlings using specific stable

Burja, A.M., Radianingtyas, H., Windust, A. and Barrow, C.J. (2006). Isolation and characterization of polyunsaturated fatty acid producing *Thraustochytrium* species: Screening of strains and optimization of omega-3 production. Appl. Microbiol. Biotechnol., 72: 1161–1169. https://doi.org/10.1007/s00253-006-0419-1.

Cahu, C., Salen, P. and De Lorgeril, M. (2004). Farmed and wild fish in the prevention of cardiovascular diseases: Assessing possible differences in lipid nutritional values. Nutr. Metab. Cardiovasc. Dis., 14: 34–41. https://doi.org/10.1016/S0939-4753(04)80045-0.

Campanati, C., Arantzamendi, L., Zorita, I., Briaudeau, T., Lekube, X. et al. (2023). Nutritional effect of substituting microalgal concentrates with microencapsulated diets in *Mytilus galloprovincialis* spat. Aquaculture, 576: 739879. https://doi.org/10.1016/j.aquaculture.2023.739879.

Carter, C.G., Lewis, T.E. and Nichols, P.D. (2003). Comparison of cholestane and yttrium oxide as digestibility markers for lipid components in Atlantic salmon (*Salmo salar* L.) diets. Aquaculture, 225: 341–351. https://doi.org/10.1016/S0044-8486(03)00300-4.

Colombo, S.M., Rodgers, T.F.M., Diamond, M.L., Bazinet, R.P. and Arts, M.T. (2020). Projected declines in global DHA availability for human consumption as a result of global warming. Ambio, 49: 865–880. https://doi.org/10.1007/s13280-019-01234-6.

Ellenbogen, B.B., Aaronson, S., Goldstein, S. and Belsky, M. (1969). Polyunsaturated fatty acids of aquatic fungi: Possible phylogenetic significance. Comp. Biochem. Physiol., 29: 805–811. https://doi.org/10.1016/0010-406X(69)91631-4.

Engell, K., Drillet, G., Hansen, B.W., Echabe, J., Evjemo, J. et al. (2006). Intensive production of calanoide copepods and possible preservation of copepod eggs for fish farming: Presentation of overall results from the FP5-craft-project, POCEFF. *In*: Joint International Conference of European Aquaculture Society and World Aquaculture Society, Firenze, Italy.

Fan, K.W., Vrijmoed, L.L.P. and Jones, E.B.G. (2002). Physiological studies of subtropical mangrove thraustochytrids. Bot. Mar., 45: 50–57. https://doi.org/10.1515/BOT.2002.006.

Galanakis, C.M. (2022). Sustainable Fish Production and Processing. Academic Press, UK, 342 pp. https://surl.li/acbjor.

Ganuza, E., Yang, S., Amezquita, M., Giraldo-Silva, A. and Andersen, R.A. (2019). Genomics, biology and phylogeny *Aurantiochytrium acetophilum* sp. nov. (Thraustrochytriaceae), including first evidence of sexual reproduction. Protist, 170: 209–232. https://doi.org/10.1016/j.protis.2019.02.004.

Gerphagnon, M., Agha, R., Martin-Creuzburg, D., Bec, A., Perriere, F. et al. (2019). Comparison of sterol and fatty acid profiles of chytrids and their hosts reveals trophic upgrading of nutritionally inadequate phytoplankton by fungal parasites. Environ. Microbiol., 21: 949–958. https://doi.org/10.1111/1462-2920.14489.

Hassett, B.T., Vonnahme, T.R., Peng, X., Jones, E.B.G. and Heuzé, C. (2020). Global diversity and geography of planktonic marine fungi. Bot. Mar., 63: 121–139. https://doi.org/10.1515/bot-2018-0113.

Ho, H.H. and Jong, S.C. (1990). *Halophytophthora*, gen. nov., a new member of the family Pythiaceae. Mycotaxon, 36: 377–382. http://www.jstor.org/stable/3760352.

Huang, J., Aki, T., Yokochi, T., Nakahara, T., Honda, D. et al. (2003). Grouping newly isolated docosahexaenoic acid-producing thraustochytrids based on their polyunsaturated fatty acid profiles and comparative analysis of 18S rRNA genes. Mar. Biotechnol., 5: 450–457. https://doi.org/10.1007/s10126-002-0110-1.

Iwasaka, H., Aki, T., Adachi, H., Watanabe, H., Kawamoto, S. and Ono, K. (2013). Utilization of waste syrup for production of unsaturated fatty acids and xanthophylls by *Aurantiochytrium*. J. Oleo. Sci., 62: 729–736. https://doi.org/10.5650/jos.62.729.

Jackson, A.J. (2006). The importance of fishmeal and fish oil in aquaculture diets. Int. Aquafeed, 9: 18–21. https://surl.li/ihnvem.

Jakobsen, A.N., Aasen, I.M., Josefsen, K.D. and Strøm, A.R. (2008). Accumulation of docosahexaenoic acid-rich lipid in thraustochytrid *Aurantiochytrium* sp. strain T66: Effects of N and P starvation and O_2 limitation. Appl. Microbiol. Biotechnol., 80: 297–306. https://doi.org/10.1007/s00253-008-1537-8.

Jaseera, K.V. and Kaladharan, P. (2019). Thraustochytrids in aquaculture: Can it replace fish meal in Aquafeed? Aquac. Spectr., 2: 25–27. http://eprints.cmfri.org.in/id/eprint/14335.

Jiang, Y., Fan, K.W., Tsz-Yeung Wong, R. and Chen, F. (2004). Fatty acid composition and squalene content of the marine microalga *Schizochytrium mangrovei*. J. Agric. Food Chem., 52: 1196–1200. https://doi.org/10.1016/j.mycres.2006.10.009.

Jones, E.B.G. and Pang, K.L. (2012). Tropical aquatic fungi. Biodivers. Conserv., 21: 2403–2423. https://doi.org/10.1007/s10531-011-0198-6.

Jones, E.B.G., Sakayaroj, J., Suetrong, S., Somrithipol, S. and Pang, K.L. (2009). Classification of marine Ascomycota, anamorphic taxa, and Basidiomycota. Fungal Divers., 35: 1–187. https://www.researchgate.net/publication/280797437.

Kraul, S., Nelson, A., Brittain, K., Ako, H. and Ogasawara, A. (1992). Evaluation of live feeds for larval and postlarval mahimahi *Coryphaena hippurus*. J. World Aquacult. Soc., 23: 299–306. https://doi.org/10.1111/j.1749-7345.1992.tb00793.x.

Kuratko, C.N. and Salem, N. Jr. (2013). Docosahexaenoic acid from algal oil. Eur. J. Lipid. Sci. Technol., 115: 965–976. https://doi.org/10.1002/ejlt.201300060.

Leaño, E.M., Jones, E.B.G. and Vrijmoed, L.L.P. (2000). Why are *Halophytophthora* species well adapted to mangrove habitats? Fungal Divers., 5: 131–151. http://hdl.handle.net/10862/1875.

Leaño, E.M., Vrijmoed, L.L.P. and Jones, E.B.G. (1998). Zoospore chemotaxis of two mangrove strains of *Halophytophthora vesicula* from Mai Po, Hong Kong. Mycologia, 90: 1001–1008. https://doi.org/10.1080/00275514.1998.12026998.

Lee Chang, K.J., Nichols, C.M., Blackburn, S.I., Dunstan, G.A., Koutoulis, A. et al. (2014). Comparison of thraustochytrids *Aurantiochytrium* sp., *Schizochytrium* sp., *Thraustochytrium* sp., and *Ulkenia* sp. for production of biodiesel, long-chain omega-3 oils, and exopolysaccharide. Mar. Biotech., 16: 396–411. https://doi.org/10.1007/s10126-014-9560-5.

Lee Chang, K.J., Parrish, C.C., Simon, C.J., Revill, A.T. and Nichols, P.D. (2020). Feeding whole thraustochytrid biomass to cultured Atlantic salmon (*Salmo salar*) fingerlings: Culture performance and fatty acid incorporation. J. Mar. Sci. Eng., 8: 207. https://doi.org/10.3390/jmse8030207.

Lee Chang, K.J., Rye, L., Dunstan, G.A., Grant, T., Koutoulis, A. et al. (2015). Life cycle assessment: Heterotrophic cultivation of thraustochytrids for biodiesel production. J. Appl. Phycol., 27: 639–647. https://doi.org/10.1007/s10811-014-0364-9.

Lewis, T.E., Nichols, P.D. and McMeekin, T.A. (1999). The biotechnological potential of thraustochytrids. Mar. Biotechnol., 1: 580–587. https://doi.org/10.1007/PL00011813.

Li, M.H., Robinson, E.H., Tucker, C.S., Manning, B.B. and Khoo, L. (2009). Effects of dried algae Schizochytrium sp., a rich source of docosahexaenoic acid, on growth, fatty acid composition, and sensory quality of channel catfish *Ictalurus punctatus*. Aquaculture, 292: 232–236. https://doi.org/10.1016/j.aquaculture.2009.04.033.

Liu, Y., Singh, P., Liang, Y., Li, J., Xie, N. et al. (2017). Abundance and molecular diversity of thraustochytrids in coastal waters of southern China. FEMS Microbiol. Ecol., 93: fix070. https://doi.org/10.1093/femsec/fix070.

Liu, Y., Singh, P., Sun, Y., Luan, S. and Wang, G. (2014). Culturable diversity and biochemical features of thraustochytrids from coastal waters of Southern China. Appl. Microbiol. Biotechnol., 98: 3241–3255. https://doi.org/10.1007/s00253-013-5391-y.

Llobet, J.M., Falco, G., Casas, C., Teixido, A. and Domingo, J.L. (2003). Concentrations of arsenic, cadmium, mercury, and lead in common foods and estimated daily intake by children, adolescents, adults, and seniors of Catalonia, Spain. J. Agric. Food Chem., 51: 838–842. https://doi.org/10.1021/jf020734q.

Maia, C., Jung, T., Engelen, A., Jung, M.H. and Custódio, L. (2023). Unravelling the lipids content and the fatty acid profiles of eight recently described *Halophytophthora* species and *H. avicenniae* from the south coast of Portugal. Mar. Drugs, 21: 227. https://doi.org/10.3390/md21040227.

Mao, X. and Chen, G. (2018). Speciation of trace mercury impurities in fish oil supplements. Food Control, 84: 221–225. https://doi.org/10.1016/j.foodcont.2017.08.001.

Marano, A.V., Pires-Zottarelli, C.L.A., de Souza, J.I., Glockling, S.L., Leaño, E. et al. (2012) Hyphochytriomycota, Oomycota and Perkinsozoa (Super-group Chromalveolata). pp. 167–214. *In*: Jones, E.B.G. and Peng, K.L. (eds.). Marine Fungi and Fungal-like Organisms. DeGruyter, Berlin. https://doi.org/10.1515/9783110264067.167.

Marchan, L.F., Chang, K.J.L., Nichols, P.D., Mitchell, W.J., Polglase, J.L. and Gutierrez, T. (2018). Taxonomy, ecology, and biotechnological applications of thraustochytrids: A review. Biotech. Adv., 36: 26–46. https://doi.org/10.1016/j.biotechadv.2017.09.003.

Marchan, L.F., Chang, K.J.L., Nichols, P.D., Polglase, J.L., Mitchell, W.J. et al. (2017). Screening of new British thraustochytrids isolates for docosahexaenoic acid (DHA) production. J. Appl. Phycol., 29: 2831–2843. https://doi.org/10.1007/s10811-017-1149-8.

Medina, E.M. and Buchler, N.E. (2020). Chytrid fungi. Curr. Biol., 30: R516–R520.

Meng, X., Yang, J., Xu, X., Zhang, L., Nie, Q. et al. (2009). Biodiesel production from oleaginous microorganisms. Renew. Energy, 34: 1–5. https://doi.org/10.1016/j.renene.2008.04.014.

Mills, G.L. and Cantino, E.C. (1974). Lipid composition of the zoospores of *Blastocladiella emersonii*. J. Bacteriol., 118: 192–201. https://doi.org/10.1128/jb.118.1.192-201.1974.

Morabito, C., Bournaud, C., Maës, C., Schuler, M., Aiese Cigliano, R. et al. (2019). The lipid metabolism in thraustochytrids. Prog. Lipid Res., 76: 101007. https://doi.org/10.1016/j.plipres.2019.101007.

Naganuma, T., Takasugi, H. and Kimura, H. (1998). Abundance of thraustochytrids in coastal plankton. Mar. Ecol. Progr. Ser., 162: 105–110. https://doi.org/10.3354/meps162105.

Nakagiri, A. (2002). *Halophytophthora* species from tropical and subtropical mangroves: A review of their characteristics. Fungal Divers., 7: 1–14.

Naylor, R.L., Goldburg, R.J., Primavera, J.H., Kautsky, N., Beveridge, M.C. et al. (2000) Effect of aquaculture on world fish supplies. Nature, 405: 1017–1024. https://doi.org/10.1038/35016500.

Nazir, Y., Halim, H., Al-Shorgani, N.K.N., Manikan, V., Hamid, A.A. et al. (2020). Efficient conversion of extracts from low-cost, rejected fruits for high-valued Docosahexaenoic acid production by *Aurantiochytrium* sp. SW1. Algal Res., 50: 101977. https://doi.org/10.1016/j.algal.2020.101977.

Neri, K.D., Devanadera, M.K.P., Watanabe, K., Bennett, R.M., Arafiles, K.H.V. et al. (2023). Fatty acid profiles of oomycetes (*Halophytophthora* and *Phytopythium* spp.) from Philippines mangroves. Lett. Appl. Microbiol., 76: ovad022. https://doi.org/10.1093/lambio/ovad022.

Neuhauser, S., Glockling, S.L., Leaño, E.M., Lilje, O., Marano, A.V. et al. (2012). An introduction to fungus-like microorganisms. pp. 137–152. *In*: Jones, E.B.G. and Peng, K.L. (eds.). Marine Fungi and Fungal-like Organisms. De Gruyter, Berlin. https://doi.org/10.1515/9783110264067.137.

Nigrelli, L. and Thines, M. (2013). Tropical oomycetes in the German Bight: Climate warming or overlooked diversity? Fungal Ecol., 6: 152–160. https://doi.org/10.1016/j.funeco.2012.11.003.

Ou, M.C., Yeong, H.Y., Pang, K.L. and Phang, S.M. (2016). Fatty acid production of tropical thraustochytrids from Malaysian mangroves. Bot. Mar., 59: 321–338. https://doi.org/10.1515/bot-2016-0031.

Pan, Y.J., Sadovskaya, I., Hwang, J.S. and Souissi, S. (2018). Assessment of the fecundity, population growth and fatty acid composition of *Apocyclops royi* (Cyclopoida, Copepoda) fed on different microalgal diets. Aquacult. Nutr., 24: 970–978. https://doi.org/10.1111/anu.12633.

Pan, Y.J., Souissi, S., Souissi, A., Wu, C.H., Cheng, S.H. et al. (2014). Dietary effects on egg production, egg-hatching rate, and female life span of the tropical calanoid copepod *Acartia bilobata*. Aquacult. Res., 45: 1659–1671. https://doi.org/10.1111/are.12113.

Pang, K.L., Chen, M.C., Chiang, M.W.L., Huang, Y.F., Yeong, H.Y. et al. (2015a). Cu(II) pollution affects fecundity of the mangrove degrader community, the Labyrinthulomycetes. Bot. Mar., 58: 129–138. https://doi.org/10.1515/bot-2015-0006.

Pang, K.L., Lin, H.J., Lin, H.Y., Huang, Y.F. and Chen, Y.M. (2015b). Production of arachidonic and eicosapentaenoic acids by the marine oomycete *Halophytophthora*. Mar. Biotechnol., 17: 121–129. https://doi.org/10.1007/s10126-014-9600-1.

Parrish, C.C. (2024). Thraustochytrids and algae as sustainable sources of long-chain omega-3 fatty acids for aquafeeds. Sustainability, 16: 9142. https://doi.org/10.3390/su16219142.

Patel, A., Mu, L., Shi, Y., Rova, U., Christakopoulos, P. et al. (2020). Novel biorefinery approach aimed at vegetarians reduces the dependency on marine fish stocks for obtaining squalene and docosahexaenoic acid. ACS Sustain. Chem. Eng., 8: 8803–8813. https://doi.org/10.1021/acssuschemeng.0c02752.

Payne, M., Rippingale, R. and Cleary, J. (2001). Cultured copepods as food for West Australian dhufish (*Glaucosoma hebraicum*) and pink snapper (*Pagrus auratus*) larvae. Aquaculture, 194: 137–150. https://doi.org/10.1016/S0044-8486(00)00513-5.

Pedroso, F.L. and Moquera, G.L. (2023). Utilization of thraustochytrid, *Schizochytrium* sp. as live food enrichment enhances growth performance, digestive enzyme activities, and stress tolerance of silver pompano (*Trachinotus blochii*) larvae. Int. J. Aquat. Biol., 11: 439–448. https://doi.org/10.22034/ijab.v11i5.2017.

Porter, D. (1990). Phylum Labyrinthulomycota. pp. 388–398. *In*: Margulis, L., Corliss, J.O., Melkonian, M. and Chapman, D.J. (eds). Handbook of Protoctista. Jones and Bartlett, Boston. https://doi.org/10.1007/978-3-319-32669-6_25-1.

Raghukumar, S. (2002). Ecology of the marine protists, the Labyrinthulomycetes (Thraustochytrids and Labyrinthulids). Eur. J. Protistol., 38: 127–145. https://doi.org/10.1078/0932-4739-00832.

Raghukumar, S. (2008). Thraustochytrid marine protists: Production of PUFAs and other emerging technologies. Mar. Biotechnol., 10: 631–640. https://doi.org/10.1007/s10126-008-9135-4.

Rasdi, N.W. and Qin, J.G. (2018). Impact of food type on growth, survival, and reproduction of the cyclopoid copepod *Cyclopina kasignete* as a potential live food in aquaculture. Aquacult. Int., 26: 1281–1295. https://doi.org/10.1007/s10499-018-0283-x.

Rateb, M.E. and Ebel, R. (2011). Secondary metabolites of fungi from marine habitats. Nat. Prod. Rep., 28: 290–344. https://doi.org/10.1039/C0NP00061B.

Rayner, T.A., Jørgensen, N.O., Blanda, E., Wu, C.H., Huang, C.C. et al. (2015). Biochemical composition of the promising live feed tropical calanoid copepod *Pseudodiaptomus annandalei* (Sewell 1919) cultured in Taiwanese outdoor aquaculture ponds. Aquaculture, 441: 25–34. https://doi.org/10.1016/j.aquaculture.2015.01.034.

Russo, G.L., Langellotti, A.L., Martín-García, B., Verardo, V., Romano, R. et al. (2023). New Biotechnological production of EPA by *Pythium irregulare* using alternative sustainable media obtained from food industry by-products and waste. Sustainability, 15(2): 1147. https://doi.org/10.3390/su15021147.

Saikia, D.K., Chikkaputtaiah, C. and Velmurugan, N. (2022). Nutritional enrichment of fruit peel wastes using lipid accumulating *Aurantiochytrium* strain as feed for aquaculture in North-East Region of India. Environ. Technol., 45: 1215–1233. https://doi.org/10.1080/09593330.2022.2139638.

Sargent, J., McEvoy, L. and Bell, J. (1997). Requirements, presentation, and sources of polyunsaturated fatty acids in marine fish larval feeds. Aquaculture, 155: 117–127. https://doi.org/10.1016/S0044-8486(97)00122-1.

Schminke, H.K. (2007). Entomology for the copepodologist. J. Plankton Res., 29: i149–i162. https://doi.org/10.1093/plankt/fbl073.

Shepon, A., Makov, T., Hamilton, H.A., Müller, D.B., Gephart, J.A. et al. (2022). Sustainable optimization of global aquatic omega-3 supply chain could substantially narrow the nutrient gap. RCR Adv., 181: 106260. https://doi.org/10.1016/j.resconrec.2022.106260.

Shields, R.J., Bell, J.G., Luizi, F.S., Gara, B., Bromagem, N.R. et al. (1999). Natural copepods are superior to enriched *Artemia nauplii* as feed for halibut larvae (*Hippoglossus hippoglossus*) in terms of survival, pigmentation, and retinal morphology: Relation to dietary essential fatty acids. J. Nutr., 129: 1186–1194. https://doi.org/10.1093/jn/129.6.1186.

Siqwepu, O., Richoux, N.B. and Vine, N.G. (2017). The effect of different dietary microalgae on the fatty acid profile, fecundity, and population development of the calanoid copepod *Pseudodiaptomus hessei* (Copepoda: Calanoida). Aquaculture, 468: 162–168. https://doi.org/10.1016/j.aquaculture.2016.10.008.

Sohedein, M.N.A., Wan-Mohtar, W.A.A.Q.I., Yeong, H.Y., Zul, I., Chang, J.S. et al. (2020). Optimisation of biomass and lipid production of a tropical thraustochytrid *Aurantiochytrium* sp. UMACC-T023 in submerged-liquid fermentation for large-scale biodiesel production. Biocatal. Agric. Biotechnol., 23: 101496. https://doi.org/10.1016/j.bcab.2020.101496.

Song, Y., Yang, X., Li, S., Luo, Y., Chang, J.S. et al. (2024). Thraustochytrids as a promising source of fatty acids, carotenoids, and sterols: Bioactive compound biosynthesis, and modern biotechnology. Crit. Rev. Biotechnol., 44: 618–640. https://doi.org/10.1080/07388551.2023.2196373.

Sparrow, F.K. (1936). Biological observations on the marine fungi of woods hole waters. Biol. Bull., 70: 236–263. https://doi.org/10.2307/1537470.

Su, C.J., Ju, W.T., Chen, Y.M., Chiang, M.W.L., Hsieh, S.Y. et al. (2021). Palmitic acid and long-chain polyunsaturated fatty acids dominate in mycelia of mangrove *Halophytophthora* and *Salispina* species in Taiwan. Bot. Mar., 64: 503–518. https://doi.org/10.1515/bot-2021-0030.

Sun, S.X., Hua, X.M., Deng, Y.Y., Zhang, Y.N., Li, J.M. et al. (2018). Tracking pollutants in dietary fish oil: From ocean to table. Environ. Pollut., 240: 733–744. https://doi.org/10.1016/j.envpol.2018.05.027.

Tedersoo, L., Sánchez-Ramírez, S., Kõljalg, U., Bahram, M., Döring, M. et al. (2018). High-level classification of the fungi and a tool for evolutionary ecological analyses. Fungal Divers., 90: 135–159. https://doi.org/10.1007/s13225-018-0401-0.

Ventura, M., Oroy, C., Llopis Monferrer, N., Le Grand, F., Long, M. et al. (2024). Hydrolysis optimization of thraustochytrid (heterotrophic protist) biomass for the production of a new and sustainable ingredient for the aquafeed industry. Sustainability, 16(14): 6052. https://doi.org/10.3390/su16146052.

Voigt, K., James, T.Y., Kirk, P.M., de Santiago, A.L.C.M., Waldman, B. et al. (2021). Early-diverging fungal phyla: Taxonomy, species concept, ecology, distribution, anthropogenic impact, and novel phylogenetic proposals. Fungal Divers., 109: 59–98. https://doi.org/10.1007/s13225-021-00480-y.

Wang, Q., Zhang, Y., Hui, R. and Zhu, Y. (2024). Marine thraustochytrid: Exploration from taxonomic challenges to biotechnological applications. Front. Mar. Sci., 11: 1371713. https://doi.org/10.3389/fmars.2024.1371713.

Wijayawardene, N.N., Pawłowska, J., Letcher, P.M., Kirk, P.M., Humber, R.A. et al. (2018). Notes for genera: Basal clades of Fungi (including Aphelidiomycota, Basidiobolomycota, Blastocladiomycota, Calcarisporiellomycota, Caulochytriomycota, Chytridiomycota, Entomophthoromycota, Glomeromycota, Kickxellomycota, Monoblepharomycota, Mortierellomycota, Mucoromycota, Neocallimastigomycota, Olpidiomycota, Rozellomycota, and Zoopagomycota). Fungal Divers., 92: 43–129. https://doi.org/10.1007/s13225-018-0409-5.

Xu, H., Cao, L., Wei, Y., Zhang, Y. and Liang, M. (2018). Lipid contents in farmed fish are influenced by dietary DHA/EPA ratio: A study with the marine flatfish, tongue sole (*Cynoglossus semilaevis*). Aquaculture, 485: 183–190. https://doi.org/10.1016/j.aquaculture.2017.11.054.

Yamasaki, T., Aki, T., Mori, Y., Yamamoto, T., Shinozaki, M. et al. (2007). Nutritional enrichment of larval fish feed with thraustochytrid producing polyunsaturated fatty acids and xanthophylls. J. Biosci. Bioeng., 104: 200–206. https://doi.org/10.1263/jbb.104.200.

Yokoyama, R. and Honda, D. (2007). Taxonomic rearrangement of the genus *Schizochytrium* sensu lato based on morphology, chemotaxonomic characteristics, and 18S rRNA gene phylogeny (Thraustochytriaceae, Labyrinthulomycetes): Emendation for *Schizochytrium* and erection of *Aurantiochytrium* and *Oblongichytrium* gen. nov. Mycoscience, 48: 199–211. https://doi.org/10.1007/S10267-006-0362-0.

3

Pigments from Marine Fungi

A Sustainable Approach for Food Applications

Maricarmen Iñiguez-Moreno

The increasing consumer demand for natural, sustainable food colorants has increased scientific interest in marine fungal pigments due to their vibrant hues and bioactivity. This chapter provides a comprehensive overview of the biotechnological potential of marine fungi as a new source of functional pigments for food applications. Marine fungi, such as *Talaromyces*, *Aspergillus*, and *Penicillium*, are widely spread in marine ecosystems and produce structurally unique pigments, including polyketides (anthraquinones and azaphilones), carotenoids, and melanins, which possess antioxidant, antimicrobial, and photoprotective properties. These metabolites not only enhance the visual appeal of foods but also help extend shelf life and promote health, aligning with the trend towards functional foods. Key findings from recent literature highlight the biosynthetic versatility of marine fungi and their capacity to produce pigments on a scalable basis through submerged and solid-state fermentation, often utilizing agro-industrial waste substrates. Omics-based approaches, particularly genome mining and transcriptomics, have identified biosynthetic gene clusters responsible for pigment production.

Department of Physics and Mathematics, Universidad de Monterrey, Avenida Morones Prieto 4500, San Pedro Garza García 66238, NL, Mexico.
Email: maricarmen.iniguez@udem.edu

DOI: 10.1201/9781003795100-4

Advances in fermentation and extraction technologies have also improved yield and purity. Regulatory pathways for food-grade fungal pigments remain challenging as they require comprehensive toxicological assessment and alignment with frameworks such as Generally Recognized as Safe and Novel Foods. Nevertheless, initial applications in dairy, beverage, and meat systems show promising sensory and functional results. Future directions include metabolic engineering and synthetic biology to optimize metabolic pathways, as well as industrial-scale-up strategies that integrate pigment production into circular economy models. Marine fungal pigments are a great alternative as next-generation colorants that meet the dual goals of naturality and functionality in the food industry.

1. Introduction

Color plays a pivotal role in consumer perception and acceptance of food products. Food-grade pigments are essential for enhancing visual appeal and indicating quality and freshness (Muñoz-Miranda and Iñiguez-Moreno, 2023). Traditionally, synthetic colorants have dominated the market owing to their vibrant hues, stability, and cost-effectiveness. However, growing health concerns and a shift towards natural ingredients have increased demand for natural colorants. Recent market analyses indicate a significant growth trajectory for natural food colorants. The global natural food colors market is projected to reach USD 6 billion by 2028, rising from USD 4.6 billion in 2023, driven by consumer preference for clean-label products and regulatory pressures on synthetic additives (Grand View Research, 2024). The increasing popularity of plant-based diets and the growing demand for sustainable, health-promoting food ingredients further support this trend.

While synthetic colorants offer advantages in terms of cost and stability, they have been associated with potential health risks, including allergic reactions and behavioral issues in children. For instance, certain azo dyes like Tartrazine (E102) and Sunset Yellow (E110) have been linked to hyperactivity and other adverse effects, leading to regulatory actions in various countries. The European Food Safety Authority (EFSA) and other regulatory bodies have imposed restrictions or bans on several synthetic dyes, prompting manufacturers to seek safer alternatives (Aguilar et al., 2009; Barciela et al., 2023).

Terrestrial sources of natural pigments, such as plants and insects, also present challenges. These include seasonal variability, limited availability, and environmental concerns related to large-scale cultivation and harvesting. Moreover, natural pigments from terrestrial sources often exhibit lower stability and color intensity compared to their synthetic counterparts, limiting their application in various food matrices (López-Cruz et al., 2023). In this sense, marine fungi have emerged as a promising and underexplored source of natural pigments. These organisms thrive in diverse and extreme marine environments, leading to unique metabolic pathways that produce a variety of bioactive compounds, including pigments (Muñoz-

Miranda and Iñiguez-Moreno, 2023). Pigments derived from marine fungi, such as polyketides and carotenoids, exhibit vibrant colors and possess functional properties like antioxidant and antimicrobial activities (Caro et al., 2017).

The use of marine fungi for pigment production offers several advantages. For example, marine fungi can be cultivated using fermentation technologies, reducing dependence on agricultural land and freshwater resources. Fermentation processes enable controlled and scalable production of pigments and ensure consistent quality and supply. Beyond coloration, marine fungal pigments can provide health benefits, which is in line with the trend towards functional foods. These properties make marine fungi a viable alternative to conventional pigment sources, meeting consumer demand and industry challenges. This chapter aims to provide a comprehensive overview of marine fungi-derived pigments, focusing on their potential as sustainable and functional colorants in food applications.

2. Marine Fungal Diversity and Ecological Role

Marine fungi comprise a diverse group of organisms adapted to various marine-like mangroves, sediments, and deep-sea environments. Genera like *Aspergillus*, *Penicillium*, and *Talaromyces* are notable for their pigment-producing capabilities (Gonçalves et al., 2022). These fungi play crucial ecological roles, including nutrient cycling and symbiotic relationships with marine flora and fauna.

2.1 *Definition and Classification of Marine Fungi*

Marine fungi are a diverse group of filamentous and unicellular fungi that either obligatorily or facultatively inhabit marine or estuarine environments. These organisms are characterized not by distinct taxonomic boundaries but rather by their ecological adaptation to saline and aquatic conditions. Marine fungi are generally categorized into two functional groups: obligate marine fungi, which grow and sporulate exclusively in marine or estuarine conditions, and facultative marine fungi, which originate from terrestrial environments but have adapted to survive in marine settings (Caro et al., 2017; Gonçalves et al., 2022).

Taxonomically, marine fungi span several major fungal phyla, including Ascomycota, Basidiomycota, Chytridiomycota, and Zygomycota, though Ascomycota comprises the majority of known marine fungal species. Genera such as *Aspergillus*, *Penicillium*, *Talaromyces*, and *Cladosporium* are frequently isolated from marine environments and are known for their ability to produce a wide range of secondary metabolites, including bioactive pigments (Sen et al., 2022). Due to the complexity and variability of marine ecosystems, defining marine fungi purely on phylogenetic grounds remains a challenge. Molecular tools, especially high-throughput sequencing and metagenomic analyses, have significantly expanded the known diversity of marine fungal taxa and revealed previously undetected lineages that contribute to ecosystem function in marine habitats.

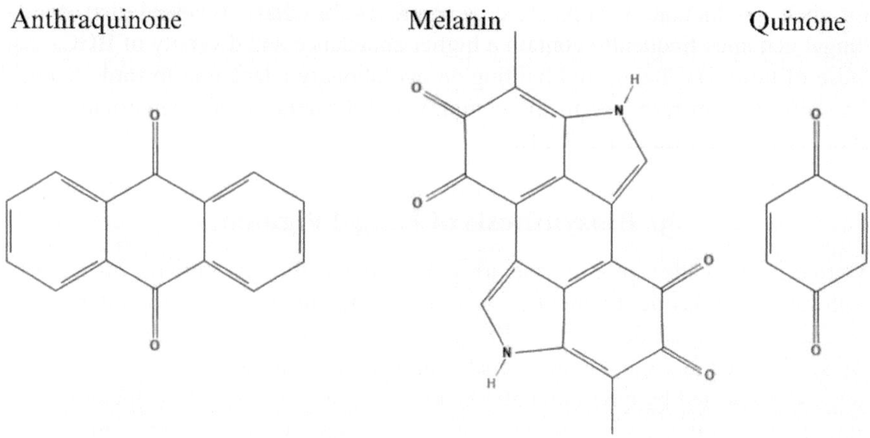

Fig. 1. General structures of representative pigments synthesized by marine-derived fungi.

On the other hand, azaphilones are another subclass of polyketides featuring a highly oxygenated pyranoquinone core. Biosynthesis of azaphilones is catalyzed by a PKS-FAS hybrid complex, and their unique chromophore results from reactions between polyketide chains and amino-containing molecules. Marine fungi such as *Talaromyces albobiverticillius* have been shown to produce yellow to red azaphilones with antioxidant and anti-inflammatory properties (de Oliveira et al., 2022). Marine fungi such as *Aspergillus*, *Penicillium*, *Talaromyces*, *Alternaria*, *Eurotium*, *Halorosellinia*, *Microsphaeropsis*, *Monodictys*, *Nigrospora*, *Paecilomyces*, *Phomopsis*, and *Stemphylium* are notable for their ability to synthesize anthraquinone-based pigments (Fouillaud et al., 2016, 2017). The production of these pigments is often upregulated in nutrient-limited environments, suggesting that ecological stressors trigger pigment biosynthesis as a form of metabolic adaptation (El-Kashef et al., 2020; Afroz Toma et al., 2023).

3.1.2 *Carotenoids*

Carotenoids are isoprenoid pigments characterized by long conjugated polyene chains that impart yellow, orange, and red colors. In fungi, their biosynthesis occurs via the mevalonate (MVA) pathway, leading to the formation of geranylgeranyl pyrophosphate (GGPP) as a key precursor. From GGPP, enzymes such as phytoene synthase, phytoene desaturase, and lycopene cyclase participate in sequential desaturation and cyclization reactions to produce common carotenoids like β-carotene, torulene, and neurosporene (Fig. 2) (Ram et al., 2020). While less commonly associated with marine fungi compared to terrestrial yeasts, certain marine-adapted strains of *Rhodosporidium* have demonstrated the ability to produce high carotenoid yields under saline conditions (Kulczyk-Małysa and Bogusławska-Wąs, 2025). Carotenoids in fungi function as antioxidants and light-harvesting molecules, helping to mitigate oxidative stress in high-UV marine environments. These pigments are particularly attractive for food applications due to their established safety and provitamin A activity (Vílchez et al., 2011).

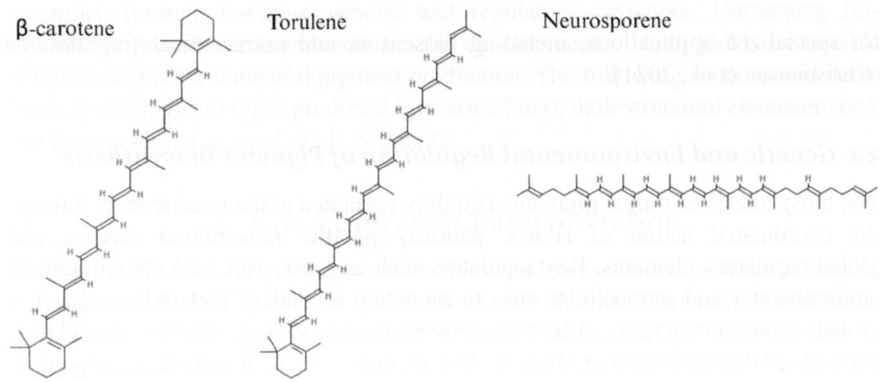

Fig. 2. Chemical structure of the most common carotenoids produced by marine fungi.

3.1.3 Melanins

Melanins are dark brown to black pigments formed by oxidative polymerization of phenolic and indolic compounds. Fungi produce three primary types of melanin: DHN-melanin (1,8-dihydroxynaphthalene pathway), DOPA-melanin (L-3,4-dihydroxyphenylalanine pathway), and pyomelanin (from homogentisic acid) (Pavan et al., 2020; Wang et al., 2020). The most prevalent fungal melanin, DHN-melanin, is synthesized via the condensation of acetate-derived units through a type I polyketide synthase. The pathway proceeds through intermediates such as 1,3,6,8-tetrahydroxynaphthalene (T4HN) and 1,8-DHN, which polymerize to form the final pigment (Pavan et al., 2020). In marine fungi such as *Antarctomyces pellizariae, Schizophyllum commune, Aspergillus nidulans, Hortaea werneckii, Colletotrichum lagenarium, Paracoccidioides brasiliensis, Magnaporthe grisea,* and *Sporothrix schenckii,* melanins have been detected in species exposed to high UV radiation or oxidative stress, where they serve as protective agents by absorbing radiation and scavenging free radicals. They may also contribute to mechanical resilience in a high-pressure environment (Ghattavi et al., 2022).

3.1.4 Other Minor Pigments

Marine fungi have also been found to produce a variety of lesser-known pigments, including flavins, quinones, and phenazine derivatives. These compounds often arise from complex biosynthetic networks involving hybrid PKS-NRPS systems or shikimate pathway intermediates (Kalra et al., 2020; Afroz Toma et al., 2023). Flavins, typically derived from riboflavin, fluoresce under UV light and may function in energy transfer or stress response (Muñoz-Miranda and Iñiguez-Moreno, 2023). Quinones function as redox-active pigments with potential roles in antimicrobial defense and cellular signaling. Although phenazines are predominantly associated with bacteria, their occurrence in certain marine fungi, particularly under co-culture conditions, suggests the possibility of interspecies metabolic exchange. While typically produced in low quantities, these pigments exhibit high bioactivity and unique chromatic characteristics, positioning them as promising candidates

2017). Emodin, a yellow anthraquinone (1,3,8-trihydroxy-6-methylanthraquinone), has been reported from various marine fungal genera, including *Eurotium chevalieri* (Vasarri et al., 2022), *Penicillium oxalicum* (Fouillaud et al., 2017). Several *Aspergillus* species (Hawas et al., 2012), and lichen-associated fungi such as *Caloplaca* (Frolov et al., 2021). Notably, emodin possesses multiple pharmacological properties, including hepatoprotective, neuroprotective, antimicrobial, and anticancer activities (Hassan et al., 2022).

While cytotoxicity raises safety concerns, controlled cytotoxic activity is beneficial in the context of anticancer applications (Table 2). Anthraquinones and polyketides produced by marine fungi have demonstrated selective cytotoxicity against human cancer cell lines, including HepG2 (liver cancer), HeLa (cervical cancer), and MCF-7 (breast cancer) (Evidente, 2024). Mechanistic studies suggest these effects are mediated through the induction of apoptosis, cell cycle arrest, and inhibition of angiogenesis. However, such findings necessitate careful evaluation when considering these pigments for food use, as high cytotoxic activity may pose risks to healthy human cells at elevated exposure levels (Hafez Ghoran et al., 2022; Evidente, 2024).

5.2 *Regulatory Landscape and Approval Prospects*

Navigating regulatory approval is a critical step toward the commercialization of fungal pigments for food applications. Agencies such as the United States Food and Drug Administration (FDA) and the EFSA require extensive dossiers detailing chemical composition, source organism, production method, stability, and toxicological profiles.

In the U.S., the GRAS designation enables substances to be used in foods without premarket approval, provided that qualified experts deem the substance safe under its intended use. Currently, few fungal pigments hold GRAS status, with most approved natural colorants derived from plants, insects, or minerals. However, pigment-producing fungi such as *Monascus purpureus* have a long history of use in Asian cuisine, laying a precedent for marine-derived analogs (Teiba et al., 2025). Applicants must submit comprehensive safety data, including toxicological studies, manufacturing details, and dietary exposure assessments. The voluntary GRAS notification program allows for industry transparency and FDA feedback.

In the European Union, marine fungal pigments would likely fall under the Novel Food Regulation (EU) 2015/2283, as they were not widely consumed within the EU before May 1997. The EFSA panel assesses identity, purity, nutritional value, metabolism, bioavailability, toxicity, allergenicity, proposed use levels, and target populations (Aguilar et al., 2009). Fungal pigments, including carotenoids and polyketide-derived compounds, are approved as food colorants by the European Union. One major advantage of producing these pigments through fungal fermentation is the consistency of production, which is not affected by seasonal changes or supply fluctuations (Aguilar et al., 2009; Meruvu and dos Santos, 2021). A positive EFSA opinion is required before authorization by the European

Commission. Given the precedent of plant-derived carotenoids and algae-based pigments, marine fungal colorants may have a viable pathway pending comprehensive data. The main challenge compared to plant- and algae-based colorants is that fungal pigments are often produced alongside mycotoxins; therefore, their purification is crucial to enable their safe application in the food industry (Darwesh et al., 2020; Lin and Xu, 2022). Additionally, fungal pigments intended for use in edible coatings, packaging films, or nutraceuticals must comply with the corresponding food contact material or supplement regulations.

6. Applications in the Food Industry

The application of marine fungal pigments in the food industry offers a natural alternative to synthetic dyes. These pigments can be incorporated into various food products, including dairy, beverages, and confectionery, to enhance visual appeal and provide health benefits (Table 3). The modern food industry is experiencing a paradigm shift driven by consumer demand for natural, health-promoting, and sustainable ingredients. Among these, natural colorants have emerged as essential alternatives to synthetic dyes, which are increasingly scrutinized for their potential health risks and environmental impact. Marine-derived fungal pigments, a promising class of natural compounds, have gained attention for their vibrant hues, stability in diverse processing conditions, and potential functional benefits such as antioxidant and antimicrobial activity. Their applicability across a wide array of food matrices, combined with their biosynthetic versatility, positions them as valuable assets in the design of next-generation food products (Fig. 4) (Venkatachalam et al., 2018; Muñoz-Miranda and Iñiguez-Moreno, 2023; Ashenafi et al., 2023).

6.1 Use as Natural Colorants in Food Matrices

Color plays a pivotal role in consumer perception, with visual appeal often influencing the initial judgment of a food product (Hassan et al., 2023; Muñoz-Miranda and Iñiguez-Moreno, 2023). Pigments from marine fungi provide a spectrum of color options ranging from bright reds and oranges to subtle yellows and purples, which are derived primarily from polyketides, azaphilones, carotenoids, and anthraquinones (Pereira et al., 2014; Hassan et al., 2023; Muñoz-Miranda and Iñiguez-Moreno, 2023).

In dairy applications, fungal pigments can be used in flavored yogurts, cheeses, and plant-based dairy analogues to impart attractive visual characteristics. Due to the high protein and fat content of dairy matrices, pigment solubility and partitioning behavior are critical. Lipophilic pigments, such as azaphilones, often integrate well in fat-rich matrices and can maintain color integrity during pasteurization and refrigeration (Xu et al., 2015; Hassan et al., 2023). For instance, azaphilone-based pigments from *Talaromyces albobiverticillius* have been successfully incorporated into cream-based desserts, demonstrating high thermal and light stability, with minimal impact on flavor or mouthfeel (Pimenta et al., 2021).

Table 3. Pigments from *Monascus* with potential applications in the food industry (GAE: Gallic Acid Equivalents; EBC: European Brewery Convention).

Microorganism, Production, and Pigment Extraction	**Pigments Used**	**Food Product**	**Main Result**	**References**
Monascus purpureus Went NRRL 1992 was cultivated under submerged conditions using a medium containing 7.81% potato waste and 2.87% peptone at 30°C and pH 6.5 for 12.82 days, followed by pigment extraction with ethyl acetate as the solvent.	Red, orange, and yellow	Ice lolly	At a pigment concentration of 0.05%, the formulation demonstrated excellent sensory performance, achieving scores higher than 9 across all evaluated attributes, including taste, color, texture, odor, and overall acceptability. The pigment contributed to a uniform coloration and proved effective even at this low concentration. Importantly, it did not negatively affect any sensory characteristics, highlighting its suitability for applications where both aesthetic appeal and sensory integrity are essential.	Monteiro et al. (2021)
M. purpureus HM188425.1 was cultured for 14 days at 30°C using solid-state fermentation with 50 g of durian seeds per batch (particle size ~1 cm^3) and a 5% (v/w) inoculum, followed by pigment extraction using a water-ethanol mixture at a 1:50 (v/v) ratio.	Red	Yogurt	The pigment was applied at concentrations of 0.15% for the powder form and 7.5% for the liquid form, whether water- or ethanol-based. It produced a strong red coloration and could be directly incorporated as a powder without compromising yield. Additionally, its inclusion led to an increase in total phenolic content (from 1.54 to 2.21 mg GAE/g), enhanced antioxidant activity, and improvements in both color intensity and flavor profile.	Gomah et al. (2017)
Monascus ruber Went AUMC 5705 was cultivated via solid-state fermentation using broken rice supplemented with 2% monosodium glutamate at pH 6.5, 30°C, and 39% moisture content. Inoculum rate of 126×10^6 spores per 10 g of fermented dried substrate, for 15 days, followed by pigment extraction using a mixture of ethanol and water.	Red (strawberry-flavored milk beverage) and yellow (banana-flavored milk beverage)	Yogurt and milk beverage from buffalo milk	The pigment was added incrementally until achieving visual standardization with a commercial reference product. The resulting formulations exhibited a uniform, attractive color, appearance, and distribution. Sensory evaluation revealed high overall acceptability, with total scores exceeding 87.8% for strawberry-flavored yogurt and milk beverage and over 83.88% for banana-flavored milk beverage.	Abdel-Raheam et al. (2022)
M. ruber CCT 3802 was cultivated via solid-state fermentation on wheat grains at 30°C for 14 days, using 5 mL of spore suspension per 100 g of dry substrate with a moisture content of 43% (m/m), followed by pigment extraction.	Red	Loaf bread	By replacing up to 15% of the flour with pigment, the formulation achieved a more intense red coloration, along with increased ash content and improved water-holding capacity. Additionally, the dough development time was reduced. However, some negative effects were observed, including decreased resistance to extension, reduced dough volume, and higher texture values, indicating a firmer consistency.	Egea Buranelo and Marcionilio de Oliveira (2022)

M. purpureus (commercial) was prepared in powder form from red yeast rice.	Red	Poultry meat products	At 0.6%, color, taste, consistency, cutting view, and inner appearance were improved.	Srianta et al. (2022)
Monascus sp. (powder of red yeast rice)	Red	Meat system as pork fat replacement	*Monascus* pigment water-in-oil-in-water emulsion at concentrations ranging from 10 to 30% showed several beneficial effects in the meat. The formulation resulted in a reduction of lipid content to below 45.37% and an increase in protein content above 7.34%. The composition of polyunsaturated fatty acids exceeded 36.27%, while saturated fatty acids remained below 19.2%. Also improved water-holding capacity and oxidative stability. Product hardness increased without negatively impacting the structure of the meat system.	Darwesh et al. (2020)
M. ruber OMNRC45; submerged fermentation; 30°C; pH 6.5; with modified rice broth medium; peptone (100 mg/L); (pH 4.5–8.5); and extraction: 95% ethanol and centrifugation.	Red	Lollipops and jelly beans	At a pigment concentration of 0.1%, the formulation exhibited an intense and persistent red color. This enhancement significantly improved the product's visual appeal, overall appearance, and consumer acceptability.	Mal'a et al. (2010)
M. ruber (10%) was used to supplement barley malt.	Red	Red beer (Pilsen)	The beer exhibited high acidity as indicated by pH and titratable acidity measurements, showed a dark brown color with a 23.64 EBC staining value, and tested positive for the presence of lovastatin.	Cordeiro de Souza (2018)
M. purpureus (Moguntia type, International AG, Europe) was used in the form of red yeast rice powder.	Red	Processed cheeses	The addition of 0.015% pigment enhanced sensory characteristics, while pigment concentrations of 0.02% and 0.05% improved both sensory attributes and color stability.	Lin and Xu (2022)

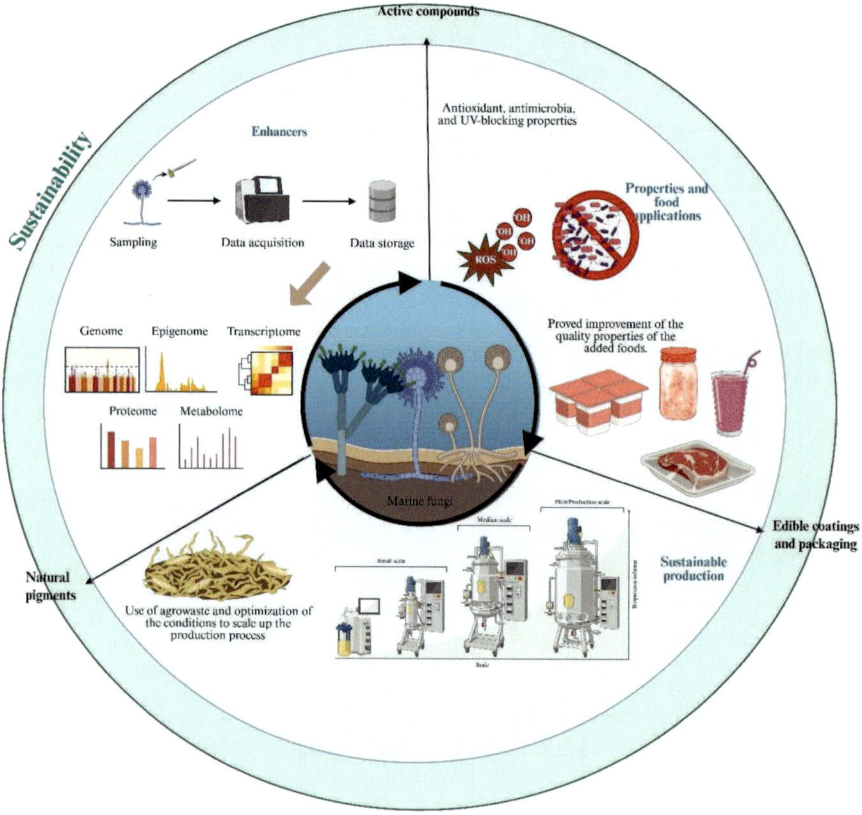

Fig. 4. Integrated biotechnological approach for the sustainable use of marine fungi in food applications.

The application of fungal pigments in beverages poses unique challenges due to the low pH and high-water activity of these systems. However, water-soluble red pigments from *Penicillium purpurogenum* and *Aspergillus sydowii* have shown promise in carbonated drinks, fruit juices, and alcoholic beverages (Méndez et al., 2011; Pimenta et al., 2021). These pigments remain stable over shelf-life periods typical of bottled beverages and have been shown to resist degradation from UV exposure, an important consideration for transparent packaging. Their vivid hues also align well with the demand for naturally colored energy drinks and functional beverages (Pimenta et al., 2021).

Marine fungal pigments are suitable for confections, such as gummies, hard candies, and jellies, where bold colors are essential (Fig. 1) (Molelekoa et al., 2023). Their thermal stability allows them to endure high-temperature processes such as sugar boiling or baking. Pigments like monascorubrin and rubropunctatin from marine-adapted *Talaromyces* strains retain their color intensity after baking in cookies and cakes, particularly when used in encapsulated form (Egea et al., 2023).

7. Challenges and Future Perspectives

Despite the growing interest and promising potential of marine fungal pigments, several scientific, technical, and regulatory hurdles must be addressed to facilitate their widespread application in the food industry. While advances in fermentation, extraction, and formulation have paved the way for industrial interest, challenges related to production scalability, regulatory compliance, and pigment consistency remain. Furthermore, the integration of synthetic biology, omics-driven strain improvement, and cross-sector collaborations is essential to accelerate innovation and enable sustainable commercialization (Jin et al., 2016; Hassan et al., 2023).

7.1 Bottlenecks in Scalability and Regulatory Hurdles

The large-scale production of marine fungal pigments faces multiple bottlenecks. These include the relatively low yields in wild-type strains, the variability of pigment output due to changes in environmental conditions, and difficulties in biomass separation and downstream purification. Additionally, the marine origin of the strains may introduce specific culture medium requirements (such as salinity, micronutrients) that complicate fermentation scale-up (Hassan et al., 2023).

In terms of regulatory approval, natural pigments from terrestrial fungi such as *Monascus* have faced a strong examination due to the co-production of mycotoxins. Similarly, marine-derived pigments must undergo comprehensive toxicological evaluations to determine their safety for human consumption. This includes assessing mutagenicity, allergenicity, and long-term metabolic impact. Current frameworks, such as those established by the EFSA and the U.S. FDA, demand extensive documentation of strain identification, production conditions, and the absence of harmful metabolites (Aguilar et al., 2009; FDA, 2023).

A key challenge lies in the limited precedent for marine fungal pigments in regulatory submissions. The lack of historical use data delays novel food approval processes and increases the costs of bringing a product to market. Establishing GRAS status, novel food applications in the EU, or equivalent approvals in the Asia-Pacific regions requires concerted efforts to generate comprehensive safety data. The development of standardized analytical protocols and metabolic profiling of pigments can support regulatory dossiers and streamline evaluation processes.

7.2 Genetic and Metabolic Engineering Opportunities

To overcome production limitations, genetic and metabolic engineering offer a powerful toolkit to enhance pigment yields, reduce unwanted byproducts, and broaden the color palette of marine fungal pigments. Gene cluster identification and characterization of biosynthetic pathways in *Talaromyces*, *Penicillium*, and *Aspergillus* species have revealed modular enzyme systems, including polyketide synthases (PKS), nonribosomal peptide synthetases (NRPS), and tailoring enzymes involved in pigment biosynthesis (Mózsik et al., 2022). Targeted overexpression of rate-limiting enzymes, disruption of competing pathways, and the integration of synthetic promoters can boost pigment biosynthesis. Heterologous expression of pigment biosynthetic gene clusters in GRAS hosts such as *Saccharomyces cerevisiae*, *Vibrio natriegens*, or *Aspergillus oryzae* represents another strategy to enhance safety and scalability (Wang et al., 2020). This platform allows decoupling pigment production from toxin-producing strains while maintaining high biosynthetic efficiency. Metabolic modeling and flux balance analysis can guide strain engineering efforts by identifying metabolic bottlenecks and redirecting precursor fluxes toward desired compounds.

7.3 Synthetic Biology and Omics Tools in Pigment Enhancement

Recent advances in synthetic biology and multi-omics technologies (genomics, transcriptomics, proteomics, and metabolomics) have revolutionized our understanding of marine fungal metabolism and opened new avenues for pigment enhancement. Genome mining has enabled the identification of cryptic biosynthetic gene clusters associated with pigment production that remain silent under standard culture conditions. Through tools such as CRISPR/Cas9 and multiplex genome editing, these clusters can be activated or reconstructed in laboratory strains to reveal novel pigments (Leal et al., 2024). Transcriptomic analyses under varying culture conditions provide insights into gene regulation dynamics and can be used to identify transcription factors and culture parameters that optimize pigment synthesis. Proteomic studies further validate enzyme expression and activity within biosynthetic pathways, while metabolomic profiling allows the quantification of pigments and their precursors. In addition, machine learning and artificial intelligence are emerging tools to analyze large omics datasets, predict pigment yields, and guide experimental design. These computational approaches accelerate the design-build-test-learn cycle in strain development and reduce the time from discovery to application. Together, these approaches create a systems biology framework to guide rational engineering strategies.

7.4 Collaboration: Academia, Industry, and Regulators

The development of marine fungal pigments for food applications requires a multidisciplinary, collaborative approach involving academic researchers, industrial partners, and regulatory authorities. Academia plays a critical role in exploring fungal biodiversity, elucidating biosynthetic pathways, and developing enabling technologies for strain improvement and bioprocessing. However, translation into commercial products necessitates engagement from industry to scale production, validate pigment functionality in food systems, and navigate market requirements. Public-private partnerships can bridge the gap between research and application. Collaborative consortia such as the European Innovation Council and the U.S. National Science Foundation's Industry-University Cooperative Research Centers can facilitate technology transfer and joint development of standards. Industrial partners can support pilot-scale validation, application testing, and investment in production infrastructure. Regulatory agencies must also be engaged early in the development process to guide safety assessments, required data packages, and submission formats. Pre-submission consultations, such as those offered by the EFSA and FDA, can help align product development with regulatory expectations and reduce approval timelines. An integrated ecosystem that fosters open communication, data sharing, and shared risk can accelerate the commercialization of marine fungal pigments. Establishing databases of characterized pigments, standardized protocols for safety evaluation, and accessible repositories of engineered strains can support global efforts toward sustainable natural colorants.

8. Conclusion

The future of marine fungi pigments in the food industry is promising, as consumer preferences for clean-label products and technological advances in the production of natural pigments increasingly coincide. Further exploration of the diversity of marine fungi, particularly from extreme habitats such as deep-sea springs and polar regions, could reveal novel pigments with unique color properties and functionalities. The development of genomic techniques to eliminate genes associated with mycotoxin production is a promising alternative to improve pigment production and increase production capacity from marine sources. Sustainable bioprocessing concepts using agro-industrial by-products as fermentation substrates can improve the environmental footprint of pigment production. Innovations in bioreactor design, process automation, and downstream purification will be critical to achieving cost-efficient production on a commercial scale. The integration of pigment production with other valuable by-products, such as enzymes, bioactive compounds, or bioplastics, could improve process economics and support circular economy models. In addition, the multifunctionality of some marine fungal pigments, which have antioxidant, antimicrobial, or UV-protective effects, extends their potential beyond coloring to active food packaging, nutraceuticals, and functional foods.

Acknowledgments

Figures were obtained by BioRender through the premium membership.

References

Abdel-Raheam, H.E., Alrumman, S.A., Gadow, S.I., El-Sayed, M.H., Hikal, D.M. et al. (2022). Optimization of *Monascus purpureus* for natural food pigments production on potato wastes and their application in ice lolly. Front. Microbiol., 13: 862080. https://doi.org/10.3389/FMICB.2022.862080/BIBTEX.

Afroz Toma, M., Rahman, M.H., Rahman, M.S., Arif, M., Nazir, K.N.H. et al. (2023). Fungal pigments: Carotenoids, riboflavin, and polyketides with diverse applications. J. Fungi, 9: 454. https://doi.org/10.3390/JOF9040454.

Aguilar, U.R., Charrondiere, B., Dusemund, P., Galtier, P., Gilbert, J. et al. (2009). Scientific opinion on the re-evaluation Tartrazine (E 102). EFSA J., 7: 1331. https://doi.org/10.2903/J.EFSA.2009.1331.

Ashenafi, E.L., Nyman, M.C., Shelley, J.T. and Mattson, N.S. (2023). Spectral properties and stability of selected carotenoid and chlorophyll compounds in different solvent systems. Food Chem. Adv., 2: 100178. https://doi.org/10.1016/J.FOCHA.2022.100178.

Barciela, P., Perez-Vazquez, A. and Prieto, M.A. (2023). Azo dyes in the food industry: Features, classification, toxicity, alternatives, and regulation. Food Chem. Toxicol., 178: 113935. https://doi.org/10.1016/J.FCT.2023.113935.

Britto Martins de Oliveira, J., Corrêa Junior, D., Parente, C.E.T. and Frases, S. (2025). Fungi in mangrove: Ecological importance, Climate change impacts, and the role in environmental remediation. Microorganisms, 13(4): 878. https://doi.org/10.3390/MICROORGANISMS13040878.

Caro, Y., Venkatachalam, M., Lebeau, J., Fouillaud, M. and Dufossé, L. (2017). Pigments and colorants from filamentous fungi. pp. 499–568. *In*: Mérillon, J.M. and Ramawat, K. (eds.). Fungal Metabolites. Reference Series in Phytochemistry. Springer, Cham. https://doi.org/10.1007/978-3-319-25001-4_26.

Christiansen, J.V., Isbrandt, T., Petersen, C., Sondergaard, T.E., Nielsen, M.R. et al. (2021). Fungal quinones: Diversity, producers, and applications of quinones from *Aspergillus*, *Penicillium*, *Talaromyces*, *Fusarium*, and *Arthrinium*. Appl. Microbiol. Biotechnol., 105: 8157–8193. https://doi.org/10.1007/S00253-021-11597-0.

Cordeiro de Souza, T.F. (2018). Cerveja com adição de arroz fermentado em estado sólido pelo fungo Monascus ruber e identificação de metabólitos secundários. https://sucupira.capes.gov.br/sucupira/public/consultas/coleta/trabalhoConclusao/viewTrabalhoConclusao.jsf?popup=true&id_trabalho=6935807.

Couttolenc, A., Medina, M.E., Trigos, Á. and Espinoza, C. (2021). Antioxidant capacity of fungi associated with corals and sponges of the reef system of Veracruz, Mexico. Electron. J. Biotechnol., 55: 40–46. https://doi.org/10.1016/J.EJBT.2021.11.002.

Darwesh, O.M., Matter, I.A., Almoallim, H.S., Alharbi, S.A. and Oh, Y.K. (2020). Isolation and optimization of Monascus ruber OMNRC45 for red pigment production and evaluation of the pigment as a food colorant. Appl. Sci., 10: 8867. https://doi.org/10.3390/APP10248867.

de Oliveira, F., Rocha, I.L., Pinto, D.C.G.A., Ventura, S.P., Dos Santos, A.G. et al. (2022). Identification of azaphilone derivatives of *Monascus* colorants from *Talaromyces amestolkiae* and their halochromic properties. Food Chem., 372: 131214. https://doi.org/10.1016/J.FOODCHEM.2021.131214.

Egea Buranelo, M. and Marcionilio de Oliveira, S.M.L. (2022). A biodiversidade como fonte de compostos bioativos: Moléculas e aplicações. https://repositorio.ifgoiano.edu.br/handle/prefix/2341.

Egea, M.B., Dantas, L.A., Sousa, T.L.D., Lima, A.G. and Lemes, A.C. (2023). The potential, strategies, and challenges of Monascus pigment for food application. Front. Sustain. Food Syst., 7: 1141644. doi: 10.3389/fsufs.2023.1141644.

El-Kashef, D.H., Youssef, F.S., Hartmann, R., Knedel, T.O., Janiak, C. et al. (2020). Azaphilones from the red sea fungus *Aspergillus falconensis*. Mar. Drugs, 18. https://doi.org/10.3390/MD18040204.

Evidente, A. (2024). Advances on anticancer fungal metabolites: Sources, chemical, and biological activities in the last decade (2012–2023). Nat. Prod. Bioprospect., 14: 31. https://doi.org/10.1007/S13659-024-00452-0.

FDA. (2023). Title 21—Food and Drugs Chapter I—Food and Drug Administration Department of Health and Human Services, Subchapter B—Food for Human Consumption, Part 172 Food Additives Permitted for Direct Addition to Food for Human Consumption. https://www.accessdata.fda.gov/scripts/cdrh/cfdocs/cfcfr/CFRSearch.cfm?CFRPart=172&showFR=1&subpartNode=21:3.0.1.1.3.3 (Accessed 29 Aug 2023).

Fouillaud, M., Venkatachalam, M., Girard-Valenciennes, E., Caro, Y., Dufossé, L. et al. (2016), Anthraquinones and derivatives from marine-derived fungi: Structural diversity and selected biological activities. Mar. Drugs, 14: 64. https://doi.org/10.3390/md14040064.

Fouillaud, M., Venkatachalam, M., Llorente, M., Magalon, H., Cuet, P. et al. (2017). Biodiversity of pigmented fungi isolated from marine environment in la réunion Island, Indian ocean: New resources for colored metabolites. J. Fungi, 3: 36. https://doi.org/10.3390/jof3030036.

Frolov, I.V., Vondrák, J., Konoreva, L.A., Chesnokov, S.V., Himelbrant, D.E. et al. (2021). Three new species of crustose Teloschistaceae in Siberia and the Far East. Lichenologist, 53: 233–243. https://doi.org/10.1017/S0024282921000177.

Gao, M., Liu, B., Li, J., Deng, Y., Zhang, Y. et al. (2024). Diversity and distribution of fungi in the marine sediments of Zhanjiang Bay, China. J. Fungi, 10: 867. https://doi.org/10.3390/JOF10120867/S1.

Ghattavi, K., Homaei, A., Kamrani, E. and Kim, S.K. (2022). Melanin pigment derived from marine organisms and its industrial applications. Dyes and Pigm., 201: 110214. https://doi.org/10.1016/J.DYEPIG.2022.110214.

Gomah, N.H., Abdel-Raheam, H.E.F. and Mohamed, T.H. (2017). Production of natural pigments from Monascus ruber by solid state fermentation of broken rice and its application as colorants of some dairy products. J. Food Dairy Sci., 8: 37–43. https://doi.org/10.21608/JFDS.2017.37112.

Gonçalves, M.F., Esteves, A.C. and Alves, A. (2022). Marine fungi: Opportunities and challenges. Encyclopedia, 2(1): 559–577. https://doi.org/10.3390/encyclopedia2010037.

Grand View Research (2024). Food Colors Market Size and Share | Industry Report, 2030. https://www.grandviewresearch.com/industry-analysis/food-colorants-market (Accessed 22 May 2025).

Hafez Ghoran, S., Taktaz, F., Ayatollahi, S.A. and Kijjoa, A. (2022). Anthraquinones and their analogues from marine-derived fungi: Chemistry and biological activities. Mar. Drugs, 2022(20): 474. https://doi.org/10.3390/MD20080474.

Hassan, S.W.M., Abdrabo, M.A.A. and Elsayis, A. (2022). Microbial pigments: Sources and applications in the marine environment. Egypt J. Aquat. Biol. Fish, 26: 99–124. https://doi.org/10.21608/ejabf.2022.215163.

Hassan, S.W.M., Abdrabo, M.A.A., Alkersh, B.M., Elsayis, A., Abdel-Latif, H.H. et al. (2023). Marine fungal pigments: From Lab to potential applications. Egypt. J. Aquat. Biol. Fish., 27(4): 1027–1046.

Hawas, U.W., El-Beih, A.A. and El-Halawany, A.M. (2012). Bioactive anthraquinones from endophytic fungus aspergillus versicolor isolated from red sea algae. Arch. Pharm. Res., 35: 1749–1756. https://doi.org/10.1007/s12272-012-1006-x.

Hebra, T., Eparvier, V. and Touboul, D. (2024). Supercritical CO_2 extraction of solid-state cultivation fungus producing azaphilone polyketides. Green Anal. Chem., 8: 100102. https://doi.org/10.1016/J.GREEAC.2024.100102.

Jin, L., Quan, C., Hou, X. and Fan, S. (2016). Potential pharmacological resources: Natural bioactive compounds from marine-derived fungi. Mar. Drugs, 14(4): 76. https://doi.org/10.3390/md14040076.

Kalra, R., Conlan, X.A. and Goel, M. (2020). Fungi as a potential source of pigments: Harnessing filamentous fungi. Front. Chem., 8: 526952. https://doi.org/10.3389/fchem.2020.00369.

Kannan, J., Pang, K.L., Ho, Y.N., Hsu, P.H., Chen, L.L. et al. (2024). A comparison of the antioxidant potential and metabolite analysis of marine fungi associated with the red algae *Pterocladiella capillacea* from Northern Taiwan. Antioxidants, 13: 336. https://doi.org/10.3390/antiox13030336.

Kaur, M., Goel, M., Mishra, R.C., Lahane, V., Yadav, A.K. et al. (2023). Characterization of the red biochromes produced by the endophytic fungus *Monascus purpureus* CPEF02 with antimicrobial and antioxidant activities. Fermentation, 9: 328. https://doi.org/10.3390/fermentation9040328.

Keem, M.J., Seo, S.W., Kim, T., Jo, B.G., Kim, S.N. et al. (2023). A High-Performance liquid chromatography with photodiode array detection method for simultaneous determination of three compounds isolated from *Wikstroemia ganpi*: Assessment of the effects on cytochrome p450-mediated metabolism *in vitro* and *in vivo*. Nutrients, 15: 4061. https://doi.org/10.3390/nu15184061.

Knowles, S.L., Raja, H.A., Roberts, C.D. and Oberlies, N.H. (2022). Fungal–fungal co-culture: A primer for generating chemical diversity. Nat. Prod. Rep., 39: 1557. https://doi.org/10.1039/D1NP00070E.

Kulczyk-Małysa, E. and Bogusławska-Wąs, E. (2025). Carotenoid yeasts and their application potential. Foods, 14: 1866. https://doi.org/10.3390/foods14111866.

Kumawat, C., Kumar, A., Parshad, J., Sharma, S.S., Patra, A. et al. (2022). Microbial diversity and adaptation under salt-affected soils: A review. Sustainability, 14: 9280. https://doi.org/10.3390/su14159280.

Leal, K., Rojas, E., Madariaga, D. et al. (2024). Unlocking fungal potential: The CRISPR-cas system as a strategy for secondary metabolite discovery. J. Fungi, 10: 748. https://doi.org/10.3390/jof10110748.

Lin, L. and Xu, J. (2022). Production of fungal pigments: Molecular processes and their applications. J. Fungi, 9: 44. https://doi.org/10.3390/jof9010044.

Li, R., Su, Z., Sun, C. and Wu, S. (2024). Antibacterial insights into alternariol and its derivative alternariol monomethyl ether produced by a marine fungus. Appl. Environ. Microbiol., 90: e00058-24. https://doi.org/10.1128/aem.00058-24.

López-Cruz, R., Sandoval-Contreras, T. and Iñiguez-Moreno, M. (2023). Plant pigments: Classification, extraction, and challenge of their application in the food industry. Food Bioproc. Tech., 16(12): 2725–2741. https://doi.org/10.1007/s11947-023-03075-4.

Maľa, P., Baranová, M., Marcinčáková, D. and Nagy, J. (2010). Organoleptic evaluation of poultry meat products with wheat protein-seitan, coloured by microbial natural pigment. Assam Univ. J. Sci. Technol. Bio. Environ. Sci., 5: 1–5.

Mansour, R.I., El Shoubaky, G.A. and Attia, E.A. (2025). Exploring the antioxidant and cytotoxicity perspective of melanin pigment produced by fungi associated with marine macroalgae, Fayed, Suez Canal, Egypt. Microb. Biosyst., 10: 181–194. https://doi.org/10.21608/mb.2025.352914.1229.

Méndez, A., Pérez, C., Montañéz, J.C., Martínez, G. and Aguilar, C.N. (2011). Red pigment production by *Penicillium purpurogenum* GH2 is influenced by pH and temperature. J. Zhejiang Univ. Sci. B, 12: 961–968. https://doi.org/10.1631/jzus.B1100039.

Meruvu, H. and dos Santos, J.C. (2021). Colors of life: A review on fungal pigments. Crit. Rev. Biotechnol., 41: 1153–1177. https://doi.org/10.1080/07388551.2021.1901647.

Molelekoa, T.B.J., da Silva, L.S., Regnier, T. and Augustyn, W. (2023). Application and stability of fungal pigments using jelly sweets as a food model system. Int. J. Food Sci. Technol., 58: 6761–6774. https://doi.org/10.1111/IJFS.16644.

Monteiro, A.B.P., Prados, C.R.M.G., Silva, M.D.L.R., Silva, E.P., Damiani, C. et al. (2021). Production of Monascus pigments by solid-state cultivation of wheat grains and application in bread formulations. Int. J. Gastron. Food Sci., 24: 100313. https://doi.org/10.1016/J.IJGFS.2021.100313.

Mózsik, L., Iacovelli, R., Bovenberg, R.A.L. and Driessen, A.J.M. (2022). Transcriptional activation of biosynthetic gene clusters in filamentous fungi. Front. Bioeng. Biotechnol., 10: 901037. https://doi.org/10.3389/FBIOE.2022.901037.

Muñoz-Miranda, L.A. and Iñiguez-Moreno, M. (2023). An extensive review of marine pigments: Sources, biotechnological applications, and sustainability. Aquat. Sci., 85: 68. https://doi.org/10.1007/s00027-023-00966-8.

Neto, F.C., Guaratini, T., Costa-Lotufo, L., Colepicolo, P., Gates, P.J. et al. (2016). Re-investigation of the fragmentation of protonated carotenoids by electrospray ionization and nanospray tandem mass spectrometry. Rapid Commun. Mass Spectrom., 30: 1540–1548. https://doi.org/10.1002/rcm.7589.

Parul, Thiyam, G., Dufossé, L. and Sharma, A.K. (2020). Characterization of *Talaromyces purpureogenus* strain F extrolites and development of production medium for extracellular pigments enriched with antioxidant properties. Food Bioprod. Process., 124: 143–158. https://doi.org/10.1016/J.FBP.2020.08.012.

Pavan, M.E., López, N.I. and Pettinari, M.J. (2020). Melanin biosynthesis in bacteria, regulation, and production perspectives. Appl. Microbiol. Biotechnol., 104: 1357–1370. https://doi.org/10.1007/s00253-019-10245-y.

Peng, X., Amend, A.S., Baltar, F., Blanco-Bercial, L., Breyer, E. et al. (2023). Planktonic marine fungi: A review. Authorea Preprints. https://doi.org/10.22541/essoar.169903594.45892424/v1.

Pereira, D.M., Valentão, P. and Andrade, P.B. (2014). Marine natural pigments: Chemistry, distribution, and analysis. Dyes Pigm., 111: 124–134. https://doi.org/10.1016/j.dyepig.2014.06.011.

Pimenta, L.P.S., Gomes, D.C., Cardoso, P.G. and Takahashi, J.A. (2021). Recent findings in azaphilone pigments. J. Fungi, 7: 541. https://doi.org/10.3390/jof7070541.

Ram, S., Mitra, M., Shah, F., Tirkey, S.R. and Mishra, S. (2020). Bacteria as an alternate biofactory for carotenoid production: A review of its applications, opportunities, and challenges. J. Funct. Foods, 67: 103867. https://doi.org/10.1016/j.jff.2020.103867.

Rengifo, L.R., Rosas, P., Méndez, N., Ludeña, Y., Sirvas, S. et al. (2022). Comparison of pigment production by filamentous fungal strains under Submerged (SmF) and Surface Adhesion Fermentation (SAF). J. Fungi, 9: 48. https://doi.org/10.3390/jof9010048.

Rokas, A., Mead, M.E., Steenwyk, J.L., Raja, H.A. and Oberlies, N.H. (2020). Biosynthetic gene clusters and the evolution of fungal chemodiversity. Nat. Prod. Rep., 37: 868. https://doi.org/10.1039/C9NP00045C.

Sen, K., Sen, B. and Wang, G. (2022). Diversity, abundance, and ecological roles of planktonic fungi in marine environments. J. Fungi, 8: 491. https://doi.org/10.3390/jof8050491.

Srianta, I., Kuswardani, I., Ristiarini, S., Kusumawati, N., Godelive, L. et al. (2022). Utilization of durian seed for Monascus fermentation and its application as a functional ingredient in yogurt. Bioresour. Bioprocess., 9: 128. https://doi.org/10.1186/s40643-022-00619-y.

Sun, X., Zhou, X., Cai, M., Tao, K. and Zhang, Y. (2009). Identified biosynthetic pathway of aspergiolide A and a novel strategy to increase its production in a marine-derived fungus *Aspergillus glaucus* by feeding of biosynthetic precursors and inhibitors simultaneously. Bioresour. Technol., 100: 4244–4251. https://doi.org/10.1016/j.biortech.2009.03.061.

Tarman, K., Lindequist, U., Wende, K., Porzel, A., Arnold, N. et al. (2011). Isolation of a new natural product and cytotoxic and antimicrobial activities of extracts from fungi of Indonesian marine habitats. Mar. Drugs, 9: 294–306. https://doi.org/10.3390/md9030294.

Teiba, I.I., Mamdouh, I., Yousef, M.I., Hussein, A. and El-Bilawy, E.H. (2025). Antimicrobial activity of *Monascus purpureus*-derived red pigments against *Salmonella typhimurium*, *Escherichia coli*, and *Enterococcus faecalis*. AMB Express, 15: 6. https://doi.org/10.1186/S13568-024-01801-5.

Vasarri, M., Vitale, G.A., Varese, G.C., Barletta, E., D'Auria, M.V. et al. (2022). Dihydroauroglaucin isolated from the Mediterranean sponge *Grantia compressa* endophyte marine fungus *Eurotium chevalieri* inhibits migration of human neuroblastoma cells. Pharmaceutics, 14: 616. https://doi.org/10.3390/pharmaceutics14030616.

Velez, P., Salcedo, D.L., Espinosa-Asuar, L., Gasca-Pineda, J., Hernandez-Monroy, A. et al. (2022). Fungal diversity in sediments from deep-sea extreme ecosystems: Insights into low- and high-temperature hydrothermal vents, and an oxygen minimum zone in the Southern Gulf of California, Mexico. Front. Mar. Sci., 9: 802634. https://doi.org/10.3389/fmars.2022.802634.

Venkatachalam, M. (2024). Pigment production using submerged fermentation. Fermentation, 10(2): 91. https://doi.org/10.3390/fermentation10020091.

Venkatachalam, M., Magalon, H., Dufossé, L. and Fouillaud, M. (2018). Production of pigments from the tropical marine-derived fungi Talaromyces albobiverticillius: New resources for natural red-colored metabolites. J. Food Compos. Anal., 70: 35–48. https://doi.org/10.1016/j.jfca.2018.03.007.

Venkatachalam, M., Mares, G., Dufossé, L. and Fouillaud, M. (2023). Scale-up of pigment production by the marine-derived filamentous fungus, *Talaromyces albobiverticillius* 30548, from shake flask to stirred bioreactor. Fermentation, 9: 77. https://doi.org/10.3390/fermentation9010077.

Vílchez, C., Forján, E., Cuaresma, M., Bédmar, F., Garbayo, I. et al. (2011). Marine carotenoids: Biological functions and commercial applications. Mar. Drugs, 9(3): 319–333. https://doi.org/10.3390/md9030319.

Wadhwa, K., Kapoor, N., Kaur, H., Abu-Seer, E.A., Tariq, M. et al. (2024). A comprehensive review of the diversity of fungal secondary metabolites and their emerging applications in healthcare and environment. Mycobiology, 52(6): 335–387. https://doi.org/10.1080/12298093.2024.2416736.

Wang, W., Liao, Y., Chen, R., Hou, Y., Ke, W. et al. (2018). Chlorinated azaphilone pigments with antimicrobial and cytotoxic activities isolated from the deep sea derived fungus *Chaetomium* sp. NA-S01-R1. Mar. Drugs, 16: 61. https://doi.org/10.3390/md16020061.

Wang, Z., Tschirhart, T., Schultzhaus, Z., Kelly, E.E., Chen, A. et al. (2020). Melanin produced by the fast-growing marine bacterium Vibrio natriegens through heterologous biosynthesis: Characterization and application. Appl. Environ. Microbiol., 86: e02749-19. https://doi.org/10.1128/AEM.02749-19.

Xu, L., Meng, W., Cao, C., Wang, J., Shan, W. et al. (2015). Antibacterial and antifungal compounds from marine fungi. Mar. Drugs, 13(6): 3479–3513. https://doi.org/10.3390/md13063479.

Zhong, M., Li, Y., Deng, L., Fang, J. and Yu, X. (2023). Insight into the adaptation mechanisms of high hydrostatic pressure in physiology and metabolism of hadal fungi from the deepest ocean sediment. Msystems, 9: e01085-23. https://doi.org/10.1128/msystems.01085-23.

Zhou, S., Wang, M., Feng, Q., Lin, Y. and Zhao, H. (2016). A study on biological activity of marine fungi from different habitats in coastal regions. Springerplus, 5: 1966. https://doi.org/10.1186/s40064-016-3658-3.

Industrial Applications

4

Marine Fungal Pigments

A Promising Source of Bioactive Hues for Industrial Applications

Mehak Kaur,[1,2] *Shubhajit Dhara*[2] and *Mayurika Goel*[2,*]

The growing demand for sustainable alternatives to non-renewable resources has driven interest in natural colorants across food, textile, cosmetic, pharmaceutical, and other industries. Among biological sources, microorganisms are highly favored due to their stability, water solubility, cost-effectiveness, and year-round availability. Marine-derived filamentous fungi have emerged as promising candidates, producing bioactive pigments such as melanin, azaphilones, and hydroxyanthroquinones that protect against UV radiation, pathogens, salinity, and environmental stressors. Moreover, marine fungal endophytes have the capacity to generate bioactive compounds and colorants that offer potential applications in food, textiles, paints, agriculture, and medicine. Their biosynthetic pathways have been systematically explored for biotransformation into cell factories for industrial-scale production. The current chapter will focus on pigments from marine fungi with significant bioactivities, biotechnological interventions utilized to enhance their production, followed by industrial avenues opened through the same. It also addresses commercialization challenges and proposes solutions to overcome these barriers.

[1] Department of Botany, University of Delhi, Delhi, 110007, India.
[2] Division of Sustainable Agriculture, The Energy and Resources Institute, Gurugram, Haryana, 122001.
* Corresponding author: mayurika.goel@teri.res.in/mayurikagoel@gmail.com

DOI: 10.1201/9781003795100-6

1. Introduction

Colorants in various shades and forms have been an inseparable part of life on Earth, serving roles ranging from photosynthesis to aesthetics. Before the 19th century, natural colorants extracted from plants, animals, rocks, insects, or microorganisms were widely used in textiles, food, cosmetics, and healing preparations. However, they were soon replaced by synthetic petroleum-based dyes and pigments with the invention of the mauvine color by Sir William Henry Perkin in 1856 (Welham, 1963). This triggered a trail of synthetic color development for a diverse range of applications in food, textiles, pharmaceuticals, and cosmetics, introducing the onset of the synthetic era. Soon after, their detrimental impacts started unfolding with occasional sightings of colored water bodies and toxicity to freshwater and marine ecosystems along with humans. Synthetic dyes have been associated with hyperactivity in children, neurodevelopmental disorders, allergenicity, endocrine disruption, and mutagenesis in humans. Moreover, the effluents generated during their production as well as application have been linked to air, water, and land pollution, reducing soil productivity and altering microbiome profiles (Singh et al., 2021; Fried et al., 2022). These findings shifted the focus of conscious consumers from synthetic colorants and their commodities to natural and sustainable alternatives. Natural colors can be extracted from plants, animals, insects, and microorganisms. However, plant-based colors often suffer from low yields and extensive extraction procedures, while insects and animal-derived colors raise ethical issues.

Amongst microorganisms, fungi stand out as prolific producers of colored compounds in the form of bioactive secondary metabolites. They exhibit robust growth, can be scaled up economically for continuous production, and allow feasible downstream processing (Kalra et al., 2020). Fungi are ubiquitous in nature, with an established presence in extreme conditions such as hot springs, glaciers, hypersaline lakes, and polar highs (Ibrar et al., 2020). One such extreme environment is the marine ecosystem, covering over 70% of earth and harboring a treasure trove of biological resources. However, the majority of it remains unexplored due to lack of detailed understanding and technological challenges in collection, cultivation, and modification. For instance, over 0.1 million fungal species have been identified across the world, with 27,500 being from India and merely 3,000 species from the aquatic habitat (Gareth Jones, 2011). Marine fungi have been isolated from seawater, mangroves, corals, sponges, plants, and other substrates, and represent an ecological classification. They can be obligate (sporulating, fast-growing, and exclusively sea-dwelling) or facultative (initially terrestrial and later on adapted to marine) (Agrawal et al., 2019).

Owing to the complex surrounding environment, marine fungi have developed unique metabolic pathways for the production of novel secondary bioactive chemical moieties. These are derived from primary metabolites and produced as a defense response against external biotic and abiotic stress conditions (Rajagopal et al., 1997). More than a thousand novel chemical moieties have been identified from marine fungi belonging to polyketides, alkaloids, terpenoids, lipids, non-ribosomal peptides, polyenes, carotenoids, melanin, monascins, and other classes. Amongst them, pigmented bioactive compounds belonging to carotenoids, melanin, polyketides, and non-ribosomal peptides classes, such as quinones, violacein, anthraquinones, naphthoquinones, flavonoids, and others have been reported to possess bioactive properties such as antimicrobial, anticancer, antioxidant, and anti-inflammatory properties (Rateb and Ebel, 2011; Kalra et al., 2020). For instance, novel azaphilones and meroterpenes were identified from *Penicillium* sp. collected from seawater in China and were identified to possess antibacterial properties (Li et al., 2014). Even though a diverse range of pigmented compounds can be isolated from marine filamentous fungi, only a handful of them have been explored, a few have been upscaled, and none has been taken to the commercial scale. This book chapter aims to shed light on the pigmented bioactive compounds isolated from marine fungi, challenges faced in their cultivation and production, biotechnological interventions utilized to overcome those challenges, and industrial avenues opened through them.

2. Pigments and their Pharmaceutical Potential

A growing number of fungi have been isolated from marine environments such as mangroves, seabeds, sponges, coral reefs, solar salterns, and hydrothermal vents as saprotrophs, parasites, and endophytes (Behera and Das, 2023). Somewhat similar to their terrestrial counterparts, these fungi can also secrete known and novel pigmented secondary metabolites with hues ranging from yellow to black and biological properties such as antioxidant, anticancer, antimicrobial, anti-inflammatory, and protein tyrosine phosphatase inhibition, as depicted in Table 1 and Fig. 1 (Xu et al., 2008; Xia et al., 2014; Kalra et al., 2020; Meruvu and dos Santos, 2021; Behera and Das, 2023). In the presence of extreme temperature, pressure, and salinity, marine fungal species such as *Eurotium rubrum*, *Halorosellinia*, and *Microsporum* sp. have been identified to synthesize biopigments, which are able to emit light and/or heat (Muñoz-Miranda and Iñiguez-Moreno, 2023).

Aspergillus, *Alternaria*, *Penicillium*, *Eurotium*, and *Talaromyces* are the major pigment-producing marine fungi, along with alterporriol, bostrycin, physcion, asperflavin, sclerotiorin, emodin, flavoglaucin, catenarin, melanin, macrosporin, bacillisporin, and chrysophanol as the major pigments (Fouillaud et al., 2016). Species of *Alternaria* have been isolated as endophytes from mangrove forests, coral reefs, and deep sea, producing bostrycin, alterporriol, and similar pigmented compounds with yellow, orange, red, and pink hues. These pigments were found to possess antibacterial, antimycobacterial, anticancer, and phytotoxic activities (Charudattan and Rao, 1982; Huang et al., 2011; Zheng et al., 2012; Xia et al., 2014).

The genus of *Aspergillus* is distributed widely in the marine ecosystem as a symbiont with coral reefs, sponges, and algae and in extreme environments such as deep seas and solar salterns. Aspergiolides, physcion, emodin, catenarin, rubrocristin, variecolorin, variecolorquinone, falconensin, bostrycin, isorhodoptilometrin-1-methyl ether, 3-(2-hydroxypropyl)-1-methyl ether-6,8-dihydroxy-9,10-anthraquinone, emodin, 1-Methyl emodin, evariquinone,

Table 1. Pigmented compounds isolated from marine filamentous fungi.

Species	Isolated From	Metabolite	Hue	Bioactivity	References
Alternaria eichhorniae	Mangrove plant	Bostrycin, 4-Deoxybostrycin	Red	Antibacterial and phytotoxic	Charudattan and Rao (1982)
Alternaria sp. ZJ-2008003	Coral reef	Tetrahydroaltersolanol F, Alterporriol N, Alterporriol O, Alterporriol P, Alterporriol Q	Pink, red, yellow	Cytotoxic	Zheng et al. (2012)
Alternaria sp. ZJ9-6B	Mangrove ecosystem	Alterporriol K, Alterporriol L, Alterporriol M,	Red	Anticancer	Huang et al. (2011)
Alternaria sp. (SK11)	Sea-derived	Alterporriol S, S-Alterporriol C	Red and orange	*Mycobacterium tuberculosis* protein tyrosine phosphatase B inhibitors	Xia et al. (2014)
Ascomycota sp. Ind19F07	Deep sea derived	Ascomycotina	Yellow	Antibacterial	Tian et al. (2015)
Aspergillus falconensis	Sea derived	Falconensin O-S	Yellow	Anticancer	El-Kashef et al. (2020)
Aspergillus glaucus	Marine derived	Aspergiolide B, Physcion, Questin, Emodin, Catenarin, Rubrocristin, (+)-Variecolorquinones A	Red, yellow	Cytotoxic	Du et al. (2008)
Aspergillus nidulans sp. SG 28	Marine water	Melanin Precursor 5,6-Dihydroxy-Indole-2-Carboxylic Acid	Orange, reddish pink	Antioxidant and UV protectant	Shanuja et al. (2018)
Aspergillus sp.	Coral reef	Bostrycin, 4-Deoxybostrycin	Red	Antimicrobial, antiprotozoal, phytotoxic, and cytotoxic activities	Xu et al. (2008)
Aspergillus sp. 16-02-1	Deep sea-derived	Epiaspinonediol	Yellow	Anticancer	Chen et al. (2014)
Aspergillus sp. SCSIO F063	Marine derived	(1′S)-6-O-Methyl-7-Chloroaverantin, (1′S)-7-Chloroaverantin-1′-Butyl Ether, 7-Chloroaverythrin, 6-O-Methyl-7-Chloroaverythrin, (1′S)-6-O-Methyl-7-Bromoaverantin	Red	Anticancer	Huang et al. (2012)

Aspergillus ustus	Sea-derived	Halimide	Yellow	Anticancer and antiseizure activities	Copmans et al. (2022)
Aspergillus versicolor	Solar salterns	Variecolorin J	Red	Antioxidant	Wang et al. (2007)
A. versicolor	Marine algal-derived	Isorhodoptilometrin-1-Methyl Ether; 3-(2-Hydroxypropyl)-1-Methyl Ether-6,8-Dihydroxy-9,10-Anthraquinone, Emodin, 1-Methyl Emodin, Evariquinone, 7-Hydroxyemodin 6,8-Methyl Ether	Red, orange	Antimicrobial activity, anti-cancer activity, and inhibition of the Hepatitis C virus (HCV) protease	Hawas et al. (2012)
Aspergillus westerdijkiae SCSIO 05233	Deep sea-derived	Botryoisocoumarin A	Brown	–	Fredimoses et al. (2015)
Aspergillus niger	Marine sponge	Terphenylquinone	Deep green	Antioxidant	Hiort et al. (2004)
Aspergillus sp. strain 05F16	Coral reef	Tetrahydrobostrycin, 1-Deoxytetrahydrobostrycin, Bostrycin	Pale yellow and red	Antibacterial	Xu et al. (2008)
Chaetomium globosum	Coral reef	Chaetoglobin, Chaetoglobosin, Chaetomugilin, Cochliodinol	Maroon, purple	Antifungal	Fouillaud et al. (2017)
C. lunata	Sponge-derived	Lunatin	Orange	Antibacterial	Jadulco et al. (2002)
Chaetomium sp. NA-S01-R1	Deep sea-derived	Chaetoviridide A, Chaetoviridide B, Chaetoviridide C	Red	Antimicrobial	Wang et al. (2018)
Eurotium nigrum (strain 749)	Marine algal-derived	Acetosellin, 5′, 6′ Dihydroxyacetosellin.	Red-orange, yellow	–	Hufendiek et al. (2017)
E. album DFFSCS021	Deep sea-derived	Engyodontiumone A-J, 2-Methoxyl Cordyol C	Pale yellow to brown	Antibacterial and antiviral	Yao et al. (2014)
E. rubrum	Mangrove tree	Eurorubrin, 3-O-(Alpha-D-Ribofuranosyl)-Questin, Asperflavin, 2-O-Methyleurotinone	Red, orange, yellow, brown	Antioxidant	Li et al. (2009)
Eurotium	Mangrove tree	Tetrahydroauroglaucin, Isodihydroauroglaucin	Yellow, orange	Antioxidant	Gawas et al. (2002)

Table 1 contd. ...

...Table 1 contd.

Species	Isolated From	Metabolite	Hue	Bioactivity	References
Eurotium chevalieri MUT 2316	Sponge- derived	Dihydroauroglaucin, Physcion, Neoechinulin D, Isodihydroauroglaucin	Yellow	Antiviral and antibacterial	Bovio et al. (2019)
Eurotium cristatum	Marine sponge	2-(2′, 3-Epoxy-1′,3′-Heptadienyl)-6-Hydroxy-5-(3-Methyl-2-Butenyl) Benzaldehyde and 1,8-Dihydroxy-6-Methoxy-3-Methyl-9,10-Anthracenedione (Physcion)	Yellow	Cell growth inhibitors	Almeida et al. (2010)
Eurotium repens	Marine-derived	Physcion, Asperflavin	Red-Green	–	Smetanina et al. (2007)
Fusarium sp. ZZF60	Mangrove endophyte	6,8-Dimethoxy-1-Methyl-2-(3-Oxobutyl) Anthrakunthone	Yellow	Cytotoxic	Shao et al. (2010)
Fusarium sp. (No. b77)	Marine endophyte	5-Acetyl-2-Methoxy-1,4,6-Trihydroxy-Anthraquinone, Javanicin, Anhydrofusarubin, Fusarubin, 3-O-Methylfusarubin	Red to dark violet	Antimicrobial and anticancer	Shao et al. (2010)
Fusarium sp. (No. ZH-210)	Marine-derived	Fusaquinon C	Red	Cytotoxic	Chen et al. (2009)
Fusarium spp. PSU-F14 and PSU-F135	Coral reef	Fusarnaphthoquinone B, Fusarnaphthoquinone C	Red	–	Trisuwan et al. (2010)
Halorosellinia werneckii	Solar salterns	Melanin	Black	Antibacterial	Rani et al. (2013)
H. werneckii	Solar salterns	DHN-Melanin	Dark brown	–	Kogej et al. (2004)
Halorosellinia sp. (No. 1403)	Mangrove endophyte	1,4,5,6,7,9-Hexahydroxy-2-Methoxy-7,-Methyl-5beta,9beta,8abeta,6alpha,10aalpha-Hexahydroanthracen-10(10ah)-One, Austrocortirubin, 1,4,6-Trihydroxy-2-Methoxy-7-Methylanthracene-9, Demethoxyaustrocortirubin	Red to brown	Anticancer	Xia et al. (2007)

Microsphaeropsis	Sponge-derived	1,3,6,8-Tetrahydroxyanthraquinone congeners	Yellow-orange	Protein kinase inhibitors	Brauers et al. (2000)
Microsporum sp.	Marine sediment	Flavoglaucin, Asperflavin Ribofuranoside	Yellow	Antioxidant and antibacterial	Li et al. (2006)
M. purpureus CPEF02	Mangrove endophyte	Monasfluore B, rubropunctine, monaphilones B and C, purpureusone, monascin and rubropunctamine	Yellow, red, red-purple	Antioxidant, antibacterial and antifungal	Kaur et al. (2023)
Monascus ruber BB5	Marine shellfish	Monauspurpurone	Yellow	Anticancer and antioxidant	Ran et al. (2020)
Nigrospora	Marine-derived	Nigrosporin B, 4-Deoxybostrycin, Bostrycin, 3-Acetoxy-4-Deoxybostrycin, 3-Acetoxybostrycin,3-Acetoxy-5,8-Dihydroxy-7-Methoxy-2-Methylanthracene9,10-Dione	Red, orange	Antibacterial	Yang et al. (2012)
Paecilomyces variotii	Marine algal-derived	Varioxepine A	Yellow	Antifungal	Zhang et al. (2014)
Penicillium	Mangrove sediment	2-(4-Acetyl Phenyl) Acetic Acid	Red	–	Chintapenta et al. (2014a)
Penicillium bilaii	Marine water	Citromycetin, Citromycin, (–)-2,3-Dihydrocitromycetin, (–)-2,3-Dihydrocitromycin	Yellow	Cytotoxic	Capon et al. (2007)
Penicillium crustosum PRB-2	Deep sea-derived	Penilactone A	Pale yellow	–	Wu et al. (2012)
Penicillium sclerotiorum FS50	Marine-derived	Sclerotiorin	Orange	Antifungal	Xin et al. (2019)
Penicillium sp. JP-1	Mangrove endophyte	Penicillinone	Red	Cytotoxic	Lin et al. (2008)
Penicillium sp. NIOM-02	Marine sediment	Unknown	Red	Antioxidant	Dhale and Vijay-Raj (2009)

Table 1 contd. ...

...Table 1 contd.

Species	Isolated From	Metabolite	Hue	Bioactivity	References
Penicillium sp. Y-50-10	Sea hydrothermal vents	Methyl Isoverrucosidinol	Light yellow	antibacterial	Pan et al. (2016)
Penicillium commune DFFSCS026	Deep sea-derived	Cyclopiamides B-J	Yellow	–	Xu et al. (2015)
Penteagramma? triangularis	Solar salterns	DHN-Melanin	Dark brown	–	Kogej et al. (2004)
Periconia	Deep sea-derived	Pericosine A	Violet	Antioxidant and anti-cancer	Meruvu and dos Santos (2021)
Periconia	Solar salterns	Unknown	Blue	–	Cantrell et al. (2006)
Stagonospora sp. SYSU-MS7888	Sponge-derived	Stasorbicillinoids A–C	Yellow	Anti-inflammatory	Chen et al. (2022)
Stemphylium globuliferum	Marine endophyte	Acetylalterporriol D, Acetylalterporriol E	Red	–	Liu et al. O(2015)
Stemphylium sp. 33231	Mangrove endophyte	2-O-Acetylaltersolanol B, Alterporriol T, Alterporriol U, Alterporriol W	Orange, red	Antibacterial, anticancer	Zhou et al. (2014)
Talaromyces albobiv (AQerticillius 30548)	Coral reef	6-[(Z)-2-Carboxyvinyl]-N-GABA-PP-V, PP-O (Compound Red Orange), N-Glutarylrubropunctamine, N-Threonine-Monascorubramine	Purple-red to red-orange	–	Venkatachalam et al. (2018)
Talaromyces assiutensis	Mangrove forest	1,5-diamino-2-(4-((4-(dibutylamino)-6-fluoro-1,3,5-triazin-2-yl) oxy) phenyl) -4,8-dihydroxyanthraquinone, p-benzoquinone, 2,5-bis((3-morpholinopropyl) amino	Red	Antibacterial, antioxidant, anticancer	Mishra et al. (2021a)
Talaromyces salinum	Solar salterns	DHN-Melanin	Dark brown	–	Kogej et al. (2004)

Talaromyces sp. 30570	Coral reef	PP-R, 6-[(Z)-2-Carboxyvinyl]-*N*-GABA-PP-V, *N*-Threonine-Monascorubramin, *N*-Glutaryl-Rubropunctamin, Monascorubramin, *N*-Threonyl-Rubropunctamin	Dark-red to light-red	–	Lebeau et al. (2020)
Talaromyces sp. BTBU20213036	Marine-derived	Bacillisporin K, Bacillisporin M, Rugulosin D	Light yellow, brown	Antibacterial	Song et al. (2022)
Talaromyces spp.	Marine sediment	*N*-threonine rubropunctamine	Red	–	Lebeau et al. (2017)
Talaromyces sp. WHUF04072	Deep-sea sediment	Talaromenes A–D, Prugosene A1, F and G, Rugulosin A and B, Skyrin	Yellow, red	Antibacterial	Sibero et al. (2017); Huo et al. (2025)
Trichoderma parareesei MT23	Marine sponge	Unknown	Yellow	Antimicrobial	Sibero et al. (2017)
Xylaria sp. 2508	Marine-derived	Xylanthraquinone, Bostrycin, Deoxybostrycin, Altersolanol A	Red	Antimycobacterial	Huang et al. (2014)
Unidentified fungus ZSUH-36	Marine-derived	6,8,1-Tri-O-Methyl Averantin, 1-Tri-O-Methyl Averantin, 6,8-Di-O-Methyl Averufanin	Red	–	Shao et al. (2008)

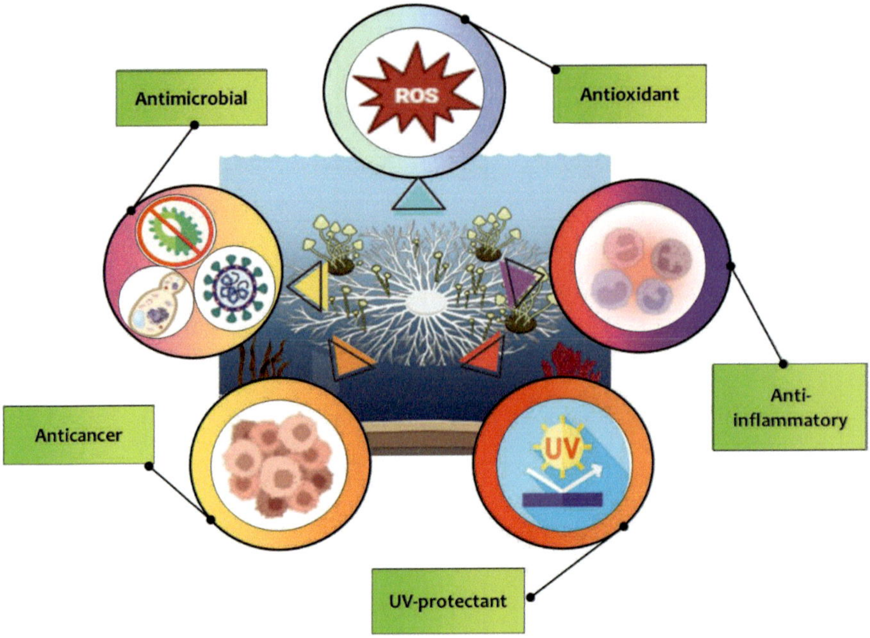

Fig. 1. Bioactive properties of pigmented compounds from marine fungi.

7-hydroxyemodin 6,8-methyl ether, epiaspinonediol, halimide, terphenylquinone, botryisocoumarin, (1′S)-6-O-methyl-7-chloroaverantin, (1′S)-7-chloroaverantin-1′-butyl ether, 7-chloroaverythrin, 6-O-methyl-7-chloroaverythrin, (1′S)-6-O-methyl-7-bromoaverantin, and melanin precursor 5,6-dihydroxy-indole-2-carboxylic acid have been isolated from the marine *Aspergillus* species. These pigments have been identified to have anticancer, antioxidant, UV-protectant, antioxidant, antimicrobial, antiseizure, antiprotozoal, phytotoxic, and cytotoxic activities with yellow, orange, reddish pink, red, deep green, and brown shades (Hiort et al., 2004; Du et al., 2008; Xu et al., 2008; Hawas et al., 2012; Huang et al., 2012; Chen et al., 2014; Fredimoses et al., 2015; El-Kashef et al., 2020; Copmans et al., 2022). *Chaetomium globosum* and *Chaetomium* sp. NA-S01-R1, isolated from a coral reef and deep sea, contained red, maroon, and purple-colored chaetoglobin, chaetoglobosin, chaetomugilin, cochliodinol, and chaetoviridides with antimicrobial properties (Fouillaud et al., 2016; Wang et al., 2018). *Eurotium* species isolated as mangrove and sponge endophytes contained eurorubrin, 3-O-(alpha-D-riboofuranosyl)-questin, asperflavin, 2-O-methyleurotinone, physcion, dihydroauroglaucin, physcion, neoechinulin D, isodihydroauroglaucin, tetrahydroauroglaucin, and 2-(2′, 3-epoxy-1′,3′-heptadienyl)-6-hydroxy-5-(3-methyl-2-butenyl) benzaldehyde emitting yellow, orange, red, reddish-green, and brown hues with antimicrobial and antioxidant activities (Gawas et al., 2002; Smetanina et al., 2007; Li et al., 2009; Almeida et al., 2010; Bovio et al., 2019).

The genus *Fusarium* has also been identified as an endophyte in mangroves and coral reefs, producing shades ranging from yellow and red to violet through

6,8-dimethoxy-1-methyl-2-(3-oxobutyl) anthrakunthone, 5-acetyl-2-methoxy-1,4,6-trihydroxy-anthraquinone, javanicin, anhydrofusarubin, fusarubin, 3-O-methylfusarubin, fusaquinon, and fusarnaphthoquinones. These compounds are reported to possess antimicrobial, anticancer, and antioxidant activities (Chen et al., 2009; Shao et al., 2010; Trisuwan et al., 2010). Halophilic *Hortaea werneckii*, which has been isolated from solar salterns in different studies, contains black and brown shaded melanin and DHN-melanin with antibacterial, UV-protectant, and antioxidant activities (Kogej et al., 2004; Rani et al., 2013). Melanin is known to be produced as a defense mechanism in several animals and microorganisms for its diverse functions as a UV protectant, antimicrobial, antioxidant, and virulence enhancer (Rani et al., 2013).

Penicillium has been isolated from several niche habitats, such as plants from mangrove endophytes, mangrove sediments, marine waters, deep sea waters, marine sediments, and sea hydrothermal vents. With a limited hue of yellow, orange, and red, 2-(4-acetyl phenyl) acetic acid, citromycetin, citromycin, (–)-2, 3-dihydrocitromycetin, penilactone A, sclerotiorin, penicillenone, methyl isoverrucosidinol, and cyclopiamides isolated from the genus possess antibacterial, anticancer, antifungal, and antioxidant activities (Capon et al., 2007; Lin et al., 2008; Dhale and Vijay-Raj, 2009; Wu et al., 2012; Chintapenta et al., 2014a; Xu et al., 2015; Pan et al., 2016; Xin et al., 2019). Deep sea derived *Periconia* produced a violet-colored pericosine A with antioxidant and anticancer properties, while solar saltern-derived *Periconia* had been identified to produce an unknown blue-colored compound (Cantrell et al., 2006; Meruvu and dos Santos, 2021). The marine endophyte *Stemphylium* produced red-hued alterporriol and its derivatives, exhibiting antibacterial and anticancer properties (Zhou et al., 2014; Liu et al., 2015). *Talaromyces* has also been isolated from coral reefs, marine sediment, and mangrove forests for the production of non-toxic *Monascus*-like pigments that can be easily employed commercially, such as novel 6-[(Z)-2-carboxyvinyl]-N-GABA-PP-V, PP-O, N-glutarylrubropunctamine, N-threonine-monascorubramine, PP-R, *N*-glutaryl-rubropunctamin, monascorubramin, and presumed *N*-threonyl-rubropunctamin with shades ranging from light red to purple red. Other antibacterial pigments isolated from the marine counterparts of the genus include light yellow colored bacillisporins and brown colored rugulosin D (Lebeau et al., 2017, 2020; Venkatachalam et al., 2018; Mishra et al., 2021a; Song et al., 2022).

Other marine fungal pigments include red-colored monasfluore B, rubropunctine, monaphilones B and C, purpureusone, monascin, and rubropunctamine from *Monascus purpureus* CPEF02, yellow-colored ascomycotin A from *Ascomycota* sp. Ind19F07, orange-hued lunatin from *Curvularia lunata*, pale yellow engyodontiumone A-J, and brown colored 2-methoxyl cordyol C from *Engyodontium album* DFFSCS021, orange-red colored acetosellin from *Epicoccum nigrum* (strain 749), novel red colored 1,4,5,6,7,9-hexahydroxy-2-methoxy-7,-methyl-5β,9β,8aβ,6α,10aα-hexahydroanthracen-10(10aH)-one along with red to brown shaded austrocortirubin, 1,4,6-trihydroxy-2-methoxy-7-methylanthracene-9, and demethoxyaustrocortirubin from *Halorosellinia* sp. (No. 1403), DHN-melanin from *Phaeotheca triangularis* and *Trimmatostroma salinum*. The high therapeutic

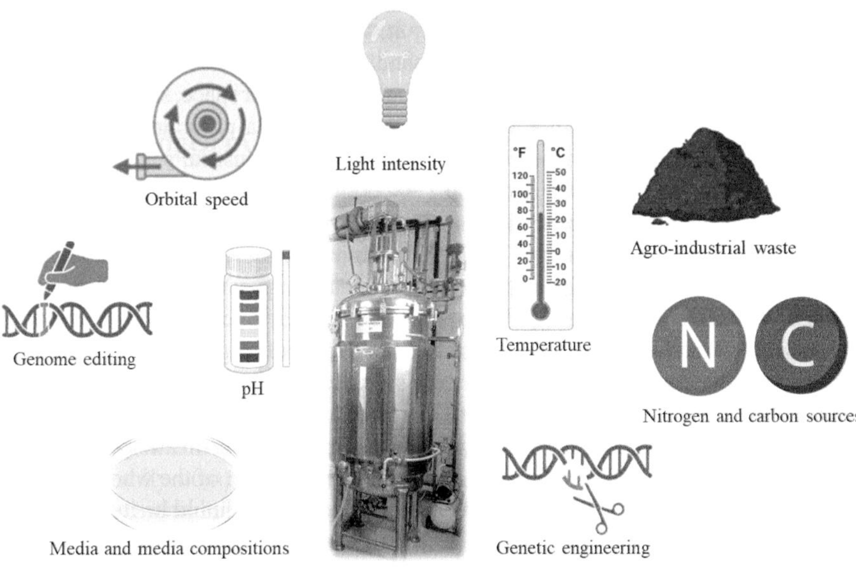

Fig. 2. Influence of key parameters on the production of fungal pigments in a bioreactor.

derived fungus *Peniophora* sp. CBMAI 1063 investigated laccase production in two types of bioreactors (Mainardi et al., 2018). The stirred tank bioreactor resulted in higher productivity along with laccase activity, whereas the air-lift bioreactor exhibited maximum growth of biomass. The purified laccase, a monomeric enzyme with ~55 kDa, showed excellent activity at pH 5.0 and thermal stability (between a range of 30–50°C). Notably, laccase is being produced exclusively in the presence of artificial seawater and copper sulfate, emphasizing the flexibility of the fungus to marine conditions and its biotechnological potency for saline processes.

Detailed work has been conducted on marine-derived *T. albobiverticillius* 30548 to improve pigment production under various scales and conditions (Venkatachalam et al., 2018, 2019, 2020, 2021, 2023). Using a one-variable-at-a-time (OVAT) approach, it has been identified that sucrose and yeast extract are optimal nitrogen and carbon sources, respectively, and this combination results in maximum production of orange and red pigments along with biomass. Additional optimization using response surface methodology (RSM) and central composite design (CCD) suggested that yeast extract (3 g/L), K_2HPO_4 (1 g/L), and $MgSO_4{\cdot}7H_2O$ (0.2 g/L) provide the best results, with pigment yields of 1.43 g/L (orange) and 2.59 g/L (red), and biomass up to 15.98 g/L. These findings underscore the importance of optimizing nutrients for enhancing scalable production of fungal pigment production. In another study, *T. albobiverticillius* 30548, acquired from marine habitat of Réunion Island, was evaluated for growth and pigment productivity, and different factors, which included variations in temperature and salinity. Maximum yield of pigment production was observed at 27°C in non-saline media, while the highest biomass was recorded under saline stress at 9% sea salts. Pigment production was reduced beyond 6% salinity along with the enhancement of fungal growth. Color variations of obtained pigments, ranging from red to yellow (CIELAB hue angles), were affected by salinity.

These findings show the effectiveness of abiotic stress in modulating the metabolic profile of marine fungi (Venkatachalam et al., 2019). Optimization of submerged fermentation (SmF) conditions is necessary for maximizing the yield of pigment and biomass for marine-derived fungi. In reference to that, *T. albobiverticillius* 30548 was evaluated for orange and red pigments along with biomass production at different processing parameters (Venkatachalam et al., 2020). The Box-Behnken Design and RSM allowed systematic evaluations of interactive effects of fermentation time (14 days), initial pH (4–9), agitation speed (100–200 rpm), and temperature (21–27°C). The mathematical models derived from the data predicted optimal yields effectively. The numerically optimized conditions, like pH 6.4, 24°C, 164 rpm, and 149 h, resulted in maximum pigment and biomass production, emphasizing the utility of statistical tools for fine-tuning SmF parameters in marine fungal bioprocesses. *T. albobiverticillius* 30548, a marine-derived fungus, has been reported to produce *Monascus*-like red and orange azaphilone pigments. A comparative study showed that pigment production was significantly higher in a 2 L bioreactor under controlled conditions (pH 5.0, 24°C, 800 rpm, 70–90% pO_2) than in shake flasks. Bioreactor cultures yielded up to 25.95 AU (orange) and 22.79 AU (red) pigments, with pellet morphology, while shake flasks showed lower yields and filamentous growth. Oxygen transfer, influenced by agitation and morphology, was identified as a key factor affecting pigment biosynthesis (Venkatachalam et al., 2023).

In one study, a marine fungus, *Penicillium sclerotiorum* M-22, was optimized for the production of penicilazaphilone C (PAC), a bioactive azaphilone pigment with antibacterial and cytotoxic properties. Use of single-factor experiments and RSM, resulted in optimal culture conditions at 24.83 days of fermentation (pH 7.0), along with precise concentrations of yeast extract, sea salt, and cornmeal. These interventions led to an increase in PAC production (1.34-fold, 19.85 mg/L), demonstrating the efficiency of biotechnological approaches in scaling up marine fungal pigment production for industrial use (Zhao et al., 2018). A red pigment-producing *Penicillium* strain (DLR-7) was isolated from mangrove soil in Andhra Pradesh, India. A stepwise optimization method demonstrated that lab-prepared potato extract provided higher yields of pigment than those exhibited by commercial media. Optimal conditions for pigment (1050 mg/L) and biomass (3.1 g/L) production included xylose (2% w/v), glycine (1% w/v), pH 3.0, and 25°C. Notably, a decrease of pigment production was observed by basic amino acids (arginine and lysine), while the opposite result was obtained in the case of acidic amino acids (aspartic and glutamic acid). The study highlighted the important role of certain carbon and nitrogen sources in regulating the biosynthesis of fungal pigment (Chintapenta et al., 2014b).

In a study on the marine-derived fungus *Alternaria tenuissima* DFFSCS013, novel hydroanthraquinones were isolated using solid fermentation (SF) at 26°C for 30 days. The extraction was completed using an ethanol-acetone mixture, and production often face challenges such as limited availability of raw materials, low pigment yields, and instability of pigments when exposed to factors like light, pH, temperature, moisture, and complex food matrices. Additionally, safety concerns arise as certain fungal strains may coproduce mycotoxins such as citrinin alongside the desired pigments.

6. Conclusion

Marine fungi have come out to be an extensive and unexplored reservoir of unique and biologically active pigments with vast commercial potential. They have an upper hand in the industrial framework owing to feasible culturing conditions, rapid upscaling and flexible downstream processes. Recent advances in genetic engineering and microbial fermentation have further accelerated exploration of marine fungi for novel pigments and their production optimization for enhanced yield. The diverse metabolic pathways, responsible for their ability to adapt to extreme environments, contribute to pigments with varied chemical structures and bioactivities. Continued exploration of marine fungal pigments promises novel compounds, expanded color profiles, and new avenues for bioprospecting and therapeutic applications.

References

Abdel-Raheam, H.E., Abdel-Mageed, W.S. and Abd El-Rahman, M.A. (2019). Optimization of production of *Monascus ruber* pigments on broth medium and their application in flavored yogurts. Egypt. J. Food Sci., 47(2): 271–283. https://doi.org/10.21608/ejfs.2019.19654.1030.

Agboyibor, C., Kong, W.B., Chen, D., Zhang, A.M. and Niu, S.Q. (2018). *Monascus* pigments production, composition, bioactivity, and its application: A review. Biocatal. Agric. Biotechnol., 16: 433–447. https://doi.org/10.1016/j.bcab.2018.09.012.

Agrawal, S., Adholeya, A., Barrow, C.J. and Deshmukh, S.K. (2018). Marine fungi: An untapped bioresource for future cosmeceuticals. Phytochem. Lett., 23: 15–20. https://doi.org/10.1016/j.phytol.2017.11.003.

Agrawal, S., Deshmukh, S.K. and Barrow, C.J. (2019). Marine fungi as a potential source of future cosmeceuticals. pp. 627–669. *In*: Satyanarayana, T., Deshmukh, S. and Deshpande, M. (eds.). Advancing Frontiers in Mycology and Mycotechnology: Basic and Applied Aspects of Fungi. Singapore: Springer Singapore. https://doi.org/10.1007/978-981-13-9349-5_25.

Almeida, A.P., Dethoup, T., Singburaudom, N., Lima, R.T. and Vasconcelos, M.H. (2010). The *in vitro* anticancer activity of the crude extract of the sponge-associated fungus *Eurotium cristatum* and its secondary metabolites. J. Nat. Pharm., 1: 25–29. https://doi.org/10.4103/2229-5119.73583.

Behera, A.D. and Das, S. (2023). Ecological insights and potential application of marine filamentous fungi in environmental restoration. Rev. Environ. Sci. Biotechnol., 22(2): 281–318. https://doi.org/10.1007/s11157-023-09655-2.

Bovio, E., Garzoli, L., Poli, A., Luganini, A. and Villa, P. (2019). Marine Fungi from the sponge grantia compressa: Biodiversity, chemodiversity, and biotechnological potential. Mar. Drugs, 17. https://doi.org/10.3390/md17040220.

Brauers, G., Edrada, R.A., Ebel, R., Proksch, P., Wray, V. et al. (2000). Anthraquinones and betaenone derivatives from the sponge-associated fungus *Microsphaeropsis* species: Novel inhibitors of protein kinases. J. Nat. Prod., 63(6): 739–745. https://doi.org/10.1021/np9905259.

Cantrell, S.A., Casillas-Martínez, L. and Molina, M. (2006). Characterization of fungi from hypersaline environments of solar salterns using morphological and molecular techniques. Mycol. Res., 110(8): 962–970. https://doi.org/10.1016/j.mycres.2006.06.005.

Capon, R.J., Stewart, M., Ratnayake, R., Lacey, E. and Gill, J.H. (2007). Citromycetins and bilains A-C: New aromatic polyketides and diketopiperazines from Australian marine-derived and terrestrial *Penicillium* spp. J. Nat. Prod., 70(11): 1746–1752. https://doi.org/10.1021/np0702483.

Caro, Y., Anamale, L., Fouillaud, M., Laurent, P., Petit, T. et al. (2012). Natural hydroxyanthraquinoid pigments as potent food grade colorants: An overview. Nat. Prod. Bioprospect., 2(5): 174–193. https://doi.org/10.1007/s13659-012-0086-0.

Charudattan, R. and Rao, K.V. (1982). Bostrycin and 4-deoxybostrycin: Two nonspecific phytotoxins produced by *Alternaria eichhorniae*. Appl. Environ. Microbiol., 43(4): 846–849. https://doi.org/10.1128/aem.43.4.846-849.1982.

Chen, S., Guo, H., Wu, Z., Jiang, M., Li, H. and Liu, L. (2022). Targeted discovery of sorbicillinoid pigments with anti-inflammatory activity from the sponge-derived fungus *Stagonospora* sp. SYSU-MS7888 using the PMG strategy. J. Agric. Food Chem., 70(48): 15116–15125. https://doi.org/10.1021/acs.jafc.2c05940.

Chen, X.W., Li, C.W., Cui, C.B., Hua, W., Zhu, T.J. et al. (2014). Nine new and five known polyketides derived from a deep sea-sourced *Aspergillus* sp. 16-02-1. Mar Drugs, 12(6): 3116–3137. https://doi.org/10.3390/md12063116.

Chen, Y., Cai, X., Pan, J., Gao, J., Li, J. et al. (2009). Structure elucidation and NMR assignments for three anthraquinone derivatives from the marine fungus *Fusarium* sp. (No. ZH-210). Magn. Reson. Chem., 47(4): 362–365. https://doi.org/10.1002/mrc.2391.

Chintapenta, L.K., Rath, C.C., Maringinti, B. and Ozbay, G. (2014a). Pigment production from a mangrove *Penicillium*. Afr. J. Biotechnol., 13(26). https://doi.org/10.5897/ajb2014.13838.

Chintapenta, L.K., Rath, C.C., Maringinti, B. and Ozbay, G. (2014b). Culture conditions for growth and pigment production of a Mangrove *Penicillium* species. J. Multidiscip. Sci., Res., 2: 1–5.

Copmans, D., Kildgaard, S., Roux, E., Partoens, M., Steurs, G. et al. (2022). From the North Sea to drug repurposing, the antiseizure activity of halimide and plinabulin. Pharma., 15(2): 247. https://doi.org/10.3390/ph15020247.

Dhale, M.A. and Vijay-Raj, A.S. (2009). Pigment and amylase production in *Penicillium* sp NIOM-02 and its radical scavenging activity. Int. J. Food Sci. Technol., 44(12): 2424–2430. https://doi.org/10.1111/j.1365-2621.2009.01983.x.

Du, L., Zhu, T., Liu, H., Fang, Y. and Zhu, W. (2008). Cytotoxic polyketides from a marine-derived fungus *Aspergillus glaucus*. J. Nat. Prod., 71(11): 1837–1842. https://doi.org/10.1021/np800303t.

Dufossé, L. (2016). Current and potential natural pigments from microorganisms (Bacteria, Yeasts, Fungi, Microalgae). pp. 419–436. *In*: Schweiggert, R. (ed.). Handbook on Natural Pigments in Food and Beverages: Industrial Applications for Improving Food Color. Woodhead Publishing. https://doi.org/10.1016/B978-0-323-99608-2.00001-X.

Dufossé, L. (2018). Red colourants from filamentous fungi: Are they ready for the food industry? J. Food Comp. Anal., 69: 156–161. https://doi.org/10.1016/j.jfca.2017.11.002.

Dufossé, L. (2019). Research, development, and production of microalgal and microbiabiocolorants. pp. 71–92. *In*: Molina, G., Gupta, V., Singh, B. and Gathergood, N. (eds.). Bioprocessing for Biomolecules Production. John Wiley & Sons, Ltd. https://doi.org/10.1002/9781119434436.ch4.

Dufosse, L., Fouillaud, M., Caro, Y., Mapari, S.A., Sutthiwong, N. et al. (2024). Filamentous fungi are large-scale producers of pigments and colorants for the food industry. Curr. Opin. Biotechnol., 26: 56–61. https://doi.org/10.1016/j.copbio.2013.09.007.

El-Kashef, D.H., Youssef, F.S., Hartmann, R., Knedel, T.O., Janiak, C. et al. (2020). Azaphilones from the red sea fungus *Aspergillus falconensis*. Mar. Drugs, 18(4): 204. https://doi.org/10.3390/md18040204.

Fouillaud, M., Venkatachalam, M., Girard-Valenciennes, E., Caro, Y., Dufossé, L. et al. (2016). Anthraquinones and derivatives from marine-derived fungi: Structural diversity and selected biological activities. Mar. Drugs, 14(4): 64. https://doi.org/10.3390/md14040064.

Fouillaud, M., Venkatachalam, M., Llorente, M., Magalon, H., Cuet, P. et al. (2017). Biodiversity of pigmented fungi isolated from marine environment in la réunion Island, Indian ocean: New resources for colored metabolites. J. Fungi, 3(3): 36. https://doi.org/10.3390/jof3030036.

Fredimoses, M., Zhou, X., Ai, W., Tian, X., Yang, B. et al. (2015). Westerdijkin A, a new hydroxyphenylacetic acid derivative from deep sea fungus *Aspergillus westerdijkiae* SCSIO 05233. Nat. Prod. Res., 29(2): 158–162. https://doi.org/10.1080/14786419.2014.968154.

Mishra, R.C., Kalra, R., Dwivedi, N. and Goel, M. (2021b). Exploring endophytes using "Omics": An approach for sustainable production of bioactive metabolites. pp. 349–376. *In*: Prasad, R., Nayak, S.C., Kharwar, R.N. and Dubey, N.K. (eds.). Mycoremediation and Environmental Sustainability (Vol. 3). Cham: Springer International Publishing.

Muñoz-Miranda, L.A. and Iñiguez-Moreno, M. (2023). An extensive review of marine pigments: Sources, biotechnological applications, and sustainability. Aquat. Sci., 85(3): 68. https://doi.org/10.1007/s00027-023-00966-8.

Nigam, P.S. and Luke, J.S. (2016). Food additives: Production of microbial pigments and their antioxidant properties. Curr. Opin. Food Sci., 7: 93–100. https://doi.org/10.1016/j.cofs.2016.02.004.

Pailliè-Jiménez, M.E., Stincone, P. and Brandelli, A. (2020). Natural pigments of microbial origin. Front. Sustain. Food Syst., 4: 590439. https://doi.org/10.3389/fsufs.2020.590439.

Pallela, R., Na-Young, Y. and Kim, S.K. (2010). Anti-photoaging and photoprotective compounds derived from marine organisms. Mar. Drugs, 8(4): 1189–1202. https://doi.org/10.3390/md8041189.

Palma, D., Oliva, V., Montanares, M., Gil-Durán, C., Travisany, D. et al. (2024). Expanding the toolbox for genetic manipulation in *Pseudogymnoascus*: RNAi-Mediated silencing and CRISPR/Cas9-mediated disruption of a polyketide synthase gene involved in red pigment production in *P. verrucosus*. J. Fungi, 10(2): 157. https://doi.org/10.3390/jof10020157.

Pan, C., Shi, Y., Auckloo, B.N., Chen, X., Chen, C.T.A. et al. (2016). An unusual conformational isomer of verrucosidin backbone from a hydrothermal vent fungus, *Penicillium* sp. Y-50-10. Mar. Drugs, 14(8): 156. https://doi.org/10.3390/md14080156.

Pan, D., Zhang, X., Zheng, H., Zheng, Z., Nong, X. et al. (2019). Novel anthraquinone derivatives as inhibitors of protein tyrosine phosphatases and indoleamine 2,3-dioxygenase 1 from the deep-sea derived fungus: *Alternaria tenuissima* DFFSCS013. Org. Chem. Front., 6(18): 3252–3258. https://doi.org/10.1039/c9qo00775j.

Qiu, X., Gong, L., Xin, X. and An, F. (2021). Enhancement of emodin production by medium optimization and KH2PO4 supplementation in submerged fermentation of marine-derived *Aspergillus flavipes* HN4-13. Mar. Drugs, 19(8): 421. https://doi.org/10.3390/md19080421.

Rajagopal, L., Sundari, C.S., Balasubramanian, D. and Sonti, R.V. (1997). The bacterial pigment xanthomonadin offers protection against photodamage. FEBS Lett., 415(2): 125–128. https://doi.org/10.1016/S0014-5793(97)01109-5.

Ran, Y.Q., Lan, W.J., Qiu, Y., Guo, Q., Feng, G.K. et al. (2020). Monarubins A-C from the marine shellfish-associated fungus *Monascus ruber* BB5. Mar. Drugs, 18(2): 100. https://doi.org/10.3390/md18020100.

Rani, M.H.S., Ramesh, T., Subramanian, J. and Kalaiselvam, M. (2013). Production and characterization of melanin pigment from halophilic black Yeast *Hortaea werneckii*. Int. J. Pharma. Res. Rev., 2(8): 9–17.

Rateb, M.E. and Ebel, R. (2011). Secondary metabolites of fungi from marine habitats. Nat. Prod. Rep., 28(2): 290–344. https://doi.org/10.1039/C0NP00061B.

Robinson, S.C., Tudor, D. and Cooper, P.A. (2012). Utilizing pigment-producing fungi to add commercial value to American beech (*Fagus grandifolia*). Appl. Microbiol. Biotechnol., 93(3): 1041–1048. https://doi.org/10.1007/s00253-011-3576-9.

Sakurai, M., Kohno, J., Yamamoto, K., Okuda, T., Nishio, M. et al. (2002). TMC-256A1 and C1, new inhibitors of IL-4 signal transduction produced by *Aspergillus niger* var *niger* TC 1629. J. Antibiot., 55(8): 685–692. https://doi.org/10.7164/antibiotics.55.685.

Shanuja, S.K., Iswarya, S., Sridevi, J. and Gnanamani, A. (2018). Exploring the UVB-protective efficacy of melanin precursor extracted from marine imperfect fungus: Featuring characterization and application studies under *in vitro* conditions. Int. Microbiol., 21(1): 59–71. https://doi.org/10.1007/s10123-018-0005-2.

Shao, C., Wang, C., Wei, M., Li, S., She, Z. et al. (2008). Structural and spectral assignments of six anthraquinone derivatives from the mangrove fungus (ZSUH-36). Magn. Reson. Chem., 46(9): 886–889. https://doi.org/10.1002/mrc.2266.

Shao, C., Wang, C., Zheng, C., She, Z., Gu, Y. et al. (2010). A new anthraquinone derivative from the marine endophytic fungus *Fusarium* sp. (No. b77). Nat. Prod. Res., 24(1): 81–85. https://doi.org/10.1080/14786410902836701.

Sibero, M.T., Triningsih, D.W., Radjasa, O.K., Sabdono, A. and Trianto, A. (2017). Evaluation of antimicrobial activity and identification of yellow pigmented marine sponge-associated fungi from Teluk Awur, Jepara, Central Java. Indones J. Biotechnol., 21: 1–11. https://doi.org/10.22146/ijbiotech.26058.

Singh, J., Gupta, P. and Das, A. (2021). Dyes from textile industry wastewater as emerging contaminants in agricultural fields. pp. 109–129. *In*: Kumar Singh, V., Singh, R. and Lichtfouse, E. (eds.). Sustainable Agriculture Reviews (Vol. 50). Springer, Cham. https://doi.org/10.1007/978-3-030-63249-6_5.

Smetanina, O.F., Kalinovskii, A.I., Khudyakova, Y.V., Slinkina, N.N., Pivkin, M.V. et al. (2007). Metabolites from the marine fungus Eurotium repens. Chem. Nat. Comp., 43(4): 395–398. https://doi.org/10.1007/s10600-007-0147-5.

Song, F., Dong, Y., Wei, S., Zhang, X., Zhang, K. et al. (2022). New antibacterial secondary metabolites from a marine-derived *Talaromyces* sp. Strain BTBU20213036. Antibiotics, 11(2): 222. https://doi.org/10.3390/antibiotics11020222.

Song, T., Cheng, F. and Sun, J. (2021). Stain capacity of three fungi on two fast-growing wood. J. For. Res. (Harbin), 32(1): 427–434. https://doi.org/10.1007/s11676-020-01103-z.

Sumbal Sajid, S.S. and Nukba Akbar, N.A. (2018). Applications of fungal pigments in biotechnology. Pesqui. Agropecu. Bras., 7(3): 922–930. https://doi.org/10.19045/bspab.2018.700111.

Tanaka, T., Shnimizu, M. and Moriwaki, H. (2012). Cancer chemoprevention by carotenoids. Molecules,17(3): 3202–3242. https://doi.org/10.3390/molecules17033202.

Tian, Y.Q., Lin, X.P., Liu, J., Kaliyaperumal, K., Ai, W. et al. (2015). Ascomycotin A, a new citromycetin analogue produced by *Ascomycota* sp. Ind19F07 isolated from deep sea sediment. Nat. Prod. Res., 29(9): 820–826. https://doi.org/10.1080/14786419.2014.988620.

Trisuwan, K., Khamthong, N., Rukachaisirikul, V., Phongpaichit, S., Preedanon, S. et al. (2010). Anthraquinone, cyclopentanone, and naphthoquinone derivatives from the sea fan-derived fungi *Fusarium* spp. PSU-F14 and PSU-F135. J. Nat. Prod., 73(9): 1507–1511. https://doi.org/10.1021/np100282k.

Venkatachalam, M., Gérard, L., Milhau, C., Vinale, F., Dufossé, L. et al. (2019). Salinity and temperature influence growth and pigment production in the marine-derived fungal strain *Talaromyces albobiverticillius* 30548. Microorganisms, 7(1): 10. https://doi.org/10.3390/microorganisms7010010.

Venkatachalam, M., Magalon, H., Dufossé, L. and Fouillaud, M. (2018). Production of pigments from the tropical marine-derived fungi *Talaromyces albobiverticillius*: New resources for natural red-colored metabolites. J. Food Comp. Anal., 70: 35–48. https://doi.org/10.1016/j.jfca.2018.03.007.

Venkatachalam, M., Mares, G., Dufossé, L. and Fouillaud, M. (2023). Scale-up of pigment production by the marine-derived filamentous fungus, *Talaromyces albobiverticillius* 30548, from Shake Flask to stirred bioreactor. Fermentation, 9(1): 77. https://doi.org/10.3390/fermentation9010077.

Venkatachalam, M., Shum-Chéong-Sing, A., Caro, Y., Dufossé, L. and Fouillaud, M. (2021). OVAT analysis and response surface methodology based on nutrient sources for optimization of pigment production in the marine-derived fungus *Talaromyces albobiverticillius* 30548 submerged fermentation. Mar. Drugs, 19(5): 248. https://doi.org/10.3390/md19050248.

Venkatachalam, M., Shum-Chéong-Sing, A., Dufossé, L. and Fouillaud, M. (2020). Statistical optimization of the physico-chemical parameters for pigment production in submerged fermentation of *Talaromyces albobiverticillius* 30548. Microorganisms, 8(5): 711. https://doi.org/10.3390/microorganisms8050711.

Vílchez, C., Forján, E., Cuaresma, M., Bédmar, F., Garbayo, I. et al. (2011). Marine carotenoids: Biological functions and commercial applications. Mar Drugs, 9(3): 319–333. https://doi.org/10.3390/md9030319.

Wang, W., Liao, Y., Chen, R., Hou, Y., Ke, W. et al. (2018). Chlorinated azaphilone pigments with antimicrobial and cytotoxic activities isolated from the deep sea derived fungus *Chaetomium* sp. NA-S01-R1. Mar. Drugs, 16(2): 61. https://doi.org/10.3390/md16020061.

Wang, W.L., Lu, Z.Y., Tao, H.W., Zhu, T.J., Fang, Y.C. et al. (2007). Isoechinulin-type alkaloids, variecolorins A– L, from halotolerant *Aspergillus variecolor*. J. Nat. Prod., 70(10): 1558–1564. https://doi.org/10.1021/np070208z.

Welham, R.D. (1963). The early history of the synthetic dye industry. J. Soc. Dyers Colour., 79(4): 146–152. https://doi.org/10.1111/j.1478-4408.1963.tb02545.x.

Wu, G., Ma, H., Zhu, T., Li, J., Gu, Q. et al. (2012). Penilactones A and B, two novel polyketides from Antarctic deep-sea derived fungus *Penicillium crustosum* PRB-2. Tetrahedron, 68(47): 9745–9749. https://doi.org/10.1016/j.tet.2012.09.038.

Xia, G., Li, J., Li, H., Long, Y., Lin, S.E. et al. (2014). Alterporriol-type dimers from the mangrove endophytic fungus, *Alternaria* sp. (SK11), and their MptpB inhibitions. Mar. Drugs, 12(5): 2953–2969. https://doi.org/10.3390/md12052953.

Xia, X.K., Huang, H.R., She, Z.G., Shao, C.L., Liu, F. et al. (2007). 1H and13C NMR assignments for five anthraquinones from the mangrove endophytic fungus *Halorosellinia* sp. (No. 1403). Magn. Reson. Chem., 45(11): 1006–1009. https://doi.org/10.1002/mrc.2078.

Xin, Q., Pan, T., Zhang, W.M. and Wang, Z. (2019). Submerged culture of marine-derived *Penicillium sclerotiorum* FS50 to produce sclerotiorin. Process Biochem., 79: 28–31. https://doi.org/10.1016/j.procbio.2018.12.009.

Xu, J., Nakazawa, T., Ukai, K., Kobayashi, H., Mangindaan, R.E. et al. (2008). Tetrahydrobostrycin and 1-deoxytetrahydrobostrycin, two new hexahydroanthrone derivatives, from a marine-derived fungus *Aspergillus* sp. J. Antibiot., 61(7): 415–419. https://doi.org/10.1038/ja.2008.57.

Xu, X., Zhang, X., Nong, X., Wei, X. and Qi, S. (2015). Oxindole alkaloids from the fungus *Penicillium commune* DFFSCS026 isolated from deep-sea-derived sediments. Tetrahedron, 71(4): 610–615. https://doi.org/10.1016/j.tet.2014.12.031.

Yang, K.L., Wei, M.Y., Shao, C.L., Fu, X.M., Guo, Z.Y. et al. (2012). Antibacterial anthraquinone derivatives from a sea anemone-derived fungus *Nigrospora* sp. J. Nat. Prod., 75(5): 935–941. https://doi.org/10.1021/np300103w.

Yao, Q., Wang, J., Zhang, X., Nong, X., Xu, X. et al. (2014). Cytotoxic polyketides from the deep-sea-derived fungus *Engyodontium album* DFFSCS021. Mar. Drugs, 12(12): 5902–5915. https://doi.org/10.3390/md12125902.

Zhang, P., Mandi, A., Li, X.M., Du, F.Y., Wang, J.N. et al. (2014). Varioxepine a, a 3H-oxepine-containing alkaloid with a new oxa-cage from the marine algal-derived endophytic fungus *Paecilomyces variotii*. Org. Lett., 16(18): 4834–4837. https://doi.org/10.1021/ol502329k.

Zhao, H.G., Wang, M., Lin, Y.Y. and Zhou, S.L. (2018). Optimization of culture conditions for penicilazaphilone C production by a marine-derived fungus *Penicillium sclerotiorum* M-22. Lett. Appl. Microbiol., 66(3): 222–230. https://doi.org/10.1111/lam.12841.

Zheng, C.J., Shao, C.L., Guo, Z.Y., Chen, J.F., Deng, D.S. et al. (2012). Bioactive hydroanthraquinones and anthraquinone dimers from a soft coral-derived *Alternaria* sp. fungus. J. Nat. Prod., 75(2): 189–197. https://doi.org/10.1021/np200766d.

Zhou, X.M., Zheng, C.J., Chen, G.Y., Song, X.P., Han, C.R. et al. (2014). Bioactive anthraquinone derivatives from the mangrove-derived fungus *Stemphylium* sp. 33231. J. Nat. Prod., 77(9): 2021–2028. https://doi.org/10.1021/np500340y.

5

Marine Fungal Hydrophobins and their Applications

Catalina Landeta-Salgado[1,2,*] and *María Elena Lienqueo*[2]

Marine and halophilic fungal hydrophobins (HFBs) offer unique biomaterials due to adaptations to extreme environments, conferring novel properties and enhanced stability. This chapter examines the significant variability in HFB gene repertoires among these fungi. This ranges from massive gene expansions, particularly of Class I HFBs rich in acidic residues, in extremophilic filamentous fungi like the basidiomycete *Wallemia ichthyophaga*, to generally reduced HFB gene numbers in many marine and halophilic ascomycetes such as *Aspergillus sydowii*, *Paradendryphiella salina*, and *Eurotium rubrum*. The chapter explores the diverse biotechnological, biomedical, and materials science applications stemming from these specialized proteins, including their use as biosurfactants and surface modifiers. However, realizing this potential is significantly challenged by difficulties in production from often extremophilic marine sources, complex purification processes (especially the co-purification of marine-derived secondary metabolites), and an evolving regulatory landscape, underscoring the need for further targeted research.

[1] Departamento de Ciencia de los Alimentos y Tecnología Química, Facultad de Ciencias Químicas y Farmacéuticas, Universidad de Chile, St. Dr. Carlos Lorca Tobar 964, Independencia, Santiago 8380000, RM, Chile.

[2] Center for Biotechnology and Bioengineering (CeBiB), Department of Chemical Engineering, Biotechnology, and Materials, Faculty of Physical and Mathematical Sciences, University of Chile, Santiago, Beauchef 851- 8370456 Chile.

* Corresponding author: catalina.landeta@ciq.uchile.cl/cmlandeta@uc.cl

DOI: 10.1201/9781003795100-7

1. Introduction

Hydrophobins (HFBs) are a unique class of small, surface-active proteins secreted exclusively by filamentous fungi. With molecular weights ranging between 5 and 20 kDa and composed of approximately 100–200 amino acids, these proteins are characterized by a conserved pattern of eight cysteine residues forming four intramolecular disulfide bonds (Sallada et al., 2018). HFBs are small extracellular proteins produced exclusively by filamentous fungi, which means that even yeasts do not contain genes for these proteins. HFBs self-assemble at hydrophobic-hydrophilic interfaces and have been associated with the morphogenesis of aerial hyphae and the adhesion of hyphae to hydrophobic surfaces (Wösten, 2001; Pérez-Llano et al., 2020). This structure confers remarkable conformational stability and chemical resistance, features that make HFBs particularly suitable for various industrial and biomedical applications (Wessels, 1996; Wösten, 2001). Based on sequence patterns (especially the spacing of eight conserved cysteine residues) and biophysical properties, HFBs are classified into Class I or Class II. Class I HFBs form highly insoluble amyloid-like rodlet layers (dissociable only in harsh solvents like 100% TFA), whereas Class II HFBs form less robust, more soluble films (De Stefano et al., 2009). These proteins have diverse roles in fungal biology—they coat spores and hyphae with a hydrophobic layer, aiding in aerial growth, spore dispersal, and adhesion to surfaces (Wösten et al., 1994; Wösten and Wessels, 1997; Sunde et al., 2008). In extreme environments (high salinity, marine habitats), HFBs may also contribute to stress adaptation, for instance, by strengthening the cell wall and regulating permeability (Plemenitaš et al., 2014).

Their most defining feature is their amphiphilic nature, derived from the spatial clustering of hydrophobic and hydrophilic amino acids. This allows HFBs to self-assemble spontaneously at interfaces such as air-water or solid-liquid, forming robust, elastic amphiphilic monolayers. In particular, Class I HFBs exhibit high resilience to environmental stresses like dehydration, osmotic stress, and proteolysis, properties that are pivotal for both fungal survival and technological utility (Linder et al., 2005; Landeta et al., 2024). Biologically, HFBs play crucial roles in fungal development. They reduce surface tension to facilitate the emergence of aerial hyphae and spore dispersal structures, and they coat conidia to promote hydrophobicity and resistance to water (Askolin et al., 2006). Beyond self-assembly, recent findings suggest that HFBs, in their monomeric form, may be involved in fungal cell wall assembly, act as signaling molecules, or even function as elicitors or effectors under stress conditions (Aimanianda et al., 2009; Bayry et al., 2012). These expanded roles position HFBs as multifunctional proteins with untapped potential in biotechnological and medical fields, including drug delivery, immunomodulation, and material sciences. Despite this growing body of research, most of the current knowledge on HFBs comes from terrestrial fungal models, leaving a significant gap regarding their diversity and function in marine environments. Marine-derived filamentous fungi, particularly those classified as halophiles or halotolerant, remain largely underexplored in this context, even though they inhabit saline conditions that require unique adaptations (Gunde-Cimerman et al., 2018).

2. Hydrophobin from Marine Fungi: A New Frontier

Extremophilic fungi, including halophilic species, have evolved to thrive under high-salinity conditions that are typically inhibitory to most eukaryotes. These fungi constitute an emerging source of novel biomolecules, including extremozymes and surface-active proteins with enhanced stability under harsh physicochemical conditions (Siglioccolo et al., 2011). Among these adaptations, the production of halophilic HFBs has been identified in marine and hypersaline fungal species such as *Wallemia ichthyophaga* and *Aspergillus sydowii* (Siglioccolo et al., 2011; Zajc et al., 2013; Pérez-Llano et al., 2020; Nowrousian and Stajich, 2024). Other marine-derived fungi, such as *Paradendryphiella salina*, have also been shown to produce surface-active proteins, including HFBs and cerato-platanins, highlighting the functional plasticity of these organisms in saline environments (Landeta et al., 2024). However, the physiological roles of HFBs in halophilic fungi remain poorly understood, and the classification of these proteins within traditional Classes I and II is often ambiguous in marine taxa. Moreover, many fungal strains of marine origin are uncultured or understudied, hindering the systematic exploration of their HFB repertoire.

Understanding the structure-function relationships and expression regulation of marine and halophilic HFBs could pave the way for novel biomaterials and antifouling technologies, salt-tolerant coatings, and stress-resilient formulations for biomedicine or agriculture (Plemenitaš et al., 2014; Pérez-Llano et al., 2020; Yovchevska et al., 2025). Thus, this chapter not only reviews the classical features and applications of fungal HFBs, but also emphasizes the urgent need to expand research on marine-derived and halophilic HFBs, highlighting their biotechnological promise and the ecological insights they may provide into fungal adaptation to extreme environments.

3. Hydrophobin Gene Counts

Researchers have identified HFB genes in various marine fungi (inhabiting sea or brackish environments) and halophilic/halotolerant fungi (thriving at high salt). Table 1 summarizes the total number of HFB genes reported for each such fungus, along with how many belong to Class I vs Class II, based on the literature.

As shown above, halophilic fungi exhibit two extremes: the basidiomycete *Wallemia ichthyophaga* encodes an exceptional number of hydrophobins (over two dozen), whereas some halotolerant ascomycetes have very few or none. *W. ichthyophaga*'s genome contains ~26 HFB genes, a dramatic expansion (the largest HFB gene family known in fungi) (Pérez-Llano et al., 2020). Its closest relative, *W. sebi* has only 12 HFB genes indicating *W. ichthyophaga* underwent a lineage-specific expansion of this gene family (Plemenitaš et al., 2014). In contrast, the extremely halotolerant black yeast *Hortaea werneckii* completely lacks HFB genes—an interesting case, possibly because *H. werneckii* often grows in yeast-like form (yeast cells typically do not require HFBs) (Kis-Papo et al., 2014; Pérez-Llano et al., 2020). Another halophilic ascomycete, *Eurotium rubrum* (a fungus from

Table 1. Hydrophobin gene counts in selected marine and halophilic fungi, with Class I vs. Class II classification and references (HFB = hydrophobin gene).

Fungi	Total HFB Genes	Class I HFBs	Class II HFBs	References
Aspergillus sydowii (marine halotolerant mold)	4	1	3	Littlejohn et al., 2012; Pérez-Llano et al., 2020
Hortaea werneckii (extremely halotolerant black yeast)	0	0	0	Lenassi et al., 2013
Eurotium rubrum (Dead Sea ascomycete)	1	1	0	Kis-Papo et al., 2014
Wallemia sebi (halotolerant relative)	12	12	0	Plemenitaš et al., 2014
Wallemia ichthyophaga (obligate halophile)	26	26	0	Pérez-Llano et al., 2020
Paradendryphiella salina (marine filamentous fungus)	2	0	2	Landeta et al., 2024

Dead Sea brines), retains only one HFB gene. Thus, in ascomycetes adapted to high salt, the number of HFBs is generally reduced (or even lost) relative to their non-halophilic relatives (Kis-Papo et al., 2014).

Among marine fungi, a similar trend of low HFB copy number is observed. For example, *Aspergillus sydowii* (a filamentous fungus isolated from marine environments) has 4 HFB genes, which is below the *Aspergillus* genus average (6–10 per species, typically ~9). Notably, genomic analysis initially found only three intact HFB genes in *A. sydowii* (sih1–3) plus one truncated gene; however, transcriptomics confirmed a fourth HFB (sih 4) is expressed, indicating *A. sydowii* does possess four HFB genes in total (Pérez-Llano et al., 2020). The marine hyphomycete *Paradendryphiella salina* also contains just 2 HFB genes (Landeta et al., 2024). In a broad screening of 23 marine fungi, only six new HFBs were isolated in total (implying many tested strains produced none detectable), underscoring that marine fungi are an underexplored source of HFBs. Interestingly, of those six new marine fungal HFBs, four were Class I and two were Class II, based on biochemical characterization (Cicatiello et al., 2016). This suggests that marine fungi can produce both classes of HFBs, though not in large numbers per species.

4. Class I vs Class II: Comparative Distribution and Trends

In the fungi above, we see varying proportions of Class I and Class II HFBs. Class I HFBs tend to dominate in certain lineages and roles. For instance, basidiomycetes *exclusively* produce Class I HFBs (Lo et al., 2014; Cicatiello et al., 2016), and indeed the Wallemia species (Basidiomycota) appear to have only Class I HFBs (Zajc et al., 2013). These Class I HFBs likely form robust rodlet layers on cell surfaces—a feature that may be beneficial for surviving extreme osmotic stress. In *W. ichthyophaga*, HFBs are hypothesized to coat the cell wall and contribute to its unique "sarcina-like" clumped morphology, possibly aiding in osmotic protection (Zalar et al., 2005).

contracted or were lost to eliminate unnecessary metabolic cost in a given niche (Kubicek et al., 2008; Cai et al., 2020). As research continues—aided by genomic and proteomic studies, including multi-omics approaches in halophiles (Plemenitaš et al., 2014; Gunde-Cimerman et al., 2018; Pérez-Llano et al., 2020) and structural analyses (Bayry et al., 2012; Landeta et al., 2024), we are gaining a clearer picture of how these "protein-amphiphiles" (Linder et al., 2005) contribute to fungal survival across the salty seas and hypersaline shores of our planet.

6. Diverse Application of Marine Fungal Hydrophobins

HFBs, with their unique self-assembly properties and amphiphilic nature, are considered promising candidates for a wide range of industrial and technological applications (Wösten, 2001; Linder et al., 2005; Cicatiello et al., 2016; Landeta et al., 2024). The exploration of marine fungal variants promises to further expand this spectrum, offering biomaterials with characteristics adapted to extreme environments (Yovchevska et al., 2025).

6.1 Biotechnological Applications

In the biotechnological field, HFBs are valued as natural biosurfactants, foaming agents, and encapsulating agents (Wösten, 2001; Bayry et al., 2012; Sánchez, 2022). Their self-assembly and non-toxic nature make them suitable for use in emulsions and personal care products (Artini et al., 2017). One of their most prominent applications is surface modification, offering a simple, rapid, and environmentally friendly method to alter chemically inert surfaces (Linder et al., 2005; Stanzione et al., 2022). This capacity to make hydrophilic surfaces hydrophobic and vice versa is crucial for preventing microbial adhesion or alternatively, for promoting fungal adhesion (Wösten, 2001; Bayry et al., 2012).

HFBs are not just biosurfactants but multifunctional ones, capable of performing various functions that would normally require different types of molecules (Deshmukh et al., 2022). This broad utility, combined with their biological origin and non-toxic nature, positions them as highly attractive alternatives to synthetic surfactants in various industries, especially those with strict biocompatibility or environmental requirements (Wösten, 2001; Linder et al., 2005; Pitocchi et al., 2022; Landeta et al., 2024). This implies a significant market opportunity if production and regulatory hurdles can be overcome. Marine fungi, in particular, are a promising and largely unexplored source of biosurfactants and bioemulsifiers (Deshmukh et al., 2022; Pitocchi et al., 2022; Yovchevska et al., 2025). Microbial bioemulsifiers are high-molecular-weight amphiphilic molecules that are efficient in emulsifying two immiscible phases, such as water and oil. Unlike synthetic emulsifiers, microbial bioemulsifiers offer unique advantages such as biocompatibility, non-toxicity, biodegradability, efficiency at low concentrations, and high selectivity under various pH, temperature, and salinity conditions (Pitocchi et al., 2022; Rubio-Ribeaux et al., 2023).

These properties make marine fungal HFBs promising candidates for applications in medical biocoatings, food industry foam stabilizers, and environmental

bioremediation (Wösten and Scholtmeijer, 2015; Landeta et al., 2024). Furthermore, HFBs enable protein immobilization in biosensors for rapid detection of bacterial contamination, such as *E. coli,* at 10 CFU/mL within 15 minutes (Pennacchio et al., 2024). They also show potential in separation technologies and as flavor stabilizers and encapsulating agents for trace ingredients in beverages (Ozdal et al., 2020). A notable advance is the possibility of fusing HFBs with nutritional proteins of interest, which allows their separation and isolation, offering a generic, rapid, and efficient protein purification system (Reuter et al., 2016).

6.2 Biomedical and Pharmaceutical Applications

The interest in HFBs in medicine and pharmaceutical formulations is considerable, primarily due to their ability to increase drug stabilization. They can enhance the biocompatibility of medical implants, increasing their chances of validation (Scholtmeijer et al., 2001). A vital need in medical device design is the development of anti-fouling coatings that prevent the adhesion of micromolecules, which can lead to complications such as thrombosis or infections (Salve et al., 2022). In this context, the HFB SC3, when coated on a nitric oxide (NO)-releasing medical-grade polymer, has shown a significant reduction in viable *Staphylococcus aureus* cell count and platelet adhesion compared to control samples (Devine et al., 2019). HFBs can also coat surfaces to repel or eliminate microbes, preventing biofilm formation, which represents an ecological strategy to reduce disease spread (Linder et al., 2005; Cicatiello et al., 2020).

For example, a fusion protein of HFB with pediocin PA1, an antibacterial peptide, showed much stronger antibacterial activity on PCL membranes, improving bacterial resistance (Wang et al., 2017). HFBs are also explored in drug delivery systems, particularly for oral administration, where they can improve bioavailability in formulations such as nanocrystals, polymeric nanocarriers, and lipid-based formulations (Wang et al., 2021). Their low immunogenicity makes them safe and functional biomaterials for medical devices (Scholtmeijer et al., 2001). Additionally, their potential in tissue engineering has been identified (Janssen et al., 2002). The diverse medical applications suggest that HFBs are not just a desirable additive but a fundamental biomaterial for the next generation of medical devices and therapies. Their ability to modify surfaces, prevent biofilm formation, and enhance drug stability, coupled with their low immunogenicity, makes them ideal candidates for addressing critical healthcare challenges, such as device-associated infections and low drug bioavailability. This implies a shift towards bio-inspired engineering using fungal proteins for advanced medical solutions. Table 2 highlights the key biomedical applications of fungal HFBs.

7. Challenges in Marine Hydrophobin Production

While HFBs, a unique class of fungal surface-active proteins, offer immense potential across diverse applications, their journey from discovery to widespread commercialization is fraught with challenges. These challenges, particularly

Table 2. Fungal hydrophobins in biomedicine: Representative examples.

Area of Application	Mechanism/Benefit	Specific Example/Evidence (from Materials)	References
Tissue Engineering	Supports tissue growth and development	Potential for application in tissue engineering; HGFI on PCL scaffolds enhances cellularization	Niu et al., 2012
Antibacterial Surfaces	Repels or eliminates microbes, prevents biofilm formation	Hydrophobin-pediocin PA1 fusion protein for antibacterial PCL membranes; fusion inhibits pathogens and biofilms	Wang et al., 2017
Anti-fouling Coatings	Prevents microbial and platelet fouling	Hydrophobin SC3 on NO-releasing polymer reduces *S. aureus* and platelet adhesion	Devine et al., 2019
Biocompatibility Enhancement	Improves implant integration	Low immunogenicity, making them safe and functional biomaterials, scaffolds for tissue engineering	Salve et al., 2022
Drug Stabilization	Increases drug stability	Improves bioavailability in nanocrystals, polymeric nanocarriers, and lipid formulations for oral administration	McGuckin et al., 2022

concerning manufacturing efficiency, purification intricacies, and navigating complex regulatory landscapes, are significantly amplified when the focus shifts to marine HFBs, especially those sourced from marine-derived fungi (Wösten, 2001; Khalesi et al., 2015b). The marine environment, with its unique biodiversity and often extreme conditions, presents a novel frontier for HFB discovery, but concurrently introduces distinct hurdles.

The general difficulties associated with HFB development are not merely replicated but are often compounded when marine organisms, and marine fungi in particular, are the source (Gonçalves et al., 2022). This intensification of challenges stems from the inherent complexities of accessing, cultivating, and processing organisms from marine ecosystems, coupled with potentially more stringent or less defined regulatory pathways for novel marine microbial products (Gladfelter et al., 2019). This section will delve into the specific challenges inherent in the production and commercialization of marine HFBs, emphasizing the additional complexities associated with utilizing marine fungal sources, from their cultivation and bioprocessing to the stringent safety and access regulations governing marine-derived compounds.

7.1 *Yield and Purification Complexities*

The direct extraction of HFBs from wild-type marine fungal strains, while offering the allure of discovering proteins with novel functionalities adapted to unique marine niches, is severely hampered by fundamental biological and bioprocess engineering challenges. These obstacles collectively contribute to low yields and make large-scale production economically unviable for many potential marine HFBs

(Khalesi et al., 2015b). A primary hurdle is the sheer difficulty in the discovery, isolation, and sustained cultivation of marine fungi. A vast proportion of marine fungal biodiversity remains largely unexplored, with estimates suggesting that less than 10% of all fungal species have been described, and of those, only a meager 1.2% originate from marine habitats (Richards et al., 2012; Gonçalves et al., 2022). This "uncultured majority" or "microbial dark matter" represents an enormous, untapped reservoir of potentially novel HFBs.

However, accessing this diversity is contingent upon overcoming significant cultivation barriers. Marine fungi frequently inhabit extreme environments such as the deep sea, hydrothermal vents, or polar regions, leading to highly specialized physiological requirements. These can include dependencies on high salinity, elevated hydrostatic pressure, specific and often scarce nutrients, or particular temperature regimes, all of which are challenging and costly to replicate under standard laboratory fermentation conditions. Furthermore, many marine fungi are slow-growing or exhibit complex symbiotic relationships with other marine organisms, making their isolation and axenic culture for consistent HFB production exceptionally difficult (Pang et al., 2016). Traditional mycological cultivation techniques, largely developed for terrestrial fungi, often prove inadequate for these specialized marine strains, necessitating innovative, though often less developed, approaches (Salgado et al., 2021). The lack of optimized and robust cultivation protocols directly translates to low biomass accumulation and, consequently, diminished native HFB yields. Even if a promising marine fungal strain is successfully isolated and cultured at the laboratory scale, the subsequent transition to industrial-scale bioreactor production introduces a new suite of challenges. The scale-up of marine fungal fermentations is often not a linear process.

Critical bioprocess parameters such as dissolved oxygen levels, agitation rates, pH control, and the resulting fungal morphology (e.g., pelletized versus filamentous growth) can differ dramatically between small-scale shake flask cultures and large-volume bioreactors (Petre and Petre, 2016). For instance, studies on the marine-derived fungus *Talaromyces albobiverticillius* demonstrated that oxygen transfer became a rate-limiting factor in bioreactors, and fungal morphology shifted from filamentous in flasks to pelleted growth in bioreactors, adversely affecting pigment production, a phenomenon that could similarly impact HFB yields (Venkatachalam et al., 2023). Filamentous marine fungi can be particularly sensitive to the shear forces generated by impellers in bioreactors, and ensuring adequate oxygen and nutrient transfer throughout viscous mycelial broths remains a persistent engineering problem. Compounding these issues is the general lack of comprehensive growth kinetic data for many marine fungal isolates; without a thorough understanding of their growth characteristics and substrate utilization preferences, the rational design and optimization of scalable and efficient fermentation processes for HFB production becomes highly problematic.

High-throughput screening methods are being explored to accelerate the acquisition of such data, but these also face technical difficulties, including data variability and the tendency of fungi to attach to surfaces in small-volume cultivation systems (Tamminen et al., 2020). The intrinsic physiology of marine fungi can also

influence hydrophobin expression levels. Many biosynthetic gene clusters (BGCs) in fungi, including marine strains, are known to be "silent" or poorly expressed under standard laboratory conditions (Tasdemir, 2017; Landeta et al., 2024). While HFBs are typically considered primary structural proteins, their expression levels can be significantly modulated by environmental cues and stresses, which are characteristic features of many marine environments. Marine fungi, for example, possess variable numbers of HFB genes, and their expression may be regulated in response to salt stress, a common adaptation in marine ecosystems (Kumar et al., 2015; Cicatiello et al., 2016). Therefore, optimizing fermentation conditions to elicit robust and sustained expression of the target HFB genes in often recalcitrant marine fungal hosts is a key hurdle for native production. Marine HFB sourcing faces a significant double bottleneck. The first arises from the biological challenges of isolating and cultivating native marine fungi, many of which remain uncultured or have specific growth requirements. The second lies in scaling up fermentation processes while ensuring product consistency and yield, both of which hinder industrial application.

7.2 *Recombinant Production Strategies*

Given the profound challenges and typically low yields associated with native production from marine fungi, recombinant DNA technology presents an alternative and often more attractive route for producing marine HFBs. This approach involves expressing the gene encoding a marine HFB in a well-characterized heterologous host system. However, this strategy is not a panacea and comes with its own distinct set of obstacles. The selection of a suitable heterologous host is crucial for functional HFB production, balancing expression efficiency, proper folding, and post-translational modifications. *Escherichia coli* is widely used due to its simplicity and cost-effectiveness. However, for HFBs, expression in *E. coli* commonly results in the accumulation of the recombinant protein as insoluble and non-functional aggregates known as inclusion bodies (Murooka and Imanaka, 2020; Ahn et al., 2021).

Pichia pastoris offers a eukaryotic alternative with a high capacity for protein secretion into the culture medium and the capability to grow to very high cell densities in bioreactors, facilitating downstream processing (Wu et al., 2025). However, it presents challenges such as low initial yields, proteolysis, hyper-glycosylation, and *in vivo* self-assembly (Werten et al., 2019; Wu et al., 2025). Consequently, achieving high yields of soluble, correctly folded, and fully functional recombinant marine HFBs remains a substantial hurdle, as underscored by research efforts focused on developing tagging strategies and optimizing expression conditions to improve soluble protein recovery.

7.3 *The Hydrophobicity Paradox*

The intrinsic hydrophobicity and amphiphilicity of HFBs, while central to their functionality, pose major challenges for purification—an issue known as the "hydrophobicity paradox" (Khalesi et al., 2015a). This problem is intensified in

marine-derived HFBs due to the complex fermentation process. The complexity of the biological matrix from marine fermentations and the general difficulties encountered in marine biotechnology downstream processing (DSP) further compound these purification challenges. Marine fungal fermentations can yield a highly complex mixture of biomolecules. In addition to the target HFB, the culture broth or fungal biomass may contain numerous other proteins, polysaccharides, pigments, lipids, and a diverse array of secondary metabolites (Tasdemir, 2017). Separating the desired HFB from this intricate soup to the high degree of purity required for many applications is a demanding task.

The general field of marine biotechnology already grapples with significant DSP challenges; it is widely recognized that DSP often constitutes the most expensive and least efficient segment of marine bioprocesses. This is largely due to the vast diversity of marine-derived products, each potentially requiring tailored and often specialized separation protocols (Muffler and Ulber, 2005). These overarching DSP difficulties directly impact the feasibility and cost-effectiveness of purifying marine HFBs. Consequently, the development and implementation of innovative purification strategies, such as advanced affinity chromatography, novel membrane separation techniques, or alternative methods like aqueous two-phase systems (ATPS) and microfluidic systems, become even more critical for marine HFBs to overcome the limitations imposed by their complex source matrix (Ahn et al., 2021). A particularly critical concern, and one that adds a significant layer of complexity to the purification of HFBs from marine fungi, is the potential for co-purification of bioactive or toxic secondary metabolites.

Marine fungi are renowned for their capacity to produce a prolific and structurally diverse array of secondary metabolites (SMs), many of which exhibit potent biological activities, including antibiotic, cytotoxic, or other toxic properties (Kempken, 2023). These SMs, which can include various classes such as polyketides, alkaloids, and terpenes, are produced by the fungus for ecological purposes but can pose serious safety risks if present as contaminants in an HFB preparation intended for food, cosmetic, pharmaceutical, or biomedical applications. During the extraction and purification of HFBs, these SMs may co-purify with the target protein due to similar solubility characteristics, adsorption properties, or direct interactions with the HFB itself. The profile and concentration of these SMs can also be highly variable, depending on the specific marine fungal strain, culture conditions, and fermentation parameters, making consistent removal challenging (Kempken, 2023). Ensuring the complete absence or reduction to safe levels of such potentially harmful co-purified SMs necessitates rigorous analytical screening and may require additional, often costly, purification steps, such as specialized chromatography or detoxification treatments. This is a crucial consideration for achieving regulatory approval and ensuring product safety. Marine fungal HFBs pose a "hidden cargo" risk during purification, as their extraction may co-purify unknown or bioactive secondary metabolites (Kamat et al., 2023). These compounds can compromise product safety and purity, adding complex analytical and regulatory challenges. This issue is less prominent in HFBs from well-characterized terrestrial fungi.

8. Future Directions

8.1 Exploration of New Marine Fungal Strain

The exploration of new marine fungal strains, particularly those from extreme environments (e.g., high salinity, pressure, or pH), represents a promising path for discovering HFBs with unique properties. Fungi adapted to extreme conditions often produce unusual bioactive compounds as survival mechanisms. For instance, a novel HFB (Sa-HFB1) isolated from the alkaliphilic fungus *Sodiomyces alkalinus*—sourced from highly basic soda soils—exhibited antifungal activity against opportunistic pathogens (Sallada et al., 2018; Kuvarina et al., 2022). This finding suggests that marine HFBs may have broader ecological functions, including antimicrobial defense. Marine ecosystems, from hypersaline estuaries to deep-sea hydrothermal vents, offer an untapped diversity of filamentous fungi. Genomic surveys have revealed that marine fungal strains can carry multiple HFB-encoding genes; some sediment-derived strains possess up to 12 distinct HFB genes. Additionally, a screening of 100 marine fungal isolates uncovered numerous novel HFB candidates, highlighting the ocean as a prolific source of unexplored HFB variants (Kempken, 2023). These HFBs may possess molecular adaptations (e.g., halotolerance) specific to their native environments, which makes them suitable for biotechnological applications under extreme conditions.

8.2 Advances in Genetic Engineering

Genetic engineering plays a key role in expanding the functional landscape of HFBs. Engineered HFBs created via site-directed mutagenesis or gene fusion techniques have enabled the generation of chimeric proteins with tailored surface-binding properties or additional biological activities. For example, HFBs have been successfully fused with functional enzymes to create bifunctional biomolecules capable of surface immobilization and catalytic action (Stanzione et al., 2022). Furthermore, novel engineered variants with enhanced surface adhesion have demonstrated superior performance compared to their wild-type counterparts. On the bioprocessing side, the large-scale production of HFBs faces challenges due to their complex folding requirements (e.g., eight cysteines forming disulfide bridges). To address this, cell-free expression systems have been developed, allowing for correct folding of HFBs outside the reducing cytosol of conventional hosts like *E. coli* (Siddiquee et al., 2020). These systems, enhanced with folding chaperones such as disulfide isomerases, have achieved HFB yields in the milligram-per-milliliter range. Additionally, new extraction and purification methods have emerged, including alcohol-based solvent systems that significantly increase recovery yields and cold-induced phase separation (CIPS) techniques that remove impurities and improve final product purity (Vereman et al., 2021).

8.3 Overcoming Commercialization Barriers

Despite scientific progress, several challenges remain before marine HFBs can be commercialized. From a regulatory perspective, HFBs would be classified as

novel ingredients in many jurisdictions, requiring thorough safety assessments. In the European Union, they would fall under the novel food category, necessitating toxicological and allergenic evaluations. In the United States, GRAS status would be required to facilitate market entry (Kale et al., 2022). Compliance with stringent food safety regulations necessitates the establishment of consistent and robust methods for obtaining highly pure HFBs, often to food-grade standards, free from undesirable contaminants (Khalesi et al., 2015a). The introduction of marine HFBs into the market, especially for food or food-contact applications, requires adherence to established regulatory pathways designed for novel ingredients and biomaterials. In the United States, the primary route for food ingredients is to achieve Generally Recognized as Safe (GRAS) status. This designation can be obtained either through a self-affirmation process, where a panel of qualified experts convenes to review the safety data and concludes that the substance is GRAS under its intended conditions of use, or through a GRAS notification submitted to the Food and Drug Administration (FDA) for review (Ziarati et al., 2024).

An example of a fungal protein achieving self-GRAS status is Quorn, derived from *Fusarium venenatum* (Denny et al., 2008). The registration of hydrophobins as edible products or food contact materials is contingent upon well-defined production, purification, and safety substantiation protocols, such as those outlined in GRAS determination processes in the USA (Hartung, 2018). This underscores that a principal barrier to market entry is not solely technical but significantly regulatory, creating a "regulatory lag" between scientific advancements and commercial viability. Addressing this challenge implies the need for a concerted effort involving industry, academia, and regulatory bodies (e.g., the FDA in the USA and EFSA in Europe) to develop clear guidelines and approval pathways for marine HFBs, especially for food-contact or ingestible applications (Coltelli et al., 2015). Technical and economic feasibility also limit industrial adoption. Although HFBs have demonstrated promising functionalities—such as stabilizing edible foams for up to four months and maintaining aerated emulsions for 45 days—their production costs remain high. Low yields, costly purification, and complex manufacturing processes pose economic constraints. Even with GRAS-compliant sources, extensive downstream processing is needed to remove pigments and impurities while preserving HFB functionality. Moreover, consumer acceptance must be considered, particularly for food applications involving fungal-derived proteins. Effective communication about the natural origin and functional benefits of HFBs will be essential. Future research and industrial investment should prioritize integrated development pipelines—from strain discovery to regulatory approval and public outreach—to ensure the successful transition of marine HGFBs from lab-scale innovation to commercial impact.

9. Conclusion

HFBs from marine and halophilic fungi represent a compelling class of proteins for biotechnology, with adaptations to extreme environments often conferring enhanced stability and novel functionalities. The genetic diversity of HFBs in these specialized fungi is striking. It includes the remarkable expansion and diversification of the

HFB gene family in the obligate halophilic filamentous basidiomycete *Wallemia ichthyophaga*, which contrasts with the trend of reduced HFB gene numbers observed in several marine and halophilic filamentous ascomycetes like *Aspergillus sydowii*, *Paradendryphiella salina*, and *Eurotium rubrum*. While the applications for these marine-derived proteins are extensive, spanning surface modification, emulsion stabilization, and biomedical uses such as drug delivery and anti-fouling coatings, significant obstacles impede their market translation. Key among these are low production yields from often difficult-to-culture marine strains and complex purification processes. A critical challenge specific to marine fungal sources is the potential co-purification of bioactive secondary metabolites, which can complicate product safety and regulatory approval. An evolving regulatory framework adds further complexity to their commercialization. To fully harness the innovative potential of these specialized fungal proteins, future research must prioritize bioprospecting in diverse and extreme marine niches, advancing genetic and bioprocess engineering tailored for marine hosts and their unique HFB characteristics, and establishing clear, supportive regulatory pathways.

The exploration of marine HFBs, particularly those sourced from the vast and largely untapped biodiversity of marine fungi, holds considerable promise for a multitude of innovative applications. However, as detailed in this chapter, the path from discovery to commercial realization is significantly more challenging than for their terrestrial counterparts. The unique conditions of marine environments and the specialized nature of marine fungi amplify existing difficulties in HFB production and introduce new layers of complexity in purification and regulatory approval.

Acknowledgements

During the preparation of this chapter, we used ChatGPT (OpenAI) to improve clarity and grammar. We reviewed and edited the content as needed and take full responsibility for the final text.

References

Ahn, S.O., Lim, H.D., You, S.H., Cheong, D.E. and Kim, G.J. (2021). Soluble expression and efficient purification of recombinant Class I hydrophobin DewA. Int. J. Mol. Sci., 22: 7843. https://doi.org/10.3390/ijms22157843.

Aimanianda, V., Bayry, J., Bozza, S., Kniemeyer, O., Perruccio, K. et al. (2009). Surface hydrophobin prevents immune recognition of airborne fungal spores. Nature, 460: 1117–1121. https://doi.org/10.1038/nature08264.

Artini, M., Cicatiello, P., Ricciardelli, A., Papa, R., Selan, L. et al. (2017). Hydrophobin coating prevents *Staphylococcus epidermidis* biofilm formation on different surfaces. Biofouling, 33: 601–611. https://doi.org/10.1080/08927014.2017.1338690.

Askolin, S., Linder, M., Scholtmeijer, K., Tenkanen, M., Penttilä, M. et al. (2006). Interaction and comparison of a Class I hydrophobin from *Schizophyllum commune* and Class II hydrophobins from *Trichoderma reesei*. Biomacromolecules, 7: 1295–1301. https://doi.org/10.1021/bm050676s.

Bayry, J., Aimanianda, V., Guijarro, J.I., Sunde, M. and Latge, J.P. (2012). Hydrophobins—Unique fungal proteins. PLoS Pathog., 8: e1002700. https://doi.org/10.1371/journal.ppat.1002700.

Cai, F., Gao, R., Zhao, Z., Ding, M., Jiang, S. et al. (2020). Evolutionary compromises in fungal fitness: Hydrophobins can hinder the adverse dispersal of conidiospores and challenge their survival. ISME J., 14: 2610–2624. https://doi.org/10.1038/s41396-020-0709-0.

Cicatiello, P., Gravagnuolo, A.M., Gnavi, G., Varese, G.C. and Giardina, P. (2016). Marine fungi as source of new hydrophobins. Int. J. Biol. Macromol., 92: 1229–1233. https://doi.org/10.1016/j.ijbiomac.2016.08.037.

Coltelli, M.B., Wild, F., Bugnicourt, E., Cinelli, P., Lindner, M. et al. (2015). State of the art in the development and properties of protein-based films and coatings, and their applicability to cellulose based products: An extensive review. Coatings, 6: 1. https://doi.org/10.3390/coatings6010001.

Crous, P.W., Shivas, R.G., Quaedvlieg, W.V., Van der Bank, M., Zhang, Y. et al. (2014). Fungal planet description sheets: 214–280. Persoonia Mol. Phyl. Evol. Fungi, 32: 184–306. https://doi.org/10.3767/003158514X682395.

De Stefano, L., Rea, I., De Tommasi, E., Rendina, I., Rotiroti, L. et al. (2009). Bioactive modification of silicon surface using self-assembled hydrophobins from *Pleurotus ostreatus*. Eur. Phys. J. E., 30: 181. https://doi.org/10.1140/epje/i2009-10481-y.

Denny, A., Aisbitt, B. and Lunn, J. (2008). Mycoprotein and health. Nutr. Bull., 33: 298–310. https://doi.org/10.1111/j.1467-3010.2008.00730.x.

Deshmukh, S.K., Deshpande, M.V. and Sridhar, K.R. (2022). Fungal Biopolymers and Biocomposites: Prospects and Avenues. Springer Nature Singapore, Singapore. https://doi.org/10.1007/978-981-19-1000-5.

Devine, R., Singha, P. and Handa, H. (2019). Versatile biomimetic medical device surface: Hydrophobin coated, nitric oxide-releasing polymer for antimicrobial and hemocompatible applications. Biomater. Sci., 7: 3438–3449. https://doi.org/10.1039/C9BM00469F.

Gladfelter, A.S., James, T.Y. and Amend, A.S. (2019). Marine fungi. Curr. Biol., 29: R191–R195. https://doi.org/10.1016/j.cub.2019.02.009.

Gonçalves, M.F.M., Esteves, A.C. and Alves, A. (2022). Marine Fungi: Opportunities and challenges. Encyclopedia, 2: 559–577. https://doi.org/10.3390/encyclopedia2010037.

Gostinčar, C., Coleine, C., Gunde-Cimerman, N. and Stajich, J.E. (2024). Fungi from extreme environments: Genome sequences and beyond. pp. 33–48. *In*: Nowrousian, M. and Stajich, J.E. (eds.). Fungal Genomics. Springer Nature Switzerland, Cham.

Gunde-Cimerman, N., Plemenitaš, A. and Oren, A. (2018). Strategies of adaptation of microorganisms of the three domains of life to high salt concentrations. FEMS Microbiol. Rev., 42: 353–375. https://doi.org/10.1093/femsre/fuy009.

Hariri Akbari, F., Song, Z., Turk, M., Gunde-Cimerman, N. and Gostinčar, C. (2024). Experimental evolution of extremotolerant and extremophilic fungi under osmotic stress. IUBMB Life, 76: 617–631. https://doi.org/10.1002/iub.2825.

Hartung, T. (2018). Rebooting the generally recognized as safe (GRAS) approach for food additive safety in the US. ALTEX, 35: 3–25. https://doi.org/10.14573/altex.1712181.

Janssen, M.I., Van Leeuwen, M.B.M., Scholtmeijer, K., Van Kooten, T.G., Dijkhuizen, L. et al. (2002). Coating with genetic engineered hydrophobin promotes growth of fibroblasts on a hydrophobic solid. Biomaterials, 23: 4847–4854. https://doi.org/10.1016/S0142-9612(02)00240-5.

Jensen, B.G., Andersen, M.R., Pedersen, M.H., Frisvad, J.C. and Søndergaard, I. (2010). Hydrophobins from *Aspergillus* species cannot be clearly divided into two classes. BMC Res. Notes, 3: 344. https://doi.org/10.1186/1756-0500-3-344.

Kale, V., Jagade, S., Dhar, H., Joshi, K. and Kulkarni, S. (2022). Hydrophobins: A revolutionary protein with potential uses in the food industry. Appl. Biol. Chem. J., 3(1): 4–10. https://doi.org/10.52679/tabcj.2022.0002.

Kamat, S., Kumar, S., Philip, S. and Kumari, M. (2023). Secondary metabolites from marine fungi: Current status and application. pp. 181–209. *In*: Kumar, A., Bilal, M., Ferreira, L.F.R. and Kmari, M. (eds.). Microbial Biomolecules. Elsevier. https://doi.org/10.1016/B978-0-323-99476-7.00001-6.

Kempken, F. (2023). Marine fungi: A treasure trove of novel natural products and for biological discovery. PLoS Pathog., 19(9): e1011624. https://doi.org/10.1371/journal.ppat.1011624.

Khalesi, M., Gebruers, K. and Derdelinckx, G. (2015a). Recent advances in fungal hydrophobin towards using in industry. Protein J., 34: 243–255. https://doi.org/10.1007/s10930-015-9621-2.

Khalesi, M., Mandelings, N., Herrera-Malaver, B., Riveros-Galan, D., Gebruers, K. et al. (2015b). Improvement of the retention of ocimene in water phase using Class II hydrophobin HFBII. Flavour Fragr. J., 30: 451–458. https://doi.org/10.1002/ffj.3260.

Kis-Papo, T., Weig, A.R., Riley, R., Peršoh, D., Salamov, A. et al. (2014). Genomic adaptations of the halophilic Dead Sea filamentous fungus *Eurotium rubrum*. Nat. Commun., 5: 3745. https://doi.org/10.1038/ncomms4745.

Klauer, R.R., Silvestri, R., White, H., Das, M., Hayes, R.D. et al. (2024). Hydrophobins from Aspergillus mediate fungal interactions with microplastics. Environ. Sci. Technol., 59(28): 14528–14539. https://doi.org/10.1021/acs.est.5c01771.

Kubicek, C.P., Baker, S., Gamauf, C., Kenerley, C.M. and Druzhinina, I.S. (2008). Purifying selection and birth-and-death evolution in the Class II hydrophobin gene families of the ascomycete *Trichoderma/Hypocrea*. BMC Evol. Biol., 8: 4. https://doi.org/10.1186/1471-2148-8-4.

Kumar, A., Henrissat, B., Arvas, M., Syed, M.F., Thieme, N. et al. (2015). *De Novo* assembly and genome analyses of the marine-derived *Scopulariopsis brevicaulis* strain LF580 unravels life-style traits and anticancerous scopularide biosynthetic gene cluster. PLoS One, 10: e0140398. https://doi.org/10.1371/journal.pone.0140398.

Kuvarina, A.E., Rogozhin, E.A., Sykonnikov, M.A., Timofeeva, A.V., Serebryakova, M.V. et al. (2022). Isolation and characterization of a novel hydrophobin, Sa-HFB1, with antifungal activity from an alkaliphilic fungus, *Sodiomyces alkalinus*. J. Fungi, 8: 659. https://doi.org/10.3390/jof8070659.

Landeta, C., Medina-Ortiz, D., Escobar, N., Valdez, I., González-Troncoso, M.P. et al. (2024). Integrative workflows for the characterization of hydrophobin and cerato-platanin in the marine fungus Paradendryphiella salina. Arch. Microbiol., 206: 385. https://doi.org/10.1007/s00203-024-04087-0.

Lenassi, M., Gostinčar, C., Jackman, S., Turk, M., Sadowski, I. et al. (2013). Whole genome duplication and enrichment of metal cation transporters revealed by *de novo* genome sequencing of extremely halotolerant black yeast *Hortaea werneckii*. PLoS One, 8: e71328. https://doi.org/10.1371/journal.pone.0071328.

Linder, M.B., Szilvay, G.R., Nakari-Setälä, T. and Penttilä, M.E. (2005). Hydrophobins: The protein-amphiphiles of filamentous fungi. FEMS Microbiol. Rev., 29: 877–896. https://doi.org/10.1016/j.femsre.2005.01.004.

Littlejohn, K.A., Hooley, P. and Cox, P.W. (2012). Bioinformatics predicts diverse Aspergillus hydrophobins with novel properties. Food Hydrocoll., 27: 503–516. https://doi.org/10.1016/j.foodhyd.2011.08.018.

Lo, V.C., Ren, Q., Pham, C.L., Morris, V.K., Kwan, A.H. et al. (2014). Fungal hydrophobin proteins produce self-assembling protein films with diverse structure and chemical stability. Nanomaterials, 4: 827–843. https://doi.org/10.3390/nano4030827.

Madern, D., Ebel, C. and Zaccai, G. (2000). Halophilic adaptation of enzymes. Extremophiles, 4: 91–98. https://doi.org/10.1007/s007920050142.

McGuckin, M.B., Wang, J., Ghanma, R., Qin, N., Palma, S.D. et al. (2022). Nanocrystals as a master key to deliver hydrophobic drugs via multiple administration routes. J. Contr. Release, 345: 334–353. https://doi.org/10.1016/j.jconrel.2022.03.012.

Muffler, K. and Ulber, R. (2005). Downstream processing in marine biotechnology. pp. 63–103. *In*: Ulber, R. and Le Gal, Y. (eds.). Marine Biotechnology II. Springer Berlin Heidelberg, Berlin, Heidelberg, pp. 63–103.

Murooka, Y. and Imanaka, T. (eds.). (2020). Recombinant Microbes for Industrial and Agricultural Applications, (1st Edn.). CRC Press, 896 pp.

Niu, B., Huang, Y., Zhang, S., Wang, D., Xu, H. et al. (2012). Expression and characterization of hydrophobin HGFI fused with the cell-specific peptide TPS in *Pichia pastoris*. Protein Expr. Purif., 83: 92–97. https://doi.org/10.1016/j.pep.2012.03.004.

Nowrousian, M. and Stajich, J.E. (eds.). (2024). Fungal Genomics. Springer Nature Switzerland, Cham, 202 pp.

Ozdal, T., Yolci-Omeroglu, P. and Tamer, E.C. (2020). Role of encapsulation in functional beverages. pp. 195–232. *In*: Grumezescu, A.M. and Holban, A.M. (eds.). Biotechnological Progress and Beverage Consumption. Elsevier.

Pang, K.L., Overy, D.P., Jones, E.G.B., da Luz Calado, M., Burgaud, G. et al. (2016). "Marine fungi" and "marine-derived fungi" in natural product chemistry research: Toward a new consensual definition. Fungal Biol. Rev., 30: 163–175. https://doi.org/10.1016/j.fbr.2016.08.001.

Pennacchio, A., Giampaolo, F., Cafaro, V., Cicatiello, P., Della Ventura, B. et al. (2024). A bacterial biosensor based on gold nanoparticles functionalized by a hydrophobin-chimera and combined with

machine learning for user-friendly detection. Sens. Actuators B Chem., 410: 135645. https://doi.org/10.1016/j.snb.2024.135645.

Pérez-Llano, Y., Rodríguez-Pupo, E.C., Druzhinina, I.S., Chenthamara, K., Cai, F. et al. (2020). Stress reshapes the physiological response of halophile fungi to salinity. Cells, 9: 525. https://doi.org/10.3390/cells9030525.

Petre, M. and Petre, V. (2016). Biotechnology of mushroom growth through submerged cultivation. pp. 1–18. *In*: Petre, M., (ed.). Mushroom Biotechnology. Elsevier.

Pitocchi, R., Piscitelli, A. and Giardina, P. (2022). Marine fungi as a source of biosurfactants and bioemulsifiers. pp. 313–327. *In*: Deshmukh, S.K., Deshpande, M.V. and Sridhar, K.R. (eds.). Fungal Biopolymers and Biocomposites. Springer Nature Singapore, Singapore.

Plemenitaš, A., Lenassi, M., Konte, T., Kejžar, A., Zajc, J. et al. (2014). Adaptation to high salt concentrations in halotolerant/halophilic fungi: A molecular perspective. Front. Microbiol., 5: 199. https://doi.org/10.3389/fmicb.2014.00199.

Reuter, L., Ritala, A., Linder, M. and Joensuu, J. (2016). Novel hydrophobin fusion tags for plant-produced fusion proteins. PLoS One, 11: e0164032. https://doi.org/10.1371/journal.pone.0164032.

Richards, T.A., Jones, M.D.M., Leonard, G. and Bass, D. (2012). Marine fungi: Their ecology and molecular diversity. Annu. Rev. Mar. Sci., 4: 495–522. https://doi.org/10.1146/annurev-marine-120710-100802.

Rubio-Ribeaux, D., Da Costa, R.A.M., Montero-Rodríguez, D., do Amaral Marques, N.S.A., Puerta-Díaz, M. et al. (2023). Sustainable production of bioemulsifiers: A critical overview from microorganisms to promising applications. World J. Microbiol. Biotechnol., 39: 195. https://doi.org/10.1007/s11274-023-03611-6.

Sahay, S. (2022). Extremophilic Fungi: Ecology, Physiology, and Applications. Springer Nature Singapore, Singapore. https://doi.org/10.1007/978-981-16-4907-3.

Salgado, C.L., Muñoz, R., Blanco, A. and Lienqueo, M.E. (2021). Valorization and upgrading of the nutritional value of seaweed and seaweed waste using the marine fungi *Paradendryphiella salina* to produce mycoprotein. Algal Res. 53: 102135. https://doi.org/10.1016/j.algal.2020.102135.

Sallada, N.D., Dunn, K.J. and Berger, B.W. (2018). A structural and functional role for disulfide bonds in a Class II Hydrophobin. Biochemistry, 57: 645–653. https://doi.org/10.1021/acs.biochem.7b01166.

Salve, A.L., Apotikar, S.B., Muddebihalkar, S.V., Joshi, K.S. and Kulkarni, S.S. (2022). Applications of hydrophobins in medical biotechnology. J. Sci. Res., 66: 86–94. https://doi.org/10.37398/JSR.2022.660312.

Sánchez, C. (2022). A review of the role of biosurfactants in the biodegradation of hydrophobic organopollutants: Production, mode of action, biosynthesis, and applications. World J. Microbiol. Biotechnol., 38: 216. https://doi.org/10.1007/s11274-022-03401-6.

Scholtmeijer, K., Wessels, J.G.H. and Wösten, H.A.B. (2001). Fungal hydrophobins in medical and technical applications. Appl. Microbiol. Biotechnol., 56: 1–8. https://doi.org/10.1007/s002530100632.

Siddiquee, R., Choi, S.S.C., Lam, S.S., Wang, P., Qi, R. et al. (2020). Cell-free expression of natively folded hydrophobins. Protein Expr. Purif., 170: 105591. https://doi.org/10.1016/j.pep.2020.105591.

Siglioccolo, A., Paiardini, A., Piscitelli, M. and Pascarella, S. (2011). Structural adaptation of extreme halophilic proteins through decrease of conserved hydrophobic contact surface. BMC Struct. Biol., 11: 50. https://doi.org/10.1186/1472-6807-11-50.

Stanzione, I., Pitocchi, R., Pennacchio, A., Cicatiello, P., Piscitelli, A. et al. (2022). Innovative surface bio-functionalization by fungal hydrophobins and their engineered variants. Front. Mol. Biosci., 9: 959166. https://doi.org/10.3389/fmolb.2022.959166.

Sunde, M., Kwan, A.H.Y., Templeton, M.D., Beever, R.E. and Mackay, J.P. (2008). Structural analysis of hydrophobins. Micron., 39: 773–784. https://doi.org/10.1016/j.micron.2007.08.003.

Tamminen, A., Happonen, P., Barth, D., Holmström, S. and Wiebe, M.G. (2020). High throughput, small scale methods to characterise the growth of marine fungi. PLoS One, 15: e0236822. https://doi.org/10.1371/journal.pone.0236822.

Tasdemir, D. (2017). Marine fungi in the spotlight: Opportunities and challenges for marine fungal natural product discovery and biotechnology. Fungal Biol. Biotechnol., 4: 5. https://doi.org/10.1186/s40694-017-0034-1.

total carbohydrate content by spectrometric techniques offers several advantages, for instance, low cost and the ability to analyze samples quickly (Rondel et al., 2013; Stauß et al., 2024).

3.2.3 *Ninhydrin Test*

The most used reagent for the spectrophotometric quantification of amino acids is ninhydrin (Stauß et al., 2024). The ninhydrin method is a rapid and sensitive colorimetric assay for detecting proteins and amino acids, optimized and applied in various fields (Marks et al., 1985; Barbehenn, 1995; Stauß et al., 2024), including biosurfactant characterization (Patowary et al., 2015). This reagent reacts with primary amino groups to form a purple compound, called Ruhemann's purple, measurable using spectrophotometry at 570 nm. The reaction is carried out at high temperature to induce the formation of this purple compound. In the literature, a wide range of temperatures and incubation times have been reported. Recently, Stauß et al. (2024) optimized the procedure by employing incubation at 90°C for 45 minutes. The optimized procedure presented an LOD and LOQ of 0.03 and 0.1 mmol L^{-1}, respectively.

3.2.4 *Anthrone Test*

The principle of this method is the conversion of carbohydrates by acid hydrolysis with concentrated sulfuric acid to monosaccharides and further dehydration to 5-hydroxymethylfurfural (HMF) and pentoses into furfural (Loewus, 1952 apud Rondel et al., 2013). These compounds subsequently react with anthracen-9(10 H)-one, often referred to as anthrone, producing green-colored compounds, which absorb light at a wavelength of around 620 nm. Anthrone is a much less toxic chemical than the classical phenol used in the phenol method. The replacement significantly decreases the safety risks related to the evaluations and the toxicity of the produced waste (Rondel et al., 2013).

3.2.5 *Saponification Test*

The classical method to estimate lipid content consists of adding 1 mL of crude biosurfactant sample, a few drops of 0.1N KOH, and a few drops of phenolphthalein indicator in a tube, then heating it in a water bath for 1–2 hours (Vijayalaxmi, 2012). Some authors simplified the method by using 2 mL of 1–2% NaOH solution and shaking (Patowary et al., 2015; Godlewska et al., 2022). A positive result is indicated by the formation of a soapy solution.

3.2.6 *Liquid Chromatography*

Liquid chromatography enables the separation, purification, and qualitative and quantitative analysis of complex mixtures, provided the samples are soluble in a suitable solvent. Liquid chromatography has various configurations concerning instrumental and separation modes (Meyer, 2005). Thin layer chromatography (TLC) and high-performance liquid chromatography (HPLC) represent distinct methods within the broader category of liquid chromatography. While TLC is primarily used for preparative purposes, HPLC is employed for analytical purposes.

3.2.7 Thin Layer Chromatography

Thin-layer chromatography is a separation and qualitative technique that can be conducted with relatively low costs (Stahl, 1969; Gupta, 2016). Standard (TLC) and high-performance (HPTLC) formats are known as versatile and high-throughput liquid chromatography techniques, with high variability of important applications (Kowalska and Sajewicz, 2022). The technique involves the movement of a liquid phase through a thin stationary phase held on a rigid support via capillary action. It is a versatile technique because a variety of solvents, stationary phases, and numerous spray reagents are available, allowing the detection of many types of compounds in life sciences (Meyer, 2005). In marine fungal biosurfactant characterization, Senthil Balan et al. (2019) and Xiaoyun et al. (2022) employed TLC for glycolipoprotein identification.

To carry out the technique, the sample solution is applied as a spot near the edge of the support. A portion of the sheet is then immersed in the mobile phase, which gradually moves into the stationary phase. Once the mobile phase has nearly reached the opposite end, the sheet is taken out of the developing chamber and allowed to dry. If the analytes are not visible due to a lack of color, the sheet is treated with a reagent to reveal the spots (Meyer, 2005). The decisive factor that considerably affects separation quality and unequivocally defines the technique (Kowalska and Sajewicz, 2022). The analysis of strongly hydrophilic compounds, such as carbohydrates, requires polar solvent systems with slow migration rates. For free amino acids and peptides, which are hydrophilic compounds slightly soluble in non-aqueous solvents, these solvents are generally employed (Stahl, 1969). Once the type of macromolecules is characterized, further advanced chemical characterization of the resultant organic extracts can be performed using a variety of specialized methods. Using TLC as a preparative technique, the purified fungal biosurfactant can be acid hydrolyzed and extracted with diethyl ether. The resulting hydrolysate is then examined for detailed biochemical composition using GC–MS analysis (Senthil Balan et al., 2019). After stepwise separation by HPLC, TLC is often performed to verify the purity of the eluates (Konishi et al., 2010).

3.2.8 High Performance Liquid Chromatography

HPLC is a technique for identifying, quantifying, and purifying the components of complex matrices such as biosurfactant crude extracts. As HPLC utilizes microparticulate packings, it requires a pump for mobile phase transport and a detector for observing eluate fractions, usually based on concentration, e.g., UV absorbance, fluorescence, or electrochemical detectors, according to the properties of the analytes (Meyer, 2005). In fungal biosurfactant characterization, mixtures present in the purified product can be separated and differentiated between various structural analogs (Patowary et al., 2017). HPLC was also used as a purification step employing reverse phase (RP)-C18 silica gel column chromatography with stepwise elution using acetonitrile (A) and water (B) as the solvent system. The resultant fractions were then collected, rotary evaporated, lyophilized, and examined individually (Konishi et al., 2010; Senthil Balan et al., 2019).

3.2.9 Gas Chromatography-Mass Spectrometry

GC-MS is an analytical technique where the mobile phase is a gas, and the sample needs to be volatile or derivatized as volatile. GC is the technique of choice for investigating volatile and very complex samples. At adequate temperatures, the sample molecules are transported by the gas due to their volatility. Retention is influenced by both the vapor pressure and the compound's affinity for the stationary phase, whereas the gaseous mobile phase does not directly affect the separation process (Meyer, 2005). For fatty acid analysis in biosurfactant extracts, pretreated samples typically undergo several extraction and purification steps before being analyzed by GC-MS. For example, Konishi et al. (2010) analyzed methyl ester surfactants, while Amaral et al. (2006) characterized a polysaccharide–protein complex with low lipid content.

3.2.10 SDS-PAGE

Electrophoresis is the migration of charged molecules in solutions in response to an electric field. As an analytical tool, it is simple, rapid, and highly sensitive. It is used to study the properties of charged species and as a preparative separation technique (Moldes et al., 2023). In Sodium Dodecyl Sulphate Polyacrylamide Gel Electrophoresis (SDS -PAGE), Sodium dodecyl sulfate (SDS), an anionic detergent, is utilized to denature proteins, conferring a negative charge to the polypeptide in proportion to its length. The sample is run in a polyacrylamide gel support matrix. This inhibits convective mixing caused by heating, separating molecules by size. At the end of the procedure, they can be stained and used for scanning, blotting, autoradiography, or storage (Rybicki, 2015). This technique was used for the analysis of biosurfactant fungal proteins. After extracting and evaluating the presence of proteins, SDS-PAGE was applied to determine the purity and molecular weight of the extracted proteins using a 15% acrylamide configuration and silver stain (Cicatiello et al., 2016), or in early steps of the characterization process after ultrafiltration dialysis purification (Pitocchi et al., 2020).

3.2.11 Fourier Transform Infrared Spectrometry

FTIR is a technology for acquiring emission spectra or infrared absorption from solid, liquid, or gas samples (Onyeaka et al., 2022). A primary application of this method is to identify functional groups present in a purified sample. For biosurfactant characterization, it is applied following standard purification steps with lyophilized samples (Amaral et al., 2006; Senthil Balan et al., 2019; Correa et al., 2020; Kaur et al., 2022). Once the main functional groups are identified, structural elucidation of the purified biosurfactant is performed using MALDI-TOF MS and NMR analyses. Figure 4 illustrates the process flow with general steps, beginning with obtaining crude extracts containing biosurfactant molecules and continuing to structural characterization.

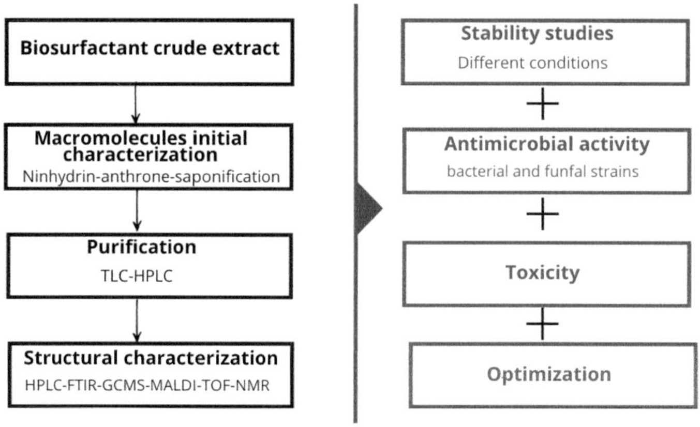

Fig. 4. Process flow scheme illustrating the general steps from biosurfactant crude extracts to structural characterization as reported in marine fungi studies.

4. Applications

A relevant aspect to discuss is the scarcity of raw materials due to the increasing development of products derived from fossil fuels, leading to intensified extraction of these natural resources, such as those used in manufacturing synthetic surfactants (Farias et al., 2021). In this context, companies' growing concern with sustainability highlights the adoption of remedial and biodegradable materials, which present an advantageous alternative for the future of the environment and the economy (Farias et al., 2021). Another important factor is the low toxicity associated with surfactants of biological origin, whose composition shows innovative and sustainable aspects (Banat et al., 2021). From this perspective, replacing synthetic surfactants with biosurfactants contributes to reducing the use of compounds that are mostly toxic to the environment and the entire ecosystem, including those present in products consumed by humans (Banat et al., 2021). Furthermore, these biologically derived compounds have wide applicability, ranging from the pharmaceutical and cosmetic industries to bioremediation processes and operations in the oil industry, highlighting their versatility and relevance in various production sectors (Jahan et al., 2020; Sarubbo et al., 2022).

4.1 Bioremediation

Currently, the use of surfactants produced by microorganisms as bioremediation agents is increasingly explored (Mapelli et al., 2017) due to their effectiveness in metabolizing petroleum compounds and degrading hydrophobic pollutants (Kaczorek and Olszanowski, 2011; Jahan et al., 2020). In the biodegradation of surfactant pollutants, microorganisms break the hydrocarbon chain, causing the surfactant pollutants to lose their amphipathic properties and dissociate into elements harmless to the environment, such as carbon dioxide (CO_2), water, and mineral salts (Lima et al., 2011).

The remediation of hydrocarbon spills in aquatic environments is crucial to preserve the local ecosystem, as they cause severe consequences for biodiversity and affect sunlight penetration, impacting marine life and disrupting the food chain. However, the methods used to disperse these oil spills often involve non-biodegradable chemical compounds with high toxicity levels for many aquatic organisms (Toledo et al., 2013; Patel et al., 2019; Feng et al., 2019). Thus, biosurfactants serve as an ecological and effective dispersant (Fenibo et al., 2019; Dang et al., 2019). A study with species *Penicillium polonicum*, *P. chrysogenum*, and *P. cyclopium* obtained from chronically polluted seawater demonstrated that the fungi were efficient degraders of pure petroleum. Particularly, these same isolates exhibited the highest biosurfactant activity in the assays. These results show the close link between biosurfactant production and hydrocarbon degradation (Maamar et al., 2020). Another example is the yeast *Yarrowia lipolytica*, which produces biosurfactants during the biodegradation process of hydrocarbons, such as asphaltene fractions and crude oil (Macedo et al., 2025).

Biosurfactants have been shown to be effective in the remediation of heavy metals, promoting their removal through processes such as sorption, desorption, and complexation, facilitating their dispersion and solubilization in the environment. Some studies indicate that the interaction of these compounds with metals may be more efficient than conventional complexation methods, as the metal desorption process occurs mainly through the formation of stable complexes and the reduction of interfacial tension between the phases (Franzetti et al., 2014). In general, biosurfactants facilitate the removal of metals through processes such as precipitation-dissolution, ion exchange, and electrostatic interactions, presenting a promising alternative for environmental remediation (Mulligan et al., 1999; Ochoa-Loza et al., 2001; Rufino et al., 2012; Franzetti et al., 2014).

4.2 Oil Industry

In the coming years, with the high extraction and use of lighter oils, a shortage of these oils is expected, making heavier oils an option in the industry (Fenibo et al., 2019; Sarubbo et al., 2022). In the recovery of residual oil through the surfactant principle, biosurfactants act to remove these oils trapped in the pores of rocks (Mazaheri-Assadi et al., 2010; Rawat et al., 2020). In the extraction of heavy oils, synthetic surfactants are used, which are harmful to the environment. On the other hand, biosurfactants are more advantageous, as they can be used in oil extraction processes with minimal impact due to their biodegradable and less toxic nature (Luna et al., 2012; Fenibo et al., 2019). In the oil industry, microbial surfactants can clean oil-contaminated vessels and transport crude oil over long distances because their high surfactant properties help the oil flow, reduce viscosity, and minimize the risk of pipeline clogging (Mazaheri-Assadi et al., 2010). Other innovative applications of biosurfactants include anticorrosives, emulsifiers/de-emulsifiers, and biocides (de Almeida et al., 2016).

4.3 *Pharmaceutical and Cosmetic Industries*

In the pharmaceutical industry, biosurfactants are utilized for their antimicrobial, anti-adhesive, and enzyme inhibitory properties (Markande et al., 2021; Banat et al., 2021; Sarubbo et al., 2022). They aid in burn treatment procedures and enhance the effectiveness of antibiotics within cells (Stipcevic et al., 2006; Sotirova et al., 2012; Joshi-Navare et al., 2013). Another recognized antimicrobial action of biosurfactants is the control of biofilms generated by pathogenic microorganisms. The attachment capacity of these organisms is reduced, causing instability in these colonies and consequently their dysfunction. Thus, the risk of infections in areas with high infectious potential decreases (Anjum et al., 2016; Sun et al., 2018; Banat et al., 2021). Furthermore, the applicability of biosurfactants has shown promise in the formulation of liquid soaps for hand use, as well as in coatings for plastic utensils, contributing to the prevention of microbial infections (Ceresa et al., 2021). Another relevant factor is the increased incidence of allergic reactions and side effects associated with synthetic surfactants in product compositions (Loranger et al., 2017; Boozalis et al., 2018). In cosmetics production, biosurfactants can replace synthetic surfactants due to their low toxicity (Vecino et al., 2017), an important consideration in personal care product formulation. Additionally, they possess properties such as wetting, solubilization, dispersion, foaming, cleansing, detergent action, and emulsion formation, which are utilized in cosmetic development (Vandana et al., 2018).

4.4 *Food Industry and Agriculture*

In agricultural production, improving planting strategies is essential to increase the efficiency and sustainability of the sector. In this context, biosurfactants have the potential to improve soil quality, promote symbiosis between microorganisms and plants, improve fertilizer adhesion, and protect against phytopathogens (Renfro et al., 2014; Liu et al., 2016). Additionally, fungal-derived biosurfactants can be effective in stimulating seeds before sowing, increasing resistance to pathogens and reducing the need for fertilizers such as biostimulants (de Vasconcelos et al., 2020). Other attributes applied in the food industry, such as antimicrobials and antibiofilms, are useful in cleaning equipment and preserving food. They also include regulating consistency, stable solubilization of ingredients like flavoring oils, and stabilizing fats. For example, Massoia lactone is known for its pleasant aroma and flavor and can be obtained from marine fungi, providing a natural alternative to chemical synthesis (Araujo et al., 2014; Mnif et al., 2016; Kiran et al., 2017; Luepongpattana et al., 2017).

4.5 *Nanotechnology*

In the nanotechnology sector, biosurfactants are notable for their ability to aid in the production of nanoparticles, especially silver (Sarubbo et al., 2022). They act as both reducers and stabilizers, making the process more sustainable than other chemical methods. Currently, there is a search for more environmentally friendly alternatives to traditional chemical processes, which require high pressure, temperature, and energy, consequently generating toxic waste (Plazar et al., 2014; Gómez-Graña et al., 2017). Biosurfactants emerge as an efficient solution, as their use does not require high energy consumption, nor does it involve substances potentially harmful to the environment and human health (Plazar et al., 2014; Gómez-Graña et al., 2017; Joanna et al., 2018).

5. Conclusion

The biotechnological potential of biosurfactants produced by marine fungi can be explored to meet the demands of various production sectors, including food, pharmaceuticals, cosmetics, oil, and agriculture. These microorganisms, adapted to extreme salinity, pressure, and temperature conditions, possess the metabolic capacity to produce biosurfactants with unique physicochemical properties and characteristics that combine functional efficiency, low toxicity, biodegradability, and stability under a wide range of conditions. These characteristics distinguish them from synthetic surfactants currently on the market. The commercial viability of biosurfactants remains constrained by their high production costs. Thus, there is a need for further investment and research in cultivation, production optimization, prospecting, purification, and structural characterization of these compounds, as well as integrated genomics, metabolomics, and metabolic engineering methods. In the current global scenario of searching for more sustainable alternatives and cleaner industrial processes, marine fungal biosurfactants represent a more natural and environmentally friendly alternative.

References

Adebajo, S., Akintokun, P.O., Ojo, A., Akintokun, A.K. and Badmos, O.A. (2020). Recovery of biosurfactant using different extraction solvent by rhizospheric bacteria isolated from rice-husk and poultry waste biochar amended soil. Egypt. J. Basic and Appl. Sci., 7: 252–266. https://doi.org/10.1080/2314808X.2020.1797377.

Amaral, P.F.F., da Silva, J.M., Lehocky, M., Barros-Timmons, A.M.V., Coelho, M.A.Z. et al. (2006). Production and characterization of a bioemulsifier from *Yarrowia lipolytica*. Process Biochem., 41: 1894–1898. https://doi.org/10.1016/j.procbio.2006.03.029.

Amend, A., Burgaud, G., Cunliffe, M., Edgcomb, V.P., Ettinger, C.L. et al. (2019). Fungi in the marine environment: Open questions and unsolved problems. mBio, 10: e01189-18. https://doi.org/10.1128/mBio.01189-18.

Anjum, F., Gautam, G., Edgard, G. and Negi, S. (2016). Biosurfactant production by *Bacillus* sp. MTCC 5877 and its multiple applications in the food industry. Bioresour. Technol., 213: 262–269. https://doi.org/10.1016/j.biortech.2016.02.091.

Araújo, S.C.D.S., Silva-Portela, R.C.B., de Lima, D.C., da Fonsêca, M.M.B., Araújo, W.J. et al. (2020). MBSP1: A biosurfactant protein derived from a metagenomic library with activity in oil degradation. Sci. Rep., 10: 1340. https://doi.org/10.1038/s41598-020-58330-x.

Asgher, M., Arshad, S., Qamar, S.A. and Khalid, N. (2020). Enhanced biosurfactant production from *Aspergillus niger* through chemical mutagenesis: Characterization and RSM optimization. SN Appl. Sci., 2: 966. https://doi.org/10.1007/s42452-020-2783-3.

Banat, I.M., Franzetti, A., Gandolfi, I., Bestetti, G., Martinotti, M.G. et al. (2010). Production of microbial biosurfactants, applications, and future potential. Appl. Microbiol. Biotechnol., 87: 427–444. https://doi.org/10.1007/s00253-010-2589-0.

Geys, R., Soetaert, W. and Van Bogaert, I. (2014). Biotechnological opportunities in biosurfactant production. Curr. Opin. Biotechnol., 30: 66–72. https://doi.org/10.1016/j.copbio.2014.06.002.

Gómez-Graña, S., Perez-Ameneiro, M., Vecino, X., Pastoriza-Santos, I., Perez-Juste, J. et al. (2017). Biogenic synthesis of metal nanoparticles using a biosurfactant extracted from corn and their antimicrobial properties. Nanomaterials, 7: 139. https://doi.org/10.3390/nano7060139.

Gonçalves, M.F.M., Hilário, S., Van de Peer, Y., Esteves, A.C. and Alves, A. (2022). Genomic and metabolomic analyses of the marine fungus *Emericellopsis cladophorae*: Insights into saltwater adaptability mechanisms and its biosynthetic potential. J. Fungi, 8: 31. https://doi.org/10.3390/jof8010031.

Gudiña, E.J., Teixeira, J.A. and Rodrigues, L.R. (2016). Biosurfactants produced by marine microorganisms with therapeutic applications. Mar. Drugs, 14: 38. https://doi.org/10.3390/md14020038.

Hanson, K.G., Desai, J.D. and Desai, A.J. (1993). A rapid and simple screening technique for potential crude oil degrading microorganisms. Biotechnol. Tech., 7: 745–748. https://doi.org/10.1007/BF00152624.

Henkel, M. and Hausmann, R. (2019). Diversity and classification of surfactant-producing microorganisms. pp. 41–63. *In*: Hayes, D.G., Solaiman, D.K.Y. and Ashby, R.D. (eds.). Biobased Surfactants: Synthesis, Properties, and Applications. Elsevier, Amsterdam. https://doi.org/10.1016/C2016-0-03179-0.

Hosseini, H., Al-Jabri, H.M., Moheimani, N.R., Siddiqui, S.A. and Saadaoui, I. (2022). Marine microbial bioprospecting: Exploitation of marine biodiversity towards biotechnological applications: A review. J. Basic Microbiol., 62: 1030–1043. https://doi.org/10.1002/jobm.202100504.

Hyde, K.D., Saleh, A., Aumentado, H.D.R., Boekhout, T., Bera, I. et al. (2024). Fungal numbers: Global needs for a realistic assessment. Fungal Diversity, 128: 191–225. https://doi.org/10.1007/s13225-024-00545-8.

Ito, T. and Masubuchi, M. (2014). Dereplication of microbial extracts and related analytical technologies. J. Antibiot., 67: 353–360. https://doi.org/10.1038/ja.2014.12.

Jahan, R., Bodratti, A.M., Tsianou, M. and Alexandridis, P. (2020). Biosurfactants, natural alternatives to synthetic surfactants: Physicochemical properties and applications. Adv. Colloid Interface Sci., 275: 102061. https://doi.org/10.1016/j.cis.2019.102061.

Jain, D.K., Collins-Thompson, D.L., Lee, H. and Trevors, J.T. (1991). A drop-collapsing test for screening surfactant-producing microorganisms. J. Microbiol. Methods, 13: 271–279. https://doi.org/10.1016/0167-7012(91)90064-W.

Jewett, M.C., Hofmann, G. and Nielsen, J. (2006). Fungal metabolite analysis in genomics and phenomics. Current Opinion in Biotechnology, 17: 191–197. https://doi.org/10.1016/j.copbio.2006.02.001.

Joanna, C., Marcin, L., Ewa, K. and Grażyna, P. (2018). A non-specific synergistic effect of biogenic silver nanoparticles and biosurfactants against environmental bacteria and fungi. Ecotoxicology, 27(3): 352–359. https://doi.org/10.1007/s10646-018-1899-3.

Joshi-Navare, K. and Prabhune, A. (2013). A biosurfactant sophorolipid acts synergistically with antibiotics to enhance their efficiency. BioMed. Res. Int., 2013: 512495.

Joye, S. and Peña-Montenegro, T.D. (2018). SnapShot: Microbial hydrocarbon bioremediation. Cell, 172(6): 1336–1336. https://doi.org/10.1016/j.cell.2018.02.059.

Kaczorek, E. and Olszanowski, A. (2011). Hydrocarbon uptake by Pseudomonas fluorescens (P1) and Pseudomonas putida (K1) strains in the presence of surfactants: A modification of the cell surface. Water, Air, Soil Pollut., 214(1-4): 451–459. https://doi.org/10.1007/s11270-010-0436-7.

Kaur, J., Kaur, S., Kumar, M., Krishnan, P., Kaur, J. et al. (2022). Studies on production, optimization and machine learning-based prediction of biosurfactant from Debaryomyces hansenii CBS767. Int. J. Environ. Sci. Technol., 19: 8465–8478. https://doi.org/10.1007/s13762-021-03639-x.

Kempken, F. (2023). Marine fungi: A treasure trove of novel natural products and for biological discovery. PLoS Pathog., 19(9): e1011624. https://doi.org/10.1371/journal.ppat.1011624.

Kiran, G.S., Hema, T.A., Gandhimathi, R., Selvin, J., Thomas, T.A. et al. (2009). Optimization and production of a biosurfactant from the sponge-associated marine fungus *Aspergillus ustus* MSF3. Colloids Surf B: Biointerfaces, 73(2): 250–256.

Kiran, G.S., Priyadharsini, S., Sajayan, A., Priyadharsini, G.B., Poulose, N., Selvin J. et al. (2017). Lipopeptide biosurfactant production by a marine *Nesterenkonia* sp. and its application in the food industry. J. App. Microbiol., 8: 1138. https://doi.org/10.3389/fmicb.2017.01138.

Kogej, T., Stein, M., Volkmann, M., Gorbushina, A.A., Galinski, E.A. et al. (2007). Osmotic adaptation of the halophilic fungus Hortaea werneckii: Role of osmolytes and melanization. Microbiology, 153(12): 4261–4273. https://doi.org/10.1099/mic.0.2007/010751-0.

Kohlmeyer, J. and Kohlmeyer, E. (1979). Marine Mycology: The Higher Fungi. Cambridge: Academic Press Inc., 690 pp.

Kopsahelis, A., Kourmentza, C., Zafiri, C. and Kornaros, M. (2018). Gate-to-gate life cycle assessment of biosurfactants and bioplasticizers production via biotechnological exploitation of fats and waste oils. J. Chem. Technol. Biotechnol., 93(10): 2833–2841. https://doi.org/10.1002/jctb.5633.

Kowalska, T. and Sajewicz, M. (2022). Thin-Layer Chromatography (TLC) in the screening of botanicals–its versatile potential and selected applications. Molecules, 27: 6607. https://doi.org/10.3390/molecules27196607.

Kubicki, S., Bollinger, A., Katzke, N., Jaeger, K.E., Loeschcke, A. et al. (2019). Marine biosurfactants: Biosynthesis, structural diversity, and biotechnological applications. Mar. Drugs, 17(7): 408. https://doi.org/10.3390/md17070408.

Leyva, A., Quintana, A., Sánchez, M., Rodríguez, E.N., Cremata, J. et al. (2008). Rapid and sensitive anthrone-sulfuric acid assay in microplate format to quantify carbohydrate in biopharmaceutical products: Method development and validation. Biologicals, 36: 134–141. https://doi.org/10.1016/j.biologicals.2007.09.001.

Lima, T.M.S., Procópio, L.C., Brandão, F.D., Carvalho, A.M.X., Totola, M.R. et al. (2011). Biodegradability of bacterial surfactants. Biodegradation, 22(3): 585–592. https://doi.org/10.1007/s10532-010-9431-3.

Liu, H., Shao, B., Long, X., Yao, Y. and Meng, Q. (2016). Enhanced foliar penetration by the biosurfactant rhamnolipid. Colloids Surf. B Biointerfaces, 145: 548–554.

Loranger, C., Alfalah, M., Ferrier Le Bouedec, M.C. and Sasseville, D. (2017). Alkyl glucosides in contact dermatitis. Dermatitis, 28(1): 5–13. https://doi.org/10.1097/DER.0000000000000240.

Lopes, J.M., Macedo, B.P., Sodré, T.S., Cayres, C.A., Ferreira, R.M. et al. (2025) Biosurfactant production by *Yarrowia lipolytica* in corn steep liquor-based media for hydrocarbon bioremediation. Processes, 13(2): 412. https://doi.org/10.3390/pr13020412.

Luepongpattana, S., Thaniyavarn, J. and Morikawa, M. (2017). Production of Massoia lactone by *Aureobasidium pullulans* YTP6-14 isolated from the Gulf of Thailand and its fragrant biosurfactant properties. J. App. Microbiol., 123(6): 1488–1497. https://doi.org/10.1111/jam.13598.

Luft, L. (2022). Fungal polysaccharides as biosurfactants and bioemulsifiers. *In*: Deshmukh, S.K., Deshpande, M.V. and Sridhar, K.R. (eds.). Fungal Biopolymers and Biocomposites. Springer, Singapore. https://doi.org/10.1007/978-981-19-1000-5_7.

Luft, L., Confortin, T.C., Todero, I., Zabot, G.L. and Mazutti, M.A. (2020). An overview of fungal biopolymers: Production of bioemulsifying and biosurfactant compounds. Crit. Rev. Biotechnol., 40: 1059–1080. https://doi.org/10.1080/07388551.2020.1752868.

Luna, J., Rufino, R., Campos, G. and Sarubbo, L. (2012). Properties of the biosurfactant produced by *Candida sphaerica* grown on low-cost substrates. Chem. Eng. Sci., 27: 67–72.

Maamar, A., Lucchesi, M.-E., Debaets, S., Nguyen van Long, N., Quemener, M., et al. (2020). Highlighting the crude oil bioremediation potential of Marine Fungi isolated from the Port of Oran (Algeria). Diversity, 12: 196. https://doi.org/10.3390/d12050196.

Mahmoud, Y.A.G., El-Halmouch, Y.H., Nasr, E.E., Al-Sodany, Y.M., El-Nogoumy, B.A. et al. (2024). Exploring the potential of biosurfactants produced by fungi found in soil contaminated with petrochemical wastes. Sci. Rep., 14: 25733. https://doi.org/10.1038/s41598-024-75865-5.

Mapelli, F., Scoma, A., Michoud, G., Aulenta, F., Boon, N. et al. (2017). Biotechnologies for oil spill cleanup in the sea: Indissoluble ties with microorganisms. Trends in Biotechnol., 35: 860–870. https://doi.org/10.1016/j.tibtech.2017.04.003.

Marks, D.L., Buchsbaum, R. and Swain, T. (1985). Measurement of total protein in plant samples in the presence of tannins. Anal. Biochem., 147: 136–143. https://doi.org/10.1016/0003-2697(85)90019-3.

Matos, M., Fernandes, M.A., Costa, I., Coelho, N., Santos, T.F. et al. (2025). Culturable yeast diversity associated with industrial cultures of the microalga *Microchloropsis gaditana* and their ability to produce lipids and biosurfactants. J. Fungi, 11: 228. https://doi.org/10.3390/jof11030228.

Mazaheri-Assadi, M. and Tabatabaee, M.S. (2010). Biosurfactants and their use in enhancing petroleum vacuum distillation residues: A review. Int. J. Environ. Res., 4: 549–572.

Mnif, I. and Ghribi, D. (2016). Glycolipid biosurfactants: Main properties and potential applications in agriculture and food industry. J. Sci. Food Agric., 96: 4310–4320. https://doi.org/10.1002/jsfa.7759.

Mohanram, R., Jagtap, C. and Kumar, P. (2016). Isolation, screening, and characterization of surface-active agent-producing, oil-degrading marine bacteria of Mumbai Harbor. Mar. Pollut. Bull., 105: 131–138. https://doi.org/10.1016/j.marpolbul.2016.02.040.

Morikawa, M., Hirata, Y. and Imanaka, T. (2000). A study on the structure-function relationship of lipopeptide biosurfactants. Biochim. Biophys. Acta, 1488: 211–218. https://doi.org/10.1016/s1388-1981(00)00124-4.

Moura, G.C.C., Ayres, Y.M., Brito, A.L.D.C., Júnior, E.F.D.S., Rocha, R.D.S. et al. (2023). Characterization of the cultivable microbiota components of marine bioaerosols in the North Tropical Atlantic. Atmosphere, 14: 1470. https://doi.org/10.3390/atmos14101470.

Mulligan, C.N. (2005). Environmental applications for biosurfactants. Environ. Pollut., 133: 183–198. https://doi.org/10.1016/j.envpol.2004.06.009.

Najmi, Z., Ebrahimipour, G., Franzetti, A. and Banat, I.M. (2018). *In situ* downstream strategies for cost-effective bio/surfactant recovery. Biotechnol. Appl. Biochem., 65: 523–532. https://doi.org/10.1002/bab.1641.

Obi, L.U., Atagana, H.I. and Adeleke, R.A. (2016). Isolation and characterisation of crude oil sludge degrading bacteria. SpringerPlus, 5: 1946. https://doi.org/10.1186/s40064-016-3617-z.

Ochoa-Loza, F.J., Artiola, J.F. and Maier, R.M. (2001). Stability constants for the complexation of various metals with a rhamnolipid biosurfactant. J. Environ. Qual., 30: 479–485.

Onyeaka, H., Passaretti, P., Miri, T. and Al-Sharify, Z.T. (2022). The safety of nanomaterials in food production and packaging. Current Research in Food Science, 5: 763–774. https://doi.org/10.1016/j.crfs.2022.04.005.

Pang, K., Overy, D.P., Jones, E.B.G., Calado, M.L., Burgaud, G. et al. (2016). Marine fungi and marine-derived fungi in natural product chemistry research: Toward a new consensual definition. Fungal Biology Reviews, 30(4): 163–175. https://doi.org/10.1016/j.fbr.2016.08.001.

Patel, S., Homaei, A. and Daverey, A. (2019). Microbial biosurfactants for oil spill remediation: Pitfalls and potentials. App. Microbiol. Biotechnol., 103(1): 27–37. https://doi.org/10.1007/s00253-018-9434-2.

Patowary, K., Saikia, R.R., Kalita, M.C. and Deka, S. (2015). Degradation of polyaromatic hydrocarbons employing biosurfactant-producing *Bacillus pumilus* KS2. Ann. Microbiol., 65: 225–234. https://doi.org/10.1007/s13213-014-0854-7.

Perfumo, A., Banat, I.M. and Marchant, R. (2018). Going green and cold: Biosurfactants from low-temperature environments for biotechnological applications. Tr. Biotechnol., 36: 277–289. https://doi.org/10.1016/j.tibtech.2018.01.005.

Pitocchi, R., Cicatiello, P., Birolo, L., Piscitelli, A., Bovio, E. et al. (2020). Cerato-Platanins from marine fungi as effective protein biosurfactants and bioemulsifiers. Int. J. Mol. Sci., 21: 2913. https://doi.org/10.3390/ijms21082913.

Pitocchi, R., Piscitelli, A. and Giardina, P. (2022). Marine Fungi as a source of biosurfactants and bioemulsifiers. pp. 313–327. *In*: Deshmukh, S.K., Deshpande, M.V. and Sridhar, K.R. (eds.). Fungal Biopolymers and Biocomposites: Prospects and Avenues. Springer Nature, Singapore.

Płaza, G.A., Zjawiony, I. and Banat, I.M. (2006). Use of different methods for detection of thermophilic biosurfactant-producing bacteria from hydrocarbon-contaminated and bioremediated soils. J. Pet. Sci. Eng., 50: 71–77. https://doi.org/10.1016/j.petrol.2005.10.005.

Plaza, G.A.P., Chojniak, J. and Banat, I.M. (2014). Biosurfactant-mediated biosynthesis of selected metal nanoparticles. Int. J. Mol. Sci., 15(8): 13720–13737. https://doi.org/10.3390/ijms150813720.

Rawat, G., Dhasmana, A. and Kumar, V. (2020). Biosurfactants: The next generation of biomolecules for diverse applications. Environ. Sustain., 3(3): 353–369. https://doi.org/10.1007/s42398-020-00128-8.

Rédou, V., Navarri, M., Meslet-Cladière, L., Barbier, G. and Burgaud, G. (2015). Species richness and adaptation of marine fungi from deep-subseafloor sediments. Appl. Environ. Microbiol., 81. https://doi.org/10.1128/AEM.04064-14.

Renfro, T.D., Xie, W., Yang, G. and Chen, G. (2014). Rhamnolipid surface thermodynamic properties and transport in agricultural soil. Colloids Surf. B Biointerfaces, 115: 317–322. https://doi.org/10.1016/j.colsurfb.2013.12.021.

Ribeiro, B.G., Guerra, J.M.C. and Sarubbo, L.A. (2020). Potential food application of a biosurfactant produced by *Saccharomyces cerevisiae* URM 6670. Front. Bioeng. Biotechnol., 8: 1–13. https://doi.org/10.3389/fbioe.2020.00434.

Rondel, C., Marcato-Romain, C.-E. and Girbal-Neuhauser, E. (2013). Development and validation of a colorimetric assay for simultaneous quantification of neutral and uronic sugars. Water Research, 47: 2901–2908. https://doi.org/10.1016/j.watres.2013.03.010.

Rosenberg, E. and Ron, E.Z. (1999). High and low molecular weight microbial surfactants. Appl. Microbiol. Biotechnol., 52(2): 154–162. https://doi.org/10.1007/s002530051502.

Rufino, R.D., Luna, J.M., Campos-Takaki, G.M., Ferreira, S.R. and Sarubbo, L.A. (2012). Application of biosurfactant produced by *Candida lipolytica* in heavy metal remediation. Chem. Eng., 27: 61–66. https://doi.org/10.3303/CET1227011.

Rybicki, E. (2015). Manual of SDS-PAGE and Immunoblotting Techniques, pp 32, Apple iBooks Store.

Santiago, I.F., Alves, T.M., Rabello, A., Sales Junior, P.A., Romanha, A.J. et al. (2012). Leishmanicidal and antitumor activities of endophytic fungi associated with Antarctic angiosperms *Deschampsia antarctica* Desv. and *Colobanthus quitensis* (Kunth) Bartl. Extremophiles, 16: 95–103. https://doi.org/10.1007/s00792-011-0403-2.

Sarubbo, L.A., Maria da Gloria, C.S., Durval, I.J.B., Bezerra, K.G.O., Ribeiro, B.G. et al. (2022). Biosurfactants: Production, properties, applications, trends, and general perspectives. Biochem. Eng. J., 181: 108377. https://doi.org/10.1016/j.bej.2022.108377.

Sen, K., Sen, B. and Wang, G. (2022). Diversity, abundance, and ecological roles of planktonic fungi in marine environments. J. Fungi, 8: 491. https://doi.org/10.3390/jof8050491.

Sen, S., Borah, S.N., Bora, A. and Deka, S. (2017). Production, characterization, and antifungal activity of a biosurfactant produced by *Rhodotorula babjevae* YS3. Microb. Cell Fact., 16: 95. https://doi.org/10.1186/s12934-017-0711-z.

Senthil Balan, S., Ganesh Kumar, C. and Jayalakshmi, S. (2019). Physicochemical, structural, and biological evaluation of Cybersan (trigalactomargarate), a new glycolipid biosurfactant produced by a marine yeast, Cyberlindnera saturnus strain SBPN-27. Process Biochem., 80: 171–180. https://doi.org/10.1016/j.procbio.2019.02.005.

Shakeri, F., Babavalian, H., Amoozegar, M.A., Ahmadzadeh, Z., Zuhuriyanizadi, S. et al. (2020). Production and application of biosurfactants in biotechnology. Biointerface Res. Appl. Chem., 11: 10446–10460. https://doi.org/10.33263/BRIAC113.1044610460.

Shaumi, A., Cheng, U.C., Guo, S.Y., Jones, E.B.G., Chan, T.Y. et al. (2023). Diversity of fungi isolated from carapace and gut of the marine crab *Portunus sanguinolentus* in northern waters of Taiwan. Bot. Mar., 66(4): 301–307. https://doi.org/10.1515/bot-2022-0081.

Sales da Silva, I.G., Gomes de Almeida, F.C., Padilha da Rocha e Silva, N.M., Casazza, A.A., Converti, A. et al. (2020). Soil bioremediation: Overview of technologies and trends. Energies, 13(18): 4664. https://doi.org/10.3390/en13184664.

Sotirova, A., Avramova, T., Stoitsova, S., Lazarkevich, I., Lubenets, V. et al. (2012). The importance of rhamnolipid-biosurfactant-induced changes in *Bacillus subtilis* membrane lipids for the antimicrobial activity of thiosulfonates. Curr. Microbiol., 65(5): 534–541. https://doi.org/10.1007/s00284-012-0191-7.

Stahl, E. (1969). Apparatus and general techniques in TLC. pp. 52–86. *In*: Stahl, E. (ed.). Thin-Layer Chromatography: A Laboratory Handbook. Springer, Berlin, Heidelberg.

Stipcevic. T., Piljac, A. and Piljac, G. (2006). Improvement in full-thickness burn wound healing using di-rhamnolipid. Burns, 32(1): 24–34.

Sun, W., Wang, Y., Zhang, W., Ying, H. and Wang, P. (2018). A novel peptide surfactant for biofilm removal. Colloids Surf. B Biointerfaces, 172: 180–186. https://doi.org/10.1016/j.colsurfb.2018.08.029.

Thies, S., Rausch, S.C., Kovacic, F., Schmidt-Thaler, A., Wilhelm, S. et al. (2016). Metagenomic discovery of novel enzymes and biosurfactants in a slaughterhouse biofilm microbial community. Sci. Rep., 6: 27035. https://doi.org/10.1038/srep27035.

Tian, Y., Li, Y.L. and Zhao, F.C. (2017). Secondary metabolites from polar organisms. Mar. Drugs, 15: 28. https://doi.org/10.3390/md15020028.

Toledo, A.V., Simurro, M.E. and Balatti, P.A. (2013). Morphological and molecular characterization of a fungus, *Hirsutella* spp., isolated from planthoppers and psocids in Argentina. J. Insect. Sci., 13(1): 18. https://doi.org/10.1673/031.013.1801.

Trindade, M., van Zyl, L.J., Navarro-Fernández, J. and Abd Elrazak, A. (2015). Targeted metagenomics as a tool to tap into marine natural product diversity for the discovery and production of drug candidates. Front. Microbiol., 6: 980. https://doi.org/10.3389/fmicb.2015.00890.

Tugrul, T. and Cansunar, E. (2005). Detecting surfactant-producing microorganisms by the drop-collapse test. World J. Microbiol. Biotechnol., 21: 851–853. https://doi.org/10.1007/s11274-004-5958-y.

Twigg, M.S., Baccile, N., Banat, I.M., Déziel, E., Marchant, R. et al. (2021). Research on microbial biosurfactants: Time to improve rigor in reporting synthesis, functional characterization, and process development. Microb. Biotechnol., 14(1): 147–170. https://doi.org/10.1111/1751-7915.13704.

Vandana, P. and Singh, D. (2018). Review on the production of biosurfactants and their applications. Int. J. Curr. Microbiol. Appl. Sci., 7(8): 4228–4241. https://doi.org/10.20546/ijcmas.2018.708.443.

Varrella, S., Barone, G., Tangherlini, M., Rastelli, E., Dell'Anno, A. et al. (2021). Diversity, ecological role, and biotechnological potential of Antarctic marine fungi. J. Fungi, 7(5): 391. https://doi.org/10.3390/jof7050391.

Vecino, X., Cruz, J.M., Moldes, A.B. and Rodrigues, L.R. (2017). Biosurfactants in cosmetic formulations: Trends and challenges. Crit. Rev. Biotechnol., 37(7): 911–923. https://doi.org/10.1080/07388551.2016.1269053.

Walter, V., Syldatk, C. and Hausmann, R. (2010). Screening concepts for the isolation of biosurfactant producing microorganisms. pp. 1–13. *In*: Sen, R. (ed.). Biosurfactants. Springer, New York, NY. https://doi.org/10.1007/978-1-4419-5979-9_1.

Walter, V., Syldatk, C. and Hausmann, R. (2013). Screening concepts for the isolation of biosurfactant producing microorganisms. *In*: Madame Curie Bioscience Database. Landes Bioscience. https://www.ncbi.nlm.nih.gov/books/NBK6189/.

Wilhelmy, L. (1863). Ueber die Abhängigkeit der Capillaritäts-Constanten des Alkohols von Substanz und Gestalt des benetzten festen Körpers. Annalen der Physik, 195: 177–217.

Williams, W., Kunorozva, L., Klaiber, I., Henkel, M., Pfannstiel, J. et al. (2019). Novel metagenome-derived ornithine lipids identified by functional screening for biosurfactants. Appl. Microbiol. Biotechnol., 103: 4429–4441. https://doi.org/10.1007/s00253-019-09768-1.

Wojciechowski, K., Borucka, K. and Mierzejewska, J. (2023). Are all yeast biosurfactants really capable of lowering surface tension below 30 mN/m? Colloids and Surfaces B: Biointerfaces, 230: 113503. https://doi.org/10.1016/j.colsurfb.2023.113503.

Youssef, N.H., Duncan, K.E., Nagle, D.P., Savage, K.N., Knapp, R.M. et al. (2004). Comparison of methods to detect biosurfactant production by diverse microorganisms. J. Microbiol. Methods, 56: 339–347. https://doi.org/10.1016/j.mimet.2003.11.001.

Zou, R., Chen, B., Sun, J., Guo, Y.W. and Xu, B. (2023). Recent advances of activation techniques-based discovery of new compounds from marine fungi. Fitoterapia, 167: 105503. https://doi.org/10.1016/j.fitote.2023.105503.

7

Marine Lichens

A Treasure Trove of Bioactive and Biotechnological Potential in Marine Fungal Symbiosis

Malshadi Ranasinghe and *Udeni Jayalal**

Marine lichens are a spectacular but poorly known category of symbiotic organisms that have developed stunning adaptations to allow them to survive in some of the most hostile marine habitats on the planet. These composite organisms can survive extreme environments such as periodic submersion, desiccation, high suprasalinity, high UV exposure, and severe temperature variations, typical of the intertidal zone. Despite comprising the world lichen diversity by about 3% (700 unknown species), marine lichens have unique biochemical characteristics that make them different from the terrestrials, offering them as lucrative products to blue biotechnology. Marine lichens are biotechnologically important, as they produce unusual secondary metabolites, especially halogenated products such as brominated and chlorinated polyketides that are rare in the terrestrial species. These metabolites are crucial to ecology, capable of reducing fouling and defending against pathogens and UV chemistry. It was especially noticed that atranorin, usnic acid, gyrophoric acid, fumarprotocetraric acid, and halogenated derivatives of

Department of Natural Resources, Faculty of Applied Sciences, Sabaragamuwa University of Sri Lanka, Belihuloya 70140, Sri Lanka.
* Corresponding author: jayalal@appsc.sab.ac.lk

DOI: 10.1201/9781003795100-9

these compounds, as well as bioactive compounds, proved to have significant antimicrobial, antioxidant, anti-inflammatory, and anticancer effects. The new pharmaceutical activities have reported more than 250 new fungal products of marine origin, which makes marine lichens of great significance in natural compounds, particularly the microorganisms resistant to antibiotics. Genomics and transcriptomics have opened up the picture of complex adaptive strategies in marine lichens, such as horizontal gene transfer and adjustment of metabolism towards survival in mobile coastal systems. The complex symbiotic design combines mycobionts that offer protective factors and stress-responding proteins and photobionts that offer photosynthetic energy as well as nitrogen fixing. Marine lichens as ecological pioneers are important to primary succession, stabilization of coastal ecosystems and have served as the guide towards environmental change. Despite the suggestions of better cultivation via a tidal-simulating bioreactor system, the elaborate symbiosis and specific environmental requirements enforce the limitation of the large-scale biotechnological use of marine lichens.

1. Introduction

Fungi spread throughout terrestrial environments, where water combines with carbon materials in the terrestrial biosphere. These organisms establish beneficial partnerships with plants and animals since they perform vital functions in ecosystem nutrient cycles that span numerous habitats. Lichens consist of symbiotic organisms that unite autotrophic algae and cyanobacteria, and heterotrophic elements developed through the coastal transition when algae and fungi adapted to land existence. These organisms show extraordinary adaptations for living in harsh environments, including coasts' littoral (intertidal) areas (Sonina and Androsova, 2020). Lichens possess strong abilities to inhabit environments that experience dramatic variations of water levels, thus proving their outstanding adaptability in border areas between land and water (Honegger, 2009).

The field of freshwater and marine lichen-forming fungi remains under investigation from both lichenologists and mycologists (Hawksworth, 2000). The study of marine fungi commonly discusses lichens because these associations form symbioses between algae that succeed in salty environments (Raghukumar, 2017). Multiple ecological and floristic research projects have studied aquatic lichen occurrences ranging from fresh and ocean-based systems. The flow of water causes fundamental alterations to the external features of freshwater and marine lichens that exist in vertically distinct ecological zones. The presentation of comparative studies in research demands evaluation of both involved groups according to the findings of Hasegawa et al. (2000).

The structural aspects and functional attributes in lichens exist as integral results of fungus-algae partnerships. Symbiotic arrangements of fungal strains include phyla Ascomycota and Basidiomycota together with photosynthetic algae or cyanobacteria species as their partners. Units developed by cyanobacteria inside lichens allow nitrogen fixation, which helps these organisms establish themselves in regions with limited nutrients. Lichen communities exist as important elements of ecosystem dynamics because they take control of areas that plants cannot survive, yet they persist in harsh environments. These organisms thrive in difficult environmental conditions through their photosynthetic capabilities, which keep most organisms away. The mutually beneficial association within lichens allows fungi to obtain photosynthetic carbohydrates, but fungi also safeguard algae and cyanobacteria along with supplying them access to moisture (Mohr et al., 2004). Recent genomic research shows the symbiotic connection between fungi and symbionts runs deeper than earlier science understood because the lichen halobiont system includes various bacterial associates (Grube et al., 2009). This chapter analyzes the area of lichens that dwell in marine conditions thoroughly; marine lichens as the distinctive symbiotic organisms living in hostile conditions along the coast, where no other organisms can live. It describes their ecological functions, distribution worldwide, evolutionary development, and the biosynthesis of bioactive components with potential pharmaceutical and industrial uses. The chapter also summarizes existing research gaps, conservation, and future possibilities of sustainable use of these underutilized but ecologically important organisms.

2. Habitats, Diversity, and Distribution

The ecological environment at rocky seacoasts ranks among the harshest conditions for organism survival. Marine lichens make up a vital aspect of coastal communities since they thrive in places that plants cannot tolerate. The lichens experience multiple abiotic elements such as solar radiation fluctuations and wave and wind actions and daily drying and moistening patterns and temperature differences and salt variations (Sonina et al., 2017). Around the middle of the 20th century, scientists started to concentrate specifically on freshwater and marine lichens as their area of research. There were approximately 200 species of lichens together with fungi that resided in freshwater lakes and river shores, while tidal zones hosted approximately 700 species during that time (Coste et al., 2016). Marine lichens currently compose 3% of worldwide lichen biodiversity, and researchers detect new species annually (Selig et al., 2014). The marine ecosystem depends on lichens because these organisms successfully colonize regions that resist biological occupation. Initial ecosystem development becomes possible through pioneering actions of lichens, which provide habitat and sustenance to marine invertebrates as well as track pollution through their environmental records (Coste et al., 2016).

The diversity of marine lichens is due to their development of various habitats in sites offering unique microclimates such as the rocky inter-tidal region as there is adequate moisture and sunshine and competition is minimal. Popular hotspots exist where marine lichens thrive in the Pacific Northwest of North America and the Mediterranean Sea and in the Southern Hemisphere locations, which include New Zealand and Chile. Such regions boast high marine lichen species diversity because they maintain climatic stability along with various substrate types. Protecting these

enables optimal survival and energy management throughout tidal changes. The study by Chrismas et al. (2023) through proteomic analysis validated these activity patterns to detect specific stress-response proteins that show timed elevations during specific tidal periods. Marine lichens exhibit geographical distributions because of climatic factors, together with limited substrates and oceanic characteristics, as Higgins et al. (2014) point out. They choose to reside in coastal zones and intertidal areas, which use rocky surfaces that exist where humid conditions and frequent fog and ocean salt exist.

Marine lichens achieve their maximum diversity in temperate areas of the Northern Hemisphere, and they have major population centers in North America's Pacific Northwest, along with the Mediterranean Basin and Atlantic coastal Europe. The Southern Hemisphere countries of southern Chile, along with New Zealand and South Africa, host different terrestrial lichen species in their marine environments. The distributions of these organisms emerged because of past glaciation patterns, together with natural ocean currents, coupled with ecological habitat availability. Marine lichens do not occur in tropical regions because this zone suffers from both ecological competition and unfavorable environmental circumstances. The choice of specific marine habitats by lichens occurs because they serve as ecological indicators that show reactions to artificial disturbances from human activities. Scientific findings indicate that climate change will modify marine lichen range distribution since populations will decline in existing areas while polar regions will become unreachable, according to Bringloe et al. (2022).

The worldwide distribution of marine lichen populations remains at risk because of climate change. The organisms lose their ecological stability when sea levels rise and storms become more powerful, while average temperatures shift accordingly. Jones et al. (2013) tracked population reductions of sensitive Atlantic European coast marine lichens, which researchers linked to the warming ocean temperatures and altered precipitation conditions. Ocean acidification negatively impacts the calcification processes in some marine lichens and could lead to their diminished capacities to attach to and stay on rocky surfaces (Hofmann et al., 2010). The protection of intertidal zones within marine protected areas has proven effective for lichen diversity conservation, according to research that combines this approach with pollution reduction efforts and coastal development monitoring, thus assuring lichen preservation (McMullin et al., 2019).

Only a few researchers study marine lichens, though scientists agree these organisms could prove important for ecological research. Current knowledge regarding marine lichen taxonomy faces significant limitations according to survey outcomes documented by Ríos et al. (2020). Lücking et al. (2020), who conducted a thorough review of coastal lichen communities, found that presently unrecognized biological diversity can be found within these communities. A vast biodiversity survey executed by Hawksworth (2000) throughout Southeast Asian coasts that had limited study revealed 47 lichen species previously unknown to marine ecosystems and potentially 12 unidentified new species because of the survey. Scientists have established that marine lichen diversity remains poorly understood, especially in tropical areas, which have received minimal study by lichenologists.

3. Bioactive Potential

The bioactive compounds produced by marine lichens show practical applications as bio-products in biotechnology. The biological compounds consist of two distinct properties: they incorporate antibiotics but also carry antioxidants together with anti-inflammatory substances. The bacteria extracted from marine lichens can serve researchers as sources to obtain compounds of scientific interest (Sonina and Androsova, 2020). The examination of marine lichens by pharmaceutical researchers yielded numerous novel compounds, which have generated new possibilities for treating cancer and fighting antimicrobial infections (Zhang and Wong, 2024).

The secondary metabolites of lichens possess many biological effects, among which are antibiotic and antioxidant behaviors, along with the abilities to protect against mycosis and cancer while displaying antiviral and anti-inflammatory effects, together with analgesic healing properties and antipyretic responses, and neuroprotective features (Ranković and Kosanić, 2020). The bioactive compounds unique to marine lichens include atranorin, fumarprotocetraric acid, usnic acid, gyrophoric acid, lecanoric acid, physodic acid, stictic acid, and protocetraric acid that display combined properties of antibacterial, antioxidant, antifungal, and antitumor capacities. Scientific interest in Atranorin has intensified because of its high capability to scavenge free radicals (Goga et al., 2020). Chen et al. (2022) demonstrated that marine lichens produce multiple derivative compounds with halogens attached because their bioactivity exceeds that of halogenated analogs. Scientific studies indicate that these compounds demonstrate potential as anti-inflammatory drugs because their IC50 values achieve the standards of contemporary anti-inflammatory medications.

Marine lichens transform secondary metabolites through acetyl-polymalonyl (polyketide) pathways while adapting these pathways for survival in marine environments. *Xanthoria parietina* produces parietin from the acetyl-polymalonyl (polyketide) pathway, which results in the formation of depsides and depsidones and anthraquinones. Research has shown that *Xanthoria parietina* creates parietin, an anthraquinone pigment, through this biosynthetic route, which guards against ultraviolet rays and toxic substances in the environment (Molnár and Farkas, 2010). The research by Molnár and Farkas (2010) through metabolomic testing demonstrated that high-stress marine lichens generate a larger assortment of polyketide derivatives in comparison to lichens residing at calmer sites, which proves that environmental stress boosts metabolic scope. Research results revealed 73 newly unrecorded metabolites, mostly consisting of halogenated polyketides, which were found in samples collected from areas experiencing intense wave action.

Marine lichens defend themselves against harmful solar radiation through the production of mycosporine-like amino acids (MAAs), which function as ultraviolet light-absorbing small secondary metabolites. The process to produce MAA requires shikimate pathway intermediates that also help make aromatic amino acids (Cardozo et al., 2006). Both terrestrial lichens produce polyketides, depsides, and depsidones (Ranković and Kosanić, 2019), but marine lichens foster unique metabolite derivatives that contain brominated and chlorinated compounds. The halogenated secondary metabolites likely serve three main functions in marine environments: they offer protection against marine fouling, marine pathogens, and heightened defense capabilities against both. The special chemical compositions between marine and terrestrial lichens demonstrate the evolutionary adaptations that occur according to their habitat type. Mandrekar, Gawas, and Majik (2018) established that brominated metabolites found in marine lichens show strong antifouling properties against barnacle larvae as well as prevent bacterial biofilm growth in marine ecosystems.

Scientists have found that the antimicrobial, anti-inflammatory, and antitumor properties of halogenated compounds found in marine Roccellaceae species position these organisms as important pharmaceutical resources (Parrot et al., 2018). Research Murugan et al. (2021) showed how marine lichen extracts enhance

Table 1. Key bioactive metabolites in marine lichens.

Compound	Species	Chemical Class	Biological Activity	Potential Applications	References
Atranorin	Various marine lichens	Depside	Antioxidant, free radical scavenging	Natural antioxidants, cosmetics	Goga et al. (2020); Ranković and Kosanić (2020)
Fumarprotocetraric acid	Marine lichens	Depsidone	Antibacterial, antitumor	Pharmaceutical development	Ranković and Kosanić (2020)
Gyrophoric acid	Marine species	Depsidone	Antioxidant, antimicrobial	Drug discovery	Ranković and Kosanić (2020)
Lecanoric acid	Various lichens	Depside	Anti-inflammatory	Therapeutic agents	Ranković and Kosanić (2020)
Parietin	*Xanthoria parietina*	Anthraquinone	UV protection, antimicrobial	Natural sunscreens, preservatives	Molnár and Farkas (2010)
Physodic acid	Marine lichens	Depsidone	Antifungal, antibacterial	Medical applications	Ranković and Kosanić (2020)
Protocetraric acid	Marine lichens	Depsidone	Antitumor, antioxidant	Cancer research	Ranković and Kosanić (2020)
Stictic acid	Various species	Depsidone	Antimicrobial	Biotechnology	Ranković and Kosanić (2020)
Usnic acid	Various species	Dibenzofuran	Antibacterial, antifungal	Antimicrobial agents	Ranković and Kosanić (2020)

biodegradable polymers by producing blends that possess antimicrobial strength and UV defensive abilities. The method presents an eco-friendly substitute for standard synthetic additives during manufacturing, which produces numerous environmental advantages throughout the complete product lifecycle.

The research by Norouzi et al. (2020) produced a secondary metabolic system model for the marine lichen *Lichina confinis* through integration of proteomic data, transcriptomic data, and metabolomic data obtained over multiple tidal cycles. The model implemented by their research group proved instrumental in forecasting metabolite genesis across different environmental scenarios, which substantially increased our comprehension of biosynthetic network management within these organisms.

4. Biotechnological Potential

Various biotechnological uses focus on marine lichens because of their distinctive metabolic compounds and ecological traits. *Xanthoria parietina* extracts prevent human parainfluenza virus Type 2 from multiplying when tested in laboratory conditions. Research indicates that this lichen can fight bacteria; therefore, scientists believe it could provide potential leads for new antimicrobial medication development (Molnár and Farkas, 2010). Due to producing parietin, *Xanthoria parietina* becomes a potential candidate to develop natural sunscreen products, as parietin protects skin from UV radiation. Fonseca et al. (2023) researched marine lichen extracts as natural preservatives in cosmetics, which revealed *Roccella phycopsis* extracts gave antimicrobial defense similar to synthetic preservatives, along with antioxidant properties. Marine lichen compounds demonstrate two effective cosmetic functions according to research, which supports reducing the usage of separate cosmetic additives.

The adaptation of marine-associated fungi to hostile marine environments leads to fundamental metabolite diversity within secondary metabolite products (Varrella et al., 2021). The identification of modern compounds with medical and pharmaceutical prospects alongside industrial uses has become possible through this discovery process. The antimicrobial and anticancer, along with antiviral and antioxidant, and anti-inflammatory activities of metabolites like terpenes, combine with steroids and polyketides, peptides, and alkaloids (Bonugli-Santos et al., 2015). The research conducted by Ji, Zha, and Zhou (2023) presented a complete review of more than 250 novel compounds obtained from marine-associated fungi during 2020–2023, which yielded about 15% of compounds from fungi found in lichen thalli. These chemical substances have diverse structures, and a few representatives are presently undergoing preclinical assessment to treat resistant bacterial infections along with inflammatory states. The researchers have established that lichenicolous fungi, which grow on lichens, contain bioactive substances that possess distinctive chemical frameworks that cannot be discovered in other areas of nature.

Researchers have focused on drug discovery through other marine-derived microorganisms because cultivating marine lichens poses serious challenges when

Table 2. Biotechnological applications and commercial potential.

Application	Target Compounds	Development Stage	Market Potential	Key Advantages	References
Anti-inflammatory drugs	Halogenated derivatives	Research	High	Comparable to synthetic drugs	Chen et al. (2022)
Antifouling agents	Brominated metabolites	Research	Medium	Eco-friendly marine coatings	Mandrekar et al. (2018)
Antimicrobial agents	Usnic acid, halogenated compounds	Research/ preclinical	High	Natural origin, broad spectrum	Parrot et al. (2018); Zhang and Wong (2024)
Antioxidant supplements	Atranorin, various depsides	Early research	Medium	Free radical scavenging	Goga et al. (2020)
Biodegradable polymers	Marine lichen extracts	Development	Medium	Antimicrobial enhancement	Kim et al. (2024)
Cancer therapeutics	CBP-115, various metabolites	Preclinical trials	High	Selective toxicity	Guzmán et al. (2025)
Natural preservatives	*Roccella* extracts	Development	Medium	Cosmetic applications	Fonseca et al. (2023)
UV protection products	Parietin, MAAs	Research/ development	Medium-high	Natural sunscreens, cosmetic appeal	Fonseca et al. (2023)

extracting sufficient compounds. The study of marine lichens trails behind terrestrial lichens in research, but investigators now demonstrate initial evidence supporting their biotechnological significance. Marine microorganisms have delivered chemical compounds consisting of Salinosporamide A (Marizomib), which acts as a potent protease inhibitor showing anticancer potential (Maldonado et al., 2005). Medical trials do not involve any compounds derived from marine lichens at present, yet marine natural products show promising therapeutic potential. The clinical assessment process has been initiated for many compounds derived from marine microorganisms. Marine lichens have yet to yield documented commercial products, but marine natural products demonstrate abundant potential therapeutic agents, according to Zhang et al. (2024). The pharmaceutical company Coastal Biopharmaceuticals began its program to develop drugs from marine lichen metabolites in 2023 via its key drug candidate CBP-115 from *Roccella fuciformis*, which is now in preclinical toxicology studies (Guzmán et al., 2025). The drug shows exclusive toxicity against triple-negative breast cancer cell types without harming regular cells, making it suitable for advanced clinical investigation.

5. Molecular Studies

Research on marine lichens through genomic analysis delivered information about symbiosis development and symbiosis-affecting genes as well as mechanisms for adapting to seawater conditions (Song et al., 2022). The process of symbiosis occurs through genetic actions related to stress resistance, alongside signaling activities and genome reprogramming elements, and nutritional and structural interactions (Armaleo et al., 2019). The lichen fungus actively adjusts its behaviors depending on environmental changes, which helps it to cope with seawater immersion pressure. The salt stress protection mechanism of certain cyanobacteria produces disaccharide sugars, which provide benefits to the lichen fungus. Specific cyanobacteria produce disaccharide sugars as a salt-stress defense against environmental stress, according to Chrismas et al. (2023). The sugars serve as energy reserves for fungi, but *Lichina pygmaea* cannot directly extract them because of their large size. RNA results showed lichen fungi activate disaccharide enzyme breakdown to transform the molecules into glucose when a high tide occurs, since cyanobacteria's adaptations help the fungal component. The research suggests this disaccharide-splitting enzyme acts outside the cell, which indicates lichen fungi decompose sugars in an environment beyond their cellular boundaries, while forest fungi release wood-degrading enzymes in their environment. The mycobionts of marine lichen release particular hydrolytic enzymes into their environment based on proteomics studies conducted by Sadowsky et al. (2016) which reveals their responses to salt stress challenges.

Genomic and transcriptomic methods in biological research on marine lichens revealed special biosynthetic techniques. Evaluations conducted by Almeida et al. (2019) established that supralittoral lichen species possess various new gene clusters that can generate halogenated secondary metabolites. The gene clusters validated their antibiotic bronodepside production capacity through heterologous expression research using model fungi when Ren et al. (2021) conducted their study. Through functional genomics research, scientists closed the genetic capability to chemical production deficiency, leading to innovative marine lichen bioprospecting techniques. Metabolomic profiling provides efficient methods to detect marine lichen compounds. According to Shin et al. (2023), the Roccellaceae family of lichens yielded multiple new brominated compounds, which demonstrated strong antimicrobial activity against marine bacterial pathogens. Norouzi et al. (2020) perfected the combination of liquid chromatography-mass spectrometry with molecular networking to generate whole metabolite spectrum maps for marine lichens extracted across different geographic locations. Their investigations showed that specific chemical patterns within these organisms corresponded with salinity conditions and wave pressures, thereby explaining the natural selection of chemical diversity in marine lichens.

The adaptability and survival of ocean lichens in diverse, extreme, and harsh environments depend on two vital mechanisms, which include horizontal gene transfer (Burmeister, 2015) and metabolic plasticity (Alster et al., 2021). Organisms use Horizontal Gene Transfer (HGT) as a natural genetic material exchange method, which avoids the conventional inheritance rules. The adaptability functions of HGT prove most significant for marine lichens with green algal photobionts. The symbiont *Trebouxia* attains protein-encoding genes from fungal partners through HGT with intron horizontal gene transfer, which enhances the effectiveness and adaptability of its symbiotic life cycle (Erokhina et al., 2004). The marine lichen symbiont *Asterochloris glomerata* transfers genetic material that includes viral DNA sequences and archaeal and bacterial genetic materials to support its survival in oceanic habitats (Pino-Bodas et al., 2015). The genomic study by (Tunjić and Korač, 2013) revealed multiple times where marine lichens share genetic material with their bacterial partners, resulting in the frequent exchange of osmotic stress response and heavy metal detoxification mechanisms. The lichen halobiont functions as a dynamic collective according to their results by obtaining genes from different sources to enhance its environmental resilience.

Marine lichens contain unique genomic traits, according to Almeida et al. (2019), that enable the synthesis of compounds that do not exist in terrestrial counterparts. The organisms developed specific survival mechanisms that let them endure periodic submersions while withstanding desiccation and salt exposure and high UV radiation levels, which produced distinctive metabolites with important biotechnological value. Zhang et al. (2024) conducted a comprehensive review, which revealed that marine lichens possess great potential to produce bioactive compounds by isolating 87 unique metabolites from these organisms throughout the last decade. Marine lichen-derived compounds demonstrate strong antibacterial effects against antibiotic-resistant pathogens, according to their analysis, and exhibit the potential to treat inflammatory disorders, which points to applications for healthcare solutions.

6. Impacts of Climate Change

Scientists need to study the different ways living beings adapt to marine conditions. The study by Gueidan et al. (2022) showed that intertidal lichens possess specialized mechanisms to survive both wet and dry ocean conditions that differ from those of land-based organisms. Through gene expression profiling during tidal exposure, Im et al. (2017) detected distinct patterns of stress response gene regulation that were dissimilar to stresses faced by lichens in dry land. The study team discovered various genes that appear essential for cellular stability during instant desiccation and submersion events. Through isotope tracer analysis, Zhang and Peterson (2024) confirmed that supralittoral cyanolichens fix nitrogen, which becomes accessible to organisms due to wave movement and rainfall, to raise the local nutrient limits in rocky shores. The scientists successfully demonstrated that lichens act as essential nutrient sources that surpass their positional importance in rocky shore food chains.

An organism demonstrates metabolic plasticity when it uses metabolic modifications to adapt its systems based on environmental changes. Marine lichens maintain plasticity to handle fluctuations in environmental parameters such as water salts and water content, and nutrient supplies (Corrêa-Junior et al., 2023). During Lichen symbiosis, photobiont and fungal partners both express adapted genes to maintain efficient nutrient exchange while improving their survival adaptability. The alga in the *A. glomerata* and *Cladonia grayi* partnership activates genes that produce extracellular hydrolases and membrane transport proteins (Singh, 2023).

A. glomerata has evolved its gene function to primarily acquire nitrogen from its fungal symbiont because of its adaptation to low-nitrate marine ecosystems (Hammer, 1997). Delmail et al. (2012) tracked significant metabolic alterations of marine lichens under salt stress conditions, showing main metabolic routes rearranged themselves in less than 48 hours. Through metabolomic analysis, scientists detected how mycelial fungi produced mannitol and arabitol quickly as compatible solutes under hypersaline conditions before using these molecules to store energy when salt conditions normalized. This showed that complex metabolic regulation occurs in lichens.

According to Rubio-Salcedo et al. (2016), marine lichen habitats encounter novel dangers because of coastal development along with climate change impacts. Their research on rocky shore lichen communities showed several conservation-concerned species experienced major range reductions. Delmail et al. (2012) conducted extensive monitoring, which showed a 30% reduction in marine lichen biodiversity spread across North Atlantic locations during the last 20 years. The scientists linked the population reductions to rising sea surface temperatures and modified storm patterns, which exposed the sensitivity of these organisms to present-day climate change conditions. Sustainable harvesting practices remain underdeveloped. A management system for sustainable lichen harvesting along Portuguese coasts came into existence through Andersen et al. (2023) as they developed a community-led approach that combines indigenous ecological understanding with scientific data collection for ecological sustainability and local economic benefit. The researchers demonstrate through their initial findings that proper management of harvesting practices supports population equilibrium while generating sufficient biomaterial quantities for pharmaceutical development as well as natural dye production. Technical challenges make it challenging to establish ex situ cultivation systems for marine lichens. The research group of Rosabal and Pino-Bodas (2024) developed modern symbiont co-culture techniques with promising conservation and biotechnology applications that require further technical refinement. The conservation approach created by Hájek et al. (2012) used whole marine lichen preservation in simulated tidal tanks as well as freezing symbiotic microbes for biobanking applications. The developed protocol enabled researchers to protect 45 marine lichen species, together with endangered species whose habitat destruction created opportunities for science-focused and conservation research.

7. Conclusion and Outlook

According to Ellis (2021), we need to study these organisms now because possible extinctions will result in undetectable losses of bioactive compounds. The sustainable collection of marine lichens requires comprehensive conservation methods, according to Tehler and Irestedt (2007), because they developed harvesting frameworks that depend on observing natural populations. The International Marine Lichen Conservation Initiative, initiated in 2023, developed a global ecological monitoring program dedicated to protecting vulnerable coastal lichen populations by combining the expertise of citizen scientists and professional researchers to monitor populations and establish protection strategies (Scheidegger and Werth, 2009). Research collaboration between experts and citizen scientists will bring better scientific insights together with practical conservation results for understudied organisms. Progress in biological research demands talent contributions from specialists who work in both lichenology and marine ecology, as well as the biochemistry and biotechnology fields. Stocker-Wörgötter (1995) stress the need for creating standardized cultivation methods for symbiotic marine lichens, which will aid sustainable bioprospecting operations. Multiple scientific experts working together in research teams ensure both complete access to the biotechnological potentials of symbiotic organisms and their preservation for future generations. The Marine Lichen Biotechnology Consortium will unite academic researchers with conservation organizations and industrial partners through its establishment in 2024 (Blackwell and Suzuki, 2024). The research program focuses on major cultivation obstacles and metabolite production levels and sustainable application methods as it creates ethical systems for partnership benefits among communities harvesting marine lichens in selected regions.

Marine lichens face substantial cultivation barriers because they require symbiosis and possess complicated ecological relationships, together with sensitive environmental conditions. Several obstacles affect marine lichens, mainly from symbiotic complexity and high salinity stress and slow growth rates, together with their dependency on specific substrates. Three essential laboratory techniques for cultivating mycobionts independently and adjusting growth media supplements, as well as co-culturing bacterial associations (Sonina and Androsova, 2020). Synthetic biology, together with metagenomics, helps researchers discover genetic parts that allow organisms to become stress resistant and produce secondary metabolites, and maintain symbiotic relationships. Both metabolic pathway engineering techniques and bacterial community research knowledge enable the optimization process for bioactive compounds extracted from lichen through synthetic symbiosis research. Interdisciplinary research must occur to develop marine lichens for biotechnology and medicine because of these challenges (Blanz, 2018). A specialized bioreactor system for marine lichen cultivation was developed by Bisht et al. (2023), resulting in better cultivation practices. Through their system with tidal cycle simulation and controlled salt spray, they produced improved growth rates and secondary metabolite production when compared to static culture. This developing method demonstrates important progress toward making sustainable marine lichen biomass accessible for biotechnological and research purposes. Marine lichens stand as an unexplored region of blue biotechnology research since their biochemical features develop from extreme coastal conditions.

Marine lichens are an exceptional synthesis of biological adaptation, biotechnological opportunity and conservation necessity which propels them into one of the most exciting and underexplored frontiers of marine biology. The extensive evolution of such extraordinary, symbiotic organisms thriving in the harshest environments on the coastal zones of the Earth has led to the development of unique biochemical phenomena resulting in creation of a range of halogenated and other compounds, mycosporine-like amino acids, and other specialized metabolites that have shown potential in pharmaceutical and antimicrobial use along with various industrial applications. Nevertheless, the huge marine lichen biotechnological potential is marred by the existing challenges including gaps in taxonomic knowledge and cultivation issues, and the impending threat of climate change, which has already been shown to decrease marine lichen biodiversity across North Atlantic sites by 30% just over the past 20 years. The creation of global conservation programs and biotechnology partnerships over the past few years is indicative of a much-needed realization that visioning marine lichen potential as fulfilled can only be achieved when a combination of sustainable bio-prospecting is aligned with the far-reaching protection of marine habitat. The problem is that as we move into a new age of blue biotechnology, marine lichens ought to take their rightful place as prototypes of how survival in highly stressful environments can lead to successful evolutionary innovations capable of providing solutions to human beings as well, providing that we act fast to discover, cultivate, and preserve these living resources before the changing climate and habitat destruction destroys the possibility of ever again learning the biochemical secrets that these living treasures had to find to survive, and of passing the answers on to a grateful society.

References

Almeida, E.L., Rincón, A.F.C., Jackson, S.A. and Dobson, A.D.W. (2019). Comparative genomics of marine sponge-derived *Streptomyces* spp. isolates SM17 and SM18 with their closest terrestrial relatives provide novel insights into environmental niche adaptations and secondary metabolite biosynthesis potential. Front. Microbiol., 10: 1713. https://doi.org/10.3389/fmicb.2019.01713.

Alster, C.J., Allison, S.D., Johnson, N.G., Glassman, S.I. and Treseder, K.K. (2021). Phenotypic plasticity of fungal traits in response to moisture and temperature. ISME Commun., 1(1): 43. https://doi.org/10.1038/s43705-021-00045-9.

Andersen, J., Ferreira, S. and Cruz, M. (2023). Community-based management of marine lichen resources: Integrating traditional knowledge with scientific monitoring in coastal Portugal. J. Coastal Conserv., 27(3): 214–229. https://doi.org/10.1139/facets-2021-0067.

Armaleo, D., Müller, O., Lutzoni, F., Andrésson, Ó.S., Blanc, G. et al. (2019). The lichen symbiosis re-viewed through the genomes of *Cladonia grayi* and its algal partner Asterochloris glomerata. BMC Genomics, 20(1): 605. https://doi.org/10.1186/s12864-019-5629-x.

Bisht, B., Verma, M., Sharma, R., Chauhan, P., Pant, K., Kim, H., Vlaskin, M.S. and Kumar, V. (2023). Development of yeast and microalgae consortium biofilm growth system for biofuel production. Heliyon, 9(9): e19353. https://doi.org/10.1016/j.heliyon.2023.e19353.

Blackwell, M. and Suzuki, K. (2024). The Marine Lichen biotechnology consortium: A collaborative framework for sustainable bioprospecting and conservation. Blue Biotechnol. J., 2(1): 45–58.

Blanz, P. (2018). Biodiversity and ecology of Fungi, Lichens, and mosses. *In*: Verlag der österreichischen Akademie der Wissenschaften eBooks, 715 pp. https://doi.org/10.1553/0x003892e7.

Blanz, P. (2018). Symbiotic cultivation techniques for lichen-forming fungi. pp. 315–332. *In*: Druzhinina, I.S. and Kubicek, C.P. (eds.). Environmental and Microbial Relationships. Springer. http://dx.doi.org/10.1007/978-3-540-71840-6.

Bonugli-Santos, R.C., dos Santos Vasconcelos, M.R., Passarini, M.R., Vieira, G.A., Lopes, V.C. et al. (2015). Marine-derived fungi: Diversity of enzymes and biotechnological applications, Front. Microbiol., 6: 269. https://doi.org/10.3389/fmicb.2015.00269.

Bringloe, T.T., Wilkinson, D.P., Goldsmit, J., Savoie, A.M., Filbee-Dexter, K. et al. (2022). Arctic marine forest distribution models showcase potentially severe habitat losses for cryophilic species under climate change. Glob. Chang. Biol., 28(11): 3711–3727. https://doi.org/10.1111/gcb.16142.

Brodo, I.M. and Wilson, D.P. (2023). Marine lichen communities of the Pacific Northwest: Diversity and conservation implications. J. Coast. Res., 45(3): 312–328.

Burmeister, A.R. (2015). Horizontal gene transfer. Evol. Med. Public Health, 2015(1): 193–194. https://doi.org/10.1093/emph/eov018.

Callieri, C. (2017). *Synechococcus* plasticity under environmental changes. FEMS Microbiol. Lett., 364(23): fnx229. https://doi.org/10.1093/femsle/fnx229.

Cardozo, K.H., Guaratini, T., Barros, M.P., Falcão, V.R., Tonon, A.P. et al. (2006). Metabolites from algae with economical impact. Comp. Biochem. Physiol. C Toxicol. Pharmacol., 146(1-2): 60–78. https://doi.org/10.1016/j.cbpc.2006.05.007.

Chen, Y., Xu, L.C., Liu, S., Zhang, Z.X. and Cao, G.Y. (2022). Halometabolites isolated from the marine-derived fungi with potent pharmacological activities. Front. Microbiol., 13: 1038487. https://doi.org/10.3389/fmicb.2022.1038487.

Chrismas, N., Tindall-Jones, B., Jenkins, H., Harley, J., Bird, K. et al. (2023). Metatranscriptomics reveals diversity of symbiotic interaction and mechanisms of carbon exchange in the marine cyanolichen *Lichina pygmaea*. New Phytol., 241(5): 2243–2257. https://doi.org/10.1111/nph.19320.

Corrêa-Junior, D., De Andrade, I.B., Alves, V., Avellar-Moura, I., Rodrigues-Alves, T. et al. (2023). Metabolic plasticity and virulence-associated factors of *Sporothrix brasiliensis* strains related to familiar outbreaks of cat-to-human transmitted sporotrichosis. J. Fungi, 9(7): 724. http://dx.doi.org/10.3390/jof9070724.

Coste, C., Chauvet, E., Grieu, P. and Lamaze, T. (2016). Photosynthetic traits of freshwater lichens are consistent with the submersion conditions of their habitat. Ann. Limnol. Int. J. Limnol., 52: 235–242. http://dx.doi.org/10.1051/limn/2016009.

Delmail, D., Grube, M., Parrot, D., Cook-Moreau, J., Boustie, J. et al. (2012). Halotolerance in lichens: Symbiotic coalition against salt stress. pp. 115–148. *In*: Ahmad, P., Azooz, M. and Prasad, M. (eds.). Ecophysiology and Responses of Plants under Salt Stress. Springer, New York. https://doi.org/10.1007/978-1-4614-4747-4_4.

Ellis, Christopher J. (2011). Lichen epiphyte diversity: A species, community and trait-based review. pp. 131–52. *In*: Perspectives in Plant Ecology Evolution and Systematics, 14(2): 131–52. https://doi.org/10.1016/j.ppees.2011.10.001.

Erokhina, L.G., Shatilovich, A.V., Kaminskaya, O.P. and Gilichinskii, D.A. (2004). Spectral properties of the green alga *Trebouxia*, a phycobiont of cryptoendolithic lichens in the Antarctic dry valley. Microbiology, 73(4): 420–424. http://dx.doi.org/10.1023/B:MICI.0000036987.18559.1f.

Fonseca, S., Amaral, M.N., Reis, C.P. and Custódio, L. (2023). Marine natural products as innovative cosmetic ingredients. Mar. Drugs, 21(3): 170. https://doi.org/10.3390/md21030170.

Gasulla, F., Del Campo, E.M., Casano, L.M. and Guéra, A. (2021). Advances in understanding of desiccation tolerance of lichens and lichen-forming algae. Plants, 10(4): 807. https://doi.org/10.3390/plants10040807.

Goga, M., Elečko, J., Marcinčinová, M., Ručová, D., Bačkorová, M. et al. (2020). Lichen metabolites: An overview of some secondary metabolites and their biological potential. pp. 175–209. *In*: Mérillon, J.M. and Ramawat, K. (eds.). Co-Evolution of Secondary Metabolites. Reference Series in Phytochemistry. Springer Cham. https://doi.org/10.1007/978-3-319-96397-6_57.

Grube, M., Cardinale, M., de Castro Jr, J.V., Müller, H. and Berg, G. (2009). Species-specific structural and functional diversity of bacterial communities in lichen symbioses. ISME J., 3(9): 1105–15. https://doi.org/10.1038/ismej.2009.63.

Hájek, J., Váczi, P., Barták, M. and Jahnová, L. (2012). Interspecific differences in cryoresistance of lichen symbiotic algae of genus *Trebouxia* assessed by cell viability and chlorophyll fluorescence. Cryobiology, 64(3): 215–222. https://doi.org/10.1016/j.cryobiol.2012.02.002.

Hammer, S. (1997). Growth dynamics in *Cladonia grayi*. Mycologia, 89(6): 900–907. https://doi.org/10.1080/00275514.1997.12026860.

Hasegawa, P.M., Bressan, R.A., Zhu, J. and Bohnert, H.J. (2000). Plant cellular and molecular responses to high salinity. Annu. Rev. Plant Physiol. Plant Mol. Biol., 51(1): 463–499. http://dx.doi.org/10.1146/annurev.arplant.51.1.463.

Hawksworth, D.L. (2000). Freshwater and marine lichen-forming fungi. Fungal Divers., 5: 1–7. http://www.fungaldiversity.org/fdp/sfdp/FD_5_1-7.pdf.

Higgins, N.F., Connan, S. and Stengel, D.B. (2014). Factors influencing the distribution of coastal lichens *Hydropunctaria maura* and *Wahlenbergiella mucosa*. Mar. Ecol., 36(4): 1400–1414. http://dx.doi.org/10.1111/maec.12239.

Hofmann, G.E., Barry, J.P., Edmunds, P.J., Gates, R.D., Hutchins, D.A. et al. (2010) The effect of ocean acidification on calcifying organisms in marine ecosystems: An organism-to-ecosystem perspective. Annu. Rev. Ecol. Evol. Syst., 41(1): 127–147. https://doi.org/10.1146/annurev.ecolsys.110308.120227.

Honegger, R. (2009). Lichen-forming fungi and their photobionts. *In*: Deising, H. (ed.). Plant Relationships (2nd Edn.), 5(5): 305–333. Springer. https://doi.org/10.1007/978-3-540-87407-2_16.

Im, S., Lee, H.N., Jung, H.S., Yang, S., Park, E.J. et al. (2017). Transcriptome-based identification of the desiccation response genes in marine red algae *Pyropia tenera* (Rhodophyta) and enhancement of abiotic stress tolerance by PtDRG2 in *Chlamydomonas*. Mar. Biotechnol., 19(3): 232–245. https://doi.org/10.1007/s10126-017-9744-x.

Ji, R., Zha, X. and Zhou, S. (2023). Marine fungi: A prosperous source of novel bioactive natural products. Curr. Med. Chem., 32(5): 992–1006. https://doi.org/10.2174/0109298673266304231015070956.

Jones, L., Garbutt, A., Hansom, J. and Angus, S. (2013). Impacts of climate change on coastal habitats. MCCIP Sci. Rev., 2013: 167–179. https://doi.org/10.14465/2013.arc18.167-179.

Kappen, L., Sommerkorn, M. and Schroeter, B. (1995) Carbon acquisition and water relations of lichens in polar regions—potentials and limitations. Lichenologist, 27(6): 531–545. https://doi.org/10.1016/s0024-2829(95)80012-3.

Kranner, I., Beckett, R.P. and Minibayeva, F.V. (2022). Desiccation tolerance in lichens: From cellular mechanisms to ecological significance. J. Exp. Bot., 73(5): 1659–1676. http://dx.doi.org/10.1639/0007-2745-111.4.576.

Lücking, R., Moncada, B. and Bungartz, F. (2024). Stochastic versus deterministic assembly of oceanic island biota: Leaf-dwelling lichens on the Galápagos and on Cocos Island. Front. Biogeogr., 17: e133819. https://doi.org/10.21425/fob.17.133819.

Maldonado, L.A., Fenical, W., Jensen, P.R., Kauffman, C.A., Mincer, T.J. et al. (2005). *Salinispora arenicola* gen. nov., sp. nov. and *Salinispora tropica* sp. nov., obligate marine actinomycetes belonging to the family Micromonosporaceae. Int. J. Syst. Evol. Microbiol., 55(5): 1759–1766. http://dx.doi.org/10.1099/ijs.0.63625-0.

Maldonado, L.A., Stach, J.E.M., Pathom-aree, W., Ward, A.C., Bull, A.T. et al. (2005). Diversity of cultivable actinobacteria in geographically widespread marine sediments. Antonie van Leeuwenhoek, 87(1): 11–18. http://dx.doi.org/10.1007/s10482-004-6525-0.

Mandrekar, V.K., Gawas, U.B. and Majik, M.S. (2018). Brominated molecules from marine algae and their pharmacological importance. Stud. Nat. Prod. Chem., 61: 461–490. https://doi.org/10.1016/b978-0-444-64183-0.00013-0.

McMullin, R.T., Wiersma, Y.F., Newmaster, S.G. and Lendemer, J.C. (2019). Risk assessment and conservation strategies for rare lichen species and communities threatened by sea-level rise in the Mid-Atlantic Coastal Plain. Biol. Conserv., 239: 108281. https://doi.org/10.1016/j.biocon.2019.108281.

Mohr, F., Ekman, S. and Heegaard, E. (2004). Evolution and taxonomy of the marine *Collemopsidium* species (lichenized Ascomycota) in north-west Europe. Mycol. Res., 108(5): 515–532. https://doi.org/10.1017/S095375620400992X.

Molnár, K. and Farkas, E. (2010). Current results on biological activities of lichen secondary metabolites: A review. Z. Naturforsch., C., 65(3-4): 157–173. http://dx.doi.org/10.1515/znc-2010-3-401.

Murugan, M., Muthu, S., Rajendran, K. and Ponnusamy P. (2021). Review: UV protection and anticancer properties of lichen secondary metabolites. Sci. Academique, 2(2): 1–29.

Norouzi, H., Azizi, A., Gholami, M., Sohrabi, M. and Boustie, J. (2020). Chemotype variations among lichen ecotypes of *Umbilicaria aprina* as revealed by LC-ESI-MS/MS: A survey of antioxidant phenolics. Environ. Sci. Pollut. Res., 27(32): 40296–40308. https://doi.org/10.1007/s11356-020-10053-2.

Parrot, D., Antony-Babu, S., Intertaglia, L., Grube, M., Tomasi, S. et al. (2015). Littoral lichens as a novel source of potentially bioactive actinobacteria. Sci. Rep., 5: 15839. https://doi.org/10.1038/srep15839.

Pino-Bodas, R., Zhurbenko, M.P. and Stenroos, S. (2015). Phylogenetic placement within Lecanoromycetes of lichenicolous fungi associated with *Cladonia* and some other genera. Persoonia: Mol. Phylogeny Evol. Fungi, 37: 1–27. http://dx.doi.org/10.3767/persoonia.2017.39.05.
Raghukumar, C. (2017). Fungi in coastal and oceanic marine ecosystems: Marine Fungi. Springer, 378 pp. http://dx.doi.org/10.1007/978-3-319-54304-8.
Ranković, B. and Kosanić, M. (2019). Lichens as a potential source of bioactive secondary metabolites. pp. 1–29. *In*: Ranković, B. (ed.). Lichen Secondary Metabolites. Springer Cham. https://doi.org/10.1007/978-3-030-16814-8_1.
Ren, H.N., Zhuo, T.X., Bai, S.J., Bai, Y., Sun, X.Y. et al. (2021). Proteomic analysis of hydrolytic proteases in excretory/secretory proteins from *Trichinella spiralis* intestinal infective larvae using zymography combined with shotgun LC-MS/MS approach. Acta Trop., 216: 105825. https://doi.org/10.1016/j.actatropica.2021.105825.
Ríos, P., Prado, E., Carvalho, F.C., Sánchez, F., Rodríguez-Basalo, A. et al. (2020). Community composition and habitat characterization of a rock sponge aggregation (Porifera, Corallistidae) in the Cantabrian Sea. Front. Mar. Sci., 7: 578. https://doi.org/10.3389/fmars.2020.00578.
Rodnikova, I.M. (2012). Effect of environmental conditions on morphological, ecological, and geographic characteristics of lichens in coastal habitats. Russ. J. Ecol., 43(2): 97–100. https://doi.org/10.1134/s1067413612020117.
Rosabal, D. and Pino-Bodas, R. (2024). A review of laboratory requirements to culture lichen mycobiont species. J. Fungi, 10(9): 621. https://doi.org/10.3390/jof10090621.
Rubio-Salcedo, M., Psomas, A., Prieto, M., Zimmermann, N.E. and Martínez, I., (2016). Case study of the implications of climate change for lichen diversity and distributions. Biodivers. Conserv., 26(5): 1121–1141. https://doi.org/10.1007/s10531-016-1289-1.
Sadowsky, A., Mettler-Altmann, T. and Ott, S. (2016). Metabolic response to desiccation stress in strains of green algal photobionts (*Trebouxia*) from two Antarctic lichens of southern habitats. Phycologia, 55(6): 703–714. https://doi.org/10.2216/15-127.1.
Scheidegger, C. and Werth, S. (2009). Conservation strategies for Lichens: Insights from population biology. Fungal Biol. Rev., 23(3): 55–66. https://doi.org/10.1016/j.fbr.2009.10.003.
Selig, E.R., Turner, W.R., Troëng, S., Wallace, B.P., Halpern, B.S. et al. (2014). Global priorities for marine biodiversity conservation. PLoS One, 9(1): e82898. https://doi.org/10.1371/journal.pone.0082898.
Shin, H.J., Lee, M.A., Lee, H.S. and Heo, C.S. (2023). Thiolactones and Δ8,9-Pregnene steroids from the marine-derived fungus *Meira* sp. 1210CH-42 and their α-Glucosidase inhibitory activity. Mar. Drugs, 21(4): 246. https://doi.org/10.3390/md21040246.
Singh, G. (2023). Linking lichen metabolites to genes: Emerging concepts and lessons from molecular biology and metagenomics. J. Fungi, 9(2): 160. https://doi.org/10.3390/jof9020160.
Sinha, D., Borkataky, M., Chowardhara, B. and Nath, R. (2024). Taxonomy of lichen. pp. 21–37. *In*: Das, A.K., Sharma, A., Kathuria, D., Ansari, M.J. and Bhardwaj, G. (eds.). Chemistry, Biology and Pharmacology of Lichen. http://dx.doi.org/10.1002/9781394190706.ch3.
Song, H., Kim, K., Park, S., Lee, G., Choi, J. et al. (2022). A comparative genomic analysis of lichen-forming fungi reveals new insights into fungal lifestyles. Sci. Rep., 12(1): 10054. http://dx.doi.org/10.1038/s41598-022-14340-5.
Sonina, A.V. and Androsova, V.I. (2020). Coastal Lichens. pp. 1–22. *In*: Grigore, M.N. (ed.). Handbook of Halophytes. Springer Cham. https://doi.org/10.1007/978-3-030-17854-3_34-1.
Sonina, A.V., Tsunskiy, I.I. and Androsova, V.I. (2017). The effect of environmental factors on morphological parameters and physiological state of *Verrucaria maura* Wahlenb lichens in the littoral zone of the White Sea. Russ. J. Ecol., 48(3): 250–257. http://dx.doi.org/10.5817/CPR2012-2-11.
Stocker-Wörgötter, E. (1995). Experimental cultivation of lichens and lichen symbionts. Canadian Journal of Botany, 73(S1): 579–589. https://doi.org/10.1139/b95-298.
Tanona, M. and Czarnota, P. (2022) What determines the diversity and succession of lichens inhabiting post-bark beetle snags in the Western Carpathians? Ann. For. Res., 65(1): 65–84. https://doi.org/10.15287/afr.2022.2146.

Tehler, A. and Irestedt, M. (2007). Parallel evolution of lichen growth forms in the family Roccellaceae (Arthoniales, Ascomycota). Cladistics, 23(5): 432–454. https://doi.org/10.1111/j.1096-0031.2007.00156.x.

Tunjić, M. and Korač, P. (2013). Vertical and horizontal gene transfer in lichens. Period. Biol., 115(3): 321–329. https://hrcak.srce.hr/110652.

Varrella, S., Barone, G., Tangherlini, M., Rastelli, E., Dell'Anno, A. et al. (2021). Diversity, ecological role and biotechnological potential of Antarctic marine fungi. J. Fungi, 7(5): 391. https://doi.org/10.3390/jof7050391.

Zhang, Q. and Wong, G. (2024). Novel bioactive compounds from marine lichens: Pharmaceutical applications and biotechnological potential. Mar. Drugs, 22(1): 45–62. http://dx.doi.org/10.1016/j.mib.2005.04.008.

Zhang, Q., Lv, L., Wang, W., Wei, M., Gu, Y. et al. (2024). Recent advances of bioactive marine natural products in drug discovery. J. Ocean Univ. China, 23(5): 1297–1318. http://dx.doi.org/10.1007/s11802-024-5975-4.

8

The Untapped Potential of Arctic and Antarctica Fungi

Bhagyashree Bora,[1] *Takam Akash,*[1] *Channakeshavaiah Chikkaputtaiah*[2,3] and *Natarajan Velmurugan*[1,2,*]

The Arctic and Antarctica are home to some of the harshest but biologically intriguing habitats. Fungi are among the resilient species that live in these frigid environments. They are renowned for their great physiological adaptability, ecological relevance, and untapped biotechnology promise. This chapter gives an integrated account of psychrophilic and psychrotolerant fungi from polar habitats, focusing on their diversity, mechanisms of cold adaptation, and functions in nutrient cycling and ecosystem stability. In spite of harsh cold, poor nutrient availability, and high UV radiation, polar fungi have an enormous range of enzymatic and metabolic characteristics that facilitate survival and ecological function. Fungi from polar regions are also demonstrated to be bio-indicators for climate change and can also be potential candidates for sustainable innovations in the fields of food, agriculture, pharmaceuticals, and green industry. By integrating recent discoveries, the chapter highlights the imperative for combined molecular, ecological, and biotechnological studies

[1] Biological Sciences Division, CSIR-North East Institute of Science and Technology (CSIR-NEIST), Branch Laboratory-Itanagar, Naharlagun 791 110, Arunachal Pradesh, India.
[2] Academy of Scientific and Innovative Research (AcSIR), Ghaziabad 201 002, India.
[3] Biological Sciences and Technology Division, CSIR-North East Institute of Science and Technology (CSIR-NEIST), Jorhat 785 006, Assam, India.
* Corresponding author: velmmk@gmail.com/natarajan@neist.res.in

DOI: 10.1201/9781003795100-10

to unlock the buried potential of polar fungi. It demands more profound investigation into their genomic and metabolic abilities, which may result in revolutionary applications in science and industry.

1. Introduction

With a total size of almost 14 million km^2, Antarctica is the world's fifth-biggest continent and the southernmost landmass. Temperatures in Antarctica coastal regions typically vary from 5°C to 35°C, with ice and snow covering over 98%. The majority of living forms in the continental Antarctica region live in places free of snow and ice, which are in fact about less than 0.5% and possible as little as 0.18% of Antarctica's area (Convey et al., 2019; Hund, 2014). The Antarctica ice sheet covers most of this icy continent with an average thickness of 1.9 kilometers (1.2 miles) and contains almost 90% of the world's glacial ice (Turner et al., 2009). In case theAntarctic ice sheet melts entirely, sea levels might increase by around 60 meters (200 feet) (Climate Change, 2007). With the exception of the Antarctic Peninsula, which stretches into South America, its geographic configuration is primarily circular (Turner et al., 2009). Located on a polar plateau at an elevation of over 2,800 meters (9,186 feet), Antarctica boasts the greatest average elevation of any continent (National Snow and Ice Data Center, 2020). The climate is characterized by extreme cold, dryness, and strong winds, with winter temperatures below –60°C (–76°F) and summer temperatures that go over 10°C (50°F) in coastal regions (Peck et al., 2018).

The Transantarctic Mountains are around 3,400 kilometers (2,100 miles) long and divide East and West Antarctica (Bindschadler, 2006). The tallest mountain, Vinson Massif, is 4,892 meters (16,050 feet) high (Glidea et al., 2007). Additionally, the continent has a number of subglacial lakes, including Lake Vostok, which is home to Russia's Vostok Station and is renowned for its distinct environment (Siegert et al., 2001). The 18,000-kilometer (11,200-mile) coastline of Antarctica is home to a variety of landforms, such as ice walls and shelves (Bindschadler, 2006). The continent is habitat to many polar animals and macro- and microorganisms, such as penguins, seals, and extremophiles, despite its severe climate. Lichens and mosses make up the majority of the Antarctic flora (Convey, 2017). Only research stations are inhabited by humans, and seasonal populations range from 1,300 to 5,100 people (SCAR, 2022). Antarctica is an important region for scientific exploration and study because of its enormous ice deposits, which are crucial in controlling the world's climate and sea levels (Maksym, 2019).

Comparatively, the Arctic Ocean, along with its neighboring landmasses, including Greenland, portions of North America, and Eurasia, makes up approximately 20 million km^2, or around 7.7 million mi^2 (AMAP, 2023). The Arctic Ocean is surrounded by a different terrain, including tundra areas and glaciated Greenland, and is home to geological structures such as the Nansen-Gakkel Ridge and the Lomonosov Ridge (Weber, 1989). Arctic sea ice is seasonal, with coverage varying throughout the year, in contrast to Antarctica's permanent ice sheet. In January

2025, the sea ice extent was observed at about 13.1 million square kilometers (5.1 million square miles), which was less than the historical average (NSIDC, 2023, 2025). In the Arctic, interior regions can experience summer highs up to 26°C (79°F) and winter lows below –50°C (–58°F) (Bulygina et al., 2011), while Antarctica reaches coastal summer temperature maxima of 5–10°C (41–50°F) and interior winter temperature minima below –80°C (–112°F), with satellite-derived records below –98°C (–144°F) (Scambos et al., 2018). The area is home to a variety of animals, including caribou, seals, polar bears, and migratory birds, and its vegetation is mostly tundra with low shrubs and grasses (Post et al., 2019). These opposing polar areas continue to face environmental issues that are being thoroughly researched, and they are crucial to biodiversity and global climate systems (Lee et al., 2023).

Cold-loving microorganisms known as psychrophiles thrive in the severe conditions of Antarctica, with an optimal growing temperature of about 15°C or below and a maximum of 25°C. But they may also thrive at 0°C or below. Conversely, psychrotolerant plants can tolerate temperatures over 20°C, but can thrive at 5°C and below. While psychrotolerants flourish in areas with periodic low temperatures, psychrophiles predominate in permanently cold environments. Both groups are essential to biodegradation and nutrient cycling in Antarctica (Morita, 1975; Arthur and Watson, 1976).

The lowest temperatures limits for life are largely defined by microorganisms, particularly psychrophilic bacteria, yeasts, and other microbes. In this regard, Antarctica's psychrophilic yeasts and bacteria may make great model organisms for comprehending the molecular underpinnings of low-temperature survival. Thus far, to date, most biological studies in Antarctica have primarily focused on identifying and classifying bacterial species, exploring their diversity (Shivaji, 2005; Prabagaran et al., 2007), investigating their potential for biotechnological applications (Cavicchioli et al., 2002), and utilizing them as model organisms to better understand microbial adaptations to cold environments. This chapter summarizes, Arctic and Antarctica fungi having significant roles as biological resources, highlighting major ecological functions and unique ability for innovative applications that addresses urgent environmental and industrial needs.

2. Fungal Diversity

About 37% and 25% of the species in the divisions Mucoromycota and Ascomycota are shared across the two poles, respectively, whereas only roughly 3% of the species in the division Basidiomycota are shared (Grum-Grzhimaylo et al., 2024). Only a few species are known to be exclusive to the Arctic or Antarctica, and between 1–6 species have been identified as bipolar, meaning they occur at both poles but not at intermediate latitudes. Tables 1 and 2 are listing the major culturable and non-culturable fungal genera of polar regions, respectively. De Menezes et al. (2021) study found that the phylum Ascomycota dominated all fungal assemblages, followed by Basidiomycota, Mortierellomycota, and Mucoromycota.

The research revealed that the highest number of fungal amplicon sequence variants (ASVs), totaling 398, were detected in encrustation Type 2, followed by

Table 1. Some of the notable culturable fungal species isolated from polar regions (1990–2025).

Fungal Species	**Location and Region**	**Source/Host**	**References**
Antarctica			
Antarctomyces psychrotrophicus	King George & Deception Islands; South Shetland Islands	Lakes; Marine invertebrates	Gonçalves et al. (2012); Godinho et al. (2019)
Cadophora cf. luteo-olivacea	King George & Deception Islands	Lakes	Gonçalves et al. (2012)
Cadophora fastigiata	Victoria Land	Explorer huts	Arenz et al. (2006)
Cadophora malorum	Victoria Land; King George/Deception Islands	Explorer huts; Lakes	Arenz et al. (2006); Gonçalves et al. (2012)
Cladosporium sp.	King George & Deception Islands	Lakes	Gonçalves et al. (2012)
Cladosporium verrucocladosporioides	South Shetland Is. & Antarctica Peninsula	Marine invertebrates	Godinho et al. (2019)
Cryomyces antarcticus	McMurdo Dry Valleys	Sandstone cryptoendolith	Selbmann et al. (2005)
Cryomyces minteri	McMurdo Dry Valleys	Sandstone cryptoendolith	Selbmann et al. (2005)
Davidiella tassiana	King George & Deception Islands	Lakes	Gonçalves et al. (2012)
Debaryomyces hansenii	South Shetland Is. & Antarctica Peninsula	Marine invertebrates	Godinho et al. (2019)
Geomyces pannorum	King George & Deception Islands	Lakes	Gonçalves et al. (2012)
Gibberella moniliformis	King George & Deception Islands	Lakes	Gonçalves et al. (2012)
Glaciozyma martinii	South Shetland Is. & Antarctica Peninsula	Marine invertebrates	Godinho et al. (2019)
Globosphaeria jamesii	Coastal east	Lichen thalli	Hawksworth (1990)
Helgardia sp.	King George & Deception Islands	Lakes	Gonçalves et al. (2012)
Helotiales sp.	King George & Deception Islands	Lakes	Gonçalves et al. (2012)
Mortierella cf. alpina	King George & Deception Islands	Lakes	Gonçalves et al. (2012)
Mortierella sp.	King George & Deception Islands	Lakes	Gonçalves et al. (2012)

Table 1 contd. ...

...Table 1 contd.

Fungal Species	**Location and Region**	**Source/Host**	**References**
Penicillium cf. verrucosum	King George & Deception Islands	Lakes	Gonçalves et al. (2012)
Penicillium brevicompactum	South Shetland Is. & Antarctica Peninsula	Marine invertebrates	Godinho et al. (2019)
Penicillium rubens	South Shetland Is. & Antarctica Peninsula	Marine invertebrates	Godinho et al. (2019)
Penicillium tardochrysogenum	South Shetland Is. & Antarctica Peninsula	Marine invertebrates	Godinho et al. (2019)
Penicillium spp.	South Shetland Is. & Antarctica Peninsula	Marine invertebrates	Godinho et al. (2019)
Mrakia frigida	South Shetland Is. & Antarctica Peninsula	Marine invertebrates	Godinho et al. (2019)
Metschnikowia australis	South Shetland Is. & Antarctica Peninsula	Marine invertebrates	Godinho et al. (2019)
Phoma fimeti	King George & Deception Islands	Lakes	Gonçalves et al. (2012)
Phoma herbarum	King George & Deception Islands	Lakes	Gonçalves et al. (2012)
Pleosporales sp.	King George & Deception Islands	Lakes	Gonçalves et al. (2012)
Pseudeurotium sp.	King George & Deception Islands	Lakes	Gonçalves et al. (2012)
Pseudogymnoascus destructans	Antarctica Peninsula; South Shetland Islands	Soil, bat pathogen	Hoyt et al. (2015); Godinho et al. (2019)
Pseudogymnoascus sp.; *P. verrucosus*	South Shetland Is. & Antarctica Peninsula	Marine invertebrates	Godinho et al. (2019)
Rhodotorula mucilaginosa	South Shetland Is. & Antarctica Peninsula	Marine invertebrates	Godinho et al. (2019)
Saprolegniaceae sp.	King George & Deception Islands	Lakes	Gonçalves et al. (2012)
Skeletocutis friabilis (friata)	King George Island	Driftwood	Niemela (1998)
Thelebolus cf. microsporus	King George & Deception Islands	Lakes	Gonçalves et al. (2012)
Thelebolus sp.	King George & Deception Islands	Lakes	Gonçalves et al. (2012)
Thelebolus cf. globosus	South Shetland Is. & Antarctica Peninsula	Marine invertebrates	Godinho et al. (2019)
Tolypocladium tundrense	South Shetland Is. & Antarctica Peninsula	Marine invertebrates	Godinho et al. (2019)

Arctic			
Aspergillus niger	Greenland Ice Sheet	Supraglacial ice	Cantrell et al. (2011)
Aspergillus pseudoglaucus	Werenskiold Glacier, Svalbard; Svalbard, Norway	Cryoconite holes	Borzęcka et al. (2022, 2025)
Bjerkandera adusta	Werenskiold Glacier, Svalbard	Cryoconite holes	Borzęcka et al. (2025)
Botrytis cinerea	Svalbard (Ny-Ålesund)	Subglacial/melt/lake/tap water	Perini et al. (2019)
Cladosporium allicinum	Werenskiold Glacier, Svalbard; Svalbard, Norway	Cryoconite holes	Borzęcka et al. (2022, 2025)
Cladosporium sp.	Svalbard (Ny-Ålesund)	Meltwater	Perini et al. (2019)
Collybia cirrhata	Svalbard (Ny-Ålesund)	Water sources	Perini et al. (2019)
Guehomyces pullulans	Alaska	Glacial meltwater	De Garcia et al. (2012); De Souza et al. (2022)
Filobasidium magnum	Svalbard (Ny-Ålesund)	Water sources	Perini et al. (2019)
Geomyces sp.	Svalbard (Ny-Ålesund)	Water sources	Perini et al. (2019)
Glaciozyma watsonii	Svalbard (Ny-Ålesund)	Water sources	Perini et al. (2019)
Rhodotorula psychrophila	Alaska	Glacial meltwater	De Garcia et al. (2012); De Souza et al. (2022)
Hortaea werneckii	Svalbard (Ny-Ålesund)	Water sources	Perini et al. (2019)
Lecanicillium sp.	Svalbard (Ny-Ålesund)	Water sources	Perini et al. (2019)
Leptosphaeria sp.	Svalbard (Ny-Ålesund)	Water sources	Perini et al. (2019)
Malassezia restricta	Svalbard (Ny-Ålesund)	Water sources	Perini et al. (2019)
Mortierella elongata	Siberian Tundra	Peat soils	Ozimek and Hanaka (2020)
Mortierella globalpina	Siberian Tundra	Peat soils	Ozimek and Hanaka (2020)
Mortierella sp.	Svalbard (Ny-Ålesund)	Water sources	Perini et al. (2019)
Mrakia hoshinonis	Ellesmere Island, Canada	Proglacial sediments	Tsuji et al. (2019)
Mrakia sp.	Svalbard (Ny-Ålesund)	Water sources	Perini et al. (2019)

Table 1 contd. ...

...Table 1 contd.

Fungal Species	**Location and Region**	**Source/Host**	**References**
Naganishia albida	Svalbard (Ny-Ålesund)	Water sources	Perini et al. (2019)
Paraleptosphaeria dryadis	Svalbard (Ny-Ålesund)	Water sources	Perini et al. (2019)
Penicillium bialowiezense	Svalbard (Ny-Ålesund)	Water sources	Perini et al. (2019)
Penicillium crustosum	Greenland Ice Sheet; Svalbard (Ny-Ålesund)	Supraglacial ice; Water sources	Cantrell et al. (2011); Perini et al. (2019)
Penicillium svalbardense	Midre Lovénbreen Glacier, Svalbard	Glacial ice	Sonjak et al. (2007)
Penicillium sumatraense	Werenskiold Glacier, Svalbard; Svalbard, Norway	Cryoconite holes	Borzęcka et al. (2022, 2025)
Penicillium velutinum	Werenskiold Glacier, Svalbard	Cryoconite holes	Borzęcka et al. (2025)
Pheoliferia psychrophenolica	Svalbard (Ny-Ålesund)	Water sources	Perini et al. (2019)
Phialocephala sp.	Svalbard (Ny-Ålesund)	Water sources	Perini et al. (2019)
Rhodotorula psychrophila	Alaska	Glacial meltwater	De Garcia et al. (2012); De Souza et al. (2022)
Rhodotorula svalbardensis	Austre Brøggerbreen Glacier, Svalbard; Svalbard (Ny-Ålesund)	Cryoconite holes; Water sources	Singh et al. (2014); Perini et al. (2019)
Thelebolus sp.	Svalbard (Ny-Ålesund)	Water sources	Perini et al. (2019)
Trametes versicolor	Werenskiold Glacier, Svalbard	Cryoconite holes	Borzęcka et al. (2025)
Tetracladium sp.	Svalbard (Ny-Ålesund)	Water sources	Perini et al. (2019)
Venturia sp.	Svalbard (Ny-Ålesund)	Water sources	Perini et al. (2019)
Vishniacozyma ellesmerensis	Ellesmere Island, Canada	Proglacial sediments	Tsuji et al. (2019)

Table 2. Unculturable fungal diversity identified in Arctic and Antarctica regions based on amplicon sequencing studies.

Region	Sample Type	Number of ASVs Detected	Dominant Fungal Phyla	References
Antarctica	Soil (Deception Island)	346	Ascomycota, Basidiomycota, Chytridiomycota, Mortierellomycota	Rosa et al. (2020)
Antarctica	Air and Snow (Livingston Island)	171	Ascomycota, and Basidiomycota	Rosa et al. (2020)
Antarctica	Marine Sediments (South Shetland Islands)	133	Ascomycota, Basidiomycota, Chytridiomycota, Mortierellomycota	De Silva et al. (2022)
Antarctica	Various Habitats (Fildes Region)	320 species	Ascomycota, Basidiomycota, Mortierellomycota, and Mucoromycota	Zhang et al. (2022)
Antarctica	Lake Sediments (James Ross Island)	132	Ascomycota, Basidiomycota, and Mortierellomycota	Gonçalves et al. (2023)
Arctic and Antarctica	Water and Sediment (Various Lakes)	5,980	Ascomycota, Basidiomycota, Chytridiomycota, Rozellomycota	Marchetta et al. (2023)

281 ASVs in encrustation Type 1, and 77 ASVs in the carbonate vein. In a separate investigation, De Souza et al. (2021) reported distinct fungal communities in two lakes located on the South Shetland Islands, identifying 27 fungal ASVs in Hennequin Lake and 13 in Crater Soto Lake, with only six taxa common to both sites. Their work demonstrated that the high-throughput sequencing (HTS) meta-barcode approach effectively uncovered hidden freshwater fungal diversity in Antarctic lakes, including taxa that were not detectable through conventional culturing methods. Additionally, the study recorded a significant number of sequences corresponding to known aquatic saprotrophic *hyphomycetes*, such as *Tetradadium marchalianum* and *Tetracladium* species. Complementing these findings, Zhang et al. (2016) reported that Ascomycota was the most diverse fungal phylum in their study, with 641 operational taxonomic units (OTUs) and 43,061 sequence reads identified across 13 water samples. Additionally, the study found that the four aquatic environments in the Ny-Ålesund Region, Svalbard, had unique aquatic fungal communities. The richness and structure of these communities were mostly determined by temperature, salinity, and conductivity. The composition and diversity of fungal communities were shown to be strongly correlated with environmental factors, including temperature, conductivity, and salt.

Additionally, the study discovered that fungal communities varied throughout the four aquatic environments, with distinct genera and OTUs linked to each habitat. Another latest study by Rosa et al. (2025) has shown that fungal assemblages varied along the transect, with the subtropical and temperate South Atlantic having the most diversity, whereas the Drake Passage as well as King George Island have the least. Out of 643,325 fungal DNA readings, 176 fungal ASVs were found. The Drake Passage contained the least diverse fungal assemblages, whereas the South

Atlantic coast of South America had the most diverse. Only a few species, including *Aspergillus versicolor* and *Chaetomiaceae* sp., were shared between the Antarctic Peninsula and the South Atlantic region.

3. Fungal Adaptation

Across Arctic and Antarctic locales, there have been documented instances of fungal growth, activity, and survival at low temperatures. Fifty-four isolates from arid soils in Franz Josef Land, located between latitudes 79.83°N and 81.83°N, mostly mitosporic fungus, grew at 0 to 25°C on malt extract agar and had psychrotrophic traits (Bergero et al., 1999). *Phoma herbarum*, obtained from soil in the Truelove Lowland region of Canada (approximately 75.55°N, 84.67°W), was cultured on a sand and glucose-mineral agar medium exhibited psychrotrophic activity, growing more quickly at 2.5°C than at 5°C (Widden and Parkinson, 1978). On tomato agar and synthetic media, a total of 31 fungal isolates, primarily mitosporic species, were obtained from plant and soil samples collected on Macquarie Island (around 55.5°S, 158.95°E) and at Casey Station (approximately 64.28°S, 100.53°E) in Antarctica and thrived at 4°C, with the majority showing optimal growth at 15–20°C (Kerry, 1990). On malt extract and mycological peptone agar, *Humicola marvinii*, which was isolated from fellfield soils on Signy Island (60°43′S, 45°38′W), was shown to be psychrophilic. It grows at –2.5°C, reaching its maximum at 15°C, but not growing at 20–22°C (Weinstein et al., 1997).

Furthermore, although higher temperatures were not investigated, Latter and Heal (1971) reported that 14 fungal species isolated from soils on Signy Island were capable of growing at 1°C. In a separate study conducted in Victoria Land (approximately 72.5°S–77.87°S), 35 mitosporic fungal isolates demonstrated the ability to grow across a temperature range of 0°C–35°C. Among these, 33 isolates showed active growth at 5°C, and 31 were able to grow at 0°C, with optimal growth observed between 15–20°C on potato dextrose agar (Zucconi et al., 1996). Psychrotrophic *Phialophora hoffmannii* and *Cladosporium cladosporioides* were found in soil and litter in Point Barrow, in Alaska (approximately 71.28°N, 156.68°W), certain fungal strains displayed pectinase activity at 1°C, with peak enzymatic performance occurring between 18°C–20°C. Meanwhile, the psychrophilic fungus *Geomyces pannorum* demonstrated cellulase activity even at –4°C, with optimal activity recorded at 5°C (Flanagan and Scarborough, 1974). *Verticillium lecanii* was shown to have chitinase activity in continental Antarctica moss samples at temperatures of 5, 15, and 25°C (Fenice et al., 1998). A viability experiment on soil fungi from six Signy Island sites revealed that plates incubated at 10°C rather than 25°C produced more fungal colonies, indicating psychrotrophic tendencies (Bailey and Wynn-Williams, 1982).

Arctic soil organisms with high winter survival have a variety of ecophysiological modifications that make them resilient to challenging conditions (Hodkinson and Wookey, 1999). These organisms have a high survival rate as they undergo processes like dehydration, antifreeze synthesis, high supercooling activity, and chill tolerance. Many have freeze tolerance, which means they can survive ice formation inside their tissues. Additionally, behavioral modifications, such as choosing microhabitats with insulating snow cover, improve survival. Some species can also withstand anoxic conditions when coated in ice, facilitating their survival in severe Arctic habitats.

There were notable variations in the generation of sugar and polyol after eight weeks of incubation at 15°C at a 99.8% confidence level between *Humicola marvinii* isolated from soils of maritime Antarctica and *Humicola fuscoatra*, which was originally isolated from Gossypium (Weinstein et al., 1997). In contrast to *H. fuscoatra*, which only generated 0.51 mg/100 mg dry weight of mannitol and 4.51 mg/100 mg dry weight of trehalose, *H. marvinii* produced larger levels of both substances (41.07 mg/100 mg d.wt and 7.76 mg/100 mg d.wt, respectively). In contrast, *H. fuscoatra* generated just 0.18 mg/100 mg d.wt of glucose and did not syntheses detectable quantities of fructose, whereas *H. fuscoatra* had much greater amounts of both glucose (8.07 mg/100 mg d.wt) and fructose (5.03 mg/100 mg d.wt). Furthermore, in comparison, although *H. fuscoatra* generated just 0.11 mg/100 mg d.wt of erythritol and no detectable arabitol, *H. marvinii* showed higher amounts of both erythritol (0.27 mg/100 mg d.wt) and arabitol (0.21 mg/100 mg d.wt). On the other hand, *H. fuscoatra* produced more glycerol (0.80 mg/100 mg d.wt) than *H. marvinii* (0.35 mg/100 mg d.wt), indicating that the two fungi had different metabolic adaptations.

Fungi may thrive in Arctic and Antarctic environments by avoiding the cold rather than developing a tolerance to it. In freezing Arctic and maritime Antarctic regions, fungi rebuild from spores every spring or summer after the coldest season of the year. This is one way that fungi resist the cold. While soil temperatures in continental Antarctica can temporarily exceed air temperatures, instances of above-freezing air temperatures are uncommon. The average daily air temperature at inland locations seldom climbs above –10°C (Convey, 1996). As a result, this adaptation strategy may not be very effective there.

4. Ecological Perspective and Importance

A study by Newsham et al. (2016) predicted that fungal species richness is expected to increase by 20–27% by 2100 as a result of the present rapid warming in maritime Antarctica. The largest and most reliable predictor of soil fungal diversity was surface air temperature, which explained 23% and 39% of the variation between the predicted and actual OTU counts, respectively. Additionally, albeit to a lesser degree, the study found that the C:N ratio and soil potassium level were significant predictors of soil fungal diversity. The study found that soil fungal diversity is related to latitude, MASAT (Mean Annual Surface Air Temperature), and soil physicochemical characteristics, with complicated and indirect interactions between these factors.

Another study by Malard and Pearce (2018) discovered that Arctic soils have diverse microbial community compositions, with various locations having distinctive phyla and classes. Of them, cyanobacteria and planctomycetes plays a significant role in carbon fixation and nitrogen cycling due to the identification of N_2O production hotspots in the Arctic. This can be rated up to 34 mg N m^{-2} day^{-1}, and the significance of comprehending the distribution of methanotrophs and archaea in the area. The findings reveal that Arctic microbial communities are complex and changeable, with various locations having distinct community compositions, and that variables like pH, vegetation, and temperature have a major influence on community structure. The study's findings indicate that Arctic soils are rich in microbial diversity, with certain locations potentially having more diversity than other regions globally. Furthermore, the study revealed extremophiles and psychrotrophs as critical components of Arctic ecosystems, with potential implications in bioprospecting and industry.

Another study by Arenz and Blanchette (2011) shows that fungal abundance in Antarctic mineral soils is substantially impacted by nutrient availability, specifically carbon and nitrogen, with positive correlations of 0.55 and 0.46, respectively. Both soil moisture and the C/N ratio have a favorable effect on fungal abundance, although pH has a substantial negative connection ($r = -0.48$). Although there is no direct association between salinity and fungal communities, it may have an indirect effect on primary producers, which are important sources of bioavailable carbon in non-ornithogenic soils. Species richness is greatest on the Antarctic Peninsula, with the Ross Sea region and the McMurdo Dry Valleys following, reflecting the diversity of environmental conditions across these areas. These findings support the notion that limitations in plant-derived nutrients are a key factor influencing the distribution as well as abundance of fungi in Antarctic soils, aligning with patterns observed in other arid ecosystems.

Altogether, fungi inhabiting Arctic and Antarctic environments play vital ecological roles, including facilitating nutrient cycling, decomposing organic matter, and forming symbiotic associations with other organisms. Their distribution and variety are influenced by environmental conditions like temperature, carbon, and nitrogen availability, moisture, and pH. As climate change accelerates, particularly in the polar areas, fungal populations are projected to grow and diversify, impacting soil health, primary production, and ecosystem stability. Their versatility and ecological capabilities make them critical components of arctic ecosystems as well as useful environmental change indicators.

5. Fungal Applications

5.1 *Enzyme Production and Bioremediation*

As polar fungal strains adapted to extreme environmental conditions, these strains can potentially produce industrially important components, especially cold-adapted enzymes (Fig. 1; Table 3). In general, fungal enzymes act as nature's efficient recyclers, breaking down waste and pollutants to support both environmental health and industrial processes. Cellulase and xylanase convert plant waste into biofuels and improve sustainable textile and paper production (Abdulhussein et al., 2025; Chatterjee et al., 2015; Fendrihan et al., 2017; Ruggaber et al., 2006). Chitinase decomposes insect and crustacean waste, offering eco-friendly pest control (Krishnaven et al., 2014; Unuofin et al., 2024). Urease helps reduce nitrogen pollution by breaking down urea, while α-amylase cleans starchy residues from

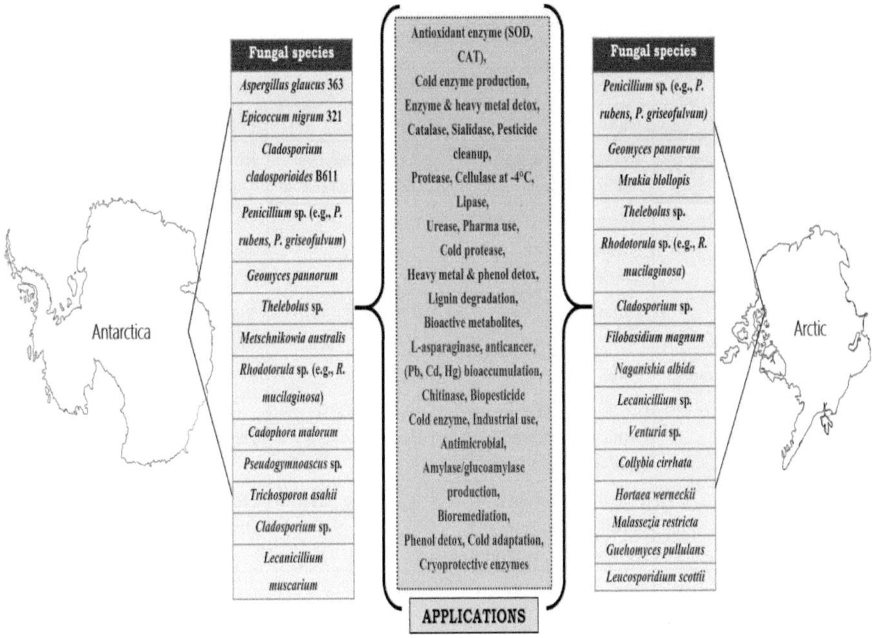

Fig. 1. Schematic representation of major industrially important fungal species isolated from polar regions.

food industry wastewater (Karimi et al., 2016). Fungal lignin peroxidase degrades wood-based pollutants and synthetic dyes, supporting cleaner industrial processes (Falade et al., 2017). Fungal keratinase targets tough proteins like those in feathers, hair, and wool, helping recycle poultry waste and textile fibers (Dipankar et al., 2019). Fungal lipase breaks down fats and oils, proving essential for treating oil spills, industrial wastewater, and grease accumulation (Kreling et al., 2020). For synthetic pollutants, fungal esterase dismantles plastics, insecticides, and herbicides, aiding in detoxifying contaminated soil (Bhatt et al., 2020).

Fungal species obtained from Livingston Island in Antarctica, such as *Aspergillus glaucus* 363, *Epicoccum nigrum* 321, *Cladosporium cladosporioides* B611, *C. oxysporum* 212, and *Penicillium* sp. strains were reported to produce cold-active antioxidant enzymes such as superoxide dismutase (SOD) and catalase (CAT). *A. glaucus* 363 showed the highest SOD activity at 25°C and CAT at 30°C, highlighting its adaptation to cold and potential for biotechnological use in oxidative stress applications (Tosi et al., 2010). A study by Krishnan et al. (2011) examined 28 fungal strains isolated from Antarctic habitats to evaluate their extracellular enzyme activities. Among these, 25% of the isolates showed good amylase activity, 25% displayed cellulase activity, and 43% demonstrated protease activity. Notably, *Mrakia frigida* exhibited the highest cellulase activity, while *Geomyces pannorum* was the most potent protease producer. Most sexual-stage ascomycetes lacked enzymatic activity, but *Ascomycete* sp. and *Thelebolus* sp. were exceptions, showing both amylase and protease activity. These results highlight the enzymatic diversity of Antarctic fungi and their potential role in cold-environment biotechnological

Table 3. Some of the notable enzymes extracted from Arctic and Antarctica fungal strains.

Fungal Species	**Source**	**Enzyme(s)**	**References**
Arthrobotrys ferox	Antarctica	Urease	Fenice et al. (1998)
Cadophora malorum	Arctic tundra and Antarctic soils	Keratinase, lignin peroxidase	Robicheau et al. (2019); Duarte et al. (2018)
Candida antarctica	Antarctica ice cores, permafrost, and marine samples	Lipase	Wentzel et al. (2018); Martorell et al. (2019); Tsuji et al. (2013)
Candida spencermartinsiae	Antarctica	Cellulase	Vaca et al. (2012); Vaz et al. (2011); Wang et al. (2013)
Cladosporium cladosporioides	Antarctica	Urease	Fenice et al. (1998)
Cladosporium sp. F09-T12-1	Antarctic marine sources	Pectinase	Henriquez et al. (2013); Vaz et al. (2011)
Cladosporium spp.	Antarctic lichens, marine sponges, continental Antarctica	Xylanase, urease	Del-Cid et al. (2013); Bradner et al. (1999)
Cryptococcus victoriae	Antarctic marine sources	Pectinase	Henriquez et al. (2013); Vaz et al. (2011)
Cystobasidium larynges	Mayo Island and Antarctica	Esterases	Martorell et al. (2017)
Cystofilobasidium infirmominiatum	Antarctic seawater	α-amylase	Bharudin et al. (2018); Vaca et al. (2013)
Geomyces spp.	Antarctica ice cores, permafrost, and marine samples	Lipase	Wentzel et al. (2018); Martorell et al. (2019); Tsuji et al. (2013)
Glaciozyma antarctica PI12	Antarctic seawater	α-amylase	Bharudin et al. (2018); Vaca et al. (2013)
Lecanicillium muscarium CCFEE 500	Polar soils	Chitinases, proteases	Fenice et al. (2016)
Leucosporidiella creatinivora	Antarctica	Cellulase, Pectinase	Vaca et al. (2012); Vaz et al. (2011); Henriquez et al. (2013)
Leucosporidiella muscorum	Antarctic marine sources	Pectinase	Henriquez et al. (2013); Vaz et al. (2011)
Leucosporidiella sp.	Antarctic seawater	α-amylase	Bharudin et al. (2018); Vaca et al. (2013)
Leucosporidiella spp.	Antarctic marine sources	Protease	Wentzel et al. (2018); Vaz et al. (2011)
Leucosporidium scottii	Antarctica	Cellulase, Pectinase	Vaca et al. (2012); Vaz et al. (2011); Henriquez et al. (2013)
Metschnikowia australis	Antarctica	Cellulase, Pectinase	Vaca et al. (2012); Vaz et al. (2011); Wang et al. (2013); Henriquez et al. (2013)

Table 3 contd. ...

...Table 3 contd.

Fungal species	**Source**	**Enzyme(s)**	**References**
Metschnikowia spp.	Antarctica ice cores, permafrost, and marine samples	Lipase	Wentzel et al. (2018); Martorell et al. (2019); Tsuji et al. (2013)
Mrakia blollopis	Antarctica ice cores, permafrost, and marine samples	Lipase	Wentzel et al. (2018); Martorell et al. (2019); Tsuji et al. (2013)
Neosetophoma samarorum	Arctic tundra and Antarctic soils	Keratinase, lignin peroxidase	Robicheau et al. (2019); Duarte et al. (2018)
Penicillium expansum	Antarctic lichens, marine sponges, continental Antarctica	Xylanase	Del-Cid et al. (2013); Bradner et al. (1999)
Penicillium chrysogenum	Antarctic lichens, marine sponges, continental Antarctica	Xylanase	Del-Cid et al. (2013); Bradner et al. (1999)
Phoma sp. 2	Antarctic lichens, marine sponges, continental Antarctica	Xylanase	Del-Cid et al. (2013); Bradner et al. (1999)
Pichia caribbica (*P. caribbica*)	Antarctica ice cores, permafrost, and marine samples	Lipase	Wentzel et al. (2018); Martorell et al. (2019); Tsuji et al. (2013)
Pseudogymnoascus sp.	Antarctic marine sources	Pectinase	Henriquez et al. (2013); Vaz et al. (2011)
Pseudogymnoascus sp. CRM1533	Antarctic marine sources	Protease	Wentzel et al. (2018); Vaz et al. (2011)
Rhodotorula glacialis	Antarctica	Protease, Cellulase, Pectinase	Wentzel et al. (2018); Vaz et al. (2011); Vaca et al. (2012); Wang et al. (2013); Henriquez et al. (2013)
Rhodotorula mucilaginosa L7	Antarctic marine sources	Protease	Wentzel et al. (2018); Vaz et al. (2011)
Verticillium cfr. lecanii A3	Antarctica	Chitinase, urease	Fenice et al. (1998)
Verticillium spp.	Antarctica	Cellulase	Vaca et al. (2012); Vaz et al. (2011); Wang et al. (2013)

applications. Cryoconite holes in Svalbard glaciers host cold-adapted fungi with enzymatic potential. The isolates comprised *Rhodotorula* sp., *Thelebolus* sp., and a newly identified strain of *Articulospora tetracladia*, exhibiting 88.7–89.4% similarity in the 5.8S rDNA region. The *Articulospora* isolate demonstrated enzymatic activities, including amylase, cellulase, lipase, and protease. *Rhodotorula* sp. showed both amylase and cellulase activity, while *Thelebolus* sp. exhibited protease and urease activity. Growth was optimal at 10°C, highlighting their relevance in cold-environment biotechnology (Singh et al., 2016). Antarctic freshwater fungi exhibit notable cold-active enzymatic activity.

From Hennequin and Kroner Lakes, 154 culturable isolates were obtained, spanning *Ascomycota*, *Basidiomycota*, and *Mortierellomycota*. Dominant species included *Thelebolus globosus*, *Gofeauzyma* sp., *Pseudogymnoascus verrucosus*, and *Metschnikowia australis*. These fungal isolates produced diverse cold-active enzymes, notably protease and pectinase. In addition, a variety of enzymes, including amylase, cellulase, lipase, invertase, agarase, carrageenase, esterase, inulinase, and gelatinase were also identified in polar fungal isolates. Further, *Leucosporidium fragarium* (inulinase), *Holtermanniella wattica* (invertase, esterase), *Mrakia blollopis* (cellulase, pectinase), *Metschnikowia australis* (protease), *Pseudohyphozyma verrucosa* (agarase, carrageenase), and *Goffeauzyma* sp. (amylase) were found to produce significant quantities of the extracellular enzymes (De Souza et al., 2023). Protease degrades proteins from organic wastes such as decaying plants and food scraps, making it valuable for marine cleanup, particularly in cold environments (Razzaq et al., 2019).

Ashok et al. (2019) reported L-asparaginase production in 30 filamentous fungal isolates from the Schirmacher Hills region of Antarctica. Of them, *Trichosporon asahii* IBBLA1 demonstrated the highest level of enzymatic activity. Duarte et al. (2018a) conducted a study on King George Island in Maritime Antarctica, where they identified 129 fungal isolates from both marine and terrestrial sources. Of these, 69% were assigned to 12 different genera, including *Penicillium*, *Cladosporium*, *Geomyces*, *Cosmospora*, and *Cadophora*. *Cosmospora* was exclusive to terrestrial samples, while others were found in both habitats. Enzyme screening revealed that *Cadophora* produced lignin peroxidase and laccase at 15°C, while *Penicillium* and unidentified fungi produced xylanase and L-asparaginase at 20°C. In total, 29 isolates produced ligninolytic enzymes, 19 xylanase, and 74 L-asparaginase. These findings highlight the biotechnological potential of cold-adapted Antarctic fungi. The fungal community of Arctic moss *Tetraplodon mimoides* was studied using fresh thallus and adhered soil, yielding 46 isolates from 20 species across 12 genera, including seven non-sporulating forms.

Species such as *Botrytis verrucosa*, *Mortierella simplex*, *Mrakia schmücken*, *Penicillium frequentans*, *Penicillium rugulosum*, and *Cladosporium chlorocephalum* were newly recorded in the region. Enzymatic screening revealed cold-adapted activity in seven isolates for amylase, 25 for pectinase, 28 for cellulase, 13 for esterase, 16 for protease, 26 for phosphatase, and 25 for urease. Notably, *P. citrinum* strain 2 and NSM 1 exhibited activity for all enzymes tested, while *P. citrinum* strain 5 was positive for all except urease (Sakhalkar et al., 2011). Da Silva et al. (2024) investigated culturable fungal communities from deep-sea sediments in Maritime Antarctica, revealing significant taxonomic diversity and biotechnological potential. Among the 68 Ascomycota isolates identified, *Penicillium chrysogenum*, *Antarctomyces psychrotrophicus*, *Cladosporium* sp., *Pseudeurotium cf. bakeri*, *Meyerozyma guilliermondii*, *Thelebolus balaustiformis*, and *Penicillium cf. palitans* were among the most abundant. Functional screening showed the production of various hydrolytic exoenzymes, with the most frequent being cellulases, followed by proteases, lipases, pectinases, invertases, agarases, esterases, and carrageenases.

Penicillium rubens strain III11-2, isolated from Antarctica, was identified as an efficient producer of extracellular catalase, an important antioxidant enzyme involved in protecting filamentous fungi from low-temperature stress (Koleva et al., 2024). *Cladosporium* sp. FAR18, isolated from Antarctic marine sediments, produced laccase, a ligninolytic enzyme (de Melo Carlos et al., 2024). Dolashki et al. (2024) found that *Penicillium griseofulvum* P29, isolated from Terra Nova Bay soil samples, is a notable producer of sialidase. Filamentous fungal genera such as *Pseudogymnoascus*, *Pseudeurotium*, and *Antarctomyces* from Boeckella Lake sediments of Hope Bay had exhibited significant enzymatic activity, predominantly producing esterase, cellulase, and protease, indicating their potential for industrial applications (Teixeira et al., 2024). Further, Antarctica strains of *Hyphozyma variabilis* and *Thelebolus ellipsoideus* showed high esterase activity, while *Ascochyta pisi* exhibited strong cellulolytic activity (Nikitin et al., 2022). These findings clearly demonstrate the potential of Arctic and Antarctica fungi to produce industrially important enzymes.

In harsh conditions, fungi demonstrate significant resilience. Their ability to metabolize toxic compounds highlights their potential for environmental cleanup and biotechnological exploitation. The ability of the local psychrophilic microbial community in cold climates to degrade organic pollutants at low temperatures biodegradation of these toxins (Margesin and Schinner, 2011). Fungi from Arctic and Antarctic regions offer a valuable reservoir of cold-adapted enzymes with applications in bioremediation (Hassan et al., 2016; Duarte et al., 2018).

According to the Arrhenius equation, reaction rates in low temperatures are significantly reduced because of higher viscosity and lower kinetic energy (Marx et al., 2007), which necessitates enzymes with higher catalytic efficiency (lower activation energy) under cold conditions. Enzymes that are cold-adapted reduce undesirable side effects and make it possible to process heat-sensitive substrates (Jeon et al., 2008). Bioremediation employs organisms or their enzymes to detoxify or remove environmental pollutants. The primary goal of bioremediation is the complete mineralization of pollutants to CO_2 and H_2O. However, transformation to less toxic intermediates is also acceptable (Margesin et al., 2001). By harnessing polar-origin enzymes, we can turn waste into resources, reduce pollution, and create more sustainable industrial practices.

5.2 *Biopharmaceuticals*

Polar fungal isolates were also reported for the production of several biologically important pharmaceutical molecules. Antarctic and Arctic fungi are emerging as promising sources of such novel antimicrobials. A study by Shi et al. (2021) identified terpenoids and steroids from the Antarctic fungus *Pseudogymnoascus* sp. HSX2#-11. This particular strain produces unique secondary metabolites, several steroidal compounds and six exclusive tremulane sesquiterpenoid pseudotremulanes. Among these, compounds have shown cytotoxic action against cell lines of hepatoma (HepG2), colorectal cancer (HCT116), and human breast cancer (MDAMB-231). Whereas harmful pathogen *Aeromonas salmonicida* was susceptible to the antibacterial action of some compounds. Antarctic fungal strains, including *Acrostalagmus luteoalbus* CH-6, *Chrysosporium* sp. HSXSD-11-1, and *Cladosporium* sp. HSXSD-12, exhibited notable antimicrobial activity, with CH-6 showing the strongest inhibition

(72.3–84.8%) against *Candida albicans* and *Aeromonas salmonicida*. Through bioactivity-guided fractionation and molecular networking of *Acrostalagmus luteo labus* CH-6, two new α-pyrones, acrostalapyrones A (1) and B (2), along with one known analog (3) and three known indole diketopiperazines, were identified. Two of the compounds demonstrated potent antimicrobial activity against *Candida albicans* and *Aeromonas salmonicida*, with one compound proving to be twice as effective as ciprofloxacin against *A. salmonicida* (Shi et al., 2022). Fungal samples from the Svalbard Arctic region (73–80°N) yielded 22 OTUs from 33 isolates, predominantly belonging to Ascomycota. Of them, 10 showed antibacterial activity, and nine may be novel species. These findings demonstrate that the Arctic is a potential source of marine fungi that are able to produce bioactive components (Hagestad et al., 2020). Soil fungi from Greenwich Island, Antarctica, including *Antarctomyces* sp., *Thelebolus* sp., *Penicillium* sp., and *Cryptococcus gilvescens*, showed varied antibacterial activity. *C. gilvescens* and *Penicillium* sp. inhibited the growth of several clinically pathogenic strains, including *Escherichia coli*, *Klebsiella pneumoniae*, *Enterococcus faecalis*, and *Staphylococcus aureus*. These findings demonstrate the potential of Antarctica as a source of new antibacterial substances (Enireb et al., 2022). Under the MICROMAT research, fungi from Antarctica's benthic microbial mats showed promising medicinal properties. Of 160 filamentous fungi, 29% exhibited antimicrobial activity, mainly against *S. aureus*, *E. coli*, *C. albicans*, and *C. neoformans*. LC–MS (Liquid Chromatography-Mass Spectrometry) studies identified rugulosin and skyrin from Antarctic *Penicillium chrysogenum*, while *Aspergillus* sp. yielded chemically novel antimicrobial compounds (Brunati et al., 2009).

5.3 *Heavy Metal Detoxification*

Several fungal and yeast strains have been reported for very strong heavy metal detoxification characteristics. Notably, *Cladosporium* sp., isolated from Antarctic rocks and lichens, was reported to detoxify heavy metals such as lead (Pb), cadmium (Cd), and mercury (Hg) and was found to form stable fungal-metal complexes (Cudowski et al., 2019). The Antarctica fungal strains *Penicillium* cf. *oxalicum* LAMAI 2400 and LAMAI 2402 exhibited a remarkable capacity to decolorize sulphur indigo blue dye even at elevated concentrations at both 15°C and 28°C, underscoring their potential for application in textile dye bioremediation under low to moderate temperature conditions (Kita et al., 2022). *Penicillium* strains isolated from Antarctic glaciers and soil samples were reported for the production of esterases and peroxidases, which break down chlorinated pesticides in cold soil.

Cold-active yeasts such as *Metschnikowia*, *Candida*, *Cryptococcus*, and *Rhodotorula* were reported to assimilate phenol at 2.5 mM concentrations, which in fact offers a potential solution to phenol contamination in industrial wastewater (Martorell et al., 2019). Another polar yeast strain, *Rhodotorula mucilaginosa* YR29, had exhibited high tolerance to heavy metals and functions in biosorption, microcapsulation, and bioaccumulation of $Pb(NO_3)_2$ (Angeles et al., 2023). In addition, psychrotrophic yeasts isolated from Antarctica have shown potential as biocontrol agents for post-harvest disease management. *Leucosporidium scottii* was recognized as an effective agent against blue and gray mold in two apple cultivars (Vero et al., 2013). Additionally, *Acremonium fusidioides* and *Penicillium verrucosus*, obtained from deep-sea sediments near the South Shetland Islands, Antarctica, display strong selective herbicidal properties (Ogaki et al., 2020).

6. Future Perspectives

Enzymes of polar fungi, which are active at low temperatures, can revolutionize industries such as food processing, detergents, leather processing, and textile production by facilitating processes at low temperatures, saving energy, and conserving heat-sensitive substrates. Pulp and paper industries can use lignin-degrading enzymes like laccases and peroxidases for eco-friendly bleaching and in production. Cold-active lipases and esterases find applications in biodiesel synthesis and organic synthesis within pharmaceutical companies. Polar fungal biopolymers and biosurfactants have potential applications in cosmetics, food packaging, and biodegradable plastics (Hamid et al., 2014; Kumar et al., 2023; Holguín-Salas et al., 2025; Margesin and Schinner, 2001; Marx et al., 2007). Polar fungi are also of immense biotechnological interest as bioremediation tools for cold-contaminated ecosystems, e.g., oil spills in polar ecosystems, textile dye contaminants, and heavy metals in soil and water (Margesin and Schinner, 2001; Martorell et al., 2019). They can be investigated as biocontrol agents and biofertilizers in alpine and cold agriculture, enhancing nutrient uptake, plant growth, and disease resistance. Chitinases and proteases may be employed in pest management and composting acceleration under cold stress (Nogi, 2017; Unuofin et al., 2024).

New secondary metabolites like tremulane sesquiterpenoids and diketopiperazines from Antarctic fungi provide lead compounds for antimicrobial and anticancer drug discovery. Cold-adapted fungi are able to synthesize extremolytes (stress protectants such as trehalose and mannitol) that can be utilized in cosmetics, cryopreservation, or therapeutic preparations (Ordóñez-Enireb et al., 2022; Shi et al., 2022).

In addition, polar fungi can be used as bioindicators of environmental change since their abundance and diversity are directly related to temperature and soil conditions (Newsham et al., 2016). Investigation of their adaptive mechanisms is insightful determining the way life could survive in extreme environments on Earth and other planets, contributing to astrobiological studies.

There is a need for further comprehensive genomic and transcriptomic analyses of unknown fungal species to determine gene clusters accountable for stress tolerance and secondary metabolism (Rosa et al., 2020). Functional metagenomics and synthetic biology can assist in utilizing polar fungal enzymes without culturing the organism. Long-term observation of fungal diversity in polar ecosystems will be key to appreciating the effects of global warming (Post et al., 2019). Interdisciplinary strategies, combining microbiology, ecology, climatology, and biotechnology, are necessary to unlock the full potential of these extremophilic fungi.

7. Conclusion

Despite harsh climates and presence of the least amount of primary producers, polar regions host diverse amounts of fungal species as reported through extensive isolation records and amplicon sequencing data from both Arctic and Antarctic environments. Tables provided in this chapter document the widespread presence of different genera like *Cadophora*, *Cladosporium*, *Penicillium*, *Mortierella*, *Mrakia*, *Metschnikowia*, *Rhodotorula*, and *Thelebolus* found across lakes, glacial habitats, soil, marine invertebrates, and cryptoendolithic niches.

Sequencing studies also reported dominance of Ascomycota and Basidiomycota dominance in polar fungal communities, but notable significant diversity is also found to be contributed by Chytridiomycota, Mortierellomycota, Mucoromycota, and occasionally Rozellomycota groups. ASVs in large amounts were found in studies from over 100 amplicons to 6,000 for each study, showing the vast diversity hidden in such extreme environments.

Functionally, the capacity of producing different ranges of oxidative and hydrolytic enzymes (cellulases, lipases, ureases, xylanases, lignin peroxidase) active at low temperatures is exhibited by culturable polar fungi. These genera: *Cadophora*, *Candida*, *Cladosporium*, *Geomyces*, *Glaciozyma*, *Lecanicillium*, *Leucosporidium*, *Metschnikowia*, *Penicillium*, *Pseudogymnoascus*, *Rhodotorula*, and *Verticillium* are majorly found for producing above mentioned enzymes. It underscores the potential of these strains in the biotechnological field and their industrial applications.

In conclusion, it can be confirmed that polar fungal communities are diverse taxonomically, significant ecologically, functionally adaptable well documented culturable forms and a broad uncultured component identified through sequencing. Diversity and proven enzyme production positions polar fungi as important contributors to nutrient cycling as well as important sources for novel biotechnological innovations focusing on cold environments.

Acknowledgement

The authors wish to thank the Director, CSIR-NEIST, Jorhat, Assam, for his kind encouragement in carrying out this work. CSIR-NEIST Manuscript Approval No.: CSIR-NEIST/PUB/2025/090. The selection of Dr. Natarajan Velmurugan for participation in the 43rd Indian Scientific Expedition to Antarctica (43rd ISEA; 43-BES/ETN-01(3)) by NCPOR (MoES)-Goa is highly appreciated.

References

Abdulhussein, H.M., Muslim, S.N. and Muslem, W.H. (2025). Bioremediation of petroleum hydrocarbon contaminated soil by xylanase enzyme. Adv. Life Sci., 12: 255–259. https://doi.org/10.62940/als.v12i1.3125.

AMAP (2023). Arctic Monitoring and Assessment Program: Climate change update. Retrieved from: https://www.amap.no.

Angeles de Paz, G., Martínez-Gutierrez, H., Ramírez-Granillo, A., López-Villegas, E.O., Medina-Canales, M.G. et al. (2023). Rhodotorula mucilaginosa YR29 is able to accumulate Pb2+ in vacuoles: A yeast with bioremediation potential. World J. Microbiol. Biotechnol., 39: 238. https://doi.org/10.1007/s11274-023-03675-4.

Arenz, B.E. and Blanchette, R.A. (2011). Distribution and abundance of soil fungi in Antarctica at sites on the Peninsula, Ross Sea region and McMurdo dry valleys. Soil Biol. Biochem., 43: 308–315. https://doi.org/10.1016/j.soilbio.2010.10.016.

Arenz, B.E., Held, B.W., Jurgens, J.A., Farrell, R.L. and Blanchette, R.A. (2006). Fungal diversity in soils and historic wood from the Ross Sea region of Antarctica. Soil, Biol. Biochem., 38: 3057–3064. https://doi.org/10.1016/j.soilbio.2006.01.016.

Arthur, H. and Watson, K. (1976). Thermal adaptation in yeast: Growth temperatures, membrane lipid, and cytochrome composition of psychrophilic, mesophilic, and thermophilic yeasts. J. Microbiol., 128: 56–68. https://doi.org/10.1128/jb.128.1.56-68.1976.

Ashok, A., Doriya, K., Rao, J.V., Qureshi, A., Tiwari, A.K. et al. (2019). Microbes producing L-asparaginase free of glutaminase and urease isolated from extreme locations of Antarctica soil and moss. Sci. Rep., 9: 1423. https://doi.org/10.1038/s41598-018-38094-1.

Bailey, A.D. and Wynn-Williams, D.D. (1982). Soil microbiological studies at Signy Island, South Orkney islands. Brit. Antarct. Surv. Bullet., 51: 167–191. https://nora.nerc.ac.uk/id/eprint/524258.

Bergero, R., Girlanda, M., Varese, G.C., Intili, D. and Luppi, A.M. (1999). Psychrotolerant fungi from Arctic soils of Franz Joseph Land. Polar Biol., 21: 361–368. https://doi.org/10.1007/s003000050374.

Bharudin, I., Bakar, M.F.A., Hashim, N.H.F., Isa, M.N.M., Alias, H. et al. (2018). Unravelling the adaptation strategies employed by *Glaciozyma Antarctica* PI12 on Antarctica sea ice. Mar. Environ. Res., 169–176. https://doi.org/10.1016/j.marenvres.2018.03.007.

Bhatt, P., Bhatt, K., Huang, Y., Lin, Z. and Chen, S. (2020). Esterase is a powerful tool for the biodegradation of pyrethroid insecticides. Chemosphere, 244: 125507. https://doi.org/10.1016/j.chemosphere.2019.125507.

Bindschadler, R. (2006). The environment and evolution of the West Antarctica ice sheet: Setting the stage. Philos. Trans. R. Soc. A, and Philos. Transact. A Math. Phys. Eng. Sci., 364: 1583–1605. https://doi.org/10.1098/rsta.2006.1790.

Borzęcka, J., Suchodolski, J., Dudek, B., Matyaszczyk, L., Spychała, K. et al. (2022). The first comprehensive biodiversity study of culturable fungal communities inhabiting cryoconite holes in the Werenskiold Glacier on Spitsbergen (Svalbard Archipelago, Arctic). Biol., 11: 1224. https://doi.org/10.3390/biology11081224.

Bradner, J.R., Sidhu, R.K., Gillings, M. and Nevalainen, K.M.H. (1999). Hemicellulase activity of Antarctica microfungi. J. Appl. Microbiol., 87: 366–370. https://doi.org/10.1046/j.1365-2672.1999.00827.x.

Brunati, M., Rojas, J.L., Sponga, F., Ciciliato, I., Losi, D. et al. (2009). Diversity and pharmaceutical screening of fungi from benthic mats of Antarctica lakes. Mar. Genomics, 2: 43–50. https://doi.org/10.1016/j.margen.2009.04.002.

Bulygina, O.N., Groisman, P.Y., Razuvaev, V.N. and Korshunova, N.N. (2011). Changes in snow cover characteristics over Northern Eurasia since 1966. Environ. Res. Lett., 6: 045204. https://doi.org/10.1088/1748-9326/6/4/04520.

Cantrell, S.A., Dianese, J.C., Fell, J., Gunde-Cimerman, N., Zalar, P. et al. (2011). Unusual fungal niches. Mycol., 103: 1161−1174. https://doi.org/10.3852/11-108.

Cavicchioli, R., Siddiqui, K.S., Andrews, D. and Sowers, K.R. (2002). Low-temperature extremophiles and their applications. Curr. Opin. Biotechnol., 13: 253–261. https://doi.org/10.1016/S0958-1669(02)00317-8.

Climate Change (2007). Climate Change 2007: The physical science basis. Agenda 6: 333. https://www.ipcc.ch/site/assets/uploads/2020/02/ar4-wg1-sum_vol_en.pdf.

Chatterjee, S., Sharma, S., Prasad, R.K., Datta, S., Dubey, D. et al. (2015). Cellulase enzyme based biodegradation of cellulosic materials: An overview. Cellulose, 5: 271–282. https://doi.org/10.38150/sajeb.

Convey, P. (1996). The influence of environmental characteristics on life history attributes of Antarctica terrestrial biota. Biol. Rev., 71: 191–225. https://doi.org/10.1111/j.1469-185X.1996.tb00747.x.

Convey, P. (2017). Antarctica ecosystems. pp. 179–188. *In*: Encyclopedia of Biodiversity (2nd Edn.) Waltham, MA, Academic Press. https://www.bas.ac.uk/data/our-data/publication/antarctic-ecosystems/.

Convey, P. and Peck, L.S. (2019). Antarctic environmental change and biological responses. Science Advances, 5(11): eaaz0888. https://doi.org/10.1126/sciadv.aaz0888.

Cudowski, A. and Pietryczuk, A. (2019). Biochemical response of *Rhodotorula mucilaginosa* and *Cladosporium herbarum* isolated from aquatic environment on iron (III) ions. Sci. Rep., 9: 19492. https://doi.org/10.1038/s41598-019-56088-5.

da Silva, M.K., Barreto, D.L.C., Vieira, R., Neto, A.A., de Oliveira, F.S. et al. (2024). Diversity and enzymatic, biosurfactant and phytotoxic activities of culturable Ascomycota fungi present in marine sediments obtained near the South Shetland Islands, maritime Antarctica. Extremophiles, 28: 20. https://doi.org/10.1007/s00792-024-01336-4.

da Silva, M.K., de Souza, L.M.D., Vieira, R., Neto, A.A., Lopes, F.A. et al. (2022). Fungal and fungal-like diversity in marine sediments from the maritime Antarctica assessed using DNA metabarcoding. Sci. Rep., 12: 21044. https://doi.org/10.1038/s41598-022-25310-2.

da Silva, T.H., Câmara, P.E., Pinto, O.H.B., Carvalho-Silva, M., Oliveira, F.S. et al. (2022). Diversity of fungi present in permafrost in the South Shetland Islands, Maritime Antarctica. Microb. Ecol., 83(1): 58–67. https://doi.org/10.1007/s00248-021-01735-6.

de Garcia, V., Brizzio, S. and van Broock, M.R. (2012). Yeasts from glacial ice of Patagonian Andes, Argentina. FEMS Microbiol. Ecol., 82: 540–550. https://doi.org/10.1111/j.1574-6941.2012.01470.x.

de Melo Carlos, L., Camacho, K.F., Duarte, A.W., de Oliveira, V.M., Boroski, M. et al. (2024). Bioprospecting the potential of the microbial community associated to Antarctica marine sediments for hydrocarbon bioremediation. Braz. J. Microbiol., 55: 471–485. https://doi.org/10.1007/s42770-023-01199-5.

de Menezes, G.C.A., Câmara, P.E., Pinto, O.H.B., Carvalho-Silva, M., Oliveira, F.S. et al. (2021). Fungal diversity present on rocks from a polar desert in continental Antarctica assessed using DNA metabarcoding. Extremophiles, 25: 193–202. https://doi.org/10.1007/s00792-021-01221-4.

de Souza, L.M., Ogaki, M.B., Câmara, P.E., Pinto, O.H., Convey, P. et al. (2021). Assessment of fungal diversity present in lakes of Maritime Antarctica using DNA metabarcoding: A temporal microcosm experiment. Extremophiles, 25: 77–84. https://doi.org/10.1007/s00792-020-01212-x.

de Souza, L.M.D., Convey, P., Lirio, J.M. and Rosa, L.H. (2022). Diversity of freshwater fungi in polar and alpine lakes. Freshw. Mycol., 2022: 37–58. https://doi.org/10.1016/B978-0-323-91232-7.00013-1.

de Souza, L.M.D., Ogaki, M.B., Teixeira, E.A.A., De Menezes, G.C.A., Convey, P. et al. (2023). Communities of culturable freshwater fungi present in Antarctica lakes and detection of their low-temperature-active enzymes. Braz. J. Microbiol., 54: 1923–1933. https://doi.org/10.1007/s42770-022-00834-x.

Del-Cid, A., Ubilla, P. and Ravanal, M.C. (2014). Cold-active xylanase produced by fungi associated with Antarctica marine sponges. Appl. Biochem. Biotechnol., 72: 524–532. https://doi.org/10.1007/s12010-013-0551-1.

Dipankar, P. and Bhan, C. (2019). Role of keratinase in bioremediation of feathers and hairs. pp. 83–98. *In*: Bhatt, P. (ed.). Smart Bioremed Tech. Academic Press. https://doi.org/10.1016/B978-0-12-818307-6.00005-6.

Dolashki, A., Abrashev, R., Kaynarov, D., Krumova, E., Velkova, L. et al. (2024). Structural and functional characterization of cold-active sialidase isolated from Antarctica fungus *Penicillium griseofulvum* P29. Biochem. Biophys. Rep., 37:101610. https://doi.org/10.1016/j.bbrep.2023.101610.

Duarte, A.W.F., Barato, M.B., Nobre, F.S., Polezel, D.A., de Oliveira, T.B. et al. (2018). Production of cold-adapted enzymes by filamentous fungi from King George Island, Antarctica. Polar. Biol., 41: 2511–2521. https://doi.org/10.1007/s00300-018-2387-1.

Duarte, A.W.F., Dos Santos, J.A., Vianna, M.V., Vieira, J.M.F., Mallagutti, V.H. et al. (2018a). Cold-adapted enzymes produced by fungi from terrestrial and marine Antarctica environments. Crit. Rev. Biotechnol., 38: 600–619. https://doi.org/10.1080/07388551.2017.1379468.

Falade, A.O., Nwodo, U.U., Iweriebor, B.C., Green, E., Mabinya, L.V. et al. (2017). Lignin peroxidase functionalities and prospective applications. Microbiol. Open, 6: e00394. https://doi.org/10.1002/mbo3.394.

Fendrihan, S. and Negoiţă, T.G. (2017). Psychrophilic microorganisms as important source for biotechnological processes. pp. 147–199. *In*: Stan-Lotter, H. and Fendrihan, S. (eds.). Adaption of Microbial Life to Environmental Extremes. Springer, Cham. https://doi.org/10.1007/978-3-319-48327-6_7.

Fenice, M. (2016). The psychrotolerant Antarctica fungus *Lecanicillium muscarium* CCFEE 5003: A powerful producer of cold-tolerant chitinolytic enzymes. Molecules, 21: 447–460. https://doi.org/10.3390/molecules21040447.

Fenice, M., Selbmann, L., Di Giambattista, R. and Federici, F. (1998). Chitinolytic activity at low temperature of an Antarctica strain (A3) of *Verticillium lecanii*. Microbiol. Res., 149: 289–300. https://doi.org/10.1016/S0923-2508(98)80304-5.

Flanagan, P.W. and Scarborough, A.M. (1974). Physiological Groups of Decomposer Fungi on Tundra Plant Remains. Soil Organisms and Decomposition in Tundra. Tundra Biome Steering Committee, Stockholm, pp.159–181.

Gawas-Sakhalkar, P. and Singh, S.M. (2011). Fungal community associated with Arctic moss, *Tetraplodon mimoides*, and its rhizosphere: Bioprospecting for production of industrially useful enzymes. Curr. Sci., 100(11): 1701–1705. https://www.jstor.org/stable/24077775.

Gildea, D. and Splettstoesser, J. (2007). Craddock Massif and Vinson Massif remeasured. In Antarctica: A Keystone in a Changing World–Online Proceedings of the 10th ISAES, edited by AK Cooper and CR Raymond et al., USGS Open-File Report (Vol. 1047).

Godinho, V.M., de Paula, M.T.R., Silva, D.A.S., Paresque, K., Martins, A.P. et al. (2019). Diversity and distribution of hidden cultivable fungi associated with marine animals of Antarctica. Fungal Biol., 123: 507–516. https://doi.org/10.1016/j.funbio.2019.05.001.

Gonçalves, V.N., de Souza, L.M.D., Lirio, J.M., Coria, S.H., Lopes, F.A.C. et al. (2022). Diversity and ecology of fungal assemblages present in lake sediments at Clearwater Mesa, James Ross Island, Antarctica, assessed using metabarcoding of environmental DNA. Fungal Biol., 126: 640–647. https://doi.org/10.1016/j.funbio.2022.08.002.

Gonçalves, V.N., Vaz, A.B., Rosa, C.A. and Rosa, L.H. (2012). Diversity and distribution of fungal communities in lakes of Antarctica. FEMS Microb. Ecol., 82: 459–471. https://doi.org/10.1111/j.1574-6941.2012.01424.x.

Grum-Grzhimaylo, O.A., Shurigina, A.A., Debets, A.J. and Aanen, D.K. (2024). Biogeography and uniqueness of filamentous terrestrial fungi in the polar regions. Fungal Biol. Rev., 49: 100382. https://doi.org/10.1016/j.fbr.2024.100382.

Hagestad, O.C., Andersen, J.H., Altermark, B., Hansen, E. and Rämä, T. (2020). Cultivable marine fungi from the Arctic Archipelago of Svalbard and their antibacterial activity. Mycol., 11: 230–242. https://doi.org/10.1080/21501203.2019.1708492.

Hamid, B., Rana, R.S., Chauhan, D., Singh, P., Mohiddin, F.A. et al. (2014). Psychrophilic yeasts and their biotechnological applications—A review. Afr. J. Biotechnol., 13(22). https://doi.org/10.4314/ajb.v13i22.

Hassan, N., Rafiq, M., Hayat, M., Shah, A.A. and Hasan, F. (2016). Psychrophilic and psychrotrophic fungi: A comprehensive review. Rev. Environ. Sci. Biotechnol., 15: 147–172. https://doi.org/10.1007/s11157-016-9395-9.

Hawksworth, D.L. (1990). *Globosphaeria*, a remarkable new pyrenomycete on *Normandina* from Tasmania. Lichenologist, 22: 301–305. https://doi.org/10.1017/S0024282990000330.

Henríquez, M., Vergara, K., Norambuena, J., Beiza, A., Maza, F. et al. (2014). Diversity of cultivable fungi associated with Antarctica marine sponges and screening for their antimicrobial, antitumoral, and antioxidant potential. World J. Microbiol. Biotechnol., 30: 65–76. https://doi.org/10.1007/s11274-013-1418-x

Hodkinson, I.D. and Wookey, P.A. (1999). Functional ecology of soil organisms in tundra ecosystems: Towards the future. Appl. Soil Ecol., 11: 111–126. https://doi.org/10.1016/S0929-1393(98)00142-5.

Holguín-Salas, A., Enríquez-Núñez, C.A., Sáenz-Marta, C.I. and Nevárez-Moorillón, G.V. (2025). Biosurfactants produced by yeasts: Environmental roles and biotechnological applications. Encyclopedia, 5: 172. https://doi.org/10.3390/encyclopedia5040172.

Hoyt, J.R., Cheng, T.L., Langwig, K.E., Hee, M.M., Frick, W.F. et al. (2015). Bacteria isolated from bats inhibit the growth of *Pseudogymnoascus destructans*, the causative agent of white-nose syndrome. PLoS One, 10: e0121329. https://doi.org/10.1371/journal.pone.0121329.

Hund, A.J. (ed.). (2014). Antarctica and the Arctic Circle: A Geographic Encyclopedia of the Earth's Polar Regions (Vol. 2). Bloomsbury Publishing USA.

Jeon, J.H., Kim, J.T., Kang, S.G., Lee, J.H. and Kim, S.J. (2009). Characterization and its potential application of two esterases derived from the arctic sediment metagenome. Mar. Biotechnol., 11: 307–316. https://doi.org/10.1007/s10126-008-9145-2.

Karimi, M. and Biria, D. (2016). The synergetic effect of starch and alpha amylase on the biodegradation of n-alkanes. Chemosphere, 152: 166–172. https://doi.org/10.1016/j.chemosphere.2016.02.120.

Kerry, E. (1990). Effects of temperature on growth rates of fungi from subantarctic Macquarie Island and Casey, Antarctica. Polar Biol., 10: 293–299. https://doi.org/10.1007/BF00238428.

Kita, D.M., Giovanella, P., Yoshinaga, T.T., Pellizzer, E.P. and Sette, L.D. (2022). Antarctica fungi applied to textile dye bioremediation. An. Acad. Bras. Ciênc., 94: 20210234. https://doi.org/10.1590/0001-3765202220210234.

Koleva, Z., Abrashev, R., Angelova, M., Stoyancheva, G., Spassova, B. et al. (2024). A novel extracellular catalase produced by the antarctica filamentous fungus *Penicillium rubens* III11-2. Fermentation, 10(1): 58. https://doi.org/10.3390/fermentation10010058.

Kreling, N.E., Simon, V., Fagundes, V.D., Thomé, A. and Colla, L.M. (2020). Simultaneous production of lipases and biosurfactants in solid-state fermentation and use in bioremediation. J. Environ. Eng., 146: 04020105. https://doi.org/10.1061/(ASCE)EE.1943-7870.0001785.

Krishnan, A., Alias, S.A., Wong, C.M.V.L., Pang, K.L. and Convey, P. (2011). Extracellular hydrolase enzyme production by soil fungi from King George Island, Antarctica. Polar Biol., 34: 1535–1542. https://doi.org/10.1007/s00300-011-1012-3.

Krishnaveni, B. and Ragunathan, R. (2014). Chitinase production from seafood wastes by plant pathogen *Bionectria* CBNR BKRR sps and its application in bioremediation studies. Int. J. Res. Med. Sci., 2: 15–19.

Kumar, A., Verma, V., Dubey, V.K., Srivastava, A., Garg, S.K. et al. (2023). Industrial applications of fungal lipases: A review. Front. Microbiol., 14: 1142536. https://doi.org/10.3389/fmicb.2023.1142536.

Latter, P.M. and Heal, O.W. (1971). A preliminary study of the growth of fungi and bacteria from temperate and Antarctica soils in relation to temperature. Soil Biol. Biochem., 3: 365–379. https://doi.org/10.1016/0038-0717(71)90047-2.

Lee, C.C., Smith, M., Kibblewhite-Accinelli, R.E., Williams, T.G., Wagschal, K. et al. (2006). Isolation and characterization of a cold-active xylanase enzyme from *Flavobacterium* sp. Curr. Microbiol., 52: 112–116. https://doi.org/10.1007/s00284-005-4583-9.

Lee, H., Calvin, K., Dasgupta, D., Krinner, G., Mukherji, A. et al. (2023). Climate Change 2023: Synthesis Report. Contribution of working groups I, II, and III to the sixth assessment report of the intergovernmental panel on climate change. The Australian National University. https://doi.org/10.59327/IPCC/AR6-9789291691647.

Maksym, T. (2019). Arctic and Antarctic sea ice change: Contrasts, commonalities, and causes. Annu. Rev. Mar. Sci., 11(1): 187–213. https://doi.org/10.1146/annurev-marine-010816-060610.

Malard, L.A. and Pearce, D.A. (2018). Microbial diversity and biogeography in Arctic soils. Environ. Microbiol. Rep., 10: 611–625. https://doi.org/10.1111/1758-2229.12680.

Marchetta, A., Papale, M., Rappazzo, A.C., Rizzo, C., Camacho, A. et al. (2023). A deep insight into the diversity of microfungal communities in Arctic and Antarctica lakes. J. Fungi, 9: 1095. https://doi.org/10.3390/jof9111095.

Margesin, R. and Miteva, V. (2011). Diversity and ecology of psychrophilic microorganisms. Res. Microbiol., 162: 346–361. https://doi.org/10.1016/j.resmic.2010.12.004.

Margesin, R. and Schinner, F. (2001). Biodegradation and bioremediation of hydrocarbons in extreme environments. Appl. Microbiol. Biotechnol., 56: 650–663. https://doi.org/10.1007/s002530100701.

Martorell, M.M., Ruberto, L.A.M., de Castellanos, L.I.F. and Cormack, W.P.M. (2019). Bioremediation abilities of Antarctica fungi. pp. 517–534. *In*: Tiquia-Arashiro, S. and Grube, M. (eds.). Fungi in Extreme Environments: Ecological Role and Biotechnological Significance. Springer, Cham. https://doi.org/10.1007/978-3-030-19030-9_26.

Martorell, M.M., Ruberto, L.A.M., Fernández, P.M., Castellanos de Figueroa, L.I. et al. (2017). Bioprospection of cold-adapted yeasts with biotechnological potential from Antarctica. J. Basic Microbiol., 57: 504–516. https://doi.org/10.1002/jobm.201700021.

Marx, J.C., Collins, T., D'Amico, S., Feller, G. and Gerday, C. (2007). Cold-adapted enzymes from marine Antarctica microorganisms. Mar. Biotechnol., 9: 293–304. https://doi.org/10.1007/s10126-006-6103-8.

Morita, R.Y. (1975). Psychrophilic bacteria. Bacteriol. Rev., 39: 144–167. https://doi.org/10.1128/br.39.2.144-167.1975.

Mousavi, S.M., Hashemi, S.A., Iman Moezzi, S.M., Ravan, N., Gholami, A. et al. (2021). Recent advances in enzymes for the bioremediation of pollutants. Biochem. Res. Int., 2021: 5599204. https://doi.org/10.1155/2021/5599204.

NSIDC (National Snow and Ice Data Center) (2020). Antarctica Ice Sheet. Retrieved from: https://nsidc.org.

NSIDC (National Snow and Ice Data Center) (2023). Antarctica Ice Sheet: Overview and statistics: https://nsidc.org.

NSIDC (National Snow and Ice Data Center) (2025). Arctic Sea Ice Extent: January 2025update: https://nsidc.org.

Newsham, K.K., Hopkins, D.W., Carvalhais, L.C., Fretwell, P.T., Rushton, S.P. et al. (2016). Relationship between soil fungal diversity and temperature in the maritime Antarctica. Nat. Clim. Change, 6: 182–186. https://doi.org/10.1038/nclimate2806.

Niemela, T. (1998). The *Skeletocutis subincarnata* complex (Basidiomycetes), a revision. Acta Bot. Fenn., 161: 1–35.

Nikitin, D.A., Sadykova, V.S., Kuvarina, A.E., Dakh, A.G. and Biryukov, M.V. (2022). Enzymatic and antimicrobial activities in polar strains of microscopic soil fungi. Dokl. Biol. Sci., 507: 380–393. https://doi.org/10.1134/S0012496622060151.

Nogi, Y. (2017). Microbial life in the deep sea: Psychropiezophiles. In Psychrophiles from Biodiversity to Biotechnology (pp. 133–152). Cham: Springer International Publishing.

Ogaki, M.B., Coelho, L.C., Vieira, R., Neto, A.A., Zani, C.L. et al. (2020). Cultivable fungi present in deep-sea sediments of Antarctica: Taxonomy, diversity, and bioprospecting of bioactive compounds. Extremophiles, 24: 227–238. https://doi.org/10.1007/s00792-019-01148-x.

Ordóñez-Enireb, E., Cucalón, R.V., Cárdenas, D., Ordóñez, N., Coello, S. et al. (2022). Antarctica fungi with antibiotic potential isolated from Fort William Point, Antarctica. Sci. Rep., 12: 21477. https://doi.org/10.1038/s41598-022-25911-x.

Ozimek, E. and Hanaka, A. (2020). *Mortierella* species as the plant growth-promoting fungi present in the agricultural soils. Agriculture, 11: 7. https://doi.org/10.3390/agriculture11010007.

Peck, L.S. (2018). Antarctica marine biodiversity: Adaptations, environments, and responses to change. Oceanogr. Mar. Biol., 56: 2–133. https://doi.org/10.1201/9780429454455.

Perini, L., Gostinčar, C. and Gunde-Cimerman, N. (2019). Fungal and bacterial diversity of Svalbard subglacial ice. Sci. Rep., 9: 20230. https://doi.org/10.1038/s41598-019-56290-5.

Post, E., Alley, R.B., Christensen, T.R., Macias-Fauria, M., Forbes, B.C. et al. (2019). The polar regions in a 2°C warmer world. Sci. Adv., 5: eaaw9883. https://doi.org/10.1126/sciadv.aaw9883.

Prabagaran, S.R., Manorama, R., Delille, D. and Shivaji, S. (2007). Predominance of roseobacter, sulfitobacter, glaciecola, and psychrobacter in seawater collected off Ushuaia, Argentina, Sub-Antarctica. FEMS Microbiol. Ecol., 59: 342–355. https://doi.org/10.1111/j.1574-6941.2006.00213.x.

Razzaq, A., Shamsi, S., Ali, A., Ali, Q., Sajjad, M. et al. (2019). Microbial proteases applications. Front. Bioeng. Biotechnol., 7: 110. https://doi.org/10.3389/fbioe.2019.00110.

Robicheau, B.M., Adams, S.J., Provencher, J.F., Robertson, G.J., Mallory, M.L. et al. (2019). Diversity and keratin degrading ability of fungi isolated from Canadian Arctic marine bird feathers. Arctic, 72: 347–359.

Rosa, L.H., Bones, F.L., Lopes, F.A., Santos, K.C., Silva, M.C. et al. (2025). Bioconnection of airborne fungal diversity across the South Atlantic Ocean, sub-Antarctica and South Shetland Islands assessed using DNA metabarcoding. Polar Biol., 48: 11. https://doi.org/10.1007/s00300-024-03340-y.

Rosa, L.H., da Silva, T.H., Ogaki, M.B., Pinto, O.H.B., Stech, M. et al. (2020). DNA metabarcoding uncovers fungal diversity in soils of protected and non-protected areas on Deception Island, Antarctica. Sci. Rep., 10: 21986. https://doi.org/10.1038/s41598-020-78934-7.

Ruggaber, T.P. and Talley, J.W. (2006). Enhancing bioremediation with enzymatic processes: A review. Pract. Period. Hazard., Toxic, Radioact. Waste Manag., 10: 73–85. https://doi.org/10.1061/(ASCE)1090-025X(2006)10:2(73).

Scambos, T.A., Campbell, G.G., Pope, A., Haran, T., Muto, A. et al. (2018). Ultralow surface temperatures in East Antarctica from satellite thermal infrared mapping: The coldest places on Earth. Geophys. Res. Lett., 45: 6124–6133. https://doi.org/10.1029/2018GL078133.

SCAR (Scientific Committee on Antarctica Research) (2022). Antarctica research stations and logistics. Retrieved from: https://www.scar.org.

Selbmann, L., De Hoog, G.S., Mazzaglia, A., Friedmann, E.I. and Onofri, S. (2005). Fungi at the edge of life: Cryptoendolithic black fungi from Antarctic desert. Stud. Mycol., 51: 1–32.

Shi, T., Li, X.Q., Zheng, L., Zhang, Y.H., Dai, J.J. et al. (2021). Sesquiterpenoids from the Antarctic fungus Pseudogymnoascus sp. HSX2#-11. Front. Microbiol., 12: 688202. https://doi.org/10.3389/fmicb.2021.688202.

Shi, T., Li, X.Q., Wang, Z.M., Zheng, L., Yu, Y.Y. et al. (2022). Bioactivity-guided screening of antimicrobial secondary metabolites from Antarctica cultivable fungus *Acrostalagmus luteo labus* CH-6 combined with molecular networking. Mar. Drugs, 20: 334. https://doi.org/10.3390/md20050334.

Shivaji, S., Reddy, G.S., Suresh, K., Gupta, P., Chintalapati, S. et al. (2005). *Psychrobacter vallis* sp. nov. and *Psychrobacter aquaticus* sp. nov., from Antarctica. Int. J. Syst. Evol. Microbial., 55: 757–762. https://doi.org/10.1099/ijs.0.03030-0.

Siegert, M.J., Carter, S.P., Tabacco, I.E., Popov, S. and Blankenship, D.D. (2001). A revised inventory of Antarctica subglacial lakes. Antarct. Sci., 13: 9–16. https://doi.org/10.1017/S0954102005002889.

Singh, P., Roy, U. and Tsuji, M. (2016). Characterisation of yeast and filamentous fungi from Brøggerbreen glaciers, Svalbard. Polar Rec., 52: 442–449. https://doi.org/10.1017/S0032247416000085.

Singh, P., Singh, S.M., Tsuji, M., Prasad, G.S. and Hoshino, T. (2014). *Rhodotorula svalbardensis* sp. nov., a novel yeast species isolated from cryoconite holes of Ny-Ålesund, Arctic. Cryobiol., 68: 122–128. https://doi.org/10.1016/j.cryobiol.2014.01.006.

Sonjak, S., Uršič, V., Frisvad, J.C. and Gunde-Cimerman, N. (2007). *Penicillium svalbardense*, a new species from Arctic glacial ice. Ant van Leeuwen., 92: 43–51. https://doi.org/10.1007/s10482-006-9133-3.

Teixeira, E.A.A., de Souza, L.M.D., Vieira, R., Lirio, J.M., Coria, S.H. et al. (2024). Enzymes and biosurfactants of industrial interest produced by culturable fungi present in sediments of Boeckella Lake, Hope Bay, north-east Antarctica Peninsula. Extremophiles, 28: 30. https://doi.org/10.1007/s00792-024-01345-3.

Tosi, S., Kostadinova, N., Krumova, E., Pashova, S., Dishliiska, V. et al. (2010). Antioxidant enzyme activity of filamentous fungi isolated from Livingston Island, Maritime Antarctica. Polar Biol., 33: 1227–1237. https://doi.org/10.1007/s00300-010-0812-1.

Tsuji, M. (2023). Survey on Fungi in Antarctica and high arctic regions, and their impact on climate change. Climate, 11: 195. https://doi.org/10.3390/cli11090195.

Tsuji, M., Tanabe, Y., Vincent, W.F. and Uchida, M. (2019). *Mrakia hoshinonis* sp. nov., a novel psychrophilic yeast isolated from a retreating glacier on Ellesmere Island in the Canadian High Arctic. Int. J. Syst. Evol. Microbiol., 69: 944–948. https://doi.org/10.1099/ijsem.0.003216.

Tsuji, M., Tsujimoto, M. and Imura, S. (2016). *Cystobasidium tubakii* and *Cystobasidium ongulense*, new basidiomycetous yeast species isolated from East Ongul Island, East Antarctica. Mycoscience, 58: 103–110. https://doi.org/10.1016/j.myc.2016.11.002.

Tsuji, M., Yokota, Y., Shimohara, K., Kudoh, S. and Hoshino, T. (2013). An application of wastewater treatment in a cold environment and stable lipase production of Antarctica basidiomycetous yeast *Mrakia blollopis*. PLoS One, 8: 59376. https://doi.org/10.1371/journal.pone.0059376.

Turner, J., Bindschadler, R., Convey, P., Di Prisco, G., Fahrbach, E. et al. (2009). Antarctica Climate Change and the Environment. Scientific Committee on Antarctica Research, Cambridge, 395 pp.

Unuofin, J.O., Odeniyi, O.A., Majengbasan, O.S. and Okoh, A.I. (2024). Chitinases: Expanding the boundaries of knowledge beyond routinized chitin degradation. Environ. Sci. Poll. Res., 31: 38045–38060. https://doi.org/10.1007/s11356-024-33728-6.

Vaca, I., Faúndez, C., Maza, F., Paillavil, B., Hernández, V. et al. (2013). Cultivable psychrotolerant yeasts associated with Antarctica marine sponges. World J. Microbiol. Biotechnol., 29: 183–189. https://doi.org/10.1007/s11274-012-1159-2.

Vaz, A.B., Rosa, L.H., Vieira, M.L., Garcia, V.D., Brandão, L.R. et al. (2011). The diversity, extracellular enzymatic activities, and photoprotective compounds of yeasts isolated in Antarctica. Braz. J. Microbiol., 42: 937–947. https://doi.org/10.1590/S1517-83822011000300012.

Vero, S., Garmendia, G., González, M.B., Bentancur, O. and Wisniewski, M. (2013). Evaluation of yeasts obtained from Antarctica soil samples as biocontrol agents for the management of postharvest diseases of apple (*Malus domestica*). FEMS Yeast Res., 13: 189–199. https://doi.org/10.1111/1567-1364.12021.

Wang, N., Zang, J., Ming, K., Liu, Y. et al. (2013). Production of cold-adapted cellulase by *Verticillium* sp. isolated from Antarctica soils. Electron. J. Biotechnol., 16: 10–10. http://dx.doi.org/10.2225/vol16-issue4-fulltext-12.

Weber, J.R. (1989). Physiography and bathymetry of the Arctic Ocean seafloor. pp. 797–828. *In*: Herman, Y. (ed.). The Arctic Seas. Springer, Boston, MA. https://doi.org/10.1007/978-1-4613-0677-1_29.

Weinstein, R.N., Palm, M.E., Johnstone, K. and Wynn-Williams, D.D. (1997). Ecological and physiological characterization of *Humicola marvinii*, a new psychrophilic fungus from fellfield soils in the maritime Antarctica. Mycologia, 89: 706–711. https://doi.org/10.1080/00275514.1997.12026836.

Wentzel, L.C.P., Inforsato, F.J., Montoya, Q.V., Rossin, B.G., Nascimento, N.R. et al. (2019). Fungi from admiralty bay (King George Island, Antarctica) soils and marine sediments. Microb. Ecol., 77: 12–24. https://doi.org/10.1007/s00248-018-1217-x.

Widden, P. and Parkinson, D. (1978). The effects of temperature on growth of some soil fungi from the Truelove Lowland, Devon Island, Canada. Can. J. Bot., 56: 1153–1158. https://doi.org/10.1139/m78-068.

Zhang, T., Wang, N.F., Zhang, Y.Q., Liu, H.Y. and Yu, L.Y. (2016). Diversity and distribution of aquatic fungal communities in the Ny-Ålesund Region, Svalbard (High Arctic) Aquatic Fungi in the Arctic. Microb. Ecol., 71: 543–554. https://doi.org/10.1007/s00248-015-0689-1.

Zhang, T., Yan, D., Ji, Z., Chen, X. and Yu, L. (2022). A comprehensive assessment of fungal communities in various habitats from an ice-free area of maritime Antarctica: Diversity, distribution, and ecological trait. Environ. Microbiome., 17: 54. https://doi.org/10.1186/s40793-022-00450-0.

Zucconi, L., Pagano, S., Fenice, M., Selbmann, L., Tosi, S. et al. (1996). Growth temperature preferences of fungal strains from Victoria Land, Antarctica. Polar Biol., 16: 53–61. https://doi.org/10.1007/BF01876829.

Bioremediation

9

Advances in Bioremediation Using Marine Fungi

Zarimah Mohd Hanafiah,[1,*] *Tuan Noraida Tuan Hamzah,*[2,3] *Wan Hanna Melini Wan Mohtar*[2] **and** *Wan Abd Al Qadr Imad Wan-Mohtar*[4,*]

This chapter explores the advances in bioremediation using marine fungi, emphasizing their emerging significance in mitigating diverse marine pollutants such as plastics, petroleum hydrocarbons, and excessive organic biomass. Marine fungi, equipped with unique enzymatic systems and physiological adaptability to extreme saline environments, have demonstrated substantial potential as eco-friendly bioremediation agents. Their ability to degrade recalcitrant pollutants is facilitated by a dual enzymatic mechanism of extracellular enzymes like laccases, peroxidases, and hydrolases that catalyze polymer breakdown, and intracellular systems dominated by cytochrome P450 monooxygenases that mediate detoxification. Notably, species such as *Aspergillus* spp., *Penicillium* spp., and *Zalerion maritimum* have exhibited high efficacy in degrading plastic polymers, converting them into benign end-products. Similarly, marine fungi play a pivotal role in the degradation of hydrocarbons from oil spills through oxidative and non-enzymatic

[1] Chemical Engineering Department, Faculty of Engineering, Universiti Malaya, Kuala Lumpur 50603, Malaysia.
[2] Civil Engineering Department, Faculty of Engineering and Built Environment, Universiti Kebangsaan Malaysia, UKM Bangi 43600 Bangi, Selangor, Malaysia.
[3] Environmental Management Centre, Institute of Climate Change, Universiti Kebangsaan Malaysia, UKM Bangi 43600 Bangi, Selangor, Malaysia.
[4] Functional Omics and Bioprocess Development Laboratory, Institute of Biological Sciences, Faculty of Science, Universiti Malaya, Kuala Lumpur 50603, Malaysia.
* Corresponding authors: zarimahhanafiah@um.edu.my; qadyr@um.edu.my

DOI: 10.1201/9781003795100-12

pathways, including biosorption and biomineralization. Their activity spans various marine environments, from coastal zones and mangroves to deep-sea sediments, indicating a broad ecological tolerance. Furthermore, marine fungi mitigate nutrient over-enrichment and heavy metal contamination by decomposing excess biomass and sequestering toxic metals through biosorption, bioaccumulation, and siderophore production. As pollution challenges intensify, marine fungi are gaining recognition for their integrative role in supporting biogeochemical cycles and restoring ecological balance. The chapter also outlines strategic directions for optimizing fungal bioremediation, including genomic advancements, immobilized biomass systems, and multi-organism consortia. Field trials and policy integration are highlighted as crucial for transitioning from laboratory success to scalable, *in situ* applications. In conclusion, marine fungi represent a versatile, sustainable, and underutilized resource for marine pollution remediation, offering promising avenues for biotechnological innovation and ecosystem restoration. This work underscores the urgent need to harness their full potential through interdisciplinary research and environmental management frameworks.

1. Introduction

Marine environments are called aquatic ecosystems with high dissolved salt, similar to those found in and near oceans. They encompass the interaction between living (biotic) and non-living (abiotic) components (Dolbeth and Arenas, 2021). Marine regions have a wide range of characteristics and habitats, including the deep sea, open seas, and coastal areas, and are exposed to pollution as the receiving bodies of inland runoff. Marine pollution refers to the introduction of harmful substances to the marine environment primarily from human activities, including chemicals, plastics, oil, industrial and agricultural wastes, and residential trash, as shown in Fig. 1. Marine pollutants cause deleterious effects on aquatic ecosystems, as well as on the health and economic structures of the countries (Dolbeth and Arenas, 2021). Chemical pollution, also known as nutrient pollution, can be caused by contaminants such as fertilizer runoff and industrial runoff into rivers and oceans (Akinnawo, 2023). The term "marine trash" indicates man-made items such as plastics and bottles that find their way into the sea (Abu Qamar et al., 2023). Littering, stormwater runoff, and ineffective waste management can also harm the marine ecosystem.

A treatment method that manipulates naturally occurring microorganisms to eradicate contamination is known as bioremediation, a process that has recently been projected (Abu Qamar et al., 2023). Microorganisms such as fungi, which are eukaryotes that can be either unicellular (yeasts) or multicellular (moulds), have recently gained attention as promising candidates for the bioremediation of various pollutants. They are capable of degrading complex chemicals and breaking them down into simpler forms using their unique enzymes (Deshmukh et al., 2016; Hanafiah et al., 2025). While terrestrial fungi have been extensively studied for their role in pollutant removal, marine fungi have received comparatively less

Fig. 1. Several factors contribute to marine pollution and some of the impact on marine life (STP- sewage treatment plant).

attention (Vaksmaa et al., 2023). Marine fungi are emerging as promising agents in bioremediation due to their distinctive enzymatic capabilities and remarkable adaptability to the harsh conditions of marine environments. They are widely distributed throughout the ocean, including in coastal zones, mangroves, the water column, and marine sediments. Notably, marine fungi have also been detected in extreme habitats such as the deep-sea biosphere, Arctic regions, and oxygen minimum zones (Dudhagara et al., 2023). Recent studies highlight their potential to degrade various aquatic pollutants, particularly plastics, oil spills, and excess nutrients.

Certain species, such as *Aspergillus sydowii* and *Penicillium* spp., have demonstrated the ability to break down synthetic polymers like polyethylene and polystyrene through enzymatic oxidation and hydrolysis, offering a sustainable biological alternative to conventional plastic waste management (Nasrabadi et al., 2023; Ojha et al., 2017). In the context of oil spill remediation, marine fungi produce lipases and other oxidative enzymes that can degrade complex hydrocarbons into simpler, less toxic compounds, thereby contributing to the natural attenuation of marine oil pollution (Saravanan et al., 2023). Additionally, these fungi aid in nutrient cycling by decomposing organic matter and assimilating nitrogen and phosphorus, which helps to mitigate eutrophication in coastal waters. Beyond their bioremediation capabilities, marine fungi play essential ecological roles similar to those of fungi in terrestrial ecosystems. Parasitic fungi in aquatic environments have been shown to significantly influence pelagic food webs and drive biogeochemical cycles (Breyer and Baltar, 2023). Various symbiotic interactions have also been observed in marine settings, particularly with phytoplankton, seaweed, and sponges. Coastal and surface marine fungi are capable of degrading structural biopolymers such as lignin, cellulose, and hemicellulose, as well as the remains of marine animals. Laboratory studies further reveal that some marine fungi can decompose complex algal compounds like agar (Balabanova et al., 2018).

Marine fungi are also dominant members of the microbial communities in bathypelagic marine snow, where they likely function as saprotrophs (Breyer and Baltar, 2023). In both terrestrial and marine environments, fungi contribute to nitrogen cycling through processes like nitrification and denitrification. Furthermore, marine fungi produce siderophores that mobilize metals and act as biosorbents, reducing metal toxicity in the environment. These processes influence the biogeochemical cycling of elements such as iron (Fe), manganese (Mn), mercury (Hg), nickel (Ni), zinc (Zn), silver (Ag), copper (Cu), cadmium (Cd), and lead (Pb). Some species are even involved in the breakdown of rocks and minerals to access essential nutrients (Pecoraro et al., 2022). As research into marine fungal genomics and metabolic pathways advances, the potential for their integration into sustainable and effective bioremediation strategies is becoming increasingly recognized and impactful. This section highlights the potential of marine fungi to bioremediate major pollutants commonly found in the ocean, namely plastics, oil spills, and nutrient enrichment. The underlying mechanisms by which these fungi operate, along with future directions for their application in environmental management, are also discussed at the end of this section.

2. Plastic-Inhabiting and Degrading Marine Fungi

2.1 Plastic Pollution in Water

Plastics are generally synthesized from inorganic and organic substances such as hydrogen, oxygen, silicon, carbon, and chloride, which could be derived primarily from sources like coal, natural gas and oil (Shah et al., 2008, Sanchez, 2021). In the contemporary world, plastics are used extensively, whether in the commercial, industrial, medicinal or municipal field (Wright et al., 2013). Plastics are widely used in our daily lives for various purposes, leading to the use of a variety of additives in their production to meet specific end-user requirements and ensure optimal quality for each application (Green, 2020). The continuous growth of the global human population has led to an exponential rise in plastic demand, with more than 300 million tons of plastic consumed annually (Isangedighi et al., 2018). Previous socioeconomic studies have proposed that the application of single-use plastics by society is driven by cost and convenience (Jambeck and Walker-Franklin, 2023). With its vital role in daily life and massive demand, global plastic production has significantly increased, rising from 299 million tons in 2013 to 360 million tons in 2018 (Sharuddin et al., 2016; Yao et al., 2021).

Due to their unique molecular and complex structure, plastics require many years to fully decompose naturally, thereby underscoring the need for effective and sustainable plastic waste management and treatment strategies to minimize the plastic impact on the environment (Ng et al., 2023). Despite the implementation of rigorous strategies to address this issue, only 9% of the plastic waste is recycled globally, while 79% piling up in limited landfills, of plastic waste ends up at the landfill, resulting an estimated 4–12 million metric tons of plastic waste, accumulating in the oceans (Geyer et al., 2017; Jambeck et al., 2015; Wen et al., 2021). Attributable to its reduced density and durability characteristics, plastic has emerged as a predominant pollutant and a pervasive component of marine litter. Several litter studies have suggested that 60% to 80% of the marine litter found in abundance is plastics (Moore, 2008; Thompson et al., 2009). Plastic is more than just marine litter; it is debris that often carries various harmful polymers and additives, some of which are known to be toxic and endocrine disruptors along with it.

After being introduced into the water system, plastic could be weathered and undergo degradation, facilitated by factors such as microbiological activity, radiation, and mechanical stress, breaking down into smaller fragments and various polymers (e.g., polyvinyl chloride, polyurethane and polyethylene terephthalate) that impact multiple systems, including marine life, the ocean's ability to sequester carbon and human health (Jambeck and Walker-Franklin, 2023; Silva et al., 2018). The resulting small polymer fragments float around the ocean and are mistaken by aquatic life such as fish and turtles as food, which are later ingested by them, while smaller fragments are retained in the sea as marine sediment (Isangedighi et al., 2018; Thompson et al., 2004). Many plastic fragments have been detected in the intestines of dead birds, fish and turtles, indicating plastic's critical impact on living organisms (Isangedighi et al., 2018). Due to its persistent characteristics, plastic could not be digested and accumulated in their body (Andrady, 2011; Gallo et al., 2018).

2.2 *Adverse Effects of Plastics on Ecosystems*

The disposed plastic in the ocean is not mere plastic pollution, but poses risks to ecosystem functions, potentially causing shifts in species composition and, in extreme cases, even leading to species extinction when the floating debris is being used by certain species to reach other regions (Aliani et al., 2003). When plastics are ingested by marine life, the contaminants, including the additives and chemicals used in the plastics production, will be transferred as well (Teuten et al., 2009). Previous technical reports have revealed that around 80% of documented interactions between marine organisms and debris involved plastic, while 11% of the reported cases specifically concerned microplastics 80% (GEF, 2012). Plastics that are pulled and remain on the seafloor in the sediment might disrupt the ecosystem process. Previous work has suggested that plastic sheets on the seafloor work like a blanket, disrupting and inhibiting the gas exchange, and resulting in conditions such as anoxia or hypoxia (Goldberg, 1997). Confused marine organisms could misunderstand the plastic waste on the seafloor as artificial hard ground that will be used by the burying creatures to hide or as a habitat, which could be suffocated due to the oxygen limitation under the plastic (Gregory, 2009).

The fragmentation and degradation incurred on the plastics generate smaller fragments materials, referred to as microplastics, which are characterized by sizes of less than 5 mm in diameter (Hammer et al., 2012; Wright et al., 2013). Compared to macroplastics, microplastics pose greater threats, as they can be transferred through the food chain, causing severe toxicity to the organisms that consume them (Guerranti et al., 2019). Microplastics sponge characteristics, attributed by their large surface area-to-volume ratio, non-polar features, and hydrophobic characteristics, allows them to absorb and accumulates harmful substances such as organic and inorganic chemical contaminants, to their surface (Selvam et al., 2021). Previous studies have suggested that microplastics may act as an antibiotic resistance gene and harmful organisms' reservoir, exerting a disease threat to the exposed organisms (Wang et al., 2020). Upon ingestion, microplastics fragments are retained in the digestive tract, triggering conditions such as mechanical disruption, leading to adverse effects such as intestinal blockage and penetration of the intestinal wall (Wang et al., 2020; Yu et al., 2020). The introduced polymer in the organisms' system cannot be digested by the organisms' enzymes, leading to the gastrointestinal blockage, subsequently causing reduction in the organisms' feeding activity (De-la-Torre, 2020; Nelms et al., 2018). This condition results in weight loss, depletion of energy reserves, reproductive impairments, alterations in cholesterol ratio and distribution, nutritional deficiencies, and stunted growth (Eltemsah and Bøhn, 2019; Wang et al., 2021; Weber et al., 2020). Previous studies have demonstrated that, to consume a similar number of zooplankton, fish exposed to microplastics exhibited a twofold increase in feeding time relative to unexposed controls (Carbery et al., 2018). The stunted growth caused by the mechanical disruption could also lead to a fertility reduction (Yu et al., 2020). The presence of microplastics in the digestive tract may induce inflammation, stimulate immune response, and alter metabolic profiles (Wang et al., 2020). While, for the guts with acidic conditions, toxic and harmful chemicals could be released by the plastics to the organism and exert prolonged and severe effects (Carbery et al., 2018; Bosker et al., 2019).

Microplastic pollution poses a growing threat not only due to its physical characteristics and chemical composition but also because of the additives and adsorbed contaminants that can disrupt hormonal synthesis and cause developmental and reproductive issues in organisms. Once considered a marine-only threat, microplastics are now recognized as a human health risk, having been detected in human stool, saliva, and even placentas (Barnes et al., 2009). This contamination is mainly linked to seafood consumption, particularly when the digestive tracts of marine organisms are not removed. The small size of microplastics allows them to penetrate cellular structures such as cell membranes, the blood-brain barrier, and even the placenta, potentially causing oxidative stress, inflammation, and organ damage. They can also travel through the lymphatic and circulatory systems, leading to liver dysfunction and other health issues such as skin irritation, respiratory problems, cardiovascular diseases, and cancer (Smith et al., 2018; Yu et al., 2020). On a broader ecological scale, microplastics floating in oceans reduce light penetration, affecting phytoplankton photosynthesis and subsequently disrupting the marine food web. Since phytoplankton produce most of the Earth's oxygen and play a key role in carbon sequestration, their decline due to microplastic exposure also affects zooplankton, which are vital in transferring carbon to the ocean floor through fecal pellets. Microplastics can fragment these pellets, slow their sinking and reduce carbon storage in the ocean floor, thus impairing the ocean's role as a carbon sink. This chain of disruptions poses a serious threat to the global carbon cycle and climate regulation.

2.3 Application of Marine Fungi to Degrade Plastics

Plastic pollution is recognized as a global environmental challenge owing to the extremely slow degradation rate of plastics, leading to their accumulation across various ecosystems (Kale et al., 2015; Webb et al., 2013). In response, researchers have explored numerous degradation methods, though many involve environmental drawbacks such as heavy metal accumulation, toxic by-products, and high operational costs (Soundararajan, 2021). Biodegradation offers a more sustainable alternative, being eco-friendly, cost-effective, and less energy-intensive. To date, over 400 microbial species have been reported to exhibit plastic-degrading abilities (Lear et al., 2021). Among them, fungi have gained attention for their unique ability to colonize and degrade both macro- and microplastic substrates using enzymatic systems that detoxify pollutants and facilitate depolymerization (Sánchez, 2020). Extensive studies have shown species like *Aspergillus niger* and *Aspergillus terreus* to accelerate low-density polyethylene (LDPE) degradation even in the absence of co-substrates or photothermal treatment (Sáenz et al., 2019).

The enzymatic degradation of plastics by fungi begins when spores adhere to the plastic surface, initiating colonization through the secretion of specific enzymes (Gao et al., 2024, 2022). Fungal hyphae can physically penetrate plastic surfaces, enhancing enzymatic access to the polymer interior and maximizing degradation. Fungi utilize plastic both as a carbon source and as a structural substrate, with degradation rates varying by plastic type (Ahmed et al., 2018). Their secreted enzymes break down polymers into simple end-products like water, carbon dioxide, and methane (Montazer et al., 2019). These processes are facilitated by two major enzymatic systems: intracellular enzymes, primarily cytochrome P450 monooxygenases, involved in detoxification and xenobiotic adaptation (Črešnar and Petrič, 2011), and extracellular enzymes, including hydrolases and oxidoreductases like laccases and peroxidases, especially prominent in lignin-degrading fungi (Sanchez, 2009). Notably, structural changes such as pitting and erosion on polyethylene films have confirmed the degradation activity of *Aspergillus* and *Penicillium* species (Mahalakshmi et al., 2012; Sowmya et al., 2015).

Marine fungi are of particular interest due to their adaptation to saline environments and their potential in mitigating oceanic plastic pollution. Marine-derived species such as *Aspergillus flavus*, *A. niger*, and *Alternaria alternata,* have been reported capable of degrading polyethylene (PE), polyvinyl chloride (PVC), and polyurethane (PU) through the secretion of enzymes including manganese peroxidase, lignin peroxidase, and laccase (Bholay et al., 2012; Gao et al., 2024). Additional enzymes such as cutinases, lipases, and esterases have also been identified in marine fungi, enabling them to degrade bioplastics like polybutylene succinate (PBS), poly(butylene succinate-co-adipate) (PBSA), polylactic acid (PLA), and polyethylene terephthalate (PET) (Thirunavukarasua et al., 2008; Furukawa et al., 2019). The efficiency of the enzymes in the degradation activity are strongly affected by physicochemical conditions such as salinity, pH, and temperature (Srikanth et al., 2022). For instance, lipase-producing *Cryptococcus* species and cutinase-like enzymes from marine isolates have shown promise in polymer degradation, particularly in saline and low-nutrient conditions.

Several marine fungal isolates stand out in the literature (Table 1). *Zalerion maritimum*, isolated from the sea, demonstrated a 43% reduction in microplastic biomass within 14 days of incubation, indicating its capability to metabolize polyethylene as energy source (Paço et al., 2017; Drzyzga and Prieto, 2018). Other fungi, including *Cladosporium cladosporioides* and *Penicillium ochrochloron*, have been reported to degrade PU and PE, although environmental factors appear to influence the degradation outcome (Alshehrei, 2017). Enzymatic profiling has shown that these fungi produce elevated levels of lignin-degrading enzymes, including laccases, manganese peroxidases, and lignin peroxidases, which plays a crucial role in polymer degradation (Ameen et al., 2015; Sangale et al., 2019). In mangrove ecosystems, fungal consortia have shown robust PE degradation activity, emphasizing the role of natural consortia in marine biodegradation.

Recent research has also emphasized the importance of previously overlooked fungi in PU degradation. *Xepiculopsis graminea* and *Leptosphaeria* sp. were identified as PU degraders in studies focused on marine plastic debris (Brunner et al., 2018). Similarly, *Aspergillus tubingensis* VRKPT1 and *Aspergillus flavus* VRKPT2 have emerged as potent marine-derived plastic degraders, largely due to their broad enzymatic repertoire, including a large number of carbohydrate-active enzymes (CAZymes) (Brunner et al., 2018). The genus *Aspergillus*, widely used in industrial biotechnology for enzyme production, appears especially well-equipped for plastic degradation, highlighting its value in circular plastic economy initiatives (Danso et al., 2019; Tournier et al., 2020). Marine fungi represent a promising, eco-friendly solution to plastic pollution. Their unique enzymatic systems, environmental adaptability, and proven degradation capabilities position them as valuable agents in future bioremediation strategies, particularly in marine and coastal environments, where plastic accumulation is most severe.

2.4 Mechanism of Plastic Degradation

Once plastic debris enters aquatic environments, it rapidly becomes colonized by microorganisms, leading to the development of biofilms on its surface due to the high microbial density present in these ecosystems (Wright et al., 2020). Biofilm formation is a dynamic, multistage process that begins with microbial adhesion, followed by the secretion of extracellular polymeric substances (EPS) and subsequent microbial proliferation (Tu et al., 2020; He et al., 2022). The composition and diversity of microbial communities within these plastic-associated biofilms are highly dependent on surrounding environmental conditions (Marsay et al., 2023). Enzymatic biocatalysis holds a crucial role in the plastics biodegradation, providing a sustainable and eco-friendly strategy for plastic recycling (Lai et al., 2022). This

Table 1. Marine fungal strains with plastic-degrading potential.

Strain	**Source of Isolation**	**Polymer**	**Observation**	**References**
Alternaria alternata	Mangrove	LDPE	• High biomass yield in culture with LDPE • CO_2 emissions • Enhanced laccase, MnP, and lignin peroxidase production	Ameen et al. (2015)
Aspergillus caespitosus	Mangrove	LDPE	• High biomass yield in culture with LDPE • CO_2 emissions • Enhanced laccase and MnP production	Ameen et al. (2015)
Aspergillus flavus	Seawater	PE	16.2% biomass reduction of polyethylene	Alshehrei (2017)
Aspergillus flavus VRKPT2	Coastal area	HDPE	Biomass reduction of 8.5±0.1%	Devi et al. (2015)
Aspergillus fumigatus	Seawater	PE	20.5% biomass reduction of polyethylene	Alshehrei (2017)
Aspergillus niger	Seawater	PE	19.5% biomass reduction of PE	Alshehrei (2017)
Aspergillus terreus	Seawater	PE	21.8% biomass reduction of polyethylene	Alshehrei (2017)
Aspergillus terreus	Mangrove	LDPE	• High biomass yield in culture with LDPE • CO_2 emission • Enhanced laccase, MnP, and lignin peroxidase production	Ameen et al. (2015)
Aspergillus tubingensis VRKPT1	Coastal area	HDPE	Biomass reduction of 6±0.2%	Devi et al. (2015)
Cladosporium cladosporioides	Lake shoreline	PU	Clear zones suggesting potential plastic degradation activity	Brunner et al. (2018)
Eupenicillium hirayamae	Mangrove	LDPE	• High biomass yield in culture with LDPE • CO_2 emissions • Enhanced laccase and MnP production	Ameen et al. (2015)
Leptosphaeria sp.	Lake shoreline	PE and PU	Clear zones suggesting potential plastic degradation activity	Brunner et al. (2018)
Paecilomyces variotii	Mangrove	LDPE	• High biomass yield in culture with LDPE • CO_2 emission • Enhanced laccase, MnP and lignin peroxidase production	Ameen et al. (2015)
Penicillium griseofulvum	Lake shoreline	PE and PU	Clear zones suggesting potential plastic degradation activity	Brunner et al. (2018)
Penicillium sp.	Seawater	PE	43.4% biomass reduction of PE	Alshehrei (2017)
Phialophora alba	Mangrove	LDPE	• High biomass yield in culture with LDPE • CO_2 emissions • Enhanced laccase and MnP production	Ameen et al. (2015)
Xepiculopsis graminea	Lake shoreline	PE and PU	Clear zones suggesting potential plastic degradation activity	Brunner et al. (2018)
Zalerion maritimum ATTC 34329	Collection of marine strain cultures	PE	FTIR analysis (carbonyl index) result showed biomass loss reduction of the plastic (56.7±2.9%)	Paço et al. (2017)

process typically begins with the enzymes being adsorbed onto the plastic surface through surface-binding domains, followed by enzymatic cleavage of the polymeric bonds.

The enzymatic degradation of plastics by fungi begins when fungal spores attach to the plastic surface and initiate colonization by secreting specific enzymes (Fig. 2). Plastic-degrading fungal mycelia can penetrate the polymer surface and reach internal regions inaccessible to other microorganisms, enhancing microplastic degradation (Gao et al., 2022; Gao et al., 2024). Plastics are being metabolized by the fungi, not only as a carbon source, but as the growth substrate as well. The plastic's degradation rate differs for different kinds of the plastic, as different plastics possess distinct and varying chemical structures (Ahmed et al., 2018). Over time, enzymatic activity leads to depolymerization of the polymers and eventually mineralization into simple end-products such as water (H_2O), carbon dioxide (CO_2), and methane (CH_4) (Montazer et al., 2019). Fungi demonstrate a distinct ability of secreting enzymes that not only break down complex polymers but also detoxify harmful intermediates. They also produce surface-active proteins like hydrophobins, improving adhesion to hydrophobic plastic surfaces. Furthermore, the physical growth of fungal hyphae can cause mechanical stress on the plastic, leading to localized swelling or rupture (Basheer, 2024).

Fungal plastic biodegradation involves two major enzymatic systems: intracellular and extracellular. The intracellular system holds a crucial role in detoxifying and adapting to xenobiotic compounds. It is largely mediated by

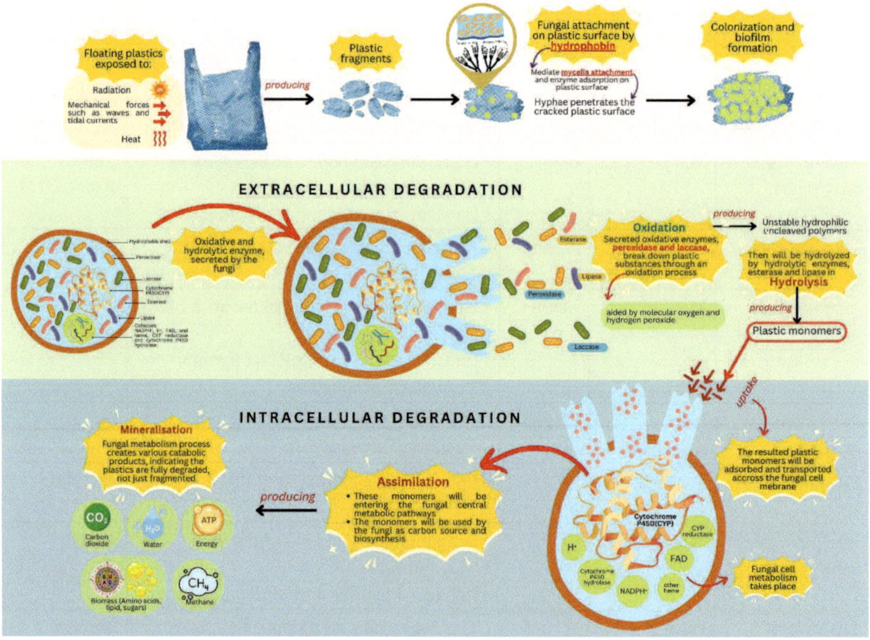

Fig. 2. General mechanism of fungal plastic degradation through extracellular and intracellular enzymatic systems.

the cytochrome P450 (CYP) enzyme family, a group of heme-containing monooxygenases involved in the oxidative metabolism of plastic-derived compounds. CYP enzymes also help maintain hyphal wall integrity and are embedded in the endoplasmic reticulum membrane, allowing access to both cytosolic and membrane-bound substrates (Črešnar and Petrič, 2011; Šrejber et al., 2018). These enzymes function with cofactors such as $NADPH^+$, H^+, FAD, and heme, along with supporting enzymes like NADPH: CYP reductase and cytochrome P450 hydrolase.

The extracellular enzymatic system comprises two types of enzymes, which are hydrolytic and oxidative enzymes. Hydrolytic enzymes break down polysaccharides and plastic-like bonds, while oxidative enzymes include non-specific oxidoreductases such as class II peroxidases (manganese peroxidase, lignin peroxidase, and versatile peroxidase), laccases, and unspecific peroxygenases. These enzymes catalyze redox reactions using molecular oxygen or hydrogen peroxide, aiding in the breakdown of complex synthetic polymers (Karich et al., 2017; Sanchez, 2009). This enzymatic system is especially prevalent in marine fungi that degrade lignin-like structures, which share chemical similarities with plastics. For instance, degradation of HDPE and LDPE has been observed through structural changes in plastics exposed to marine-derived fungal isolates (Ojha et al., 2017).

Marine fungi are increasingly recognized for their ability to degrade plastic materials such as PE, PVC, and PU by secreting various extracellular enzymes. For example, *Aspergillus flavus* and *Aspergillus niger*, both isolated from marine environments, produce manganese peroxidase and lignin peroxidase, contributing to the biodegradation of PE materials like plastic carry bags (Bholay et al., 2012). Similarly, the marine fungus *Alternaria alternata* effectively colonizes and degrades PE, producing diglycolamine as a breakdown product, primarily due to laccase and peroxidase enzyme activity (Gao et al., 2024). These enzymes, commonly involved in lignin degradation, are crucial for initiating oxidative depolymerization in marine settings. In addition to oxidative enzymes, hydrolytic enzymes such as cutinases, lipases, and esterases were detected in marine fungi, further enhancing their ability to degrade plastics. For example, lipase from *Cryptococcus* sp. exhibited activity against biodegradable plastics like PBS and PBSA (Thirunavukarasua et al., 2008). Cutinase-like enzymes from marine fungal isolates have also shown potential in degrading polymers with high molecular weight plastics including PLA and PET (Gemeren et al., 1998; Furukawa et al., 2019). The enzymatic degradation efficiency of these marine fungi is highly affected by physicochemical factors such as temperature, pH, and moisture, which in turn influence enzyme activity and the overall degradation rate (Srikanth et al., 2022).

3. Degradation of Oil Spills by Marine Fungi

3.1 Oil Spillages into the Environment

Crude oil, known as a naturally occurring liquid form of petroleum, is a complex mixture with varying compositions, primarily consisting of hydrocarbons (compounds with carbon and hydrogen only) and minor constituents such as sulfur, nitrogen, oxygen, and helium. The most common hydrocarbon components in crude oil are alkanes (or paraffins), cycloalkanes, aromatics, polycyclic aromatic hydrocarbons (PAHs) and asphaltenes, with PAHs appearing to pose the greatest toxicity to the environment compared to the other hydrocarbon compounds (Vignier et al., 2019). As the process of handling and production process is involved with the crude oil products, including extraction, refining process, transportation, and consumer distribution, hence spills and release of the oil in any other process stages occur. Oil spills often because of accidents involving broken pipelines, oil tankers, drilling rig blowouts, and storage facilities have become one of the most significant and urgent environmental challenges of the last century. High-profile spill events often draw widespread concern to the public because their visible and detrimental impacts on the environment, such as oil-covered coastlines and the death or distress of wildlife, especially seabirds and marine mammals coated in oil.

In marine environments, oil spills pose significant threats to biodiversity, causing both acute and chronic harm to a wide range of species (Almeida, 2023). These impacts include increased mortality, inhibited growth and reproduction at various trophic levels, and the alteration of natural habitat structures such as coral reefs and mangroves. Furthermore, oil spills also significantly affect the coastal communities by disrupting their socio-economic activities such as fishery and tourism, leading to long-term losses (Alvernia et al., 2021). As environmental pollution from petroleum hydrocarbons remains a critical concern, comprehensive and sustainable remediation strategies are urgently needed. Bioremediation, particularly using fungi, holds considerable promise in addressing this challenge. Some fungal species have shown the capacity to use hydrocarbons as their exclusive carbon source, particularly those isolated from marine habitats (Velez et al., 2020). Toxic long-chain n-alkanes can be broken down by these fungi into shorter-chain alkanes that are more easily volatilized and less persistent. As such, fungal biodegradation represents a vital component of an integrated approach to petroleum hydrocarbon remediation and merits further investigation as a viable solution for environmental cleanup efforts. Hence, the purpose of this section is to be the subset of marine fungi in degrading petroleum hydrocarbons.

3.2 *Role of Marine Fungi in Degrading Oil Spills*

Marine fungi play a significant role in the biodegradation of petroleum hydrocarbons, including *n*-alkanes and PAHs, due to their ability to use these compounds as a sole carbon source, offering an eco-friendly alternative to chemical or mechanical cleanup methods. Marine fungi naturally inhabit diverse marine environments, including sediments, seawater, mangrove areas, and deep-sea ecosystems, where they have adapted to survive in extreme conditions and degrade complex organic compounds such as hydrocarbons from their surroundings. Thus, their metabolic versatility and enzymatic machinery allow them to utilize oil components such as carbon and energy sources, making them valuable for marine environmental restoration. For instance, fungi oxidize hydrocarbons and convert them into intermediates such as acids, alcohols, and aldehydes, which then enter central metabolic pathways like β-oxidation, glycolysis, and the tricarboxylic acid (TCA) cycle to generate ATP as an energy source.

Marine fungal genera from phyla Ascomycota, Basidiomycota, and Mucoromycota, including *Acremonium*, *Aspergillus*, *Aureobasidium*, *Candida*, *Cladosporium*, *Cryptococcus*, *Fusarium*, *Mollisia*, *Mortierella*, *Penicillium*, *Phialocephala*, *Phoma*, *Scedosporium*, *Scopulariopsis*, and *Trichoderma*, have all been reported to include potential degraders of oil spill products (Table 2). Some species have extracellular enzymes like laccase and peroxidases that can oxidize contaminants like phenols and aromatic amines. Fungal intercellular enzymatic systems are permeable to several types of hydrocarbons and eventually break down the pollutant, such as cytochrome P450, dehalogenase and nitro reductases, to simpler organic compounds (Abu Qamar et al., 2023; Simister et al., 2015). This simpler form of organic compounds was further metabolized through processes such as β-oxidation and entry into the TCA cycle (Li et al., 2020).

Research by Barnes et al. (2018) highlights the bioremediation potential of hydrocarbon-degrading fungi belonging to genera *Penicillium*, *Acremonium*, *Fusarium*, and *Aspergillus*, isolated from various marine habitats across India, are capable of metabolizing crude oil components, with some exhibiting robust degradation activity even in saline and low-nutrient conditions. Analysis of marine fungi isolated in isolated polar regions also shows promising oil-degradation capacities, despite the extremely cold environment. Hughes et al. (2007) reported that Antarctic soil fungi possess significant tolerance to hydrocarbons, suggesting that cold-adapted marine fungi such as *Phoma herbarum*, *Mortierella* sp., *Penicillium commune*, *Mollisia* sp., *Trichoderma koningii*, and *Trichoderma* sp. could be applied to spill sites in high-latitude regions. The psychrotolerant nature of these fungi enables bioremediation processes to occur even at sub-zero temperatures, widening the scope for their use in harsh or remote marine environments. The fungi that are isolated on beaches also have shown some promising results in bioremediating hydrocarbon spills (Simister et al., 2015) utilizing those fungi collected from oil-contaminated beaches in the Gulf of Mexico, where there were dedicated areas of the Deepwater Horizon oil spill and had isolated *Aspergillus terreus* and *Fusarium solani*. In the study, the fungi were able to degrade various hydrocarbons, including PAHs, which are among the most persistent and toxic oil components. The study provided evidence that indigenous fungal species can naturally respond to oil contamination and play a vital role in the long-term degradation process, particularly when bacterial activity declines due to environmental constraints or nutrient limitations.

Deep-sea oil reserves also host cultivable fungi such as *Phialocephala* sp., *Aureobasidium* sp., *Cladosporium* sp. 1 and 2, with genetic adaptations that enhance hydrocarbon degradation, as shown by (Velez et al., 2020). These fungi exhibit unique gene expressions when exposed to petroleum compounds, indicating an evolved biochemical toolkit for breaking down complex hydrocarbons. Their resilience and specialized enzymatic pathways offer exciting opportunities for developing bioaugmentation strategies tailored for deep-sea oil spill remediation, where access and intervention are often limited. Other marine fungi isolated from mangrove areas are also promising species in degrading hydrocarbons. Lotfinasabasl et al. (2012) conducted a study along the Maharashtra coast, India in the mangrove forests of Alibaug and Akshi, where they isolated fungal strains from soil and tarball samples.

Table 2. Marine fungal strains, isolation location and performance on oil spill.

Strain	Isolation Location	Type of Oil	Performance	References
Acremonium sp., *Aspergillus* sp., *Fusarium* sp., and *Penicillium* sp.	Mangrove sediments, Arabian Sea sediments, and tarballs (Panaji, Goa, India)	n-alkanes	Long-chain n-alkanes in crude oil were more easily broken down by fungi than short-chain ones. *Penicillium citrinum* (#NIOSN-M126), a mangrove fungus, has shown remarkable efficacy in purifying crude oil, lowering its overall oil content by 77% and decomposing roughly 95.37% of the n-alkanes.	Barnes et al. (2018)
Aspergillus niger, A. terreus, Penicillium sp. and *Rhizopus* sp.	Soil and tarballs from mangrove forest in the Alibaug and Akshi coastal areas, Maharashtra, India	Kerosene	In 5% kerosene, *Rhizopus* sp. displayed the largest growth diameter, while *Aspergillus niger* displayed the largest growth diameter in 20% kerosene. Across all kerosene concentrations, *Penicillium* sp. displayed the smallest growth diameter.	Lotfinasabasl et al. (2012)
Aspergillus sclerotiorum CRM 348 and *Cryptococcus laurentii* CRM 707	Isolated from marine invertebrates (São Paulo north coast, Brazil and King George Island, Maritime Antarctica)	n-alkanes	3–65% degradation by natural attenuation, 72–92% degradation under biostimulation (BS)/bioaugmentation (BA) for short alkanes (C12–C19). Long-chain alkanes (C20–C24) under BS/BA show high degradation capabilities.	Giovanella et al. (2023)
Aspergillus sclerotiorum CBMAI 849 and *Mucor racemosus* CBMAI 847	São Sebastião, on the northern coast of the state of São Paulo, Brazil	Pyrene, benzo[a]pyrene	After 8 and 16 days, *A. sclerotiorum* performed the best in terms of pyrene (99.7%) and benzo[a]pyrene (76.6%) depletion. By *M. racemosus*, benzo[a]pyrene ($>$ 50.0%) was produced. Cytochrome P-450 monooxygenase mediates the hydroxylation process, which is then followed by conjugation with sulphate ions.	Passarini et al. (2011)
Aspergillus sydowii (AMF15, AMF77), *A. versicolor* (AMF75, AMF97), *Penicillium chrysogenum* AMF4, *P. cyclopium* (AMF40, AMF47, AMF74), *P. mononematosum* AMF79, *P. olsonii* AMF72, *P. polonicum* AMF16, *P. steckii*, and AMF25	Port of Oran, Algeria	Arabian Light crude oil	Crude oil was the only carbon source that 12 fungal strains could use. Based on a screening test that used 2,6-dichlorophenol indophenol to identify their oil-degrading capability, four of them were found to be the most efficient at biodegrading crude oil.	Maamar et al. (2020)

Aspergillus terreus, *Fusarium solani*, and *Scopulariopsis brevicaulis*	Oil-soaked sand patties, Gulf of Mexico beaches	n-alkanes	On average, 47% of all fungal isolates can thrive on crude oil. In contrast to their branched counterparts, pristane and phytane (70–98%), fungi prefer degraded short (< C18: 90–99%) chain n-alkanes over long (C19–C36: 7–87%) and straight-chain C17- and C18-n-alkanes (91–99%). The fungal isolates decomposed PAHs (42–84% total destroyed), preferring low molecular weight over high molecular weight.	Simister et al. (2015)
Aureobasidium sp., *Cladosporium* sp. 1, *Cladosporium* sp. 2, *Penicillium brevicompactum, Penicillium* sp., and *Phialocephala* sp.	Deep-sea sediment, Gulf of Mexico	Hexadecane 1-hexadecene	Genes involved in transmembrane transport, six-carbon carbohydrate metabolism, and nitric oxide pathways are up-regulated in the hydrocarbonoclastic fungus *Penicillium* sp. when hexadecane and 1-hexadecene are used as the only carbon sources.	Velez et al. (2020)
Candida tropicalis	Petroleum hydrocarbon-contaminated soil	n-hexadecane	An energy-dependent mechanism drives the transmembrane transport of n-hexadecane, a common petroleum hydrocarbon, in Candida tropicalis cells.	Li et al. (2021)
Hyphomonas adhaerens, *Marinobacter mobilis*, and *Scedosporium* sp. ZYY	Marine sediment from intertidal mudflat, Zhoushan Island, Zhejiang Province, China	Total petroleum hydrocarbon (TPH)	7% of the TPH is eliminated. By reducing oil toxicity through adsorption, increasing microbial closeness to the oil, and giving the microorganisms a favorable surface area for interfacial uptake of hydrocarbons, biochar, and the exogenous fungus performed synergistic activities. The fungus simultaneously enhanced the bio-accessibility of the target pollutants to the relevant species and encouraged their growth.	Atakpa et al. (2024)
Mollisia sp., *Mortierella* sp., *Penicillium commune, Phoma herbarum, Trichoderma koningii*, and *Trichoderma* sp.	Adelaide Island, Antarctic Peninsula	Dodecane, Hexadecane, Benzoic acid, p-hydroxybenzoic acid, toluene, phenol biphenyl, naphthalene, m- and p-xylenes with ethylbenzene	Dodecane might be the only source of carbon and energy for *Mortierella* sp. However, the presence of hydrocarbons inhibits other fungal hyphal.	Hughes et al. (2007)

Among the isolates, *Aspergillus terreus* and *Rhizopus* species were identified from tarball samples, with *Rhizopus* sp. demonstrating the most significant growth, particularly in the presence of 5% kerosene.

3.3 The Mechanism of Marine Fungi Degrading Oil Spills

Fungi play a pivotal role in the biodegradation of petroleum hydrocarbons through both enzymatic and non-enzymatic mechanisms. Their ability to secrete extracellular enzymes enables the breakdown of complex hydrocarbons into compounds that are less toxic and simpler, which facilitates further assimilation or mineralization. Apart from enzymatic breakdown, fungi also take role in non-enzymatic processes such as biosorption, biomineralization, and transmembrane stripping, which contribute to the removal of hydrocarbons from contaminated environments (Barnes et al., 2018). Marine fungi, in particular, have demonstrated significant potential in hydrocarbon degradation, especially in marine and coastal ecosystems where they contribute to the natural remediation of oil spills. These fungi produce a wide range of extracellular enzymes, notably ligninolytic enzymes such as laccases and peroxidases, which are capable of breaking down structurally complex hydrocarbon compounds. Additionally, hydrolases, including esterases, cutinases, lipases, and proteases, facilitate the degradation of hydrocarbon polymers, rendering them more accessible for microbial uptake and subsequent use as carbon sources (Okal et al., 2023).

Fungal peroxidases, in particular, can oxidize hydrocarbons both directly and indirectly, enhancing their breakdown. The non-enzymatic pathways of fungal remediation also play a significant role. Biosorption allows fungal cells to adsorb hydrocarbons onto their cell walls, effectively sequestering pollutants from the environment. In biomineralization, fungi can transform hydrocarbons into mineral forms that are less harmful and easier to remove. Moreover, fungi are capable of transmembrane stripping, a process in which hydrocarbons are separated from the contaminated medium and transported through the fungal cell membrane, contributing further to pollutant removal (Negi and Das, 2023). Several marine fungi exhibit ligninolytic activity, meaning they produce enzymes that can degrade lignin, a highly recalcitrant polymer found in plant cell walls and woody tissues. These fungi are commonly found in coastal environments such as mangrove ecosystems and are considered valuable sources of novel enzymes with unique properties, including salt tolerance (Luo et al., 2005). In recent years, considerable attention has been given to ligninolytic fungi, particularly white rot fungi, for their ability to degrade PAHs, a class of persistent organic pollutants. These fungi secrete extracellular peroxidases, which catalyze the initial oxidation of PAHs (Kadri et al., 2017; Zhang et al., 2020).

The lignin-degrading enzyme system, which includes lignin-modifying enzymes (LMEs) with limited substrate specificity, is the primary mechanism of fungal biodegradation (Dashtban et al., 2010). This allows them to target a broad spectrum of structurally diverse and recalcitrant organic compounds. The enzymes that produce hydrogen peroxide (H_2O_2), laccase, manganese peroxidase (MnP), and

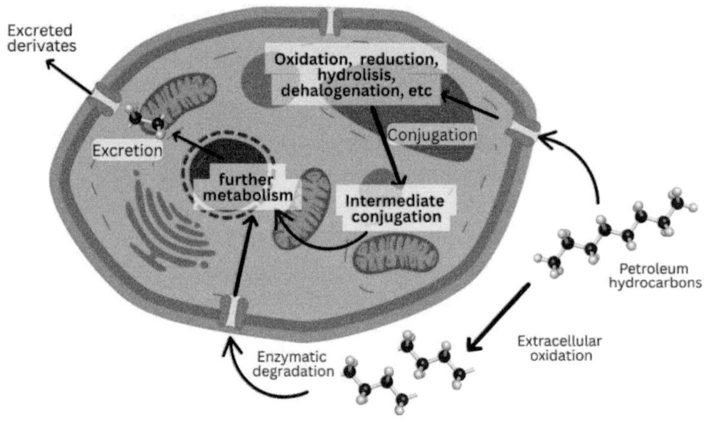

Fig. 3. Simplified diagram of fungal interactions with petroleum hydrocarbons (modified from Li et al., 2020).

lignin peroxidase (LiP) are essential parts of the LME system. Nevertheless, not all ligninolytic fungi have all of these enzymes (Kumar and Chandra, 2020). In addition to extracellular mechanisms, fungi also employ intracellular enzymatic pathways for hydrocarbon degradation (Fig. 3). The breakdown of substances like 1-hexadecene and hexadecane has shed light on these molecular mechanisms, demonstrating the role of epoxide hydrolases and the cytochrome P450 system in mycoremediation (Velez et al., 2020). Under aerobic conditions, cytochrome P450 monooxygenases and alkane-oxygenase enzymes usually mediate an oxidation step that starts the breakdown of hydrocarbons into intermediates that are easier to metabolize (Asemoloye et al., 2019).

4. Marine Biomass Degradation

4.1 Marine Biomass

Nutrient enrichment and heavy metal contamination are two major forms of marine pollution that significantly contribute to the generation and accumulation of biomass pollutants in marine ecosystems. Excessive inputs of nitrogen and phosphorus, primarily from agricultural runoff, wastewater discharge, and aquaculture activity, often lead to nutrient overloading in coastal waters (Devlin and Brodie, 2023). This enrichment stimulates the rapid growth of phytoplankton and cyanobacteria, resulting in harmful algal blooms (HABs) that increase the biomass load within the water column. Upon decomposition, the organic matter from these blooms sinks and is broken down by microorganisms, a process that consumes large amounts of dissolved oxygen, leading to hypoxic or anoxic conditions (Lan et al., 2024). These conditions not only reduce biodiversity but also create zones known as dead zones, where most marine life cannot survive. The excessive buildup of organic material from dead algal blooms constitutes biomass pollution, which can smother benthic habitats and disrupt the balance of marine ecosystems.

Simultaneously, heavy metals such as mercury (Hg), cadmium (Cd), lead (Pb), nickel (Ni), and arsenic (As) enter marine environments through industrial discharge, mining runoff, and urban effluents (Oladimeji et al., 2024). These metals often bind to particulate organic matter or accumulate in living biomass such as algae, plankton, and biofilms, turning them into vectors for toxic metal transfer through the food web. Contaminated biomass not only affects marine organisms but also poses serious health risks to humans through seafood consumption (Ali et al., 2019). In this context, marine fungi have emerged as crucial agents in mitigating both nutrient- and metal-related biomass pollution. They play a pivotal role in breaking down excess organic matter, thus preventing the long-term buildup of biomass pollutants and facilitating nutrient recycling. Certain fungi contribute to nitrogen cycling processes such as nitrification and denitrification, helping to alleviate eutrophication in nutrient-rich waters (Lan et al., 2024).

Moreover, marine fungi exhibit remarkable potential in heavy metal remediation through various mechanisms. These include biosorption, where fungal cell walls bind and immobilize metal ions; bioaccumulation within fungal cells; and the secretion of siderophores and organic acids that chelate and detoxify metals (Ramya et al., 2020). By immobilizing toxic metals and reducing their bioavailability, marine fungi help prevent further contamination of marine biomass and support the restoration of ecological balance in polluted habitats. As such, marine fungi represent an underexplored but highly promising biological tool for addressing the dual challenges of nutrient enrichment and heavy metal pollution, both of which are closely tied to the occurrence of biomass pollution in marine environments.

4.2 Marine Fungi Degradation of Excessive Marine Biomass

Through both active and passive processes, marine-derived fungi may efficiently remove marine material, including pollutants from heavy metals and nutrients. Both live and non-living fungal biomass have the ability to detoxify and degrade pollutants present in contaminated settings (Dudhagara et al., 2023). A review of past studies shows that research on marine fungal-based nutrient remediation focusing on heavy metals and organic degradations is as widespread as that involving algae or bacteria; however, the range of fungal strains investigated remains limited, as shown in Tables 3 and 4.

The heat-inactivated *Aspergillus flavus* strain EGY11 has shown promise as a biosorbent for removing heavy metals (Cd, Hg, and Pb) from aqueous solutions with a high removal percentage (> 90%), according to earlier research like Mahmoud et al. (2017). The finding also highlights that using dead fungal biomass offers several advantages, including reusability of the biomaterial, ease of storage and handling, suitability for treating large volumes of wastewater with low metal concentrations, shorter operation times, no production of potentially toxic secondary compounds, and no limitations related to enzymatic activity during adsorption. Another study, by Dell'anno et al. (2022), demonstrated that marine-derived fungi (*Aspergillus niger* and *Trichoderma* sp.) are more effective than bacteria in bioleaching heavy metals (Zn, Cd, As) from highly contaminated marine sediments in Portman Bay.

Table 3. The adsorption capacities and removal efficiencies of heavy metals by different species of marine fungi and their sorption capacity.

Taxon	Isolation/Location	Heavy Metals	Removal (%)	Sorption Capacity, q	References
Aspergillus flavus strain EGY11	Different sites of the sediment samples from Red Sea, Egypt	Cd Hg Pb	92.7–99.0 91.3–95.6 95.3–98.2	1200.0 –4000.0 μmol g^{-1}	Mahmoud et al. (2017)
Aspergillus niger	Seawater and sediment samples, Gujarat coast, India	Hg	–	40.53 mg/g	Khambhaty et al. (2008)
Aspergillus niger and *Trichoderma* sp.	Sediments of the Portman Bay, Northwest Mediterranean Sea	As Cd Zn	20–80	–	Dell'anno et al. (2022)
Cryptococcus sp. (NIOCC #PY13)	Sediments of Polymetallic nodule-bearing from Central Indian Basin	Zn Cd Pb Cu	30–90	–	Singh et al. (2013)
Penicillium fellutanum	Pakistan	Ni Zn	–	161 mg/g 78.5 mg/g	Rashid et al. (2016)

Table 4. The degradation efficiency of several types of nutrients by different species of marine fungi and their brief mechanism.

Taxon	Isolation/ Location	Nutrients	Removal (%)	Mechanism	References
Aspergillus piperis	Sediment from mangrove forest (Leizhou, China)	Chloride (Cl^-)	74.11	*A. piperis* actively transports chloride ions into their cells	Han et al. (2023)
Pestalotiopsis sp. J63 and *Penicillium janthinellum* P1	Samples of sediment taken from China's Nanji Island	Decolorized dye Azure B and paper mill wastewater	99% capability for biodegradation in 10 hours of treatment	Whole immobilized pellets biodegrade contaminants through the extracellular enzyme	Chen et al. (2014)
Tinctoporellus sp. CBMAI 1061	Brazilian marine sponges	Dye Remazol Brilliant Blue R (RBBR)	90%	Fungus maximum decolorization of the dye in liquid culture over 12 days driven by extracellular oxidative enzymes	Rodriguez et al. (2015)
Talaromyces atroroseus, *Penicillium crustosum*, *Trichoderma harzianum*, *Rhodotorula mucilaginosa*, and *Trichoderma deliquescens*	Marine sediment from Chilean fjord, Chile	Oxytetracycline	72–92	Optimal degradation is achieved at 21 days of treatment	Ahumada-Rudolph et al. (2016)

Fungal treatments not only outperformed bacterial ones in terms of solubilization efficiency, but also uniquely enabled the mobilization of metals bound to otherwise non-leachable sediment fractions, highlighting their potential as a promising strategy for marine sediment bioremediation.

A research group investigated chloride biodegradation originating from wastewater generated during crustacean shell processing, utilizing the potential of mangrove-derived fungi. Their study found that fungal mycelium from *Aspergillus piperis* showed the most promising results by actively transporting Cl^- into its cells, removing over 70% of the chloride from the wastewater within three days, with the most rapid biomass increase (2-fold) and chloride removal occurring between days one and two (Han et al., 2023). Meanwhile, Ahumada-Rudolph et al. (2016) studied the biodegradation of another nutrient pollutant, oxytetracycline, an antibiotic used in salmon farming, by several marine fungal species, including *Trichoderma deliquescens*, *Talaromyces atroroseus*, *Trichoderma harzianum*, *Rhodotorula mucilaginosa*, and *Penicillium crustosum*, achieving up to 92% reduction within 21 days of treatment. Other types of contaminants such as dyes and industrial effluent are also being studied and promising results have been reported, up to 99% removal of the contaminant by marine fungi species of *Tinctoporellus* sp. CBMAI 1061, *Pestalotiopsis* sp. J63, and *Penicillium janthinellum* P1 (Chen et al., 2014; Rodriguez et al., 2015).

4.3 *Mechanism of Degradation*

Marine fungi play a multifaceted role in degrading both organic and inorganic components of biomass pollutants associated with nutrient enrichment and heavy metal contamination. In nutrient-rich environments, marine fungi contribute significantly to the decomposition of excess organic matter generated by algal blooms and detritus. Through the secretion of extracellular enzymes such as cellulases, ligninases, and proteases, these fungi break down complex polymers into simpler compounds, facilitating the mineralization and recycling of nutrients (Damare et al., 2006). This enzymatic activity not only reduces the volume of accumulated biomass but also helps prevent the formation of hypoxic conditions by limiting the oxygen demand typically caused by bacterial degradation of organic matter (Raghukumar, 2008). Furthermore, by metabolizing dissolved and particulate organic nitrogen and phosphorus, marine fungi help rebalance nutrient levels in the water column, which is essential for maintaining ecological stability in eutrophic zones.

In the context of heavy metal pollution, marine fungi employ several defence and detoxification strategies that also serve to remediate contaminated environments (Fig. 4). One primary function is biosorption, where fungal cell walls rich in chitin, glucans, and other polysaccharides bind heavy metal ions through functional groups such as carboxyl (COOH), hydroxyl (OH), and amino groups (NH_2) (Alabssawy and Hashem, 2024). This process immobilizes metals, reducing their bioavailability and toxicity. In addition, some marine fungi, such as *Aspergillus fumigatus* and

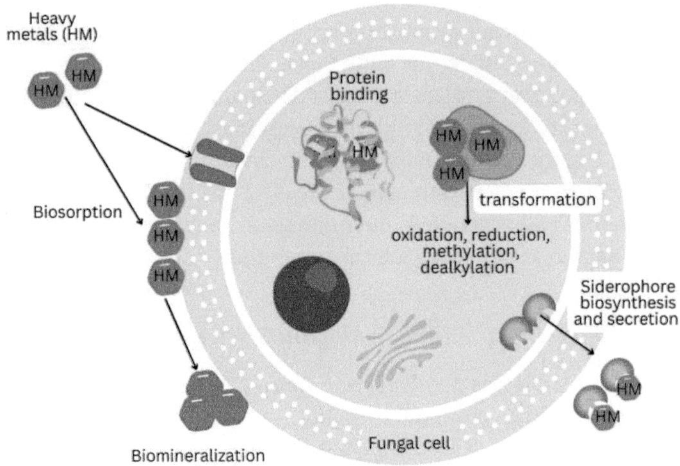

Fig. 4. Simplified fungal cell mechanism in heavy metal bioremediation.

Aspergillus nidulans, produce siderophores, which are metal-chelating compounds that scavenge and sequester metals like iron (Fe), zinc (Zn), and copper (Cu), making them less harmful in the aquatic environment (Pecoraro et al., 2022). Other marine fungi engage in bioaccumulation, actively transporting metal ions into their cells for sequestration within vacuoles or binding to intracellular proteins (Greeshma et al., 2022). These mechanisms are particularly effective in detoxifying metals such as cadmium (Cd), lead (Pb), and mercury (Hg), which are often associated with biomass in polluted waters. In more extreme environments, such as metal-rich sediments or industrial discharge zones, certain fungi have evolved to tolerate and even thrive under high metal concentrations, showcasing their adaptive capacity for bioremediation in harsh conditions.

Collectively, these functions position marine fungi as key biological agents in breaking down biomass pollutants while simultaneously mitigating the toxic effects of nutrient and metal overloading in marine ecosystems. Their enzymatic versatility, metabolic adaptability, and metal-handling capabilities make them invaluable not only for natural ecosystem processes, but also for the development of future biotechnological applications in marine pollution control. Through physico-chemical and biological processes, such as extracellular binding by metabolites and biopolymers, binding to certain polypeptides, and metabolism-dependent accumulation, fungi can accumulate metal and radionuclide species. In sum, marine fungal biomass serves as both a catalyst and a sink for various marine pollutants. Through the enzymatic breakdown of organic matter, biosorption and chelation of heavy metals, and redox modifications, marine fungi facilitate the degradation and detoxification of diverse contaminants. These mechanisms not only support the resilience of marine ecosystems under pollution stress but also position marine fungi as valuable agents in the development of eco-friendly bioremediation strategies.

5. Strategies and Future Directions

Marine fungi hold significant promise as natural bioremediation agents capable of addressing a wide range of marine pollutants, including oil spills, plastic waste, and excessive organic and nutrient biomass. Their internal and external enzymatic versatility, adaptability to extreme marine conditions, and ability to interact with both organic and inorganic pollutants make them uniquely positioned to support sustainable remediation practices in aquatic ecosystems. As scientific understanding of their metabolic pathways and environmental interactions deepens, several strategies and future research directions are emerging.

5.1 Integrated Enzymatic Biodegradation Strategies

One of the primary strategies for deploying marine fungi in pollution mitigation is the exploitation of their enzymatic toolkit, which includes lipases, laccases, peroxidases, and hydrolases. These enzymes facilitate the degradation of:

- Crude oil and hydrocarbons through oxidative cleavage of aliphatic and aromatic chains;
- Synthetic plastics, particularly PE and polystyrene (PS), via extracellular oxidative and hydrolytic enzymes;
- Excessive organic biomass, such as algal detritus, through cellulolytic and ligninolytic enzyme systems and heavy metal contamination through their biosorption properties.

Future strategies should focus on metagenomic screening and genetic engineering of high-performing fungal strains to enhance enzyme production and substrate specificity (Bonugli-Santos et al., 2015). Additionally, combining fungal bioremediation with bacterial or algal consortia may yield synergistic effects, improving overall degradation efficiency.

5.2 Development of Immobilized Fungal Biomass and Biofilms

To increase the effectiveness and stability of marine fungi in bioremediation applications, researchers are exploring immobilization techniques, where fungal biomass is attached to inert carriers (e.g., alginate beads, activated carbon, or marine substrates) for continuous treatment of pollutants (Harms et al., 2011). Immobilized fungal systems can be deployed in contaminated coastal zones or estuarine environments, enabling long-term degradation of oil residues and microplastics. Fungal biofilms are also promising, as they naturally form on marine debris, helping to colonize and degrade pollutant surfaces.

5.3 Biotechnological Tools and Genomic Insights

Advancements in omics technologies, particularly genomics, transcriptomics, and proteomics, are revolutionizing our understanding of marine fungal physiology and pollutant response mechanisms. By decoding the genomes of key species such as *Aspergillus sydowii*, *Penicillium* spp., and *Cladosporium* spp., researchers can identify and optimize gene clusters responsible for pollutant degradation (Raghukumar, 2017). Future directions include synthetic biology approaches to engineer fungi with enhanced pollutant specificity and tolerance to marine stressors like salinity and low oxygen.

5.4 *Field Trials and Scalable Bioremediation Systems*

Despite promising laboratory results, field-scale applications of marine fungi remain limited. Future efforts must focus on pilot studies and environmental field trials to evaluate fungal efficacy in real-world marine conditions. These trials can help assess survival, pollutant degradation rates, ecological impacts, and the resilience of fungal strains in variable marine environments. Establishing fungal-based bioreactors or deploying floating fungal mats in oil-contaminated or eutrophic zones may offer scalable solutions for continuous remediation.

5.5 *Policy Integration and Ecosystem-Based Management*

The integration of marine fungal bioremediation into national and international marine pollution management frameworks remains a key challenge. Future directions should involve multi-stakeholder collaboration among scientists, environmental agencies, policymakers, and the blue economy sector. Promoting fungal bioremediation as a nature-based solution could complement traditional mechanical and chemical cleanup methods and reduce ecological harm while promoting biodiversity.

6. Conclusion

Marine fungi offer a potent, yet underutilized, solution to pressing marine pollution issues, namely oil spills, plastic debris, and excess biomass linked to eutrophication. Strategic approaches combining biotechnology, ecological field studies, and policy support will be essential to unlock their full bioremediation potential. As research accelerates and practical deployment models are tested, marine fungi may become a cornerstone in sustainable marine ecosystem restoration and pollution mitigation.

References

AbuQamar, S.F., Abd El-Fattah, H.I., Nader, M.M., Zaghloul, R.A., Abd El-Mageed, T.A. et al. (2023). Exploiting fungi in bioremediation for cleaning up emerging pollutants in aquatic ecosystems. Mar. Environ. Res., 190: 106068. https://doi.org/10.1016/j.marenvres.2023.106068.

Ahmed, T., Shahid, M., Azeem, F., Rasul, I., Shah, A.A. et al. (2018). Biodegradation of plastics: Current scenario and future prospects for environmental safety. Environ. Sci. Pollut. Res., 25: 7287–7298. https://doi.org/10.1007/s11356-017-9339-x.

Ahumada-Rudolph, R., Novoa, V., Sáez, K., Martínez, M., Rudolph, A. et al. (2016). Marine fungi isolated from Chilean fjord sediments can degrade oxytetracycline. Environ. Monit. Assess., 188: 8. https://doi.org/10.1007/s10661-016-5475-0.

Akinnawo, S.O. (2023). Eutrophication: Causes, consequences, physical, chemical, and biological techniques for mitigation strategies. Environ. Chall., 12. https://doi.org/10.1016/j.envc.2023.100733.

Alabssawy, A.N. and Hashem, A.H. (2024). Bioremediation of hazardous heavy metals by marine microorganisms: A recent review. Arch. Microbiol., 206: 3. https://doi.org/10.1007/s00203-023-03793-5.

Ali, H., Khan, E. and Ilahi, I. (2019). Environmental chemistry and ecotoxicology of hazardous heavy metals: Environmental persistence, toxicity, and bioaccumulation. J. Chem., 2019: 1–14. https://doi.org/10.1155/2019/6730305.

Aliani, S. and Molcard, A. (2003). Hitch-hiking on floating marine debris: Macrobiotic species in the Western Mediterranean Sea. Hydrobiologia, 503: 59–67.

Almeida, R. (2023). The effects of oil spills on marine life and coastal communities. Int. J. Geogr. Geol. Environ., 5: 263–265.

Alshehrei, F. (2017). Biodegradation of low density polyethylene by fungi isolated from Red Sea water. Int. J. Curr. Microbiol. Appl. Sci., 6: 1703–1709. https://doi.org/10.20546/ijcmas.2017.608.204.

Alvernia, P., Utomo, S., Soesilo, T.E.B. and Herdiansyah, H. (2021). Studies of fishermen's economic loss due to oil spills. *In*: IOP Conf. Ser.: Earth Environ. Sci., 802: 1. https://doi.org/10.1088/1755-1315/802/1/012002.

Ameen, F., Moslem, M., Hadi, S. and Al-Sabri, A.E. (2015). Biodegradation of low density polyethylene (LDPE) by mangrove fungi from the Red Sea coast. Prog. Rubber Plast. Recycl. Technol., 31: 125–143. https://doi.org/10.1177/147776061503100204.

Andrady, A.L. (2011). Microplastics in the marine environment. Mar. Pollut. Bull., 62: 1596–1605. https://doi.org/10.1016/j.marpolbul.2011.05.030.

Asemoloye, M.D., Jonathan, S.G. and Ahmad, R. (2019). Synergistic plant-microbes interactions in the rhizosphere: A potential headway for the remediation of hydrocarbon polluted soils. Int. J. Phytorem., 21: 71–83. https://doi.org/10.1080/15226514.2018.1474437.

Atakpa, E.O., Yan, B., Okon, S.U., Liu, Q., Zhang, D. et al. (2024). Asynchronous application of modified biochar and exogenous fungus *Scedosporium* sp. ZYY for enhanced degradation of oil-contaminated intertidal mudflat sediment. Environ. Sci. and Pollut. Res., 31: 20637–20650. https://doi.org/10.1007/s11356-024-32419-6.

Balabanova, L., Slepchenko, L., Son, O. and Tekutyeva, L. (2018). Biotechnology potential of marine fungi degrading plant and algae polymeric substrates. Front. Microbiol., 9. https://doi.org/10.3389/fmicb.2018.01527.

Barnes, N.M., Khodse, V.B., Lotlikar, N.P., Meena, R.M. and Damare, S.R. (2018). Bioremediation potential of hydrocarbon-utilizing fungi from select marine niches of India. 3 Biotech, 8: 1. https://doi.org/10.1007/s13205-017-1043-8.

Barnes, D.K., Galgani, F., Thompson, R.C. and Barlaz, M. (2009). Accumulation and fragmentation of plastics debris in global environment. Philos. Trans. R. Soc. Lond B Biol. Sci., 364: 1985–1998. https://doi.org/10.1098/rstb.2008.0205.

Basheer, S.M., Akshatha, B.S., Anupama, K.P., Arun Kumar, G. and Sreeja Chellappan. (2024). Hydrophobins from Marine Fungi: A comprehensive review. Af. J. Biomed. Res., 27(4s): 10344–10350. https://doi.org/10.53555/AJBR.v27i4S.5655.

Bholay, A.D., Borkhataria, B.V., Jadhav, P.U., Palekar, K.S., Dhalkari, M.V. et al. (2012). Bacterial lignin peroxidase: A tool for bioleaching and biodegradation of industrial effluents. Univers. J. Environ. Res. Technol., 2: 58–64.

Bonugli-Santos, R.C., dos Santos Vasconcelos, M.R., Passarini, M.R.Z., Vieira, G.A., Lopes, V.C. et al. (2015). Marine-derived fungi: Diversity of enzymes and biotechnological applications. Front. Microbiol., 6. https://doi.org/10.3389/fmicb.2015.00269.

Bosker, T., Bouwman, L.J., Brun, N.R., Behrens, P. and Vijver, M.G. (2019). Microplastics accumulate on pores in seed capsule and delay germination and root growth of the terrestrial vascular plant *Lepidium sativum*. Chemosphere, 226: 774–781. https://doi.org/10.1016/j.chemosphere.2019.03.163.

Breyer, E. and Baltar, F. (2023). The largely neglected ecological role of oceanic pelagic fungi. Tr. Ecol. Evol., 38: 870–888. https://doi.org/10.1016/j.tree.2023.05.002.

Brunner, I., Fischer, M., Rüthi, J., Stierli, B. and Frey, B. (2018). Ability of fungi isolated from plastic debris floating in the shoreline of a lake to degrade plastics. PLoS One, 13: e0202047. https://doi.org/10.1371/journal.pone.0202047.

Carbery, M., O'Connor, W. and Thavamani, P. (2018). Trophic transfer of microplastics and mixed contaminants in the marine food web and implications for human health. Environ. Int., 115: 400–409. https://doi.org/10.1016/j.envint.2018.03.007.

Chen, H., Wang, M., Shen, Y. and Yao, S. (2014). Optimization of two-species whole-cell immobilization system constructed with marine-derived fungi and its biological degradation ability. Chin. J. Chem. Eng., 22: 187–192. https://doi.org/10.1016/S1004-9541(14)60024-0.

Črešnar, B. and Petrič, Š. (2011). Cytochrome P450 enzymes in the fungal kingdom. Biochim. Biophys. Acta (BBA)—Proteins and Proteomics, 1814: 29–35. https://doi.org/10.1016/j.bbapap.2010.06.020.

Damare, S., Raghukumar, C. and Raghukumar, S. (2006). Fungi in deep-sea sediments of the Central Indian Basin. Deep Sea Res. I: Oceanogr. Res. Pap., 53: 14–27. https://doi.org/10.1016/j.dsr.2005.09.005.

Danso, D., Chow, J. and Streit, W.R. (2019). Plastics: Environmental and biotechnological perspectives on microbial degradation. Appl. Environ. Microbiol., 85: e01095-19. https://doi.org/10.1128/AEM.01095-19.

Dashtban, M., Schraft, H., Syed, T.A. and Qin, W. (2010). Fungal biodegradation and enzymatic modification of lignin. Int. J. Biochem. Mol. Biol., 1: 36–50.

De-la-Torre, G.E. (2020). Microplastics: An emerging threat to food security and human health. J. Food Sci. Technol., 57: 1601–1608. https://doi.org/10.1007/s13197-019-04138-1.

Dell'Anno, F., Rastelli, E., Buschi, E., Barone, G., Beolchini, F. et al. (2022). Fungi can be more effective than bacteria for the bioremediation of marine sediments highly contaminated with heavy metals. Microorganisms, 10: 5. https://doi.org/10.3390/microorganisms10050993.

Deshmukh, R., Khardenavis, A.A. and Purohit, H.J. (2016). Diverse metabolic capacities of fungi for bioremediation. Indian J. Microbiol., 56: 247. https://doi.org/10.1007/S12088-016-0584-6.

Devi, R.S., Kannan, V.R., Nivas, D., Kannan, K., Chandru, S. et al. (2015). Biodegradation of HDPE by *Aspergillus* spp. from marine ecosystem of Gulf of Mannar, India. Mar. Pollut. Bull., 96: 32–40. https://doi.org/10.1016/j.marpolbul.2015.05.050.

Devlin, M. and Brodie, J. (2023). Nutrients and eutrophication. pp. 75–100. *In*: Reichelt-Brushett, A. (eds.). Marine Pollution: Monitoring, Management, and Mitigation. Springer Textbooks in Earth Sciences, Geography and Environment. Springer. https://doi.org/10.1007/978-3-031-10127-4_4.

Dolbeth, M. and Arenas, F. (2021). Marine Ecosystems: Types, their importance, and main impacts. pp. 1–17. *In*: Filho, W.L., Azul, A.M. Brandli, L. Salvia, A.L., Wall, T. et al. (eds.). Life below water (Encyclopedia of the UN Sustainable Development Goals). Springer Nature Switzerland AG. https://doi.org/10.1007/978-3-319-71064-8_38-1.

Drzyzga, O. and Prieto, A. (2018). Plastic waste management, a matter for the "community". Microb. Biotechnol., 12: 66–68. https://doi.org/10.1111/1751-7915.13328.

Dudhagara, D.R., Javia, B.M. and Vala, A.K. (2023). Exploiting marine fungi in the removal of hazardous pollutants and biomass valorisation. pp. 117–146. *In*: Encarnação, T. and Canelas Pais, A. (eds.). Marine organisms: A solution to environmental pollution? Uses in bioremediation and in biorefinery. Environmental Challenges and Solutions. Springer, Cham. https://doi.org/10.1007/978-3-031-17226-7_7.

Eltemsah, Y.S. and Bøhn, T. (2019). Acute and chronic effects of polystyrene microplastics on juvenile and adult Daphnia magna. Environ. Pollut., 254: 113070. https://doi.org/10.1016/j.envpol.2019.07.087.

Furukawa, M., Kawakami, N., Tomizawa, A. and Miyamoto, K. (2019). Efficient degradation of Poly (ethylene terephthalate) with *Thermobifida fusca* cutinase exhibiting improved catalytic activity generated using mutagenesis and additive-based approaches. Sci. Rep., 9(1): 16038. doi: 10.1038/s41598-019-52379-z.

Gao, R., Liu, R. and Sun, C. (2022). A marine fungus *Alternaria alternata* FB1 efficiently degrades polyethylene. J. Hazard. Mater., 431: 128617. https://doi.org/10.1016/j.jhazmat.2022.128617.

Gao, W., Xu, M., Zhao, W., Yang, X., Xin, F. et al. (2024). Microbial degradation of (micro)plastics: Mechanisms, enhancements, and future directions. Fermentation, 10: 441. https://doi.org/10.3390/fermentation10090441.

Gallo, F., Fossi, C., Weber, R., Santillo, D., Sousa, J. et al. (2018). Marine litter plastics and microplastics and their toxic chemical components: The need for urgent preventive measures. Environ. Sci. Eur., 30: 13. https://doi.org/10.1186/s12302-018-0139-z.

Gemeren, I.A.V., Beijersbergen, A., Van den Hondel, C.A.M.J.J. and Verrips, C.T. (1998). Expression and secretion of defined cutinase variants by *Aspergillus awamori*. Appl. Environ. Microbiol., 64: 2794–2799. https://doi.org/10.1128/AEM.64.7.2794-2799.1998.

GEF (2012). Secretariat of the Convention on Biological Diversity and Scientific and Technical Advisory Panel - Global Environment Facility Impacts of marine debris on biodiversity: Current status and potential solutions (CBD Technical Series No. 67) [Montreal. May 2025]. https://www.cbd.int/doc/publications/cbd-ts-67-en.pdf.

Geyer, R., Jambeck, J.R. and Law, K.L. (2017). Production, use, and fate of all plastics ever made. Sci. Adv., 3: e1700782. https://doi.org/10.1126/sciadv.1700782.

Giovanella, P., Taketani, R.G., Gil-Solsona, R., Saldanha, L.L., Naranjo, S.B.E. et al. (2023). A comprehensive study on diesel oil bioremediation under microcosm conditions using a combined microbiological, enzymatic, mass spectrometry, and metabarcoding approach. Environ. Sci. Pollut. Res., 30: 101250–101266. https://doi.org/10.1007/s11356-023-29474-w.

Goldberg, E.D. (1997). Plasticizing the sea floor: An overview. Environ. Technol., 18: 195–202. https://doi.org/10.1080/09593331808616527.

Green, D.S. (2020). Biological and ecological impacts of plastic debris in aquatic ecosystems. Riv. Biol./Biol. Forum, 113: 219–232.

Greeshma, K., Kim, H.S. and Ramanan, R. (2022). The emerging potential of natural and synthetic algae-based microbiomes for heavy metal removal and recovery from wastewaters. Environ, Res., 215. https://doi.org/10.1016/j.envres.2022.114238.

Gregory, M.R. (2009). Environmental implications of plastic debris in marine settings: Entanglement, ingestion, smothering, hangers-on, hitch-hiking, and alien invasions. Philos. Trans. R. Soc. Lond., B Biol. Sci., 364: 2013–2025. https://doi.org/10.1098/rstb.2008.0285.

Guerranti, C., Martellini, T., Perra, G., Scopetani, C. and Cincinelli, A. (2019). Microplastics in cosmetics: Environmental issues and needs for global bans. Environ. Toxicol. Pharmacol., 68: 75–79. https://doi.org/10.1016/j.etap.2019.03.007.

Hammer, J., Kraak, M.H.S. and Parsons, J.R. (2012). Plastics in the marine environment: The dark side of a modern gift. Rev. Environ. Contam. Toxicol., 220: 1–44. https://doi.org/10.1007/978-1-4614-3414-6_1.

Han, Z., Moh, E.S., Santos, A.L., Barcellos, I.C., Peng, Y. et al. (2023). Dechlorination of wastewater from shell-based glucosamine processing by mangrove wetland-derived fungi. Front. Microbiol., 14. https://doi.org/10.3389/fmicb.2023.1271286.

Hanafiah, Z.M., Mohtar, W.H.M.W., El Messaoudi, N., Miyah, Y., Maged, A. et al. (2025). Wastewater treatment using Ganoderma species via ganoremediation bioreactor: A comprehensive review. Bioresour. Technol. Rep., 30. https://doi.org/10.1016/j.biteb.2025.102105.

Harms, H., Schlosser, D. and Wick, L.Y. (2011). Untapped potential: Exploiting fungi in bioremediation of hazardous chemicals. Nat. Rev. Microbiol., 9: 177–192. https://doi.org/10.1038/nrmicro2519.

He, S., Jia, M., Xiang, Y., Song, B., Xiong, W. et al. (2022). Biofilm on microplastics in aqueous environment: Physicochemical properties and environmental implications. J. Hazard. Mater., 424: 127286. https://doi.org/10.1016/j.jhazmat.2021.127286.

Hughes, K.A., Bridge, P. and Clark, M.S. (2007). Tolerance of Antarctic soil fungi to hydrocarbons. Sci. Total Environ., 372: 539–548. https://doi.org/10.1016/j.scitotenv.2006.09.016.

Isangedighi, I.A., David, G.S. and Obot, O.I. (2018). Plastic waste in the aquatic environment: Impacts and management. Int. J. Environ. Sci. Nat. Resour., 26: 556195. https://doi.org/10.31058/j.envi.2018.21001.

Jambeck, J.R., Geyer, R., Wilcox, C., Siegler, T.R., Perryman, M. et al. (2015). Plastic waste inputs from land into the ocean. Science, 347: 768–771. https://doi.org/10.1126/science.1260352.

Jambeck, J.R. and Walker-Franklin, I. (2023). The impacts of plastics' life cycle. One Earth, 6: 582–587. https://doi.org/10.1016/j.oneear.2023.05.015.

Kadri, T., Rouissi, T., Brar, S.K., Cledon, M., Sarma, S. et al. (2017). Biodegradation of polycyclic aromatic hydrocarbons (PAHs) by fungal enzymes: A review. J. Environ. Sci., 51: 52–74. https://doi.org/10.1016/j.jes.2016.08.023.

Kale, S.K., Deshmukh, A.G. and Dudhare, M.S. (2015). Microbial degradation of plastic: A review. J. Biochem. Technol., 6: 952–961.

Karich, A., Ullrich, R., Scheibner, K., Hofrichter, M. et al. (2017). Fungal unspecific peroxygenases oxidize the majority of organic EPA priority pollutants. Front. Microbiol., 8: 1463. https://doi.org/10.3389/fmicb.2017.01463.

Khambhaty, Y., Mody, K., Basha, S. and Jha, B. (2008). Hg(II) removal from aqueous solution by dead fungal biomass of marine *Aspergillus niger*: Kinetic studies. Separation Science and Technology, 43: 1221–1238. https://doi.org/10.1080/01496390801888235.

Kumar, A. and Chandra, R. (2020). Ligninolytic enzymes and its mechanisms for degradation of lignocellulosic waste in environment. Heliyon, 6: 2. https://doi.org/10.1016/j.heliyon.2020.e03170.

Lai, J., Huang, H., Lin, M., Xu, Y., Li, X. and Sun, B. (2022). Enzyme catalyzes ester bond synthesis and hydrolysis: The key step for sustainable usage of plastics. Front. Microbiol., 13: 1113705. https://doi.org/10.3389/fmicb.2022.1113705.

Lan, J., Liu, P., Hu, X. and Zhu, S. (2024). Harmful algal blooms in eutrophic marine environments: Causes, monitoring, and treatment. Water, 16: 17. https://doi.org/10.3390/w16172525.

Li, J., Xu, Y., Song, Q., Zhang, S., Xie, L. et al. (2021). Transmembrane transport mechanism of n-hexadecane by *Candida tropicalis*: Kinetic study and proteomic analysis. Ecotoxicol. Environ. Saf., 209. https://doi.org/10.1016/j.ecoenv.2020.111789.

Li, Q., Liu, J. and Gadd, G.M. (2020). Fungal bioremediation of soil co-contaminated with petroleum hydrocarbons and toxic metals. Appl. Microbiol. Biotechnol., 104: 8999–9008. https://doi.org/10.1007/s00253-020-10854-y/Published.

Lotfinasabasl, S., Gunale, V.R. and Rajurkar, N.S. (2012). Assessment of petroleum hydrocarbon degradation from soil and tarball by fungi. Biosci. Discov., 3: 186–192. http://www.biosciencediscovery.com.

Luo, W., Vrijmoed, L.L.P. and Jones, E.B.G. (2005). Screening of marine fungi for lignocellulose-degrading enzyme activities. Bot. Mar.. 48: 379–386. https://doi.org/10.1515/BOT.2005.051.

Maamar, A., Lucchesi, M.E., Debaets, S., Nguyen van Long, N., Quemener, M. et al. (2020). Highlighting the crude oil bioremediation potential of marine fungi isolated from the Port of Oran (Algeria). Diversity, 12(5). https://doi.org/10.3390/D12050196.

Mahalakshmi, V., Siddiq, A. and Niren, A.S. (2012). Analysis of polyethylene degrading potentials of microorganisms isolated from compost soil. Int. J. Pharm. Biol. Arch., 3: 1190–1196.

Mahmoud, M.E., El Zokm, G.M., Farag, A.E. and Abdelwahab, M.S. (2017). Assessment of heat-inactivated marine Aspergillus flavus as a novel biosorbent for removal of Cd(II), Hg(II), and Pb(II) from water. Environ. Sci. Pollut. Res., 24: 18218–18228. https://doi.org/10.1007/s11356-017-9323-8.

Marsay, K.S., Ambrosino, A.C., Koucherov, Y., Davidov, K., Figueiredo, N. et al. (2023). The geographical and seasonal effects on the composition of marine microplastic and its microbial communities: The case study of Israel and Portugal. Front. Microbiol., 14: 1089926. https://doi.org/10.3389/fmicb.2023.1089926.

Montazer, Z., Najafi, M.B.H. and Levin, D.B. (2019). Microbial degradation of low-density polyethylene and synthesis of polyhydroxyalkanoate polymers. Can. J. Microbiol., 65: 1–11. https://doi.org/10.1139/cjm-2018-0386.

Moore, C.J. (2008). Synthetic polymers in the marine environment: A rapidly increasing, long term threat. Environ. Res. https://doi.org/10.1016/j.envres.2008.07.025.

Nasrabadi, A.E., Ramavandi, B. and Bonyadi, Z. (2023). Recent progress in biodegradation of microplastics by Aspergillus sp. in aquatic environments. Colloids Interface Sci. Commun., 57. https://doi.org/10.1016/j.colcom.2023.100754.

Negi, B.B. and Das, C. (2023). Mycoremediation of wastewater, challenges, and current status: A review. Bioresour. Technol. Rep., 22. https://doi.org/10.1016/j.biteb.2023.101409.

Nelms, S.E., Galloway, T.S., Godley, B.J., Jarvis, D.S. and Lindeque, P.K. (2018). Investigating microplastic trophic transfer in marine top predators. Environ. Pollut., 238: 999–1007. https://doi.org/10.1016/j.envpol.2018.02.016.

Ng, C.H., Mistoh, M.A., Teo, S.H., Galassi, A., Ibrahim, A. et al. (2023). Plastic waste and microplastic issues in Southeast Asia. Frontiers in Environmental Science Front. Environ. Sci., 11: 1142071. https://doi.org/10.3389/fenvs.2023.1142071.

Ojha, N., Pradhan, N., Singh, S., Barla, A., Shrivastava, A. et al. (2017). Evaluation of HDPE and LDPE degradation by fungus, implemented by statistical optimization. Sci. Rep., 7. https://doi.org/10.1038/srep39515.

Okal, E.J., Heng, G. and Magige, E.A. (2023). Insights into the mechanisms involved in the fungal degradation of plastics. Ecotoxicol. Environ. Saf., 262. https://doi.org/10.1016/j.ecoenv.2023.115202.

Oladimeji, T.E., Oyedemi, M., Emetere, M.E., Agboola, O., Adeoye, J.B. et al. (2024). Review on the impact of heavy metals from industrial wastewater effluent and removal technologies. Heliyon, 10: 23. https://doi.org/10.1016/j.heliyon.2024.e40370.

Paço, A., Duarte, K., da Costa, J.P., Santos, P.S., Pereira, R. et al. (2017). Biodegradation of polyethylene microplastics by the marine fungus *Zalerion maritimum*. Sci. Total Environ., 586: 10–15. https://doi.org/10.1016/j.scitotenv.2017.02.017.

Passarini, M.R., Rodrigues, M.V., da Silva, M. and Sette, L.D. (2011). Marine-derived filamentous fungi and their potential application for polycyclic aromatic hydrocarbon bioremediation. Mar. Pollut. Bull., 62: 364–370. https://doi.org/10.1016/j.marpolbul.2010.10.003.

Pecoraro, L., Wang, X., Shah, D., Song, X., Kumar, V. et al. (2022). Biosynthesis pathways, transport mechanisms, and biotechnological applications of fungal siderophores. J. Fungi, 8: 1. https://doi.org/10.3390/jof8010021.

Raghukumar, C. (2008). Marine fungal biotechnology: An ecological perspective. Fungal Diver. http://drs.nio.org/drs/handle/2264/1392.

Raghukumar, S. (2017). Fungi in Coastal and Oceanic Marine Ecosystems. Springer International Publishing. https://doi.org/10.1007/978-3-319-54304-8.

Ramya, D., Kiruba, N.J.M. and Thatheyus, A.J. (2020). Biosorption of heavy metals using fungal biosorbents: A review. pp. 331–352. *In*: Sharma, V.K., Shah, M.P., Parmar, S. and Kumar, A. (eds.). Fungi Bio-prospects in Sustainable Agriculture, Environment, and Nano-technology: (Volume 2): Extremophilic Fungi and Myco-mediated Environmental Management. Elsevier. https://doi.org/10.1016/B978-0-12-821925-6.00015-0.

Rashid, A., Bhatti, H.N., Iqbal, M. and Noreen, S. (2016). Fungal biomass composite with bentonite efficiency for nickel and zinc adsorption: A mechanistic study. Ecol. Eng., 91: 459–471. https://doi.org/10.1016/j.ecoleng.2016.03.014.

Rodriguez, J.P., Williams, D.E., Sabater, I.D., Bonugli-Santos, R.C., Sette, L.D. et al. (2015). The marine-derived fungus Tinctoporellus sp. CBMAI 1061 degrades the dye Remazol Brilliant Blue R producing anthraquinones and unique tremulane sesquiterpenes. RSC Adv., 5: 66360–66366. https://doi.org/10.1039/C5RA13580J.

Sáenz, M., Borodulina, T., Diaz, L. and Banchon, C. (2019). Minimal conditions to degrade low density polyethylene by Aspergillus terreus and *A. niger*. J. Ecol. Eng., 20: 40–51. https://doi.org/10.12911/22998993/100685.

Sanchez, C. (2009). Lignocellulosic residues: Biodegradation and bioconversion by fungi. Biotechnol. Adv., 27: 185–194. https://doi.org/10.1016/j.biotechadv.2008.11.001.

Sánchez, C. (2020). Fungal potential for the degradation of petroleum-based polymers: An overview of macro- and microplastics biodegradation. Biotechnol. Adv., 40: 107501. https://doi.org/10.1016/j.biotechadv.2020.107501.

Sanchez, C. (2021). Microbial capability for the degradation of chemical additives present in petroleum-based plastic products: A review on current status and perspectives. J. Hazard. Mater., 402: 123534. https://doi.org/10.1016/j.jhazmat.2020.123534.

Sangale, M.K., Shahnawaz, M. and Ade, A.B. (2019). Potential of fungi isolated from the dumping sites mangrove rhizosphere soil to degrade polythene. Sci. Rep., 9: 1. https://doi.org/10.1038/s41598-019-41448-y.

Saravanan, A., Karishma, S., Kumar, P.S. and Rangasamy, G. (2023). Biodegradation of oil-contaminated aqueous ecosystem using an immobilized fungi biomass and kinetic study. Environ. Res., 220: 115252. https://doi.org/10.1016/j.envres.2023.115252.

Selvam, S., Jesuraja, K., Venkatramanan, S., Roy, P.D. and Kumari, V.J. (2021). Hazardous microplastic characteristics and its role as a vector of heavy metal in groundwater and surface water of coastal south India. J. Hazard. Mater., 402: 123786. https://doi.org/10.1016/j.jhazmat.2020.123786.

Shah, A.A., Hasan, F. and Hameed, A. (2008). Biological degradation of plastics: A comprehensive review. Biotechnol. Adv., 26: 246–265. https://doi.org/10.1016/j.biotechadv.2007.12.005.

Sharuddin, S.D.A., Abnisa, F., Daud, W.M.A.W. and Aroua, M.K. (2016). A review on pyrolysis of plastic wastes. Energy Convers. Manag., 115: 308–326. https://doi.org/10.1016/j.enconman.2016.02.037.

Silva, A.B., Bastos, A.S., Justino, C.I., da Costa, J.P., Duarte, A.C. et al. (2018). Microplastics in the environment: Challenges in analytical chemistry: A review. Anal. Chim. Acta, 1017: 1–19. https://doi.org/10.1016/j.aca.2018.02.043.

Simister, R.L., Poutasse, C.M., Thurston, A.M., Reeve, J.L., Baker, M.C. et al. (2015). Degradation of oil by fungi isolated from Gulf of Mexico beaches. Mar. Pollut. Bull., 100: 327–333. https://doi.org/10.1016/j.marpolbul.2015.08.029.

Singh, P., Raghukumar, C. and Parvatkar, R. (2013). Heavy metal tolerance in the psychrotolerant *Cryptococcus* sp. isolated from deep-sea sediments of the Central Indian Basin. Yeast, 30: 93–101. https://doi.org/10.1002/yea.2943.

Smith, M., Love, D.C., Rochman, C.M., Neff, R.A. et al (2018). Microplastics in seafood and the implications for human health. Curr. Environ. Health Rep., 5: 375–386. https://doi.org/10.1007/s40572-018-0206-z.

Soundararajan, N. (2021). Pune researchers discover fungi that can break up polythene. Research Matters. https://researchmatters.in/news/pune-researchers-discover-fungi-can-break-polythene (accessed, 1 June 2025].

Sowmya, H.V., Ramalingappa, B., Nayanashree, G., Thippeswamy, B. and Krishnappa, M. (2015). Polyethylene degradation by fungal consortium. Int. J. Environ. Res., 9: 823–830.

Šrejber, M., Navrátilová, V. and Paloncýová, M. (2018). Membrane-attached mammalian cytochromes P450: An overview of the membrane's effects on structure, drug binding, and interactions with redox partners. J. Inorg. Biochem., 183: 117–136. https://doi.org/10.1016/j.jinorgbio.2018.03.002.

Srikanth, M., Sandeep, T.S.R.S. and Sucharitha, K. (2022). Biodegradation of plastic polymers by fungi: A brief review. Bioresour. Bioprocess., 9: 42. https://doi.org/10.1186/s40643-022-00532-4.

Teuten, E.L., Saquing, J.M., Knappe, D.R.U., Barlaz, M.A., Jonsson, S. et al. (2009) Transport and release of chemicals from plastics to the environment and to wildlife. Philos. Trans. R. Soc. B Biol. Sci., 364: 2027–2045. https://doi.org/10.1098/rstb.2008.0284.

Thirunavukarasu, K., Edwinoliver, N.G., Anbarasan, S.D., Gowthaman, M.K., Iefuji, H. et al. (2008). Removal of triglyceride soil from fabrics by a novel lipase from *Cryptococcus* sp. Process Biochem., 43: 701–706. https://doi.org/10.1016/j.procbio.2007.12.011.

Thompson, R.C., Olsen, Y., Mitchell, R.P., Davis, A., Rowland, S.J. et al. (2004). Lost at sea: Where is all the plastic? Science, 304: 838. https://doi.org/10.1126/science.1094559.

Thompson, R.C., Moore, C.J. and Vom Saal, F.S. (2009) Plastics, the environment, and human health: Current consensus and future trends. Philos. Trans. R. Soc. Lond. B Biol. Sci., 364: 2153–2166.

Tournier, V., Topham, C.M. and Gilles, A. (2020). An engineered PET depolymerase to break down and recycle plastic bottles. Nature, 580: 216–219. https://doi.org/10.1038/s41586-020-2149-4.

Tu, C., Chen, T., Zhou, Q., Liu, Y., Wei, J. et al. (2020). Biofilm formation and its influences on the properties of microplastics as affected by exposure time and depth in the seawater. Sci. Total Environ., 734: 139237. https://doi.org/10.1016/j.scitotenv.2020.139237.

Vaksmaa, A., Guerrero-Cruz, S., Ghosh, P., Zeghal, E., Hernando-Morales, V. et al. (2023). Role of fungi in bioremediation of emerging pollutants. Front. Mar. Sci., 10: https://doi.org/10.3389/fmars.2023.1070905.

Velez, P., Gasca-Pineda, J. and Riquelme, M. (2020). Cultivable fungi from deep-sea oil reserves in the Gulf of Mexico: Genetic signatures in response to hydrocarbons. Mar. Environ. Res. 153. https://doi.org/10.1016/j.marenvres.2019.104816.

Vignier, J., Volety, A., Soudant, P., Chu, F.L., Loh, A.N. et al. (2019). Evaluation of the toxicity of the deepwater horizon oil and associated dispersant on early life stages of the Eastern Oyster, *Crassostrea virginica*. Sep. Sci. Technol., 11: 169–198. https://doi.org/10.1016/B978-0-12-815730-5.00008-9.

Wang, W., Ge, J. and Yu, X. (2020). Bioavailability and toxicity of microplastics to fish species: A review. Ecotoxicol. Environ. Saf., 189: 109913. https://doi.org/10.1016/j.ecoenv.2019.109913.

Wang, C., Zhao, J. and Xing, B. (2021). Environmental source, fate, and toxicity of microplastics. J. Hazard. Mater., 407: 124357. https://doi.org/10.1016/j.jhazmat.2020.124357.

Webb, H.K., Arnott, J., Crawford, R.J. and Ivanova, E.P. (2013). Plastic degradation and its environmental implications with special reference to poly(ethylene terephthalate). Polymers, 5: 1–18. https://doi.org/10.3390/polym5010001.

Weber, A., von Randow, M., Voigt, A.L., von der Au, M., Fischer, E. et al. (2020). Ingestion and toxicity of microplastics in the freshwater gastropod Lymnaea stagnalis: No microplastic-induced effects alone or in combination with copper. Chemosphere, 128040. https://doi.org/10.1016/j.chemosphere.2020.128040.

Wen, Z., Xie, Y., Chen, M. and Dinga, C.D. (2021). China's plastic import ban increases prospects of environmental impact mitigation of plastic waste trade flow worldwide. Nat. Commun., 12: 425. https://doi.org/10.1038/s41467-020-20741-9.

Wright, S.L., Thompson, R.C. and Galloway, T.S. (2013). The physical impacts of microplastics on marine organisms: A review. Environ. Pollut., 178: 483–492. https://doi.org/10.1016/j.envpol.2013.02.031.

Wright, R.J., Erni-Cassola, G. and Zadjelovic, V. (2020). Marine plastic debris: A new surface for microbial colonization. Environ. Sci. Technol., 54: 11657–11672. https://doi.org/10.1021/acs.est.0c03778.

Yao, D., Li, H., Dai, Y. and Wang, C.H. (2021). Impact of temperature on the activity of Fe-Ni catalysts for pyrolysis and decomposition processing of plastic waste. Chem. Eng. J., 408: 127268. https://doi.org/10.1016/j.cej.2020.127268.

Yu, Q., Hu, X., Yang, B., Zhang, G., Wang, J. et al. (2020). Distribution, abundance, and risks of microplastics in the environment. Chemosphere, 249: 126059. https:// doi.org/10.1016/j.chemosphere.2020.126059.

Zhang, X., Wang, X., Li, C., Zhang, L., Ning, G. et al. (2020). Ligninolytic enzyme involved in removal of high molecular weight polycyclic aromatic hydrocarbons by Fusarium strain ZH-H2. Environ. Sci. Pollut. Res., 27: 42969–42978. https://doi.org/10.1007/s11356-020-10192-6.

10

Marine Fungi in Bioremediation for a Sustainable Ecosystem

Sabrin R.M. Ibrahim,[1,2,*] *Hagar M. Mohamed,*[3,4] *Mahmoud A. Elfaky*[5,6] and *Gamal A. Mohamed*[5,*]

Marine ecological systems make up two-thirds of the planet and host over 80% of global biodiversity, although they are still understudied in comparison to terrestrial life. Among marine-derived microorganisms, fungi provide a sustainable pathway for the synthesis of biometabolites and enzymes with environmental and/or industrial importance because of their distinct metabolic characteristics and ability to thrive in harsh environments, like low nutrients, high pressure, and extreme salt conditions. Marine fungi have shown great promise in bioremediation as a sustainable, biological and environmentally responsible approach to converting potentially hazardous, recalcitrant pollutants into products that are safe for the environment. This study discusses the role of marine fungi in elimination and converting a variety of contaminants, including plastics, dyes, hydrocarbons, and persistent organic contaminants.

[1] Preparatory Year Program, Department of Chemistry, Batterjee Medical College, Jeddah 21442, Saudi Arabia.
[2] Department of Pharmacognosy, Faculty of Pharmacy, Assiut University, Assiut 71526, Egypt.
[3] Department of Medical Laboratory Analysis, College of Medical and Health Sciences, Liwa University, Abu Dhabi 41009, United Arab of Emirates.
[4] Department of Applied Medical Chemistry, Medical Research Institute, Alexandria University, Alexandria, Egypt.
[5] Department of Natural Products and Alternative Medicine, Faculty of Pharmacy, King Abdulaziz University, Jeddah 21589, Saudi Arabia.
[6] Centre for Artificial Intelligence in Precision Medicines, King Abdulaziz University, Jeddah 21589, Saudi Arabia.
* Corresponding authors: sabrin.ibrahim@bmc.edu.sa; gahussein@kau.edu.sa

DOI: 10.1201/9781003795100-13

This chapter highlights the untapped potential of marine fungi that assists in understanding fungal-based strategies for promoting global sustainability, particularly in reducing pollution and protecting the environment.

1. Introduction

Mycoremediation is a type of bio-remediation strategy that employs fungi to remove contaminants and waste from the contaminated areas (Akpasi et al., 2023). The conversion of hazardous pollutants to less hazardous products has become more common. Fungi are capable of developing in different harsh environments through spreading their spores in the air, including marine environments, fresh water, and industrialized wastewater treatment plants. Additionally, their ability to secrete various enzymes, making them an optimal choice for bioremediation (Geris et al., 2024). Marine microorganisms have drawn marked attention due to their richness of different natural metabolites with appealing bioactivities. Currently, 20 approved drugs for commercial applications are obtained from marine sources, in addition, more than thirty drugs are at different stages of clinical trials (Haque et al., 2022). Furthermore, investigations currently have emphasized on the enzymes derived from marine sources, which possess unique characteristics and industrial and biotechnological significance (Birolli et al., 2019). Marine fungi face difficulties in the marine environment, such as pressure gradients, fluctuating temperatures, and high salinity (Barzkar and Sohail, 2020). To adapt to these conditions, marine fungi create special enzyme machinery to operate efficiently in the sea. The fungi separated from various marine substrates, including seawater, sediments, algae, sponges, and mangroves, have been proven as manufacturers of a variety of enzymes, i.e., cellulase, laccase, amylase, chitinase, alginate lyase, glucosidase, keratinase, inulinase, xylanase, and ligninase (Bonugli-Santos et al., 2015).

These enzymes can be active in sizable temperature ranges, elevated salt concentrations, metal ions, pH, surfactants, organic solvents, and great incidences of pressure and light, as well as extremely low temperatures (Duarte et al., 2018). Consequently, they might endure vital under diverse working conditions, allowing efficiency and competitiveness in several industrial procedures (Fernandes, 2014), such as in the food, biofuel, pharmaceutical, and fine chemical industries (Dalmaso et al., 2015; Bonugli-Santos et al., 2015). The marine fungal enzymes market includes four sectors: technical enzymes, which are primarily used in the textile, paper, leather, pulp, and biofuel industries; animal feed enzymes; beverage and food enzymes; enzymes with ecological uses; and enzymes with cosmetic and pharmaceutical applications (Bonugli-Santos et al., 2015).

These fungi are also renowned as a pool of bioactive constituents that reveal marked bioactivities, including anti microbial, anti-thrombotic, anti-inflammatory, anti viral, anti cancer, and anti oxidant, with enormous applications in cosmetic, agriculture, medical, and pharmaceutical sectors (Gonçalves et al., 2022; Gao et al., 2022a). Interestingly, the development of biotechnological processes using marine fungal enzyme production promotes sustainable industrial practice that support a number of Sustainable Development Goals (SDGs), such as SDG7 (Affordable and

Clean Energy) by encouraging the use of renewable biofuels, SDG12 (Responsible Consumption and Production) by making effective usage of marine resources, SDG13 (Climate Action) by lowering greenhouse-gas emissions, and SDG14 (Life Below Water) by encouraging the sustainable use of marine resources without interfering with crops grown on land. Marine fungi have been shown as potential options for biotechnological uses, bioremediation, and biomass valorization because of their capability to decompose various synthetic compounds and biomasses, e.g., polycyclic-aromatic hydrocarbons, insecticides, toluene, benzene, xylenes, and ethylbenzene, as well as persistent organic contaminants (Vaksmaa et al., 2023a; Zeghal et al., 2021). The primary target of this chapter is to show the fungal potential in the bioremediation processes. It discusses the crucial role of marine fungi in degrading various hazardous, persistent, and resistant pollutants. A number of recognized pollutant classes and the published literature on the marine fungi potential to mitigate their negative environmental effects were reviewed.

2. Marine Fungi for Micro-embolization

Thrombosis is a leading reason for death worldwide, as it is linked to cardiac illnesses like arrhythmias, acute myocardial infarction, high blood pressure, stroke, and peripheral vascular, valvular heart, and ischemic heart conditions. Thrombi or blood clots are created due to catalyzing fibrinogen transformation into fibrin by thrombin, resulting in fibrin clot formation. According to the WHO, cardiac illnesses accounted for 17.9 million fatalities worldwide yearly, with strokes and heart attacks accounting for more than four out of five deaths from these illnesses. Cardiac illnesses are becoming an international health issue and financial strain (Mensah et al., 2023).

Coagulation and fibrinolysis are two vital operations for controlling the blood vessel's patency. Any inappropriate homeostasis in these operations results in unrestrained fibrinolysis (i.e., grievous bleeding) and coagulation activation (i.e., thrombi) (Jeevithan et al., 2024). In addition to thrombolytic agents, fibrinolytic enzymes are thought to be the most beneficial remedy for the clinical management of thrombosis (Kotb, 2013). However, the use of thrombolytic agents such as alteplase, urokinase, streptokinase, and their analogs is restricted because of their high cost and intolerance risks, such as hypersensitivity reactions, blood loss, and substantial hemorrhage in important organs (Zhang et al., 2023a). Thus, scientists have directed their efforts to discover natural metabolites for achieving safety and marked potential in thrombolysis. Further, the need for microbial thrombolytic agents is rising dramatically and the finding of new microbial strains with elevated levels of fibrinolysis enzyme manufacturing is required. Diverse metabolites from marine sources, particularly fungi, have revealed marked potential as thrombolytic agents. Additionally, marine fungi have the potential for producing fibrinolytic enzymes that are critically important for treating thrombosis-linked conditions.

FGFCs, including FGFC1 to FGFC7, are isoindolone-class metabolites that are separated from marine-derived *S. microspora* IFO-30018 and *Stachybotrys longispora* FG216 (Yan et al., 2015; Guo et al., 2016; Hang et al., 2022) (Fig. 1). These compounds

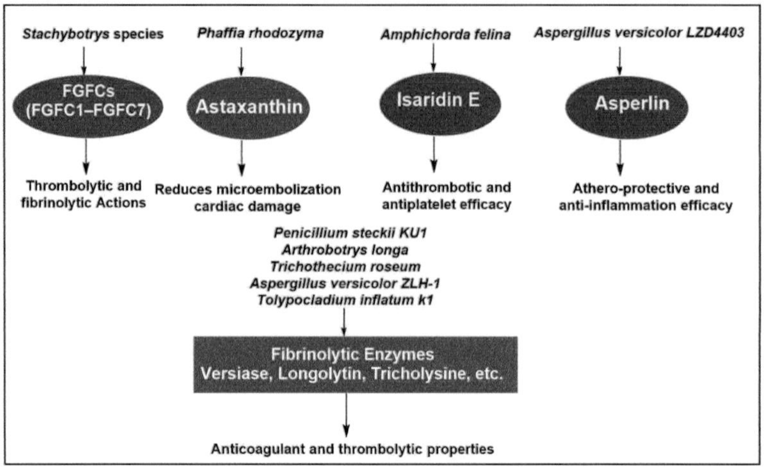

Fig. 1. Role of marine fungi and their metabolites and enzymes in micro-embolization.

revealed valuable impacts for treating various illnesses such as thrombolysis, hemorrhage, cerebral and myocardial infarctions, pulmonary embolism, embolic or ischemic stroke, and cancer (Bailly, 2023). Among these metabolites, FGFC1 is the most investigated compound. Wang et al. (2015) stated that FGFC1 (Conc. 0.1–0.4 M) increased plasmin action by 2.05–11.44-fold via activating Glu-Plg (Glu-plasminogen) and Lys-Plg (Wang et al., 2015).

Yan et al. (2015) established its thrombolytic effectiveness for treating thrombus without the notable possibility of hemorrhagic action through pharmacodynamic investigations of FGFC1 in both thrombolysis and hemolysis by fibrinogenolysis. Evaluating FGFC1 on acute-pulmonary thromboembolism *in vivo* and FITC (fluorescein isothiocyanate)-fibrin degradation *in vitro* revealed that FGFC1 (conc. 5–25 μM) degenerated FITC-fibrin and dissolved extrinsic FITC-fibrin-produced blood clots (doses 5–25 mg/kg) lysis (Yan et al., 2015).

Guo et al. (2016) reported that the bio-guided separation of the *S. longispora* FG216 fibrinolytic EtOAc fraction resulted in FGFC2 and ergosterol. These compounds had moderately fibrinolytic activities *in vitro* (EC_{50}s 108.16 and 156.30 μM/L, respectively), in comparison to FGFC1 (EC_{50} 47.00 μM/L) (Guo et al., 2016). Another study by Gao et al. (2023) proved that FGFC1 augmented the p-nitroaniline origination rate in plasma and interfered with fibrinopeptide B liberation, resulting in a rising protofibril number and clot vulnerability to fibrinolysis by changing its composition (Gao et al., 2023).

Yin et al. (2017) stated that *S. longispora* FG216 culture supply with l-threonine, (±)-3-amino-2-piperidinone, and glycine resulted in new isoindolinone analogs: FGFC4–GFC7 with high yields (Fig. 2). Among them, FGFC6 and FGFC7 possessed prominent fibrinolytic capacities through prourokinase activation to generate plasmin`s active form from the zymogen protease plasminogen, which aided in blood clot deconstruction (Yin et al., 2017).

Astaxanthin is among the xanthophyll-type metabolites that give characteristic marine organism pigmentation, and it was reported from *Phaffia rhodozyma*

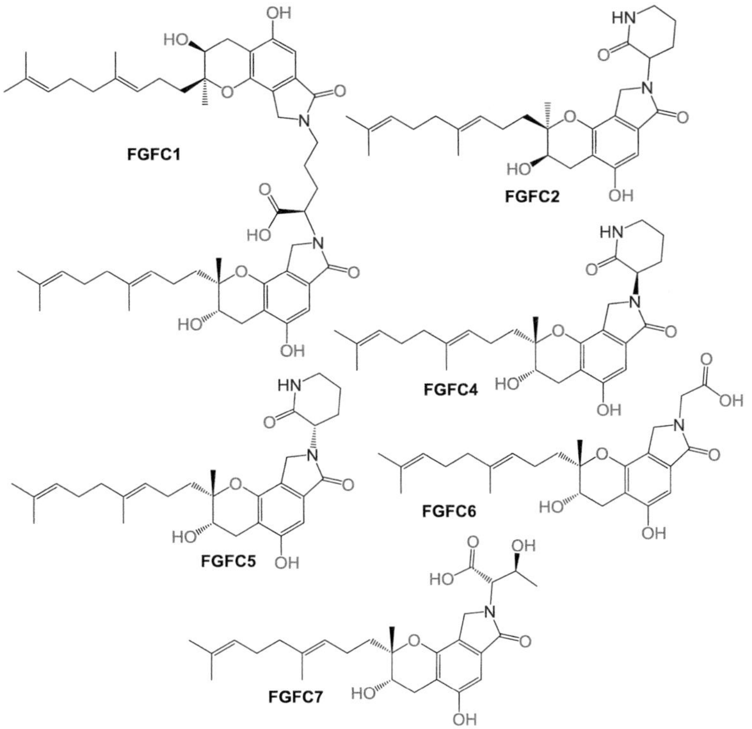

Fig. 2. Chemical structures of FGFCs separated from *Stachybotrys* species.

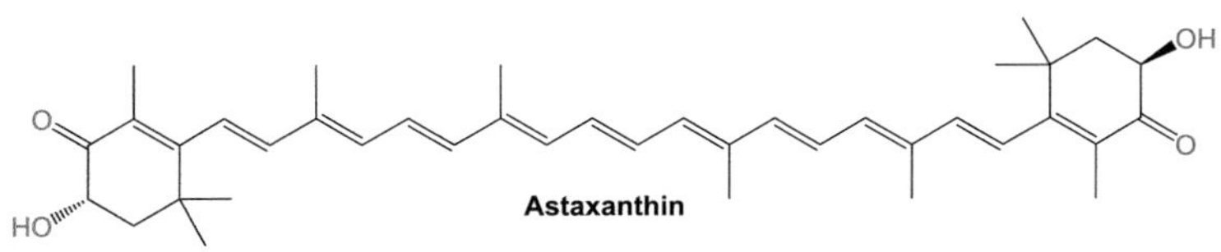

Fig. 3. Chemical structures of astaxanthin separated from *Phaffia rhodozyma.*

yeast (Fig. 3). It was examined for preventing coronary microembolization and hemeoxygenase-1/erythroid 2-related factor-2 pathway activation in rats. The compound alleviated cardiomyocyte cell deaths following coronary micro-embolization and cardiac function impairment by reactive oxygen species repression through hemeoxygenase-1/erythroid 2-related factor-2 pathway activation. Therefore, astaxanthin could have marked potential in cardiovascular safeguarding in rats (Pereira et al., 2021).

The cyclodepsipeptide, isaridin E, biosynthesized by *Amphichorda felina* SYSUMS7908, was examined *in vitro* for its efficacy on platelet functions (Fig. 4). It was found to concentration-dependently repress ADP (adenosine-diphosphate)-caused mouse platelet activation, which is a critical step in forming arterial thrombosis and platelet hyperactivity in cardiovascular illnesses linked to an elevated probability of thrombosis. It was demonstrated that isaridin E especially diminished ADP-caused platelet aggregation, ATP (adenosine triphosphate) releasing, and

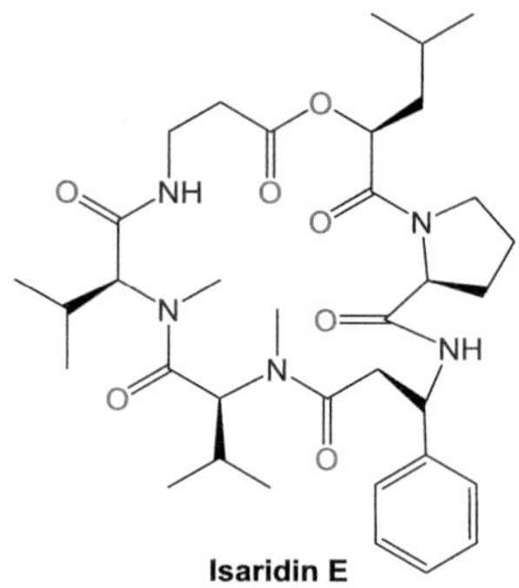

Fig. 4. Chemical structures of isaridin E separated from *Amphichorda felina*.

platelet activation with no influence on aggregation caused by thrombin or collagen. Hence, isaridin E revealed strong antithrombotic and antiplatelet efficacy without significant alteration of coagulation parameters, platelet counts, or bleeding time (Pan et al., 2022).

Aasperlin is a polyketide isolated from *Aspergillus versicolor* LZD4403 accompanied by *Pseudopterogorgia* sp. gorgonian (Fig. 5). Asperlin notably prohibited LPS-caused foam cell creation, boosted cholesterol efflux, repressed pro-inflammation factor production (iNOS/IL-1β/TNFα), and raised IL-10/IL-4 expression. *In vivo*, it substantially impeded atherosclerotic plaque initiation in the aorta by reducing atherosclerotic lesion regions and aortic dilatation. Asperlin might possess its athero-protective efficacy by lowering inflammation instead of improving dyslipidemia (Zhou et al., 2017).

Sharkova et al. (2015) revealed that *Tolypocladium inflatum* k1 produced extracellular proteinases with strong collagenolytic, plasmin-like, and plasminogen-activating properties. It was demonstrated to have 20% greater specific plasminogen-activating ability than its plasmin-like function (Sharkova et al., 2015).

Tolypocladium inflatum 62a, acquired from the White Sea soil, proved to be a promising thrombolytic agent producer. It revealed the utmost growth rate and fibrinolytic and caseinolytic (proteolytic) capacities among the examined strains, compared with *T. inflatum* k1. It also had a high specificity for fibrillar proteins and enzymatic index with fibrinogen and fibrin in media (Fokichev et al., 2022).

Zhao et al., identified versiase, a fibrinolytic bifunctional enzyme from *Callyspongia* sp.-associated *Aspergillus versicolor* ZLH-1 that had optimal potential at pH 5.0/40°C. This enzyme possessed anticoagulant and thrombolytic efficiency that effectively dissolved blood clots (Zhao et al., 2022).

The proteinase enzymes longolytin and tricholysine were characterized from *Arthrobotrys longa* and *Trichothecium roseum* cultures. They displayed fibrinolytic

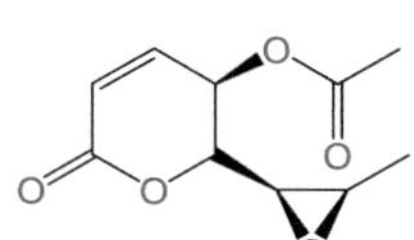

Fig. 5. Chemical structures of Asperlin separated from *Aspergillus versicolor*.

and anticoagulant characteristics after adding them to the animals' blood plasma. Additionally, a slight fibrinogen concentration reduction, insufficient prolongation of thrombin and recalcification times, and a rise in fibrinolytic action were noted upon their intravenous administration into animals. Tricholysine preparation indicated minimal toxicity at I.V. doses exceeding 80 times the thrombolytic dose resulted in 34% mortality in mice (Serebriakova et al., 1984).

An experiment by Podorol'skaya et al. (2007) proved that longolitin externally administered to the thrombotic venous section raised the thrombolysis rate by 4–4.5 times and reduced the clot lysis time by 2 times, whereas its effect was boosted by heparin (Podorol'skaya et al., 2007).

Kunhiraman et al. (2024) established that *Penicillium steckii* KU1 produced fibrinolytic enzymes on a massive scale. In contrast to the expected value (441.42 U/mL), the optimal circumstances (5.75 million spore inoculum/mL, casein 2.7% w/v, pH 7, –30°C, incubation 115 h) achieved a 10-times rise (454.49 U/mL) in the enzyme synthesis (Kunhiraman et al., 2024).

3. Decolorization of Dye by Marine Fungi

Numerous industrial wastes are poured into freshwater and marine environments (Steffen et al., 2015). Ten to 15% of discarded dyes are eventually released into wastewater because of the dyeing and subsequent washing procedures (Przystas et al., 2015). Exposure to dyes can cause a variety of health issues, such as nausea, mucous, and skin membrane ulcers, bleeding, and even serious harm to the liver, kidney, brain, and reproductive and central nervous systems (Vala and Dave, 2017). The majority of the yearly output of synthetic colors comprises triphenylmethane, heterocyclic polymeric, azo, and anthraquinone dyes that are harmful, carcinogenic, and mutagenic. Most of the traditional physicochemical techniques for color removal have drawbacks, including confined applicability, high expense, and using many chemical components. This demands the discovery of eco-friendly biological techniques (Bittencourt et al., 2023).

Much research has been directed on using fungi to develop ecological wastewater remediation techniques, as they have an exceptional capability for producing extracellular enzymes (Ghosh et al., 2023). Marine fungi are employed to remove dye color from coastal waters and wastewater treatment facilities. Their decolorization capability is due to extracellular enzyme secretion (Chin et al., 2024). The process of employing marine-derived fungi to degrade dyes include immobilized cells, growing cultures, and biosorbents made from cultures.

Laccase, lignin, and manganese peroxidases are among the main lignin-degrading enzymes with the most promise in industrial usage (Kumar and Chandra, 2020). These enzymes can break down a variety of xenobiotics, such as colored industrial effluents from textiles, leather, pulp and paper mills, dye manufacturing, and alcohol distillery industries, as well as lignocellulosic materials (Ali and Sreekrishnan, 2001).

Laccase is an effective bioremediation enzyme that can oxidize and degrade different contaminants, including dyes, insecticides, aromatic hydrocarbons, chlorophenols, and medications because of its wide substrate specificity (Dong et al., 2023b).

Out of 26 fungal strains separated by Torres et al. (2010) from seagrasses and sea sediments, 21 strains possessed full to partial decolorizing capacities for crystal violet. Interestingly, *Cladosporium* sp., *Phialophora* sp., and *Penicillium* sp. demonstrated full decolorization capacities for Congo red, while *Cladosporium* sp. and *Penicillium* sp. revealed decolorizing potential for crystal violet (87–91%) with no influence on brilliant green (Torres et al., 2011).

The central-composite design and Plackett-Burman design were utilized to optimize circumstances for producing laccase by *Alternaria tenuissima* KM-651985. It was found that this fungus effectively and its enzyme decolorized Congo Red and Crystal Violet (Abd El Aty et al., 2016). Further, its covalent-immobilized laccase in Ca2+ (AlgChG) beads possessed improved decolorizing properties for Congo red and crystal violet with minimizing their toxicity (Abd El Aty et al., 2016).

Toker et al., revealed that *Alternaria* sp. D-21, with high laccase capability, decolorized dyes featuring anthraquinone, azo, heterocyclic, and triarylmethane chemical groups. It possessed decolorization percentages of 98%, 58%, 71%, 93%, 53%, and 78%, for Methyl Orange, Crystal Violet, Remazol Brilliant Blue, Acid Green-3, Congo Red (Conc and Methylene Blue (Conc. 100 mg/L, after 144 hr). The fungi's ability to eliminate contains various contaminants (Toker et al., 2021).

Wikee et al. (2019) specified two laccase genes (PsLac-1 and -2) from *Pestalotiopsis* sp. that were cloned in *Aspergillus niger* (Wikee et al., 2019). The PsLac-2 and PsLac-1 enzymes demonstrated remarkable stability at up to 50°C/pH 4–6, maintaining 100% efficiency after three hours at 50°C. They exhibited the strongest catalytic efficiency values against 2,6-dimethoxyphenol and syringaldazine, respectively. PsLac1 more effectively decolorized Bromo Cresol Purple, Acid Yellow, Reactive Black-5, and Nitrosulfonazo-III in existence of hydroxybenzotriazole (redox mediator-1). It was noted that PsLac2 appeared to be more adaptable to the existence of sea salt than PsLac1 (Wikee et al., 2019).

Aspergillus nidulans, *Trichoderma asperellum*, and *Stemphylium lucomagnoense* collected from the Tunisian coast were estimated for their oxidase capacities. It is noteworthy that *T. asperellum* was found to have the greatest amount of oxidative functions. *T. asperellum* optimally demonstrated laccase efficiency at $CuSO_4$ (1.8 mM, inducer), NaCl 1%, and sucrose 3% (carbon source). Several synthetic dyes of various chemical categories were successfully bleached by *T. asperellum* culture supernatant that was boosted by HBT (1-hydroxybenzotriazole) addition (Ben Ali et al., 2020). *Trematosphaeria mangrovei* secreted laccase that revealed higher capacity at pH 4.0/65°C with Vmax = 184.84 U/mg protein and Km = 1.4 mM (Atalla et al., 2013) that was totally repressed by $FeSO_4$. Laccase's substrate selectivity was greatly impacted by type and location of the phenolic ring substituents (Atalla et al., 2013). The lignin-degrading, hypersaline-tolerant *Phlebia* sp. MG-60 separated from mangrove forests, was proved to be implicated in decolorizing dyes and pulp bio-bleaching at varying sea salts concentrations (Li et al., 2002).

According to D'souza-Ticlo et al. (2009) the marine fungus *Cerrena unicolor* NIOCC #2a purified laccase maintained 75% of its function in discoloring a number of artificial dyes in half-strength seawater and was unaffected by NaCl up to 0.3 M (D'Souza-Ticlo et al., 2009). It produced more laccase on exposure to synthetic dyes and textile effluent (~ 85,000 U/L). Besides, a variety of colored effluents, including textile mill wastewater, alcohol distillery molasses spent wash, and black liquor from the pulp and paper industries were decolorized by the culture supernatant and the live culture of this fungus (D'Souza et al., 2006). Several *Aspergillus* species have remarkable dye-decolorizing properties (Hareeri et al., 2022; Ibrahim et al., 2023). Assess et al., stated that *A. niger* at pH 5 decolorized 97% of 200 milligrams per liter of Congo red in 6 days while shaking due to the production of manganese peroxidase and lignin peroxidase (Asses et al., 2018).

A study by Chin et al. (2024) on *Biatriospora* sp., *Exserohilum rostratum*, *Aspergillus ochraceus*, *A. chevalieri*, and *A. iizukae* separated from Mauritius brown algae and sponges revealed that *A. chevalieri* and *A. ochraceus* produced lipase and DNAse with maximal decolorizing capacity (79.40 and 87.18%, respectively) for Congo Red and Methylene Blue, respectively. While *Biatriospora* sp. moderately produced protease, amylase, and DNAse with 56.29% decolorizing potential for Crystal Violet, suggesting the potential industrial and pharmaceutical applications of Mauritian fungi (Chin et al., 2024).

4. Degradation of Pesticides by Marine Fungi

Pesticides are synthetic or natural substances that are employed to stop, manage, and get rid of weeds, insects, and other pests that interfere with plant growth. Commercial farming is believed to be the primary reason for the sharp and steady increase in the demand for pesticides worldwide from the mid-1940s (Ahmad et al., 2024). Pesticides include disinfectants, antiseptics, fungicides, anti-bacterials, algicides, herbicides, and rodenticides (Pathak et al., 2022). Organic synthetic pesticides are in four categories: carbamates, organophosphates, pyrethroids, and organochlorines. Natural pesticides (biopesticides) are derived from living organisms like bacteria, fungi, and plants (Zhou et al., 2025). The excessive and uncontrolled uses of pesticides, particularly carbamates, organochlorines, pyrethroids, and organophosphates, are accompanied by negative effects, including high environmental retention, bio-accumulation, long-range spread, and adverse consequences on non-target species (Zhou et al., 2025). Pesticide exposure can lead to endocrine and neurological dysfunction, teratogenic and genotoxic abnormalities, and many forms of human cancer (Ahmad et al., 2024). Pesticides can have deadly short-term effects; however, some appear to have longer-term negative health and environmental implications (Pathak et al., 2022).

Extensive studies have been promoted to address the difficulty of getting rid of pesticides from different matrices. Strategies that combine chemical and physical treatments are popular methods; however, each has advantages and disadvantages, such as releasing more poisonous substances, and both methods are expensive and dangerous. These methods are impacted by variables such as temperature, pH, pesticide properties, and cost. The major processes for removing pesticides include enzymatic biodegrading, reverse osmosis, adsorption, photolysis, advanced oxidation processes, and activated carbon, membrane, and nano-filtrations (Desisa et al., 2022).

Eco-friendly approaches, such as bioremediation techniques, are accessible for eliminating hazardous pollutants to maintain a sustainable environment with a robust and healthy ecosystem (Desisa et al., 2022). These environmentally harmless and cost-effective methods include mycoremediation, microalgae bioremediation, phytoremediation, and bacterial-pesticide degradation that involve fungi, algae, plants, bacteria, and their interactions (Singh et al., 2020).

Marine fungi create a variety of powerful enzymes that break down hazardous substances. A number of studies assessed their potential capacity to decompose numerous pesticides, such as carbamates, pyrethroids, organochlorines, and organophosphorus (Table 1). According to Oliveira et al. (2015), *Trichoderma harzianum*, *Aspergillus terreus*, *Aspergillus fumigatus*, and *Penicillium citrinum* all resisted and broke down chlorfenvinphos (Oliveira et al., 2015). The capability of *Pleurotus ostreatus*, *Pycnoporus coccineus*, *Trametes versicolor*, and *Phlebiopsis gigatea* to breakdown chlorpyrifos, linuron, and metribuzin was investigated. All species showed a remarkable level of pesticide resistance (Gouma et al., 2019). *P. ostreatus* achieved the minimal growth rate in the pesticide mixture, whereas *P. coccineus* was more effective than the other isolates. High amounts of polyphenol oxidase are produced by *T. versicolor*, *P. coccineus*, and *P. gigantea*; however, only *T. versicolor* was able to breakdown linuron when chlorpyrifos and metribuzin were added (Gouma et al., 2019).

The marine-isolated *Westerdykella* sp. CBMAI_1679, *Acremonium* sp. CBMAI_1676, *Aspergillus* sp. CBMAI_1829, and *Microsphaeropsis* sp. CBMAI 1675 enantio-selectively biodegraded (±)-lambda-cyhalothrin in malt liquid (3%) medium, containing 100 mg/L of (±)-lambda-cyhalothrin (Birolli et al., 2018b). Whilst *Aspergillus* sp. was the most effective strain and (1R,3R,αS)-γ-cyhalothrin was the preferentially biodegraded enantiomer (Birolli et al., 2018b). After 14 days of incubation, *Penicillium implicatum* and *Aspergillus viridinutans* demonstrated 100% degradation of lambda-cyhalothrin and chlorpyrifos (conc. 2.5 ppm). For both strains, the optimal concentrations of lambda-cyhalothrin were 2.5 and 2.5 ppm, respectively, while the optimal amount of chlorpyrifos was 5 ppm (Abdel-Wareth et al., 2016).

Aspergillus sydowii CBMAI_935, *Penicilliumraistrickii* CBMAI_931, *Microsphaeropsis* sp. CBMAI_1675, *Cladosporium* sp. CBMAI_1237, *Acremonium* sp. CBMAI-1676, *Cladosporium* sp. CBMAI-1678, and *Westerdykella* sp. CBMAI_1679 associated with *Chelonaplysilla erecta*, and *Dragmacidon reticulate* sponges degraded esfenvalerate with various effectiveness. It was noted that *Microsphaeropsis* sp., *Acremonium* sp., and *Westerdykella* sp. were the most powerful strains (Birolli et al., 2016). Further, *A. sydowii* CBMAI-935 generated methyltransferases and phosphoesterases appropriate for the degradation of profenofos, chlorpyrifos, and methyl parathion through condensation, methylation, and hydrolysis processes (Soares et al., 2021). *A. sydowii* CBMAI_935 and *Penicillium decaturense* CBMAI_1234 completely biodegraded methyl parathion in 20 and 30 days, respectively. Methyl parathion was converted to a more

Table 1. Examples of pesticide degradation by marine-derived fungal stains.

Pesticide	**Tested Fungal Strain**	**Result**	**References**
Chlorfenvinphos, Diuron, Atrazine, and Isoproturon	*Trichoderma harzianum*, *Aspergillus fumigatus*, *Penicillium citrinum*, and *Aspergillus terreus*	All fungi were able to break down chlorfenvinphos to undetectable levels. These fungi couldn't break down isoproturon, atrazine, or diuron.	Oliveira et al. (2015)
Dichlorodiphenyltrichloroethane	*Aspergillus sydowii* (Ce15, Ce19, and Gc12), *P. raistrickii* Ce16, *P. miczynskii* Gc5, *Bionectria* sp. Ce5, and *Trichoderma* sp. Gc1	*A. sydowii* (Ce15 and Gc12), *P. miczynskii* Gc5, *Trichoderma* sp. Gc1, and had the capacity to grow in dichlorodiphenyltrichloroethane (conc. 5.0–15.0 mg) even on a liquid or solid medium. *Trichoderma* sp. Gc1 degradation capacity was enhanced when dichlorodiphenyltrichloroethane (5.0 mg) was previously introduced to a medium in the existence of hydrogen peroxide.	Ortega et al. (2011)
Dieldrin	*Trichoderma* sp. CBMAI_932, *A. sydowii* (CBMAI_935 and CBMAI-933), and *P. miczynskii* CBMAI_930	*P. miczynskii* possessed high tolerance to dieldrin 70% and 90% of dieldrin were degraded after 4 and 14 days, respectively.	Birolli et al. (2015)
Esfenvalerate	*Cladosporium* sp. CBMAI_1678, *A. sydowii* CBMAI_935, *Cladosporium* sp. CBMAI_1237, *P. raistrickii* CBMAI-931, *Acremonium* sp. CBMAI_1676, *Microsphaeropsis* sp. CBMAI_1675, and *Westerdykella* sp. CBMAI_1679	*Microsphaeropsis* sp. CBMAI_1675, *Acremonium* sp. CBMAI_1676, *Westerdykella* sp. CBMAI_1679 were the most powerful strains. After 28 days, ↑ toxic compounds concentrations: 2-(4-chlorophenyl)-3-methylbutyric acid and 3-phenoxybenzoic acid (2.7–16.6 mg/L) (6.6–13.4 mg/L) were noted.	Birolli et al. (2016)
Methyl parathion	*A. sydowii* CBMAI_934, *A. sydowii* CBMAI_935, *A. sydowii* CBMAI_1241, *P. ecaturense* CBMAI_1234, *P. raistrickii* CBMAI_931, *P. raistrickii* CBMAI_1235, and *Trichoderma* sp. CBMAI_932	In 30 days, methyl parathion was totally broken down by *P. decaturense* CBMAI_1234 and *A. sydowii* CBMAI_935. The bioactivation of methyl parathion to the oxon form led to its biodegradation.	Alvarenga et al. (2014)

Table 1 contd. ...

...Table 1 contd.

Pesticide	**Tested Fungal Strain**	**Result**	**References**
Metribuzin, Linuron, and Chlorpyrifos	*Pycnoporus coccineus*, *Pleurotus ostreatus*, *Trametes versicolor*, and *Phlebiopsis gigatea*	All strains demonstrated a significant tolerance to all pesticides. *P. ostreatus* has a minimal growth rate *P. coccineus* was the more efficient isolate. *P. coccineus, P. gigantea* and *T. versicolor* generated high polyphenol oxidase levels. *T. versicolor* decomposed linuron in combination with chlorpyrifos and metribuzin.	Gouma et al. (2019)
Pentachlorophenol	*A. sydowii* DL6A, *T. harzianum* CBMAI_1677, *C. oxysporum* DL5G, *A. versicolor* DL5A, and *F. proliferatum* DL_11A	After 7 days, 20 mg/L of pentachlorophenol was fully degraded by *T. harzianum* CBMAI_1677. *T. harzianum* biodegraded pentachloroanisole and 2,3,4,6-tetrachloroanisole in a liquid medium.	Vacondio et al. (2015)
Profenofos, Chlorpyrifos, and Methyl parathion	*A. sydowii* (CBMAI_1241, CBMAI_934, and CBMAI_935), *P. raistrickii* (CBMAI_931 and CBMAI_1235), *P. decaturense* CBMAI_1234, and *Trichoderma* sp. CBMAI_932	*A. sydowii* CBMAI-935 breaks down profenofos, chlorpyrifos, and methyl parathion. Chlorpyrifos: 32% degradation, giving 3,5,6-trichloro-1-methylpyridin-2(1H)-one, 3,5,6-trichloropyridin-2-ol, tetraethyl dithiodiphosphate, and 2,3,5-trichloro-6-methoxypyridine. Methyl parathion 80% degradation gives methyl paraoxon, isoparathion, trimethyl phosphate, O,O,O-trimethyl phosphorothioate, O,O,S-trimethyl phosphorothioate, 4-nitrophenol, and 1-methoxy-4-nitrobenzene. Profenofos: 52% degradation, giving O,O-diethyl S-propylphosphorothioate, 4-bromo-2-chloro-1-methoxybenzene, and 4-bromo-2-chlorophenol.	Soares et al. (2021)

hazardous form, methyl paraoxon, which was further broken down to p-nitrophenol (Alvarenga et al., 2014). *A. sydowii* (Ce15, Ce19, and Gc12), *Penicillium raistrickii* Ce16, *Penicillium miczynskii* Gc5, *Bionectria* sp. Ce5, and *Trichoderma* sp. Gc1, accompanied by marine sponges *Chelonaplysylla erecta* and *Geodia corticostylifera* were examined for dichlorodiphenyltrichloroethane (Conc. 5, 10, and 15 mg) degrading capability in liquid and solid media (Ortega et al., 2011).

Penicillium miczynskii Gc5 and *Trichoderma* sp. Gc1 demonstrated a noteworthy capacity to break down dichlorodiphenyltrichloroethane up to 15.0 mg. *Trichoderma* sp. Gc1 produced 100% breakdown of dichlorodiphenyltrichloroethane in liquid culture (Ortega et al., 2011).

Vacondio et al. (2015) proved that *Trichoderma harzianum* CBMAI-1677 demonstrated the highest pentachlorophenol biodegradation effectiveness among the *Didemnum ligulum*-associated fungal strains that were examined. This fungus biodegraded pentachlorophenol and its primary metabolites: 2,3,4,6-tetrachloroanisole and pentachloroanisole (Vacondio et al., 2015).

Trichoderma sp. CBMAI-932, *A. sydowii* (CBMAI_935 and CBMAI_933), and *Penicillium miczynskii* CBMAI_930 separated from the *Chelonaplysylla erecta* and *Geodia corticostylifera* sponges were explored for their dieldrin (conc. 25, 50, and 75 mg/L)-degrading capacities. All strains grew in the existence of dieldrin, whereas *P. miczynskii* was the most tolerant fungus to dieldrin. *P. miczynskii* was noted to catalyze the dieldrin (50 mg/L) biotransformation after 14 days with 90% conversion rates in liquid medium. Accordingly, *P. miczynskii* is the first possible strain for dieldrin bioremediation (Birolli et al., 2015).

5. Degradation Biomass by Marine Fungi

Biomass is a biological material derived from plants and animals that contains diverse carbon, hydrogen, oxygen, and nitrogen, along with traces of minerals and other components. It is also defined as renewable/sustainable materials, which include forest, aquatic plants, wood, animal, and urban/industrial wastes, and agricultural residues and products (He et al., 2023). Biomass contributes to the advancement of the bioeconomy because of its sustainable supply, wide availability, low cost, ease of access, and production potential, which make this renewable biomass is an ideal source to produce bioproducts, bioenergy, and biofuels (Manzanares, 2020; Omoriyekomwan et al., 2021). It is significant that biomass waste is the most beneficial carbon source among all biomass sources, considering its advantages for the environment. For instance, the global leather industry generates 8–9 MT of skin annually along with 1.4 MT of solid waste, the majority of which is protein (Ayele et al., 2021).

The majority of this solid biomass waste is disposed of in the environment by thermal incineration or landfilling, resulting in damage to the ecosystem by polluting the air and releasing odors, as well as taking up land resources (Wang et al., 2018; Ayele et al., 2021).

Thus, innovative strategies to improve the safe and effective industrial use of these renewable resources are desperately needed to reduce the buildup of organic waste. Recently, the bioconversion of biomass into biofuels and biochemicals has attracted more interest due to the overuse of petroleum resources and the rising demand for fossil-based fuels and chemicals (Ling et al., 2025).

5.1 Degradation of Lignocellulosic Biomass by Marine Fungi

Lignocellulosic biomass is among the most auspicious and natural sustainable resources for feedstock, feed additives, biofuel, and other valuable products (Lee et al., 2012). Additionally, it can be used as a substrate for generating a number of crucial industrial enzymes (Shahzadi et al., 2014). Lignocellulosic biomass is made up of cellulose (40–50%), lignin (10–25%), and hemicellulose (20–30%) (Thapa et al., 2020).

The xylanolytic and cellulolytic enzymes catalyze the breakdown of the cellulosic biomass. The hydrolytic enzymes, such as xylanases and cellulase, convert hemicellulose and cellulose into xylose and glucose, respectively, which are the fundamental components of many industrial products. These enzymes have proven to be crucial in different industries, such as textiles, biofuel, drinks, food, feed, paper and pulp, and biopolymers (Thapa et al., 2020; He et al., 2023).

Fungi serve as crucial contributors to the nutrients` cycling and the sustainability of ecosystems by their ability to break down organic biomasses through the secretion of various enzymes. These enzymes can break down cellulose, the most prevalent organic polymer on Earth, making them particularly interesting (Dong et al., 2023a). They have potential commercial applications in different kinds of industrial processes, from the generation of biofuel to the paper and textile industries. This helps in the development of sustainable energy by converting cellulosic biomass into fermentable sugars for the manufacture of biofuel (Saini et al., 2022).

Several lignocellulolytic extracellular enzymes were produced by white rot molds, particularly those belonging to the Trametes, Phanerochaete, Pleurotus, and Bjerkandera genera (Rouches et al., 2016; Torres-Farradá et al., 2024). Overall, lignocellulosic biomass conversion involves a pretreatment phase to remove the covering lignin seal that envelopes structural polysaccharides. This is followed by enzymatic hydrolysis and fermentation of soluble sugars. Consequently, other microbes may metabolize the reducing sugars from hemicellulose and cellulose fractions hydrolysis to produce chemicals with added value, including phenolics, flavonoids, alcohols, and organic acids (Rouches et al., 2016; Torres-Farradá et al., 2024).

A study by Batista-García et al. (2017) identified *Pseudogymnoascus* sp. TS-12, *Cadophora* sp. TS2, and *Emericellopsis* sp. TS11 from the *Stelletta normani* deep-sea sponge that demonstrated high xylanase and CMCase capacities with optimal halotolerance and thermostability at pH 5–8 and 50–70°C (Batista-García et al., 2017). *Pestalotiopsis* sp. J63 was found to be moderately halotolerant and secreted 0.11 U/mL cellulase, along with laccase, under submerged fermentation using alkaline-pretreated sugarcane bagasse as substrate (Chen et al., 2011). *Trichoderma* sp. and *Aspergillus niger* obtained from the seacoasts of Karnataka and Tamil Nadu showed powerful oxidation of lignin using a variety of substrates within the first three weeks (Sethi et al., 2013). Eight fungal species associated with *Avicennia marina* obtained from the Red Sea coast could break down paper and clothing waste.

These fungi released cellulase and ligninolytic enzymes, accumulated substantially greater biomass, and emitted more CO_2 (Ameen et al., 2014).

The deep-sea-derived *Chaetomium* sp. CS1 degraded alkali lignin by 55% in 10 days (Ling et al., 2025). It secreted ligninolytic enzymes, especially those belonging to the AA3 family that exhibit monooxygenase and oxidoreductase activities during lignin degradation and that have crucial roles in lignin breakdown. The pretreatment with *Chaetomium* sp. remarkably boosted the sawdust digestibility by black-soldier fly larvae that increased the lignocellulosic material bioavailability. This fungus enabled the production of biofuel and bioproducts from lignocellulosic biomass (Ling et al., 2025).

5.2 Degradation of Plastic by Marine Fungi

The ecosystem is negatively impacted by the hazardous substances included in plastics. The annual production of plastic has exceeded 300 million tons. Since 1964, the utilization of plastics has grown 20-fold, and by 2035, it is predicted to double (Sánchez, 2020). It was estimated that around 1,800 million tons of synthetic plastic will be resulted by 2050 (Gallo et al., 2018) and by that time, there will be 12,000 million tons of plastic waste in the environment worldwide (The Future of Plastic, 2018). Plastic is a polymer consisting of hydrogen, carbon, oxygen, silicon, nitrogen, and chloride. Plastics are widely utilized extensively in almost every industrial sector, but particularly in transportation, packaging, and construction because of their low weight and cost, as well as their great durability, stability, and versatility (Li et al., 2021).

Plastics could be synthetic or bio-based plastics. Bio-based plastics are made up of natural substances like terpenes, lignin, hemicellulose, cellulose, vegetable oils, soybean proteins, carbohydrates, corn starch, potatoes, and microbes, while synthetic (non-biodegradable) plastics are mostly made of crude oil (Ekanayaka et al., 2022; Vaksmaa et al., 2023a).

Synthetic plastics pollute the environment due to their extremely low rate of degradation, which represents one of the most significant environmental issues. Even their chemical degradation, by products are more harmful and have adverse impacts. Its recycling has mostly failed to be an effective means to dispose of these plastic wastes. Thus, there is a pressing need for creative solutions for decreasing this risk to the environment as contaminants continue to pile up in oceans and rivers in the form of nano-, micro-, macro-, meso-, and plastics.

Biodegradation is a process by which organic materials are degraded by organisms, most commonly microorganisms. Biodegradation is promoted by compounds such as peroxides and acids or enzymes secreted by microorganisms. Fungal enzymes have the potential to break up the plastic polymer chemical bonds. These include lignin and manganese peroxidases, which are frequently related to lignin decomposition. These enzymes transform a variety of chemicals into polymerized or oxidized products by catalyzing processes of oxidation and reduction, including free radicals (Wei and Zimmermann, 2017). Also, laccase, a multicopper oxidase, promotes the breakdown of plastic. It is a lignin-modifying enzyme that mediates the polymers` carbon backbone oxidation (Amobonye et al., 2021).

Various marine-derived fungi belonging to the *Aspergillus*, *Phanerochaete*, *Fusarium*, *Penicillium*, and *Acremonium* genera, as well as their related enzymes, such as laccases, manganese peroxidase, and lignin peroxidases have demonstrated capability to biodegrade plastics (Restrepo-Flórez et al., 2014; Vaksmaa et al., 2021). *Aspergillus flavus* PEDX3, separate from *Galleria mellonella* (wax moth) gut contents degraded high-density polyethylene into microplastic particles and secreted laccase-like multicopper oxidases with promising polyethylene degradation properties (Zhang et al., 2020). *A. terreus* and *Penicillium* sp. possessed polyethylene-degrading capability, whereas *Penicillium* sp. reduced the weight of polyethylene by the greatest amount (43.4%) (Alshehrei, 2017). *Alternaria alternata* FB1, a marine fungus, possessed the potential to colonize polyethylene film, broke it down effectively, and created numerous visible holes in the plastic surface (Gao et al., 2022b). Paço et al. (2017) reported that *Zalerion maritimum* utilized polyethylene in a minimum growth medium within seven days, causing a decrease of the pellets' size and mass (Paço et al., 2017). *Parengyodontium album* obtained from floated plastic trash in the North Pacific Subtropical Gyre showed a capacity to decompose PE and mineralize it to CO_2 (Vaksmaa et al., 2024).

Deep sea-derived *Cladosporium halotolerans* demonstrated up to 80% degradation of polyurethane after three days of incubation at 28°C. It released hydrophobic surface binding proteins and esterase, both of which are crucial for the absorption and subsequent decomposition of polyurethane (Zhang et al., 2022). Also, *Rhodotorula mucilaginosa* mineralized and assimilated carbon from plastics (Vaksmaa et al., 2023b). Brunner et al. (2018) investigated the polyethylene and polyurethane degrading capability of *Leptosphaeria* spp., *Cladosporium cladosporioides*, *Penicillium griseofulvum*, and *Xepiculopsis graminea* separated from plastic trash found along the Lake Zurich shoreline (Brunner et al., 2018). *C. cladosporioides*, *P. griseofulvum*, and *X. graminea* were able to break down polyurethane; however, none of them degraded polyethylene (Brunner et al., 2018).

Polylactic acid (PLA), an aliphatic polyester and one of the most widely used bioplastics. *Aspergillus*, *Fusarium*, *Penicillium*, *Tritirachium*, *Cryptococcus*, and *Trichoderma* fungal species are among the reported polylactic acid-degrading fungi (Vaksmaa et al., 2023a). These genera produce cutinase and protease, which are known to hydrolyze and depolymerize polylactic acid. Further, polylactic acid breakdown by *Tritirachium album* ATCC-22563 was triggered after adding 0.1% gelatin, and the enzyme assumed to be responsible for that was protease (Jarerat and Tokiwa, 2001). Polyhydroxyalkanoates (PHAs), a class of biopolymers. have more than 100 different forms that come from microbial metabolism, including PHV (polyhydroxy valerate), PHBV (poly (3-hydroxy butyrate-co-3-hydroxyvalerate), and PHB (poly (b-hydroxy butyrate) that are more biodegradable than polylactic acid (DiGregorio, 2009). *Aspergillus ustus* and *Emericellopsis minima* W2 showed a marked capacity to break down PHB (Gonda et al., 2000).

The cutinase of *Aspergillus oryzae* revealed specific activity towards polylactic acid, polybutylenesuccinate, and butylenesuccinate-co-adipate which was determined to be as 0.067, 0.42, and 11 U/mg, respectively (Maeda et al., 2005). Miyazaki et al. purified PHB depolymerase from *Penicillium funiculosum* IFO-6345 that demonstrated notable hydrolytic efficiency against 3-hydroxybutyrate oligomers (Miyazaki et al., 2000). *Cryptococcus* spp. lipase degraded polylactic acid and other degradable plastics such as poly(3-hydroxybutyrate), polybutylene succinate, and poly (ε-caprolactone) (Masaki et al., 2005).

6. Degradation of Polycyclic Aromatic Hydrocarbons by Marine Fungi

Polycyclic aromatic hydrocarbons are organic substances with two or more connected aromatic rings. Their extensive prevalence in nature, especially in developing nations, is caused by oil spills, burning fossil fuels, vandalism, shipping, disposal and use of coal, petroleum products, and careless oil exploration (Bacosa and Inoue, 2015). PAH contamination poses a serious threat to the environment and human health due to its recalcitrant, long-lasting, and stable characteristics (Vaidya et al., 2018). The paradigm shift primarily focuses on investigating microorganisms as a viable and practical method for removing PAHs because of their affordability and ecologically friendly nature.

Various researches focus on studying the PAHs degradation capacity by marine fungi. *Mucor irregularis*/*Aspergillus aculeatus* co-culture demonstrated 98.4% fluoranthene degrading efficiency after 7 days incubation. During fluoranthene breakdown, they produced 71, 71, 93, and 85% induction of lignin and manganese peroxidases and laccase, respectively (Bankole et al., 2021). Fluoranthene is PAHs that has carcinogenic, mutagenic, teratogenic, and genotoxic properties (Hentati et al., 2016).

Mucor irregularis bpo1 possessed 81.50% fluorene decomposing effectiveness on the 5th day with 87%, 31%, and 59%, laccase, lignin peroxidase, and manganese peroxidase induction, respectively, during the degradation process (Bankole et al., 2021). This revealed the potential of marine fungal co-culture in the environmentally benign remediation of PAHs.

Marine sponge-associated *Xylaria* sp. CBMAI-1464 and *Tolypocladium* sp. CBMAI-1346 displayed benzo[a]pyrene and pyrene removal efficiency, respectively. *Tolypocladium* sp. CBMAI 1346 degrades pyrene 95% of Py after 7-day incubation. The degradation carried out through the cytochrome P450 pathway and dioxygenases and monooxygenases had a crucial role (Vasconcelos et al., 2019).

Marasmiellus sp. CBMAI_1062 revealed a great PAH-degrading capability that could be a new microbial bioremediation resource. *Marasmiellus* sp. demonstrated 100% pyrene degradation after 48 h incubation. It was suggested that fungus utilized the epoxide hydrolases and cytochrome P450 in pyrene degradation (Vieira et al., 2018).

Among the sponge-derived fungi, *Cladosporium* sp. CBMAI_1237 showed efficient anthracene biodegradation capacity with 26% and 42% biodegradation in the absence and existence of artificial seawater, respectively, suggesting that it worked faster in seawater than in a non-saline circumstance. *Cladosporium* sp. CBMAI-1237 biodegraded. After 21 days, anthrone, anthracene, acenaphthene, anthraquinone, fluorene, fluoranthene, phenanthrene, nitropyrene, and pyrene were

100%, 71%, 78%, 32%, 70%, 52%, 47%, 64%, and 62%, respectively (Birolli et al., 2018a).

7. Degradation of Oil Spills by Marine Fungi

The rapid growth of urban areas and industrial development has led to increased energy use and waste output, degrading environmental quality. This degradation poses a remarkable threat to humanity and its diverse ecosystems. As industrialization accelerates and the global energy demand rises, oil spills have emerged as a significant danger. Approximately 40,000 incidents occur each year, resulting in the release of crude oil into the environment. Notably, oil spills associated with transportation contribute to 13% of the oil pollution in oceans (Daghio et al., 2017; Chen et al., 2019).

Crude oil is a complicated blend of hydrocarbons, which also includes slight quantities of metals such as iron, nickel, and vanadium, along with traces of oxygen and sulfur. The lasting nature of these components results in varying chemical compositions among different types of oil (Saravanan et al., 2023).

An oil spill can lead to three types of outcomes: immediate lethal effects, direct sublethal effects, and indirect effects (El-Din et al., 2018; Wu et al., 2019). Additionally, human consumption of marine species can cause significant harm to their health (Bento et al., 2005; Wu et al., 2019). Traditional methods for cleaning up oil spills are often ineffective, high cost, and result in harmful byproducts. On the other hand, bioremediation is considered an effective and environmentally friendly tool to clean up these pollutants (Bento et al., 2005). Bioremediation is the process of detoxification of pollutants, including petroleum hydrocarbons, by the aid of microorganisms (Sukhdhane et al., 2019). Bioremediation is considered an environmentally friendly process for marine environment cleaning through converting complex oil hydrocarbons to harmless residues (Sukhdhane et al., 2019). Bacteria, fungi, and other oil-degrading microorganisms are widely found in different climates; however, there has been limited research in high temperate regions (Prince, 1993). Several research has been conducted in the previous century on the use of biodegrading microorganisms of oil spills (Torres et al., 2008; Liu et al., 2014). These studies revealed the diversity and the dominancy of oil-degrading microorganisms through marine ecosystems over other species (Bartha and Atlas, 1977). Over 200 species of oil-decomposing microorganisms in marine ecosystem, including bacteria, fungi, and algae, have been documented (Hara et al., 2003; Bao ct al., 2012; Prince et al., 2013; Spier et al., 2013; Jiang et al., 2013; Brakstad et al., 2014). For instance, in marine environments, bacteria play a crucial role in degrading petroleum. Notable examples include *Acinetobacter*, *Achromobacter*, *Alcaligenes*, *Bacillus*, *Archrobacter*, *Flavobacterium*, *Coryneforms*, *Microbacterium*, *Pseudomonas*, *Micrococcus*, *Actinomycetes*, *Aureobasidium*, *Nocardia*, *Candida*, *Sporobolomyces,* and *Rhodotorula*.

Additionally, certain fungal species such as *Aspergillus*, *Fusarium*, *Mucor*, and *Penicillium* also contribute to petroleum breaking down. Gordonia is widely found in terrestrial, coastal, and oceanic environments, where it has the ability to degrade different environmental pollutants, including dibenzothiophene, benzothiophene, alkanes, and PAHs, among others (Soddell et al., 2006; Pasumarthi et al., 2013). Fungi have a conclusive contribution to petroleum hydrocarbons' biodegradation in polluted aquatic ecosystems. Moreover, sediment polluted with spilled hydrocarbons is one of the favored habitats for these fungi, as they utilize carbon sourced from the hydrocarbons (Hamad et al., 2021).

As a natural decomposer, fungi have developed a distinctive technique to convert complex organic substances, such as dyes, pesticides, and aromatic hydrocarbons to harmless byproducts making them chief agents in controlling environmental pollution (Garg, 2025). Although research on oil-degrading fungi has focused on the enzymatic degradation capabilities, other research has shown the contribution of bacteria in conjunction with fungi to degrade petroleum hydrocarbons (Hamad et al., 2021). A study by Ramzan et al. (2025) revealed the remarkable effect of photosynthetic bacteria in altering the composition of the fungal community in marine ecosystems (Ramzan et al., 2025). Also, co-culture of fungi and *Acinetobacteria* utilize a wider variety of genes and metabolic capacities through their unique interactions, creating new opportunities for the development of innovative bioremediation technologies (Saez et al., 2025). Some of the previous studies of oil-degrading fungi are listed in Table 2.

7.1 Mechanisms of Oil Degradation by Marine Fungi

The mechanism of oil biodegradation by marine fungi depends mainly on the concentration and type of the mixture of the hydrocarbon present in the oil (Tomer et al., 2021). Several studies revealed that the petroleum hydrocarbons' degradation is more rapid and results are better under aerobic conditions (Rudakiya et al., 2019). Additionally, types of nutrients, temperature, level of oxygen, and salinity have made great contributions to the efficiency of biodegradation of hydrocarbons by marine fungi (Abdel-Shafy and Mansour, 2019).

Biodegradation of hydrocarbons usually occurs through enzymatic and non-enzymatic pathways (Mahmud et al., 2022). The latter one is usually mediated by a group of biochemical reactions such as, biosorption, stripping, and biomineralization (Tomer et al., 2021). Nonetheless, the mechanism that includes enzymatic processes is facilitated by specific fungal enzymes, which have been previously discussed regarding their roles in the bioremediation of petro-hydrocarbons (Nikolova and Gutierrez, 2021). The mechanisms that involve enzymes were further classified into extracellular and intracellular processes (Fig. 6), depending on the sources of the fungal enzymes. Intracellular enzymes include peroxidases and transferases from the cytochrome P450 family, and their engagement in fungal bioremediation of petroleum hydrocarbons can facilitate the activation of hydrocarbon mixtures through dehalogenation, oxidation, hydrolysis, and reduction processes (Das and Chandran, 2011). Intermediate metabolites subsequently produced from the activated hydrocarbon compounds, which are ultimately converted into excreted derivatives. On the other hand, extracellular enzymes involve an enzymatic attack by specific fungal enzymes, including laccases, peroxidases, and oxidases, which catalyze the breakdown of hydrocarbons into excreted products (Ghosal et al., 2016; Daccò et al., 2020; Tomer et al., 2021).

Table 2. Some fungal species involved in oil derivative degradation.

Pollutant	**Species**	**References**
n-alkanes and crude oil	*P. chrysogenum, A. niger*, and *A. terreus*	Elshafie et al. (2007)
Arabian light crude oil	*A. sydowii, Penicillium chrysogenum, P. mononematosum, P. steckii, P. cyclopium, P. polonicum, P. olsonii, P. cyclopium, A. versicolor*, and *P. cyclopium*	Maamar et al. (2020)
Crude oil	*Cochliobolus lutanus, A. niger, P. documbens, F. solani*	Al-Nasrawi (2012)
Crude oil	*P. commune, A. niger*, and *A. oryzae*	El-Hanafy et al. (2017)
Diesel oil	*Trichoderma reesei*	Nazifa et al. (2018)
Hydrocarbon contaminants in old drilling wastes	*Aspergillus sydowii, Cladosporium cladosporioides*, and *Phanerochaete chrysosporium*	Steliga (2012)
Petroleum hydrocarbon	*Penicillium funiculosum, A. fumigatus, A. niger*, and *F. solani*	Flayyih and Ai-Jawhari (2014)
Oil-soaked sand patty samples	*Fusarium solani, Scopulariopsis brevicaulis*, and *Aspergillus terreus*	Simister et al. (2015)
Polycyclic aromatic hydrocarbons	*Acremonium sclerotigenum, Penicillium citrinum Aspergillus versicolor, Aspergillus polyporicola, Aspergillus sydowii Fusarium equiseti, Aspergillus sydowii, Aspergillus flavus*, and *Fusarium* sp.,	Barnes et al. (2023)
SAE 5W-30 non-synthetic engine oil	*Cladosporium sphaerospermum, Alternaria alternata, Aspergillus terreus, Aspergillus nomius, Eupenicillium hirayamae, Paecilomyces variotii, Globodera pallida, Penicillium alba,* and *Penicillium vinaceum*	Ameen et al. (2015)
Used vehicle oil	*Lasiodiploidia thermobromae*	Saravanan et al. (2023)

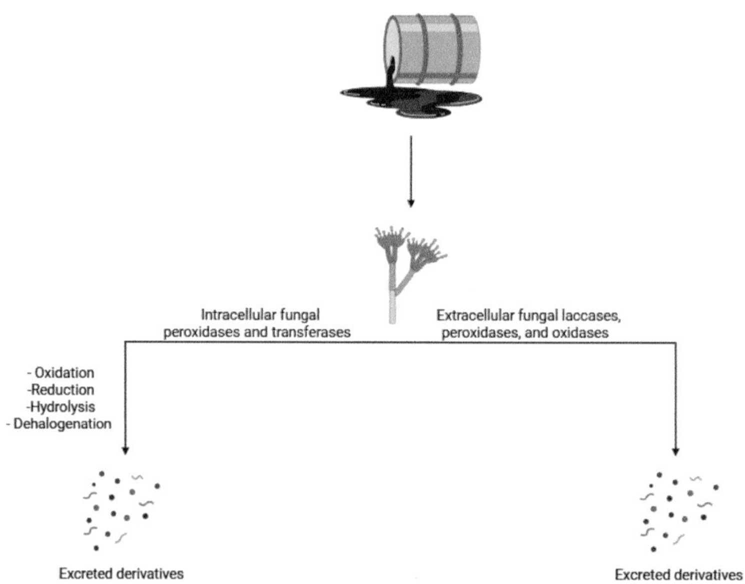

Fig. 6. Enzymatic pathways for the degradation of hydrocarbons by fungi.

7.1.1 Aerobic Degradation of Hydrocarbon Pollutants

The first step of aerobic degradation of petroleum hydrocarbons is oxidation processes, by oxygenases and peroxidases, that result in intracellular breakdown of pollutants. The pollutants then transformed into several fragments via tricarboxylic acid cycle. Finally, fungal cell masses originated from the parent metabolites, such as succinate and acetyl-CoA (Ghosal et al., 2016). The process of gluconeogenesis was utilized to synthesize sugars that are necessary for fungal biosynthesis and growth (Tomer et al., 2021).

7.1.2 Anaerobic Degradation of Hydrocarbon Pollutants

On the other hand, anaerobic degradation does not involve an oxidized functional groups addition which activates the hydrocarbon molecules, hence lowering the rate of hydrocarbon degradation (Nikolova and Gutierrez, 2021). Besides, anaerobic degradation depends on terminal electron acceptors, such as CO_2 and NO_3 during respiration process (Meckenstock and Mouttaki, 2011; Barnes et al., 2018; Al-Hawash et al., 2018; Vieira et al., 2018).

8. Conclusion and Future Directions

Marine fungi, through their enzymatic activities, resistance to adverse environmental conditions, and ability to degrade various pollutants, play an important role in bioremediation. Biodegradation of these pollutants usually occurs through enzymatic and non-enzymatic processes either aerobically or anaerobically. However, the basic mechanism of bioremediation of these hazardous pollutants is usually complicated and needs more research. Marine fungal mycoremediation is still in its infancy as a scientific and technological topic, despite its great potential. Particularly concerning the long-term effects on marine biodiversity, the ecological safety of such species, and their widespread use in actual environments, there is still a substantial knowledge gap. Additionally, the diversity of marine fungi is still mostly unknown, necessitating additional research using contemporary molecular and metagenomic methods (Zhang et al., 2023b).

Genetically engineered organisms (GMO) have been introduced for the biodegradation and bioremediation field through modification of microbial genes in order to develop new pathways to improve the process (Pieper and Reineke, 2000; Furukawa, 2003). Additionally, recombinant DNA technology is another promising technique for bioremediation (Lovley, 2003). Chakrabarty was the first to address the use of recombinant DNA technology for bioremediation (Chakrabarty, 1985). This technology can be used for bioremediation of various pollutants, such as pesticides, aplastics, heavy metals, and dyes. DNA shuffling is a reliable mutagenesis method that can produce novel enzyme activity and biocatalysts with enhanced degradation rates (Canada et al., 2002). Understanding fungal interaction with different kingdoms, such as bacteria, plants, and other kingdoms is another promising aspect in fungal biotechnology that plays an important role in environmental bioremediation. Finally, AI-enabled biotechnology can optimize fungal growth conditions, and allow monitoring of fungal interaction in various ecosystems through combining multi-omics data, and regulation of fungal genes.

References

Abd El Aty, A.A., Hamed, E.R., Ahmed, A. and El-Diwany, A.I. (2016). Induction and enhancement of the novel marine-derived *Alternaria tenuissima* KM651985 laccase enzyme using response surface methodology: Application to azo and triphenylmethane dyes decolorization. J. Appl. Pharm. Sci., 6: 006–014. https://doi.org/10.7324/JAPS.2016.60402.

Abdel-Shafy, H.I. and Mansour, M.S.M. (2019). Microbial degradation of hydrocarbons in the environment: An overview. pp. 353–386. *In*: Kumar, V., Kumar, M. and Prasad, R. (eds.). Microbial Action on Hydrocarbons. Springer Singapore. https://doi.org/10.1007/978-981-13-1840-5_15.

Abdel-Wareth, M.T.A. and Abd El-Hamid, R.M. (2016). Mycoremediation of chlorpyrifos and lambda-cyhalothrin by two species of filamentous fungi. Int. J. Environ. Stud., 73: 974–987. https://doi.org /10.1080/00207233.2016.1220721.

Ahmad, M.F., Ahmad, F.A., Alsayegh, A.A., Zeyaullah, M. and AlShahrani, A.M. (2024). Pesticides' impacts on human health and the environment with their mechanisms of action and possible countermeasures. Heliyon, 10: e29128. https://doi.org/10.1016/j.heliyon.2024.e29128.

Akpasi, S.O., Anekwe, I.M.S., Tetteh, E.K., Amune, U.O., Shoyiga, H.O. et al (2023). Mycoremediation as a potentially promising technology: Current status and prospects: A review. Appl. Sci., 13: 4978. https://doi.org/10.3390/APP13084978.

Al-Hawash, A.B., Dragh, M.A., Li, S., Alhujaily, A., Abbood, H.A. et al. (2018). Principles of microbial degradation of petroleum hydrocarbons in the environment. Egypt J. Aquat. Res., 44: 71–76. https://doi.org/10.1016/J.EJAR.2018.06.001.

Al-Nasrawi, H. (2012). Biodegradation of crude oil by fungi isolated from Gulf of Mexico. J. Bioremed. Biodegrad., 3: 1–6. https://doi.org/10.4172/2155-6199.1000147.

Ali, M. and Sreekrishnan, T.R. (2001). Aquatic toxicity from pulp and paper mill effluents: A review. Adv. Environ. Res., 5: 175–196. https://doi.org/10.1016/S1093-0191(00)00055.

Alshehrei, F. (2017). Biodegradation of low density polyethylene by fungi isolated from Red Sea Water. Int. J. Curr. Microbiol. Appl. Sci., 6: 1703–1709. https://doi.org/10.20546/IJCMAS.2017.608.204.

Alvarenga, N., Birolli, W.G., Seleghim, M.H.R., Porto, A.L. et al. (2014). Biodegradation of methyl parathion by whole cells of marine-derived fungi *Aspergillus sydowii* and *Penicillium decaturense*. Chemosphere, 117: 47–52. https://doi.org/10.1016/J.CHEMOSPHERE.2014.05.069.

Ameen, F., Hadi, S., Moslem, M., Al-Sabri, A. and Yassin, M.A. (2015). Biodegradation of engine oil by fungi from mangrove habitat. J. Gen. Appl. Microbiol., 61: 185–192. https://doi.org/10.2323/JGAM.61.185.

Ameen, F., Moslem, M.A., Hadi, S. and Al-Sabri, A. (2014). Biodegradation of cellulosic materials by marine fungi isolated from South Corniche of Jeddah, Saudi Arabia. J. Pure Appl. Microbiol., 8: 3617–3626.

Amobonye, A., Bhagwat, P., Singh, S. and Pillai, S. (2021). Plastic biodegradation: Frontline microbes and their enzymes. Sci. Total Environ., 759: 143536. https://doi.org/10.1016/j.scitotenv.2020.143536.

Asses, N., Ayed, L., Hkiri, N. and Hamdi, M. (2018). Congo red decolorization and detoxification by *Aspergillus niger*: removal mechanisms and dye degradation pathway. Biomed. Res. Int., 2018: 3049686. https://doi.org/10.1155/2018/3049686.

Atalla, M.M., Zeinab, H.K., Eman, R.H., Amani, A.Y., Abeer, A.A.E.A. et al. (2013). Characterization and kinetic properties of the purified *Trematosphaeria mangrovei* laccase enzyme. Saudi. J. Biol. Sci., 20: 373–381. https://doi.org/10.1016/J.SJBS.2013.04.001.

Ayele, M., Limeneh, D.Y., Tesfaye, T., Mengie, W. and Abuhay, A. (2021). A review on utilization routes of the leather industry biomass. Adv. Mat. Sci. Eng., 2021: 1503524. https://doi.org/10.1155/2021/1503524.

Bacosa, H.P. and Inoue, C. (2015). Polycyclic aromatic hydrocarbons (PAHs) biodegradation potential and diversity of microbial consortia enriched from tsunami sediments in Miyagi, Japan. J. Hazard. Mater., 283: 689–697. https://doi.org/10.1016/J.JHAZMAT.2014.09.068.

Bailly, C. (2023). Discovery and current developments of isoindolinone-based fungal natural products. Eur. J. Med. Chem. Rep., 9: 100112. https://doi.org/10.1016/J.EJMCR.2023.100112.

Bankole, P.O., Semple, K.T., Jeon, B.H. and Govindwar, S.P. (2021). Biodegradation of fluorene by the newly isolated marine-derived fungus, *Mucor irregularis* strain bpo1 using response surface methodology. Ecotoxicol. Environ. Saf., 208: 111619. https://doi.org/10.1016/J.ECOENV.2020.111619.

Bao, M.T., Wang, L.N., Sun, P.Y., Cao, L.X., Zou, J. et al. (2012). Biodegradation of crude oil using an efficient microbial consortium in a simulated marine environment. Mar. Pollut. Bull., 64: 1177–1185. https://doi.org/10.1016/J.MARPOLBUL.2012.03.020.

Barnes, N.M., Damare, S.R., Bhatawadekar, V.C., Garg, A., Lotlikar, N.P. et al. (2023). Degradation of crude oil-associated polycyclic aromatic hydrocarbons by marine-derived fungi. 3 Biotech, 13: 1–7. https://doi.org/10.1007/S13205-023-03753-2/METRICS.

Barnes, N.M., Khodse, V.B., Lotlikar, N.P., Meena, R.M., Damare, S.R. et al (2018). Bioremediation potential of hydrocarbon-utilizing fungi from select marine niches of India. 3 Biotech, 8: 1–10. https://doi.org/10.1007/S13205-017-1043-8/METRICS.

Bartha, R. and Atlas, R.M. (1977). The microbiology of aquatic oil spills. Adv. Appl. Microbiol., 22: 225–266. https://doi.org/10.1016/S0065-2164(08)70164-3.

Barzkar, N. and Sohail, M. (2020). An overview on marine cellulolytic enzymes and their potential applications. Appl. Microbiol. Biotechnol., 104: 6873–6892. https://doi.org/10.1007/S00253-020-10692-Y/TABLES/4.

Batista-García, R.A., Sutton, T., Jackson, S.A., Tovar-Herrera, O.E., Balcázar-López, E. et al. (2017). Characterization of lignocellulolytic activities from fungi isolated from the deep-sea sponge *Stelletta normani*. PLoS One, 12: e0173750. https://doi.org/10.1371/JOURNAL.PONE.0173750.

Ben Ali, W., Chaduli, D., Navarro, D., Lechat, C., Turbé-Doan, A. et al. (2020). Screening of five marine-derived fungal strains for their potential to produce oxidases with laccase activities suitable for biotechnological applications. BMC Biotechnol., 20: 1–13. https://doi.org/10.1186/S12896-020-00617-Y/FIGURES/7.

Bento, F.M., Camargo, F.A., Okeke, B.C., Frankenberger, W.T. et al. (2005). Comparative bioremediation of soils contaminated with diesel oil by natural attenuation, biostimulation, and bioaugmentation. Bioresour. Technol., 96: 1049–1055. https://doi.org/10.1016/J.BIORTECH.2004.09.008.

Birolli, W.G., Alvarenga, N., Seleghim, M.H.R., Porto, A.L. et al. (2016). Biodegradation of the pyrethroid pesticide esfenvalerate by marine-derived Fungi. Mar. Biotechnol., 18: 511–520. https://doi.org/10.1007/S10126-016-9710-Z/FIGURES/2.

Birolli, W.G., Lima, R.N. and Porto, A.L.M. (2019). Applications of marine-derived microorganisms and their enzymes in biocatalysis and biotransformation, the underexplored potentials. Front. Microbiol., 10: 452098. https://doi.org/10.3389/FMICB.2019.01453/PDF.

Birolli, W.G., Santos, D.D.A., Alvarenga, N., Garcia, A.C. and Romão, L.P. (2018a). Biodegradation of anthracene and several PAHs by the marine-derived fungus *Cladosporium* sp. CBMAI 1237. Mar. Pollut. Bull., 129: 525–533. https://doi.org/10.1016/J.MARPOLBUL.2017.10.023.

Birolli, W.G., Vacondio, B., Alvarenga, N., Seleghim, M.H., Porto, A.L. et al. (2018b). Enantioselective biodegradation of the pyrethroid (±)-lambda-cyhalothrin by marine-derived fungi. Chemosphere, 197: 651–660. https://doi.org/10.1016/J.CHEMOSPHERE.2018.01.054.

Birolli, W.G., Yamamoto, K.Y., de Oliveira, J.R., Nitschke, M., Seleghim, M.H. et al. (2015). Biotransformation of dieldrin by the marine fungus *Penicillium miczynskii* CBMAI 930. Biocatal. Agric. Biotechnol., 4: 39–43. https://doi.org/10.1016/J.BCAB.2014.06.002.

Bittencourt, G.A., de Souza Vandenberghe, L.P., Martínez-Burgos, W.J., Valladares-Diestra, K.K. and de Mello, A.F.M. (2023). Emerging contaminants bioremediation by enzyme and nanozyme-based processes: A review. iScience, 26: 106785.

Bonugli-Santos, R.C., dos Santos Vasconcelos, M.R., Passarini, M.R., Vieira, G.A., Lopes, V.C. et al. (2015). Marine-derived fungi: Diversity of enzymes and biotechnological applications. Front. Microbiol., 6: 128060. https://doi.org/10.3389/FMICB.2015.00269/PDF.

Brakstad, O.G., Daling, P.S., Faksness, L.G., Almås, I.K., Vang, S.H. et al. (2014). Depletion and biodegradation of hydrocarbons in dispersions and emulsions of the Macondo 252 oil generated in an oil-on-seawater mesocosm flume basin. Mar. Pollut. Bull., 84: 125–134. https://doi.org/10.1016/J.MARPOLBUL.2014.05.027.

Brunner, I., Fischer, M., Rüthi, J., Stierli, B., Frey, B. et al. (2018). Ability of fungi isolated from plastic debris floating in the shoreline of a lake to degrade plastics. PLoS One, 13: e0202047. https://doi.org/10.1371/JOURNAL.PONE.0202047.

Canada, K.A., Iwashita, S., Shim, H. and Wood, T.K. (2002). Directed evolution of toluene ortho-monooxygenase for enhanced 1-naphthol synthesis and chlorinated ethene degradation. J. Bacteriol.,

184: 344–349. https://doi.org/10.1128/JB.184.2.344-349.2002/ASSET/8610FB8B-C22E-452A-82E5-0DCDA99BAB82/ASSETS/GRAPHIC/JB0221165001.JPEG.

Chakrabarty, A.M. (1985). Genetically-manipulated microorganisms and their products in the oil service industries. Trends Biotechnol., 3: 32–39. https://doi.org/10.1016/0167-7799(85)90056-3.

Chen, H.Y., Xue, D.S., Feng, X.Y., Yao, S.J. et al. (2011). Screening and production of ligninolytic enzyme by a marine-derived fungal *Pestalotiopsis* sp. J63. Appl. Biochem. Biotechnol., 165: 1754–1769. https://doi.org/10.1007/S12010-011-9392-Y/METRICS.

Chen, J., Zhang, W., Wan, Z., Li, S., Huang, T. et al. (2019). Oil spills from global tankers: Status review and future governance. J. Clean Prod., 227: 20–32. https://doi.org/10.1016/J.JCLEPRO.2019.04.020.

Chin, J.M.W., Puchooa, D., Bahorun, T., Alrefaei, A.F., Neergheen, V.S. et al. (2024). Multigene phylogeny, bioactive properties, enzymatic and dye decolorization potential of selected marine fungi from brown algae and sponges of Mauritius. Heliyon, 10: e28955. https://doi.org/10.1016/j.heliyon.2024.e28955.

Daccò, C., Girometta, C., Asemoloye, M.D., Carpani, G., Picco, A.M. et al. (2020). Key fungal degradation patterns, enzymes and their applications for the removal of aliphatic hydrocarbons in polluted soils: A review. Int. Biodeterior. Biodegrad., 147: 104866. https://doi.org/10.1016/J.IBIOD.2019.104866.

Daghio, M., Aulenta, F., Vaiopoulou, E., Franzetti, A., Arends, J.B. et al. (2017). Electrobioremediation of oil spills. Water Res., 114: 351–370. https://doi.org/10.1016/J.WATRES.2017.02.030.

Dalmaso, G.Z.L., Ferreira, D. and Vermelho, A.B. (2015). Marine extremophiles: A source of hydrolases for biotechnological applications. Mar. Drugs, 13: 1925–1965. https://doi.org/10.3390/MD13041925.

Das, N. and Chandran, P. (2011). Microbial degradation of petroleum hydrocarbon contaminants: An overview. Biotechnol. Res. Int., 2011: 941810. https://doi.org/10.4061/2011/941810.

Desisa, B., Getahun, A. and Muleta, D. (2022). Advances in biological treatment technologies for some emerging pesticides. Pesticides Bioremediation, 259–280. https://doi.org/10.1007/978-3-030-97000-0_10.

DiGregorio, B.E. (2009). Biobased performance bioplastic: Mirel. Chem. Biol., 16: 1–2. https://doi.org/10.1016/J.CHEMBIOL.2009.01.001.

Dong, C.D., Patel, A.K., Madhavan, A., Chen, C.W., Singhania, R.R. et al. (2023a). Significance of glycans in cellulolytic enzymes for lignocellulosic biorefinery: A review. Bioresour. Technol., 379: 128992. https://doi.org/10.1016/J.BIORTECH.2023.128992.

Dong, C.D., Tiwari, A., Anisha, G.S., Chen, C.W., Singh, A. et al. (2023b). Laccase: A potential biocatalyst for pollutant degradation. Environ. Pollut., 319: 120999. https://doi.org/10.1016/J.ENVPOL.2023.120999.

D'Souza, D.T., Tiwari, R., Sah, A.K., Raghukumar, C. et al. (2006). Enhanced production of laccase by a marine fungus during treatment of colored effluents and synthetic dyes. Enzyme Microb., Technol., 38: 504–511. https://doi.org/10.1016/J.ENZMICTEC.2005.07.005.

D'Souza-Ticlo, D., Sharma, D. and Raghukumar, C. (2009). A thermostable metal-tolerant laccase with bioremediation potential from a marine-derived fungus. Mar. Biotechnol., 11: 725–737. https://doi.org/10.1007/S10126-009-9187-0.

Duarte, A.W.F., Dos Santos, J.A., Vianna, M.V., Vieira, J.M.F., Mallagutti, V.H. et al. (2018). Cold-adapted enzymes produced by fungi from terrestrial and marine Antarctic environments. Crit. Rev. Biotechnol., 38: 600–619. https://doi.org/10.1080/07388551.2017.1379468.

Ekanayaka, A.H., Tibpromma, S., Dai, D., Xu, R., Suwannarach, N. et al. (2022). A review of the fungi that degrade plastic. J. Fungi, 8: 772. https://doi.org/10.3390/JOF8080772/S1.

El-Din, G.A., Amer, A.A., Malsh, G., Hussein, M. et al. (2018) Study on the use of banana peels for oil spill removal. Alexandria Eng. J., 57: 2061–2068. https://doi.org/10.1016/j.aej.2017.05.020.

EL-Hanafy, A.A.E.M., Anwar, Y., Sabir, J.S., Mohamed, S.A., Al-Garni, S.M. et al. (2017). Characterization of native fungi responsible for degrading crude oil from the coastal area of Yanbu, Saudi Arabia. Biotechnol. Biotechnol. Equip., 31: 105–111. https://doi.org/10.1080/13102818.2016.1249407.

Elshafie, A., AlKindi, A.Y., Al-Busaidi, S., Bakheit, C. and Albahry, S.N. (2007). Biodegradation of crude oil and n-alkanes by fungi isolated from Oman. Mar. Pollut. Bull., 54: 1692–1696. https://doi.org/10.1016/J.MARPOLBUL.2007.06.006.

Fernandes, P. (2014). Marine enzymes and food industry: Insight on existing and potential interactions. Front. Mar. Sci., 1: 112823. https://doi.org/10.3389/FMARS.2014.00046/PDF.

Flayyih, I. and Ai-Jawhari, H. (2014). Ability of some soil fungi in biodegradation of petroleum hydrocarbon. J. Appl. Environ. Microbiol., 2: 46–52. https://doi.org/10.12691/JAEM-2-2-3.

Fokichev, N.S., Kokaeva, L.Y., Popova, E.A., Kurakov, A.V. and Osmolovskiy, A.A. (2022). Thrombolytic potential of micromycetes from the genus *Tolypocladium*, obtained from white sea soils: Screening of producers and exoproteinases properties. Microbiol. Res. (Pavia), 13: 898–908. https://doi.org/10.3390/MICROBIOLRES13040063.

Furukawa, K. (2003). 'Super bugs' for bioremediation. Trends Biotechnol., 21: 187–190. https://doi.org/10.1016/S0167-7799(03)00054-4.

Gallo, F., Fossi, C., Weber, R., Santillo, D., Sousa, J. et al. (2018). Marine litter plastics and microplastics and their toxic chemicals components: The need for urgent preventive measures. Environ. Sci. Eur., 30: 1–14. https://doi.org/10.1186/S12302-018-0139-Z/FIGURES/1.

Gao, C., Bao, B., Bao, C. and Wu, W. (2023). Fungi fibrinolytic compound 1 plays a core role in modulating fibrinolysis, altering plasma clot structure, and promoting susceptibility to lysis. Pharmaceutics, 15: 2320. https://doi.org/10.3390/PHARMACEUTICS15092320.

Gao, C., Tang, S., Zhang, H., Zhang, H., Zhang, T. et al. (2022a). A novel marine pyran-isoindolone compound enhances fibrin lysis mediated by single-chain urokinase-type plasminogen activator. Mar. Drugs, 20: 495. https://doi.org/10.3390/MD20080495.

Gao, R., Liu, R. and Sun, C. (2022b). A marine fungus *Alternaria alternata* FB1 efficiently degrades polyethylene. J. Hazard Mater., 431: 128617. https://doi.org/10.1016/J.JHAZMAT.2022.128617.

Garg, S. (2025). The importance of fungal biotechnology for sustainable applications. Trends Biotechnol. https://doi.org/10.1016/j.tibtech.2025.06.010.

Geris, R., Malta, M., Soares, L.A., de Souza Neta, L.C., Pereira, N.S. et al. (2024). A review about the mycoremediation of soil impacted by war-like activities: Challenges and gaps. J. Fungi, 10: 94. https://doi.org/10.3390/JOF10020094/S1.

Ghosal, D., Ghosh, S., Dutta, T.K., Ahn, Y. et al. (2016). Current state of knowledge in microbial degradation of polycyclic aromatic hydrocarbons (PAHs): A review. Front. Microbiol., 7: 208111. https://doi.org/10.3389/FMICB.2016.01369/PDF.

Ghosh, S., Rusyn, I., Dmytruk, O.V., Dmytruk, K.V., Onyeaka, H. et al. (2023). Filamentous fungi for sustainable remediation of pharmaceutical compounds, heavy metal and oil hydrocarbons. Front. Bioeng. Biotechnol., 11: 1106973. https://doi.org/10.3389/FBIOE.2023.1106973/PDF.

Gonçalves, M.F.M., Esteves, A.C. and Alves, A. (2022). Marine Fungi: Opportunities and challenges. Encyclopedia, 2: 559–577. https://doi.org/10.3390/ENCYCLOPEDIA2010037.

Gonda, K.E., Jendrossek, D. and Molitoris, H.P. (2000). Fungal degradation of the thermoplastic polymer poly-ß-hydroxybutyric acid (PHB) under simulated deep sea pressure. Hydrobiologia, 426: 173–183. https://doi.org/10.1007/978-94-011-4148-2_17.

Gouma, S., Papadaki, A.A., Markakis, G., Magan, N., Goumas, D. et al. (2019). Studies on pesticides mixture degradation by white rot fungi. J. Ecol. Eng., 20: 16–26. https://doi.org/10.12911/22998993/94918.

Guo, R., Zhang, Y., Duan, D., Fu, Q., Zhang, X. et al. (2016). Fibrinolytic evaluation of compounds isolated from a marine fungus *Stachybotrys longispora* FG216. Chin. J. Chem., 34: 1194–1198. https://doi.org/10.1002/CJOC.201600623.

Hamad, A.A., Moubasher, H.A., Moustafa, Y.M., Mohamed, N.H. and Abd-El Rhim, E.H. (2021). Petroleum hydrocarbon bioremediation using native fungal isolates and consortia. Sci. World J., 2021: 6641533. https://doi.org/10.1155/2021/6641533.

Hang, S., Chen, H., Wu, W., Wang, S., Fang, Y. et al. (2022) Progress in isoindolone alkaloid derivatives from marine microorganism: Pharmacology, preparation, and mechanism. Mar. Drugs, 20: 405. https://doi.org/10.3390/MD20060405.

Haque, N., Parveen, S., Tang, T., Wei, J. and Huang, Z. (2022). Marine natural products in clinical use. Mar. Drugs, 20: 528. https://doi.org/10.3390/MD20080528/S1.

Hara, A., Syutsubo, K. and Harayama, S. (2003). Alcanivorax which prevails in oil-contaminated seawater exhibits broad substrate specificity for alkane degradation. Environ. Microbiol., 5: 746–753. https://doi.org/10.1046/J.1468-2920.2003.00468.X.

Hareeri, R.H., Aldurdunji, M.M., Abdallah, H.M., Alqarni, A.A., Mohamed, S.G. et al. (2022) *Aspergillus ochraceus*: Metabolites, bioactivities, biosynthesis, and biotechnological potential. Molecules, 27: 6759. https://doi.org/10.3390/MOLECULES27196759.

He, H., Zhang, R., Zhang, P., Wang, P., Chen, N. et al. (2023), Functional carbon from nature: Biomass-derived carbon materials and the recent progress of their applications. Adv. Sci., 10: 2205557. https://doi.org/10.1002/ADVS.202205557.

Hentati, D., Chebbi, A., Loukil, S., Kchaou, S., Godon, J.J. et al. (2016). Biodegradation of fluoranthene by a newly isolated strain of *Bacillus stratosphericus* from Mediterranean seawater of the Sfax fishing harbour, Tunisia. Environ. Sci. Poll. Res., 23: 15088–15100. https://doi.org/10.1007/S11356-016-6648-7/TABLES/4.

Ibrahim, S.R., Mohamed, S.G., Alsaadi, B.H., Althubyani, M.M., Awari, Z.I. et al. (2023). Secondary metabolites, biological activities, and industrial and biotechnological importance of *Aspergillus sydowii*. Mar. Drugs, 21: 441. https://doi.org/10.3390/MD21080441.

Jarerat, A. and Tokiwa, Y. (2001). Degradation of poly(l-lactide) by a fungus. Macromol. Biosci., 1: 136–140. https://doi.org/10.1002/1616-5195(20010601)1:4.

Jeevithan, L., Diao, X., Hu, J., Elango, J., Wu, W. et al. (2024). Recent advancement of novel marine fungi derived secondary metabolite fibrinolytic compound FGFC in biomedical applications: A review. Front. Cell Infect. Microbiol., 14: 1422648. https://doi.org/10.3389/FCIMB.2024.1422648/BIBTEX.

Jiang, W., Pelaez, M., Dionysiou, D.D., Entezari, M.H. and Tsoutsou, D. (2013). Chromium(VI) removal by maghemite nanoparticles. Chem. Eng. J., 222: 527–533. https://doi.org/10.1016/J.CEJ.2013.02.049.

Kotb, E. (2013). Activity assessment of microbial fibrinolytic enzymes. Appl. Microbiol. Biotechnol., 97: 6647–6665. https://doi.org/10.1007/S00253-013-5052-1.

Kumar, A. and Chandra, R. (2020). Ligninolytic enzymes and its mechanisms for degradation of lignocellulosic waste in environment. Heliyon, 6: e03170. https://doi.org/10.1016/J.HELIYON.2020.E03170.

Kunhiraman, S., Kumar, S.S., Haridas, M. and Abdulhameed, S. (2024). Process optimization for the production of fibrinolytic enzyme by a novel marine fungus: *Penicillium steckii* KU1. Biocatal. Agric. Biotechnol., 58: 103207. https://doi.org/10.1016/J.BCAB.2024.103207.

Lee, C.C., Kibblewhite, R.E., Wagschal, K., Li, R., Robertson, G.H. et al. (2012). Isolation and characterization of a novel GH67 α-glucuronidase from a mixed culture. J. Ind. Microbiol. Biotechnol., 39: 1245–1251. https://doi.org/10.1007/S10295-012-1128-7.

Li, R., Espi, Wayman, C. and Niemann, H. (2021). The fate of plastic in the ocean environment: A minireview. Environ. Sci. Process. Impacts., 23: 198–212. https://doi.org/10.1039/D0EM00446D.

Li, X., Kondo, R. and Sakai, K. (2002). Biodegradation of sugarcane bagasse with marine fungus *Phlebia* sp. MG-60. J. Wood Sci., 48: 159–162. https://doi.org/10.1007/BF00767294/METRICS.

Ling, M., Zheng, P., Huang, X., Fan, G., Zhang, H. et al. (2025). Improving bioavailability of lignocellulosic biomass by pretreatment with the marine fungus *Chaetomium* sp. CS1. Int. Biodeterior. Biodegradation, 199: 106031. https://doi.org/10.1016/J.IBIOD.2025.106031.

Liu, S., Zhao, Y., Jiang, W., Wu, M. and Ma, F. (2014). Inactivation of *Microcystis aeruginosa* by electron beam irradiation. Water Air Soil Pollut., 225: 1–6. https://doi.org/10.1007/S11270-014-2093-8/FIGURES/5.

Lovley, D.R. (2003). Cleaning up with genomics: Applying molecular biology to bioremediation. Nat. Rev. Microbiol., 1: 35–44. https://doi.org/10.1038/nrmicro731.

Maamar, A., Lucchesi, M.E., Debaets, S., Nguyen van Long, N., Quemener, M. et al. (2020). Highlighting the crude oil bioremediation potential of marine fungi isolated from the Port of Oran (Algeria). Diversity (Basel), 12: 196. https://doi.org/10.3390/D12050196.

Maeda, H., Yamagata, Y., Abe, K., Hasegawa, F., Machida, M. et al. (2005). Purification and characterization of a biodegradable plastic-degrading enzyme from *Aspergillus oryzae*. Appl. Microbiol. Biotechnol., 67: 778–788. https://doi.org/10.1007/S00253-004-1853-6/METRICS.

Mahmud, T., Sabo, I.A., Lambu, Z.N., Danlami, D. and Shehu, A.A. (2022). Hydrocarbon degradation potentials of Fungi: A review. J. Environ. Bioremediat. Toxicol., 5: 50–56. https://doi.org/10.54987/JEBAT.V5I1.681.

Manzanares, P. (2020). The role of biorefinering research in the development of a modern bioeconomy. Acta Innovations, 47–56. https://doi.org/10.32933/ACTAINNOVATIONS.37.4.

Masaki, K., Kamini, N.R., Ikeda, H. and Iefuji, H. (2005). Cutinase-like enzyme from the yeast *Cryptococcus* sp. strain S-2 hydrolyzes polylactic acid and other biodegradable plastics. Appl. Environ. Microbiol., 71: 7548–7550. https://doi.org/10.1128/AEM.71.11.7548-7550.2005.

Meckenstock, R.U. and Mouttaki, H. (2011). Anaerobic degradation of non-substituted aromatic hydrocarbons. Curr. Opin. Biotechnol., 22: 406–414. https://doi.org/10.1016/J.COPBIO.2011.02.009.

Medina-Bellver, J.I., Marín, P., Delgado, A., Rodríguez-Sánchez, A., Reyes, E. et al. (2005). Evidence for in situ crude oil biodegradation after the Prestige oil spill. Environ. Microbiol., 7: 773–779. https://doi.org/10.1111/J.1462-2920.2005.00742.X.

Mensah, G.A., Fuster, V., Murray, C.J. and Roth, G.A. (2023). Global burden of cardiovascular diseases and risks, 1990–2022. J. Am. Coll. Cardiol., 82: 2350–2473. https://doi.org/10.1016/J.JACC.2023.11.007.

Miyazaki, S., Takahashi, K., Shiraki, M., Saito, T. and Tezuka, Y. (2000). Properties of a poly(3-hydroxybutyrate) depolymerase from *Penicillium funiculosum*. J. Polym. Environ., 8: 175–182. https://doi.org/10.1023/A:1015245710406/METRICS.

Nazifa, T.H., Bin Ahmad, M.A., Hadibarata, T. and Aris, A. (2018). Bioremediation of diesel oil spill by filamentous fungus *Trichoderma reesei* H002 in aquatic environment. Int. J. Integr. Eng., 10: 103–107. https://doi.org/10.30880/ijie.2018.10.09.019.

Nikolova, C. and Gutierrez, T. (2021). Biosurfactants and their applications in the oil and gas industry: Current state of knowledge and future perspectives. Front. Bioeng. Biotechnol., 9: 626639. https://doi.org/10.3389/FBIOE.2021.626639/PDF.

Oliveira, B.R., Penetra, A., Cardoso, V.V., Benoliel, M.J., Barreto Crespo, M.T. et al (2015). Biodegradation of pesticides using fungi species found in the aquatic environment. Environ. Sci. Pollut. Res., 22: 11781–11791. https://doi.org/10.1007/S11356-015-4472-0/FIGURES/5.

Omoriyekomwan, J.E., Tahmasebi, A., Dou, J., Wang, R. and Yu, J. (2021). A review on the recent advances in the production of carbon nanotubes and carbon nanofibers via microwave-assisted pyrolysis of biomass. Fuel Process Technol., 214: 106686. https://doi.org/10.1016/J.FUPROC.2020.106686.

Ortega, S.N., Nitschke, M., Mouad, A.M., Landgraf, M.D., Rezende, M.O.O. et al. (2011). Isolation of Brazilian marine fungi capable of growing on DDD pesticide. Biodegradation, 22: 43–50. https://doi.org/10.1007/S10532-010-9374-8/METRICS.

Paço, A., Duarte, K., da Costa, J.P., Santos, P.S., Pereira, R. et al. (2017). Biodegradation of polyethylene microplastics by the marine fungus *Zalerion maritimum*. Sci. Total Environ., 586: 10–15. https://doi.org/10.1016/J.SCITOTENV.2017.02.017.

Pan, N., Li, Z.C., Li, Z.H., Chen, S.H., Jiang, M.H. et al. (2022). Antiplatelet and antithrombotic effects of Isaridin E isolated from the marine-derived fungus via downregulating the PI3K/Akt signaling pathway. Mar. Drugs, 20: 23. https://doi.org/10.3390/MD20010023/S1.

Pasumarthi, R., Chandrasekaran, S. and Mutnuri, S. (2013). Biodegradation of crude oil by *Pseudomonas aeruginosa* and *Escherichia fergusonii* isolated from the Goan coast. Mar. Pollut. Bull., 76: 276–282. https://doi.org/10.1016/J.MARPOLBUL.2013.08.026.

Pathak, V.M., Verma, V.K., Rawat, B.S., Kaur, B., Babu, N. et al. (2022). Current status of pesticide effects on environment, human health and its eco-friendly management as bioremediation: A comprehensive review. Front. Microbiol., 13: 962619. https://doi.org/10.3389/FMICB.2022.962619/PDF.

Pereira, C.P.M., Souza, A.C.R., Vasconcelos, A.R. and Prado, P.S. (2021). Antioxidant and anti-inflammatory mechanisms of action of astaxanthin in cardiovascular diseases (Review). Int. J. Mol. Med., 47: 37–48. https://doi.org/10.3892/IJMM.2020.4783.

Pieper, D.H. and Reineke, W. (2000). Engineering bacteria for bioremediation. Curr. Opin. Biotechnol., 11: 262–270. https://doi.org/10.1016/S0958-1669(00)00094-X.

Podorol'skaya, L.V., Serebryakova, T.N., Sharkova, T.S., Neumyvakin, L.V., Khromov, I.S. et al. (2007). Experimental thrombosis in rabbit marginal ear vein and evaluation of the thrombolytic action of longolytin. Bull. Exp. Biol. Med., 143: 577–580. https://doi.org/10.1007/S10517-007-0184-X/METRICS.

Prince, R.C. (1993). Petroleum spill bioremediation in marine environments. Crit. Rev. Microbiol., 19: 217–240. https://doi.org/10.3109/10408419309113530.

Prince, R.C., McFarlin, K.M., Butler, J.D., Febbo, E.J., Wang, F.C. et al. (2013). The primary biodegradation of dispersed crude oil in the sea. Chemosphere, 90: 521–526. https://doi.org/10.1016/J.CHEMOSPHERE.2012.08.020.

Przystas, W., Zablocka-Godlewska, E. and Grabinska-Sota, E. (2015). Efficacy of fungal decolorization of a mixture of dyes belonging to different classes. Braz. J. Microbiol., 46: 415–424. https://doi.org/10.1590/S1517-838246246220140167.

Ramzan, M.N., Shen, D., Wei, Y., Raza, B., Yuan, H. et al. (2025). A Synergistic role of photosynthetic bacteria and fungal community in pollutant removal in an integrated aquaculture wastewater bioremediation system. Biology, 14(8): 959. https://doi.org/10.3390/biology14080959.

Restrepo-Flórez, J.M., Bassi, A. and Thompson, M.R. (2014). Microbial degradation and deterioration of polyethylene: A review. Int. Biodeterior. Biodegradation, 88: 83–90. https://doi.org/10.1016/J.IBIOD.2013.12.014.

Rouches, E., Zhou, S., Steyer, J.P. and Carrère, H. (2016). White-Rot Fungi pretreatment of lignocellulosic biomass for anaerobic digestion: Impact of glucose supplementation. Process. Biochem., 51: 1784–1792. https://doi.org/10.1016/J.PROCBIO.2016.02.003.

Rudakiya, D.M., Tripathi, A., Gupte, S. and Gupte, A. (2019). Fungal bioremediation: A step towards cleaner environment. pp. 229–249. *In*: Satyanarayana, T., Deshmukh, S. and Deshpande, M. (eds.). Advancing Frontiers in Mycology and Mycotechnology. Springer, Singapore. https://doi.org/10.1007/978-981-13-9349-5_9.

Saez, J.M., Raimondo, E.E., Costa-Gutierrez, S.B., Aparicio, J.D., Angelucci, D.M. et al. (2025). Enhancing environmental decontamination and sustainable production through synergistic and complementary interactions of actinobacteria and fungi. Heliyon, 11(3): 42135. https://doi.org/10.1016/j.heliyon.2025.e42135.

Saini, R., Kaur, A., Saini, J.K., Patel, A.K., Varjani, S. et al. (2022). Trends in lignin biotransformations for bio-based products and energy applications. Bioenergy Res., 16: 88–104. https://doi.org/10.1007/S12155-022-10434-0.

Sánchez, C. (2020). Fungal potential for the degradation of petroleum-based polymers: An overview of macro- and microplastics biodegradation. Biotechnol. Adv., 40: 107501. https://doi.org/10.1016/J.BIOTECHADV.2019.107501.

Saravanan, A., Karishma, S., Kumar, P.S. and Rangasamy, G. (2023). Biodegradation of oil-contaminated aqueous ecosystem using an immobilized fungi biomass and kinetic study. Environ. Res., 220: 115252. https://doi.org/10.1016/J.ENVRES.2023.115252.

Serebriakova, T.N., Andreenko, G.V., Maksimova, R.A., Tsymanovich, S.T. and Murashova, N.L. (1984). Anticoagulating properties of fungal proteases tricholysine (triase) and longolytin. Vopr. Med. Khim., 30: 37–40.

Sethi, R., Padmavathi, T. and Sullia, S.B. (2013). Lignocellulose biomass degradation by marine microorganisms. Pelagia Res. Lib. Eur. J. Exp. Bio., 3: 129–138.

Shahzadi, T., Mehmood, S., Irshad, M., Anwar, Z., Afroz, A. et al. (2014). Advances in lignocellulosic biotechnology: A brief review on lignocellulosic biomass and cellulases. Adv. Biosci. Biotechnol., 5: 246–251. https://doi.org/10.4236/ABB.2014.53031.

Sharkova, T.S., Kurakov, A.V., Osmolovskiy, A.A., Matveeva, E.O., Kreyer, V.G. et al. (2015). Screening of producers of proteinases with fibrinolytic and collagenolytic activities among *Micromycetes*. Microbiol. (Russian Federation), 84: 359–364. https://doi.org/10.4236/ABB.2014.53031.

Simister, R.L., Poutasse, C.M., Thurston, A.M., Reeve, J.L., Baker, M.C. et al. (2015). Degradation of oil by fungi isolated from Gulf of Mexico beaches. Mar. Pollut. Bull., 100: 327–333. https://doi.org/10.1016/J.MARPOLBUL.2015.08.029.

Singh, D.V., Ali, R., Anita, Kulsum, M. and Bhat, R.A. (2020). Ecofriendly approaches for remediation of pesticides in contaminated environs. Bioremediation and Biotechnology (Vol. 3): Persistent and Recalcitrant Toxic Substances, 3: 173–194. https://doi.org/10.1007/978-3-030-46075-4_8/TABLES/5.

Soares, P.R.S., Birolli, W.G., Ferreira, I.M. and Porto, A.L.M. (2021). Biodegradation pathway of the organophosphate pesticides chlorpyrifos, methyl parathion and profenofos by the marine-derived fungus *Aspergillus sydowii* CBMAI 935 and its potential for methylation reactions of phenolic compounds. Mar. Pollut. Bull., 166: 112185. https://doi.org/10.1016/J.MARPOLBUL.2021.112185.

Soddell, J.A., Stainsby, F.M., Eales, K.L., Seviour, R.J. and Goodfellow, M. (2006). *Gordonia defluvii* sp. nov., an actinomycete isolated from activated sludge foam. Int. J. Syst. Evol. Microbiol., 56: 2265–2269. https://doi.org/10.1099/IJS.0.64034-0/CITE/REFWORKS.

Spier, C., Stringfellow, W.T., Hazen, T.C. and Conrad, M. (2013). Distribution of hydrocarbons released during the 2010 MC252 oil spill in deep offshore waters. Environ. Poll., 173: 224–230. https://doi.org/10.1016/J.ENVPOL.2012.10.019.

Steffen, W., Richardson, K., Rockström, J., Cornell, S.E., Fetzer, I. et al. (2015). Planetary boundaries: Guiding human development on a changing planet. Science, 347(6223): 1259855. https://doi.org/10.1126/SCIENCE.1259855/SUPPL_FILE/STEFFEN-SM.PDF.

Steliga, T. (2012). Role of fungi in biodegradation of petroleum hydrocarbons in drill waste. Pol. J. Environ. Stud., 22: 471–479.

Sukhdhane, K.S., Pandey, P.K., Ajima, M.N.O., Jayakumar, T., Vennila, A. et al. (2019). Isolation and characterization of phenanthrene-degrading bacteria from PAHs contaminated mangrove sediment of Thane Creek in Mumbai, India. Polycycl. Aromat. Compd., 39: 73–83. https://doi.org/10.1080/10406638.2016.1261911.

Thapa, S., Mishra, J., Arora, N., Mishra, P., Li, H. et al. (2020). Microbial cellulolytic enzymes: Diversity and biotechnology with reference to lignocellulosic biomass degradation. Rev. Environ. Sci. Biotechnol., 19: 621–648. https://doi.org/10.1007/S11157-020-09536-Y/METRICS.

The Future of Plastic (2018). Editorial. Nat. Commun., 9: 2157. https://doi.org/10.1038/s41467-018-04565-2.

Toker, S.K., Evlat, H. and Koçyiğit, A. (2021). Screening of newly isolated marine-derived fungi for their laccase production and decolorization of different dye types. Reg. Stud. Mar. Sci., 45: 101837. https://doi.org/10.1016/J.RSMA.2021.101837.

Tomer, A., Singh, R., Singh, S.K., Dwivedi, S.A. and Reddy, C.U. (2021). Role of Fungi in bioremediation and environmental sustainability. pp. 187–200. *In*: Prasad, R., Nayak, S.C., Kharwar, R.N. and Dubey, N.K. (eds.). Mycoremediation and Environmental Sustainability. Fungal Biology. Springer, Cham. https://doi.org/10.1007/978-3-030-54422-5_8.

Torres, J.M.O., Cardenas, C.V., Moron, L.S., Guzman, A.P.A. and Dela Cruz, T.E.E. (2011). Dye decolorization activities of marine-derived fungi isolated from Manila Bay, Philippines. Phillip J. Sci., 140(2): 133–143.

Torres, M.A., Barros, M.P., Campos, S.C.G., Pinto, E., Rajamani, S. et al. (2008). Biochemical biomarkers in algae and marine pollution: A review. Ecotoxicol. Environ. Saf., 71: 1–15. https://doi.org/10.1016/J.ECOENV.2008.05.009.

Torres-Farradá, G., Thijs, S., Rineau, F., Guerra, G. and Vangronsveld, J. (2024). White rot fungi as tools for the bioremediation of xenobiotics: A review. J. Fungi, 10: 167. https://doi.org/10.3390/JOF10030167.

Vacondio, B., Birolli, W.G., Ferreira, I.M., Seleghim, M.H., Gonçalves, S. et al. (2015). Biodegradation of pentachlorophenol by marine-derived fungus *Trichoderma harzianum* CBMAI 1677 isolated from ascidian *Didemnum ligulum*. Biocatal. Agric. Biotechnol., 4: 266–275. https://doi.org/10.1016/J.BCAB.2015.03.005.

Vaidya, S., Devpura, N., Jain, K. and Madamwar, D. (2018). Degradation of chrysene by enriched bacterial consortium. Front. Microbiol., 9: 366036. https://doi.org/10.3389/FMICB.2018.01333/BIBTEX.

Vaksmaa, A., Guerrero-Cruz, S., Ghosh, P., Zeghal, E., Hernando-Morales, V. et al. (2023a). Role of fungi in bioremediation of emerging pollutants. Front. Mar. Sci., 10: 1070905. https://doi.org/10.3389/FMARS.2023.1070905/PDF.

Vaksmaa, A., Hernando-Morales, V., Zeghal, E. and Niemann, H. (2021). Microbial degradation of marine plastics: current state and future prospects. pp. 111–154. *In*: Joshi, S.J., Deshmukh, A. and Sarma, H. (eds.). Biotechnology for Sustainable Environment. Springer, Singapore. https://doi.org/10.1007/978-981-16-1955-7_5.

Vaksmaa, A., Polerecky, L., Dombrowski, N., Kienhuis, M.V., Posthuma, I. et al. (2023b). Polyethylene degradation and assimilation by the marine yeast *Rhodotorula mucilaginosa*. ISME Communications, 3(1): 68. https://doi.org/10.1038/S43705-023-00267-Z.

Vaksmaa, A., Vielfaure, H., Polerecky, L., Kienhuis, M.V.M., Van Der Meer, M.T.J. et al. (2024). Biodegradation of polyethylene by the marine fungus *Parengyodontium album*. Sci. Total. Environ., 934: 172819. https://doi.org/10.1016/J.SCITOTENV.2024.172819.

Vala, A.K. and Dave, B.P. (2017). Marine-derived fungi: Prospective candidates for bioremediation. pp. 17–37. *In*: Prasad, R. (ed.). Mycoremediation and Environmental Sustainability. Springer, Cham. https://doi.org/10.1007/978-3-319-68957-9_2.

Vasconcelos, M.R., Vieira, G.A., Otero, I.V., Bonugli-Santos, R.C., Rodrigues, M.V. et al. (2019). Pyrene degradation by marine-derived ascomycete: Process optimization, toxicity, and metabolic analyses. Environ. Sci. Pollut. Res., 26: 12412–12424. https://doi.org/10.1007/S11356-019-04518-2/FIGURES/3.

Vieira, G.A., Magrini, M.J., Bonugli-Santos, R.C., Rodrigues, M.V. and Sette, L.D. (2018). Polycyclic aromatic hydrocarbons degradation by marine-derived basidiomycetes: Optimization of the degradation process. Braz. J. Microbiol., 49: 749–756. https://doi.org/10.1016/J.BJM.2018.04.007.

Wang, G., Wu, W., Zhu, Q., Fu, S., Wang, X. et al. (2015). Identification and fibrinolytic evaluation of an isoindolone derivative isolated from a rare marine fungus *Stachybotrys longispora* FG216. Chin. J. Chem., 33: 1089–1095. https://doi.org/10.1002/CJOC.201500176.

Wang, Z., Shen, D., Wu, C. and Gu, S. (2018). State-of-the-art on the production and application of carbon nanomaterials from biomass. Green Chem., 20: 5031–5057. https://doi.org/10.1039/C8GC01748D.

Wei, R. and Zimmermann, W. (2017). Microbial enzymes for the recycling of recalcitrant petroleum-based plastics: How far are we? Microb. Biotechnol., 10: 1308–1322. https://doi.org/10.1111/1751-7915.12710.

Wikee, S., Hatton, J., Turbé-Doan, A., Mathieu, Y., Daou, M. et al. (2019). Characterization and dye decolorization potential of two laccases from the marine-derived fungus *Pestalotiopsis* sp. Int. J. Mol. Sci., 20: 1864. https://doi.org/10.3390/IJMS20081864.

Wu, C., De Visscher, A. and Gates, I.D. (2019). On naphthenic acids removal from crude oil and oil sands process-affected water. Fuel, 253: 1229–1246. https://doi.org/10.1016/J.FUEL.2019.05.091.

Yan, T., Wu, W., Su, T., Chen, J., Zhu, Q. et al. (2015). Effects of a novel marine natural product: Pyrano indolone alkaloid fibrinolytic compound on thrombolysis and hemorrhagic activities *in vitro* and *in vivo*. Arch. Pharm. Res., 38: 1530–1540. https://doi.org/10.1007/S12272-014-0518-Y/FIGURES/8.

Yin, Y., Fu, Q., Wu, W., Cai, M., Zhou, X. et al. (2017). Producing novel fibrinolytic isoindolinone derivatives in marine fungus *Stachybotrys longispora* FG216 by the rational supply of amino compounds according to its biosynthesis pathway. Mar. Drugs, 15: 214. https://doi.org/10.3390/MD15070214.

Zeghal, E., Vaksmaa, A., Vielfaure, H., Boekhout, T. and Niemann, H. (2021). The potential role of marine fungi in plastic degradation: A review. Front. Mar. Sci., 8: 738877. https://doi.org/10.3389/FMARS.2021.738877/PDF.

Zhang, J., Gao, D., Li, Q., Zhao, Y., Li, L. et al. (2020). Biodegradation of polyethylene microplastic particles by the fungus Aspergillus flavus from the guts of wax moth *Galleria mellonella*. Sci. Total Environ., 704: 135931. https://doi.org/10.1016/J.SCITOTENV.2019.135931.

Zhang, K., Hu, J., Yang, S., Xu, W., Wang, Z. et al. (2022). Biodegradation of polyester polyurethane by the marine fungus *Cladosporium halotolerans* 6UPA1. J. Hazard. Mater., 437: 129406. https://doi.org/10.1016/J.JHAZMAT.2022.129406.

Zhang, K., Jiang, Y., Zeng, H. and Zhu, H. (2023a). Application and risk prediction of thrombolytic therapy in cardio-cerebrovascular diseases: A review. Thromb. J., 21: 1–16. https://doi.org/10.1186/S12959-023-00532-0/TABLES/2.

Zhang, T., Ji, Z., Chen, X. and Yu, L. (2023b). Shotgun metagenomics reveals a diverse mycobiome in the seawater from a High Arctic fjord (Kongsfjorden, Svalbard). Environ. Res., 233: 116437. https://doi.org/10.1016/J.ENVRES.2023.116437.

Zhao, L., Lin, X., Fu, J., Zhang, J., Tang, W. et al. (2022). A novel bi-functional fibrinolytic enzyme with anticoagulant and thrombolytic activities from a marine-derived Fungus *Aspergillus versicolor* ZLH-1. Mar. Drugs, 20: 356. https://doi.org/10.3390/MD20060356/S1.

Zhou, W., Li, M. and Achal, V. (2025). A comprehensive review on environmental and human health impacts of chemical pesticide usage. Emerg. Contam., 11: 100410. https://doi.org/10.1016/J.EMCON.2024.100410.

Zhou, Y., Chen, R., Liu, D. et al. (2017). Asperlin inhibits LPS-evoked foam cell formation and prevents atherosclerosis in ApoE−/− Mice. Mar. Drugs, 15: 358. https://doi.org/10.3390/MD15110358.

11

Marine Fungal Potential for the Bioremediation of Polycyclic Aromatic Hydrocarbons and Heavy Metals

Olguin, J.,[1,2,*] *Barthelemy-Ake, S.*,[2,3] *Valdes, J.*[3,4] and *Lopez, A.*[2,3,4]

Bioremediation is a sustainable strategy for reducing, containing or eliminating pollutants by leveraging the intrinsic abilities of organisms. This review highlights the promising role of marine fungi for the bioremediation of two major contaminants: polycyclic aromatic hydrocarbons (PAHs) and heavy metals (HMs). Both contaminants originate from primary sources, such as industrial emissions and wastewater discharges. Particularly, they tend to accumulate in organisms and marine sediments. Whereas HMs are non-biodegradable elements that persist in the environment, they also bioaccumulate in organisms, and contribute to long-term ecosystem degradation. Marine fungi have shown exceptional potential in decomposing PAHs because of their production of extracellular enzymes such as laccases and peroxidases. Marine microfungi such as *Corollospora*, and *Aspergillus* have demonstrated remarkable tolerance and degradation abilities of PAH.

[1] Posgrado en Ciencias del Mar y Limnología, Universidad Nacional Autónoma de México, Mexico City, Mexico.
[2] Instituto de Biología, Universidad Nacional Autónoma de México, Mexico City, Mexico.
[3] Posgrado en Ciencias Biológicas, Universidad Nacional Autónoma de México, Mexico City, Mexico.
[4] Facultad de Estudios Superiores Zaragoza, Universidad Nacional Autónoma de México, Mexico City, Mexico.
* Corresponding author: juliooolguin@gmail.com

DOI: 10.1201/9781003795100-14

1. Introduction

Pollutants in marine ecosystems have increased every year as a result of anthropogenic activity. Bioremediation is a biotechnological technique that employs the natural ability of microorganisms such as bacteria, yeasts and microalgae to break down, modify, or eliminate toxic compounds (Aliko et al., 2022). Compared to chemical and physical strategies, bioremediation is a sustainable, safer, less invasive, and low-cost alternatives (Kuppan et al., 2024; Passarini et al., 2011). Therefore, biological techniques for environmental management can counteract the damage caused by various pollutants and prevent the degradation of soil and water (Dua et al., 2002).

Despite the fact that marine fungi play an essential role in the biogeochemical cycles of marine environments, their study remains scarce. In recent years, research efforts have focused in their biotechnological potential, discovering new metabolites (e.g., those with antibacterial and anticarcinogenic activities) and exploring their potential use in bioremediation (Gonçalves et al., 2022). In this regard, fungi have shown the ability to degrade PAHs (polycyclic aromatic hydrocarbons) or bioaccumulate HMs (heavy metals) by transforming these contaminants into nontoxic end-products through metabolic processes at the molecular level (Lamichhane et al., 2016; Dell'Anno et al., 2022; Imam et al., 2022).

The PHAs and HMs are important pollutants threatening the welfare of marine environments. In general, pollutants are studied separately, but many of them co-occur in the same environment at the same time (Fig. 1). This problem creates a complex ecological problem in marine ecosystems. The Environmental Protection Agency's (EPA) National Priority List (NPL) register that 40% in hazardous wastes sites are co-contaminated with organic pollutants and heavy metals (Ma et al., 2014). Co-existing PAHs and HMs interact through cation-π bonding, intensifying toxicity while decreasing pollutant bioavailability and microbial degradation (Wang et al., 2025).

This chapter compiles current knowledge on the application of marine fungi to the bioremediation of PAHs and HMs, alongside an overview of the ecotoxicological impacts linked to the presence of these pollutants in marine ecosystems. Additionally, we briefly discussed some perspectives and future directions for applying marine fungi in bioremediation strategies.

2. Sources of Polycyclic Aromatic Hydrocarbons

Crude oils are complex mixtures of hydrocarbons and many organic compounds with varying concentrations of trace elements like S, N, and O (Philp, 2020). Particularly, PAHs are unsaturated hydrocarbons with one or more benzene rings; they are ubiquitous, dense, extra viscous, persistent, and recalcitrant, mainly in aquatic environments (Hussein and Mona, 2016). Polycyclic aromatic hydrocarbons represent around 30% of the composition of oil (Farrington, 2020), are generally categorized as environmentally harmful pollutants, and are widely detected in marine habitats, including open sea, water columns, coastal zones, estuaries, mangroves, sediments, and coral reefs (Honda et al., 2018). Low molecular weight PAHs are

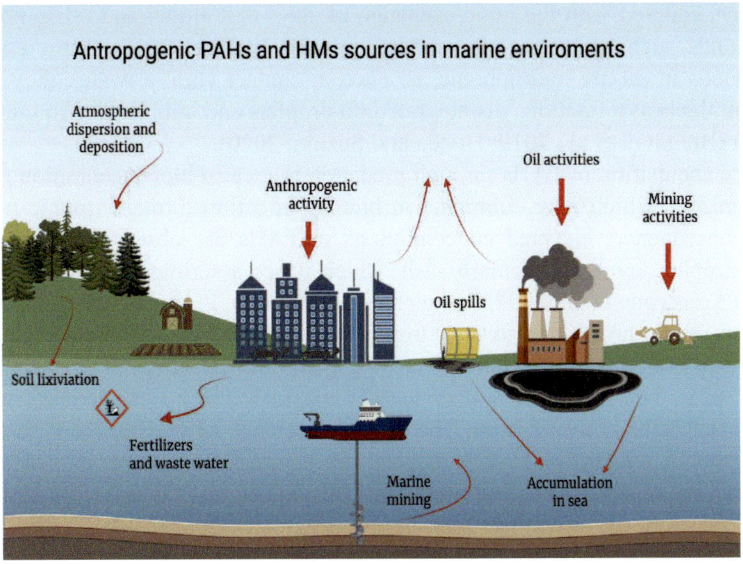

Fig. 1. Illustration of atmos pheric sources of PAHs and HMs. Both pollutants can originate from urban activities and agricultural practices, particularly through the discharge of fertilizers and wastewater into marine environments. Additionally, Deep sea mining and petroleum industries could contribute to the transport and liberation of PAHs and HMs into the ocean. Atmospheric deposition, resulting from vehicle emissions and fuel combustion, has been identified as another significant source of contamination (created in https://BioRender.com).

volatile, while high molecular weight PAHs are more persistent and toxic (Wolska et al., 2012). Contamination with PAHs occurs via inputs of industrial discharges or the incomplete combustion of organic materials (Zhang et al., 2024).

In marine systems, the presence of PAHs is also added by drilling in search of new oil sources, accidental spills, and accidents in marine platforms. Particularly in coastal areas, petrogenic and pyrogenic sources are mainly artificial and are important contributors of environmental PAH pollution (Hong et al., 1995). Light components of oil, such as aliphatic hydrocarbons and light PAHs, could recirculate in ocean layers, not reaching the deep-sea sediments, since some heavier components of oil, such as complex PAHs, resins, and asphaltenes, precipitate in benthic zones (Kostka et al., 2020). In the deep sea, marine sediments are an important reservoir of PAHs (Fig. 1).

2.1 Ecological Effects of Polycyclic Aromatic Hydrocarbons

Persistence of PAHs in the environment is influenced by environmental physicochemical conditions. In the coastal zone, particularly in areas with strong waves, low-viscosity hydrocarbons may end up being washed from the beach by the action of waves on the surface stones, where they are cleaned by abrasion, while buried hydrocarbons may persist for a longer time (Chen et al., 2022); whereas, in fine-grained sands, which are mainly found on flatter beaches, less penetration occurs (Hallin et al., 2023). In low-energy coastal areas such as estuaries and mangroves, abundant organic components increase the adsorption capacity of non-volatile oil fractions, coupled with the ionic exchange of clays that attract and retain nonpolar compounds such as hydrocarbons (Holland and Elmore, 2015). Most chemical compounds in oil are hydrophobic, so once in contact with seawater, they do not mix, but the heavy fractions disintegrate into droplets and bubbles due to interfacial tension (Brandvik et al., 2019; Honda and Suzuki, 2020).

The acquisition of PAHs through predation induces a bioconcentration process in organisms, which may culminate in biomagnification through trophic transfer. As a consequence, elevated concentrations of PAHs are observed in animals at higher trophic levels, particularly fish, which poses a serious concern for human health (Armstrong et al., 2004; Liang et al., 2007). Also, PHAs constitute a potential threat to public health due to their toxic effects, like carcinogenic and mutagenic effects and immune-suppressant effects (Zapelini de Melo et al., 2022, Manjunatha et al., 2025). The mechanism of toxicity is considered to be interference with the function of cellular membranes. The chronic presence of hydrocarbons in marine environments could cause physiological or behavioral damage in marine and terrestrial organisms, particularly in areas of fishing activity, where genetic damage and decreases in fish growth and fertility occur (Heintz, 2007).

2.2 *Bioremediation of Polycyclic Aromatic Hydrocarbons*

To mitigate the contamination with PAHs, bioremediation is a biological alternative with low threats to the environment and other organisms, high efficiency, and low costs (Azubuike et al., 2016). Fungi have significant adaptations and characteristics for nutrition; they form a large amount of biomass, release extracellular enzymes into the surrounding environments, and have a high adaptation index that allows them to spread easily in diverse media (Johnsen et al., 2005). So, some fungi contribute to biodegradation due to their ability to transform and use toxic organic compounds as a carbon source (Gadd, 2007; Peng et al., 2008; Bik et al., 2012).

In a broad sense, the degradative potential of complex compounds by fungi lies in the extracellular nature of their enzymatic system, mainly composed of catalases, laccases, lignin-peroxidases, and manganese-peroxidases (Morel et al., 2013). The metabolic pathways involved in the use of hydrocarbons by fungi as a substrate under aerobic conditions could follow an initial oxidation mediated by cytochrome P450 monooxygenases and alkane-oxygenase enzymes (Asemoloye et al., 2019). Limitation of nutrients with low carbon content, impulses towards the release of degrading enzymes such as extracellular lignin-modifying enzymes to facilitate the degradation of PAHs (Mansur et al., 2003).

For deep-sea fungi, anaerobic degradation of organic compounds predominates, which is typically slower compared to aerobic degradation (Roy et al., 2012). The proposed method for hydrocarbon degradation under anaerobic conditions starts with the addition of fumarate, and other defined substitute methods include intra-aerobic hydroxylation, oxygen-independent hydroxylation, and carboxylation (Boll et al., 2020). When oxygen is limited in benthic environments, it has been demonstrated that microorganisms produce Fe^{3+}, SO_4^{2-}, and NO_3^- for hydrocarbon biodegradation (Foght, 2008). The activation for further metabolism in the aerobic

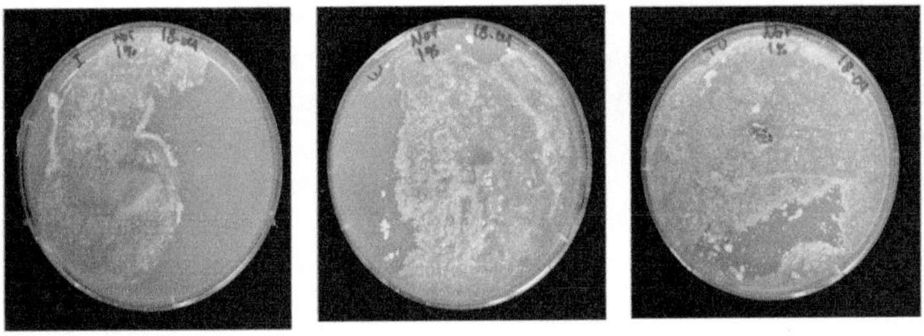

Aspergillus cejpii *Aspergillus* sp. *Corollospora maritima*

Fig. 2. Growth of three marine fungal isolates in naphthalene. After 15 days, growth of *Aspergillus cejpii*, *Aspergillus* sp., and *Corollospora maritima* in PDA medium with 1% of naphthalene.

pathway requires a terminal and subterminal reaction for aliphatic hydrocarbons and peripheral hydroxylation reactions for aromatics, while anaerobic pathways require a double reaction for this process (Sierra-García and Oliveira, 2013).

With technological advances that allow access to unique habitats in the deep-sea such as hydrothermal vents, the potential of fungal species with key metabolic characteristics can be widely exploited. The coexistence of oil and its derivatives with these microorganisms presents significant potential for their adaptation to degrade these compounds. Research conducted in recent years has shed light on the diverse microorganisms with unique adaptations inhabiting deep marine environments (Prol-Ledesma et al., 2004; Camerlenghi, 2018). While most fungal species discovered in subterranean environments thus far are known as aerobic fungi (facultative marine or facultative anaerobes), the exact metabolic pathways involved in the deep remain unknown. However, they undoubtedly possess the ability to develop anaerobic metabolism if environmental conditions demand it (Drake and Ivarsson, 2018). The quest for new marine metabolites has been a crucial biotechnology activity over the past three decades (Rangel and Falkenberg, 2015), highlighting the need for continued investigation into the potential benefits of these and other species.

2.3 *Enzymes*

One of the key advantages of fungi in bioremediation is their ability to secrete extracellular enzymes, which can effectively infiltrate and degrade pollutants in many substrates. Ligninolytic fungal species possess an inherent ability to efficiently degrade polycyclic aromatic hydrocarbons (Davis et al., 1993). This capability is linked to the production of extracellular enzymes that exhibit broad substrate specificity, which allows them to effectively degrade and mineralize various organic complexes, including recalcitrant pollutants (Kadri et al., 2017).

Obligate and facultative marine fungi, particularly those isolated from mangrove environments, produce lignocellulose-degrading enzymes (Bucher et al., 2004). Several studies have reported the presence of various lignolytic enzymes derived from marine fungi, including cellulase (Balabanova et al., 2018; Luo et al., 2005), xylanases (Qeshmi et al., 2020), and laccase (Raghukumar et al., 1994). In particular, laccases have been used *in vitro* for the depolymerization of lignin and other aromatic compounds. Investigations have demonstrated that certain mangrove fungi produce ligninolytic enzymes like cellulase and xylanase in media with seawater, indicating these enzymes exhibit tolerance to high salinity (D'Souza-Ticlo et al., 2009).

The efficiency and rate of PAH degradation mediated by fungal enzymes depend on several factors, such as oxygen availability for microbial growth, nutrient accessibility, and optimal physicochemical conditions for enzymatic activity. In the culture medium influences pH, temperature, pressure, salinity, compound structure, and distribution (Kadri et al., 2017).

3. Heavy Metals in the Marine Environment

HMs are defined as dense elements ($\geq$ 5 g/cm^3) mostly included within groups 3–16 in period 4 of the periodic table (Hawkes, 1997). HMs are insoluble and persistent, accumulating in living organisms because they cannot be degraded or destroyed (Alabssawy and Hashem, 2024). Furthermore, metals are capable of reacting with other elements, resulting in compounds that exhibit increased toxicity and mobility compared to their pristine forms (Quander and Kingscott, 1991). These pollutants cause the deterioration of marine ecosystems by reducing the diversity and abundance of species (Hosono et al., 2011; Naser, 2013). Also, biomagnification, dispersion, and trophic transfer represent a significant hazard to ecosystems and organisms, including humans (Fonti et al., 2013; Vaksmaa et al., 2023).

Coastal zones are among the most affected marine systems due to HM deposition from industrial activity, metal waste, chemical fertilizers, and other sources (Fonti et al., 2013). Through sedimentation processes, HMs tend to accumulate on the seabed via the water column (Ansari et al., 2004). Additional contamination arises from oil extraction, waste discharge, and dredging activities, further exacerbating HM concentrations (Fig. 1). Marine organisms such as algae, mollusks, and fish frequently exhibit HM concentrations significantly higher than those found in uncontaminated systems and prior studies (Naser, 2013).

3.1 *Sources of Heavy Metals*

While HMs naturally undergo biogeochemical cycling, anthropogenic activities, including industrial expansion, urbanization, and resource extraction, have led to increased environmental HM concentrations (Flegal et al., 1984). Furthermore, emerging industries such as deep-sea mining disturb marine sediments and degrade geological material, releasing Cd, Cu, Zn, Pb, and other HMs into the water column and adjacent ecosystems (Hauton et al., 2017). Additionally, the extraction and crushing of polymetallic nodules pose a significant risk of exacerbating metal release and further degrading deep-sea benthic ecosystems (El-Sharkawy et al., 2025).

Physicochemical methods (e.g., ion exchange, chemical precipitation, electrochemical treatment) reduce the concentration of HMs in soils, sediments, and water. Nevertheless, they become less effective and more costly as the concentration of HMs increases (Dermont et al., 2008). Furthermore, these approaches often generate toxic sludge that requires specialized disposal, adding to both the complexity and environmental concerns of the process. Consequently, their application is more viable in scenarios with moderate contamination levels or as a complement to other remediation techniques.

3.2 *Bioremediation of Heavy Metals*

Conventional treatments for the remediation of HMs focus on: (1) the solubilization of metals from sediments through aqueous solutions; (2) reducing the mobility of contaminants (US EPA, 2015). Acid solubilization, for instance, is a physicochemical technique that uses sulfuric acid to extract metals from an inorganic matrix (Peng et al., 2009). In marine ecosystems, bioremediation primarily targets water and sediment treatment. However, the complex dynamics of these environments and the extensive infrastructure needed for remediation often necessitate *ex situ* approaches. Fungal activity can achieve the same effect because of its intrinsic ability to withstand extreme conditions; it represents an efficient biological alternative at a lower economical cost and ecological impact (Valdes, 2025).

Fungi possess a wide range of mechanisms that favor their survival and growth in heavy metals (HMs)-contaminated sites (Gadd, 2007). The fungus-HM interaction can modify the metal through various processes: changes in oxidation state, intracellular precipitation, chelation, and ion exchange (Pande et al., 2022; Alabssawy and Hashem, 2024). The fungal mechanism to thrive in the presence of toxic metals includes the binding of metallic ions to functional groups with high affinity, metal accumulation, and complexation with different microbial biomolecules, to mention a few (Priyadarshini et al., 2021). The biosorption mechanisms in fungi are closely related to the composition of their cell wall, which contains components capable of interacting with HMs. Polysaccharides such as chitin and chitosan play a crucial role, as they feature functional groups with high affinity for these contaminants, facilitating their capture (Basil et al., 2012). The fungal cellular membrane provides protection against HM toxicity through selective entry, preventing agents that compromise the cell's interior from entering (Valdes, 2025).

Furthermore, bioaccumulation allows metal ions to precipitate within cellular organelles, thereby reducing their toxic form (Priyadarshini et al., 2021). The accumulation of metal ions by fungal cells is influenced by the age of the mycelium, composition of the culture medium, time and surface of contact, salinity, pH, chemical species of the metal, and temperature, among many other variables (Bishnoi and Garima, 2005; Khambhaty et al., 2008). Additionally, fungi can be generalists or selective during the bioaccumulation of metals (Cecchi et al., 2020). The interaction of microorganisms with certain metals can promote the formation of chemical species with increased toxicity. For instance, cosmopolitan fungi such as *Aspergillus niger* have been found to produce and accumulate methylmercury as a result of interactions with relatively low concentrations of Hg (Vonk and Sijpesteijn, 1973).

The potential of terrestrial fungi for the removal of HMs is widely studied, but the use of marine fungi is a scarcely studied field (Vala et al., 2018). Some of the genera with the most reports of HM tolerance belong to the genera *Aspergillus*, *Penicillium*, *Alternaria*, *Geotrichum*, *Fusarium*, *Rhizopus*, *Monilia*, and *Trichoderma* (Gadd, 1992). Many of these genera are also common in marine systems, although some could be considered as "marine-derived" or "facultative marine". The use of native fungi from marine environments facilitates the bioremediation of saline systems due to their tolerance to typical salinity conditions (Vala et al., 2018).

Pioneering research by Babich and Stotzky (1983) explored marine fungal capacity to thrive in the presence of HMs. They evaluated the growth of five marine isolates under progressively increasing concentrations of Ni. The growth of the isolates was affected at concentrations of ~ 50 ppm of Ni; however, the presence of a medium enriched with seawater and salts improved the tolerance of some isolates up to 500 ppm of Ni. Furthermore, it was demonstrated that Mg is the primary element enabling the growth of the tested isolates in the presence of Ni. The authors included a comparison between the performance of marine and non-marine fungi, in which Mg once again demonstrated a reduction in Ni toxicity, indicating that this is not an exclusive characteristic of marine isolates. One year later Hicks and Newell (1984) conducted microcosm bioassays inoculating the salt marsh fungus *Typhicola typharum* (reported as *Phaeospharia typharum*) to evaluate the degradation of *Spartina alterniflora* in the presence of mercury (Hg). Their findings indicate that *T. typharum* increases its biomass, metabolic activity, and glucosamine in the presence of Hg. The fungus does not exhibit sensitivity to Hg during the degradation of *S. alterniflora* at a concentration of 0.74 mg Hg/L, which may facilitate metal mobilization and cycling. The use of marine fungal microcosm experiments provides a perspective closer to real environmental conditions; however, such studies remain scarce to this day.

Research on interactions between HMs and strictly marine fungal taxa has been and continues to be limited. For instance, motivated by previous findings indicating that the presence of HMs stimulates the production of lipofuscin and malondialdehyde as stress indicators in eukaryotes, the potential of *Corollospora maritima* as a biomarker was investigated (Totáro et al., 1986). The authors of this work demonstrated its applicability in environments with Cu and Fe concentrations of up to 25 μg/L and 100 μg/L. Furthermore, although this was not their primary objective, they observed that *C. maritima* is capable of thriving in the presence of metal concentrations exceeding natural levels. Years later transition metals (Mn, Cu, V, Cr, Co, Fe, and Ni) were analyzed at concentrations of up to 0.05 μM (Pisanti et al., 1988). The production of oxidative stress molecules tends to decrease after 10 days of growth. The most noteworthy finding is the increased biomass production in the presence of Mn and Cr compared to the control and the capability of the fungus to reduce the oxidative stress induced by HMs and mitigate the cellular damage caused by free radicals.

The tolerance of *Corollospora lacera* and *Pericauda pelagica* was evaluated under progressively increasing concentrations of Cd and Pb, ranging from 10 mg/L to 500 mg/L (Taboski et al., 2005). The performance of each isolate varied depending on the fungal species and the metal tested, with optimal growth at 100 mg/L Pb for *C. lacera* and the highest Pb bioaccumulation at the maximum concentration, despite substantial inhibition of mycelial development (bioaccumulating 259 ± 36 mg Pb/g of mycelium). Conversely, in the presence of Cd, biomass production declined at 500 mg/L; however, this condition yielded the highest Cd accumulation (7.42 mg Cd/L). Similarly, *M. pelagica,* incubated in liquid medium for 30 days under static conditions, showed maximum metal accumulation at 500 mg/L of Pb (6.72 mg Pb/g of mycelium) and at 500 mg Cd/L (66.7 mg Cd/g of mycelium). One of the key findings of the study was that *C. lacera* sequestered between 93.6–95.6% of the total Pb in the extracellular fraction of the mycelium, while only 1.7–5.5% was retained intracellularly. Biomass development supports the potential use of isolates such as *C. lacera* in bioremediation, offering the possibility of establishing more efficient and rapid systems. However, it is also essential to consider isolates with a higher capacity for heavy metal bioaccumulation over extended periods.

In the past 20 years, publications on tolerance and bioaccumulation have increased, although studies on marine-derived fungi remain more frequent than those on strictly marine fungi. For instance, Khambhaty et al. (2008) demonstrated that the adsorption of Hg by a marine-derived isolate of *A. niger* is optimized at pH 3. The dead fungal biomass at 0.8 g/L was able to absorb 40.53 mg of Hg in 6 hours, under a concentration of 250 mg Hg/L. Wahab et al. (2017), using a mangrove isolate of *Penicillium citrinum,* reported growth capacity in 500 mg Pb/L, with an optimal performance at 400 mg Pb/L, and an adsorption of 329 ± 33.4 mg Pb/g in biomass. Valdes (2025) documented the utilization of marine fungal isolates from sediments collected in deep-sea hydrothermal systems, demonstrating their capacity for growth in the presence of Pb and Zn, and described structural changes and the production of extracellular matrix in the hyphae as a result of the presence of metals (Fig. 3). The isolate of *Aspergillus cejpii* showed the highest biomass development at 1,000 mg Pb/L and 10,000 mg Zn/L.

Fungi possess the genetic tools necessary to regulate metal concentrations within the cell (Behera and Das, 2023). The mechanisms of interaction and bioaccumulation of HMs have been studied in marine-derived fungi, highlighting three main key factors: siderophore production, extracellular substances, and the cell wall.

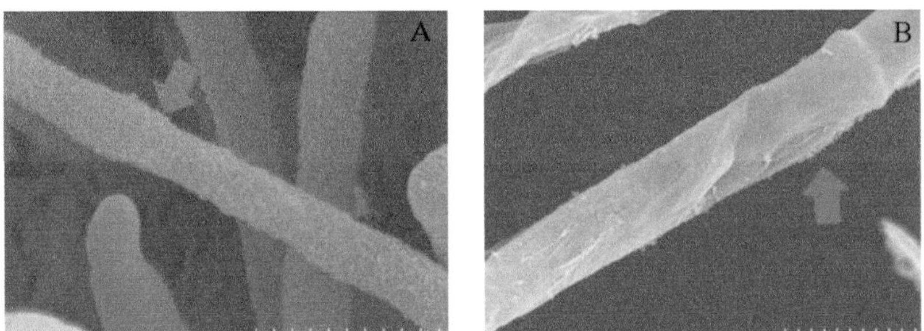

Fig. 3. Micrograph of hyphae in the presence of Pb and Zn. Micrographs were obtained using scanning electron microscopy (SEM) on *Aspergillus cejpii* isolated from the hydrothermal system at Alarcon Rise. (A) Growth in 1,000 mg Pb/L, and (B) growth in 10,000 mg Zn/L. The arrows show structural changes and the production of extracellular matrix in the hyphae.

The fungal cell possesses key proteins, like metallothioneins, located on the inner side of the cell wall involved in the regulation of metal influx and efflux (Tucker et al., 2004). Further, extracellular polymeric substances, like polysaccharides and glycoproteins, adsorb metals via ionic interactions or chelations (Alabssawy and Hashem, 2024). These extracellular substances act as a first line of defense by immobilizing metals outside the cell. Additionally the presence of carboxyl and phosphate confer the ability to capture HMs (Say et al., 2003). Marine fungal isolates, in particular, showed an excellent efflux mechanism for excluding cations that likely contributes to their ability to manage HMs exposure (Lotlikar et al., 2018). For instance, the marine fungus *A. sydowii* has demonstrated exceptional adaptation to ion homeostasis as a result of saline environments (Fernando et al., 2023).

While an extensive range of siderophores produced by terrestrial fungi are well studied, knowledge regarding marine fungi remains scarce. Holinsworth and Martin (2009) reported siderophore production in 29 marine-derived fungi isolated from the water column, mangrove plants, and marine sediments. Their study found that 25 of these 29 isolates produced pistillarin and rhizoferrin, which are carboxylate and hydroxamate siderophores, respectively. It is important to note that siderophore production is typically associated with iron-deficient systems; however, evidence suggests they also play a pivotal role in HM binding and tolerance.

4. Conclusion and Outlook

The PAHs and HMs are two major classes of pollutants threatening marine ecosystems. Fungi possess unique metabolic capabilities, especially the secretion of extracellular enzymes for PAH degradation and various mechanisms for heavy metal interaction, such as biosorption, bioaccumulation, and biotransformation. Particularly, the ability of marine fungi to thrive in diverse marine environments with high potential risk of pollution positions them as advantageous alternatives to traditional physical and chemical remediation methods, offering a potential alternative with lower costs and reduced environmental impact.

Despite the promising findings, the application of marine fungi in bioremediation, especially for HMs, remains a relatively underexplored field compared to terrestrial fungi. Maybe because the development of research on the bioremediation of HMs by fungi is limited by the generation of toxic waste, the high specificity of fungus-metal interactions, the necessary infrastructure and technology, and the high economic costs.

Investigations have demonstrated the tolerance, degradation, and bioaccumulation capabilities of obligate and facultative marine fungal genera like *Corollospora*, *Aspergillus*, and *Monodictys*. These microorganisms benefit from bioaccumulative toxic compounds, which opens up the field to further investigate the potential application of these fungi to attack specific contaminants (Vaksmaa et al., 2023). The exploration of marine fungi, particularly those from unique deep-sea habitats, could yield novel enzymes and metabolic pathways with enhanced bioremediation efficiencies. Deep investigation into the diversity, physiology, and genetic potential of marine fungi holds the key to unlocking innovative and sustainable solutions for mitigating PAH and heavy metal pollution in marine environments.

Acknowledgements

The authors thank the Programa de Apoyo a Proyectos de Investigación e Innovación Tecnológica DGAPA-PAPIIT IN203524 and the Secretaría de Ciencia, Humanidad, Tecnología, e Innovación (SECIHTI) for funding. We appreciate the use of the artificial intelligence tool ChatGPT (*OpenAI*, 2025), which served as support for the style review.

References

Alabssawy, A.N. and Hashem, A.H. (2024). Bioremediation of hazardous heavy metals by marine microorganisms: A recent review. Arch. Microbiol., 206(3): 103. https://doi.org/10.1515/botm.1994.37.6.515.

Aliko, V., Multisanti, C.R., Turani, B. and Faggio, C. (2022). Get rid of marine pollution: Bioremediation, an innovative, attractive, and successful cleaning strategy. Sustainability, 14(18): 11784. https://doi.org/10.3390/su141811784.

Ansari, T.M., Marr, I.L. and Tariq, N. (2004). Heavy metals in marine pollution perspective: A mini review. J. Appl. Sci., 4(1): 1–20.

Armstrong, B., Hutchinson, E., Unwin, J. and Fletcher, T. (2004). Lung cancer risk after exposure to polycyclic aromatic hydrocarbons: A review and meta-analysis. Environ. Health Pers., 112(9): 970–978. https://doi.org/10.1289/ehp.6895.

Asemoloye, M.D., Jonathan, S.G. and Ahmad, R. (2019). Synergistic plant-microbes interactions in the rhizosphere: A potential headway for the remediation of hydrocarbon polluted soils. Int. J. Phytorem., 21(1): 71–83. https://doi.org/10.1080/15226514.2018.1474437.

Azubuike, C.C., Chikere, C.B. and Okpokwasili, G.C. (2016). Bioremediation techniques–classification based on site of application: Principles, advantages, limitations, and prospects. World J. Microbiol. Biotechnol., (32): 1–18. https://doi.org/10.1007/s11274-016-2137-x.

Babich, H. and Stotzky, G. (1983). Nickel toxicity to estuarine/marine fungi and its amelioration by magnesium in sea water. Water, Air, Soil Poll., 19: 193–202.

Balabanova, L., Slepchenko, L., Son, O. and Tekutyeva, L. (2018). Biotechnology potential of marine fungi degrading plant and algae polymeric substrates. Front. Microbiol., 9: 1527. https://doi.org/10.3389/fmicb.2018.01527.

Basil, G., Nirmal, K., Kumar, R.N. and Sajish, P.R. (2012). Biosorption potentiality of living *Aspergillus niger* tiegh in removing heavy metal from aqueous solution. Bioremediation J., 16(4): 195–203. https://doi.org/10.1080/10889868.2012.731442.

Behera, A.D. and Das, S. (2023). Ecological insights and potential application of marine filamentous fungi in environmental restoration. Rev. Environ. Sci. Biotechnol., 22(2): 281–318. https://doi.org/10.1007/s11157-023-09655-2.

Bik, H.M., Halanych, K.M., Sharma, J. and Thomas, W.K. (2012). Dramatic shifts in benthic microbial eukaryote communities following the Deepwater Horizon oil spill. PloS One, 7 (6): e38550. https://doi.org/10.1371/journal.pone.0038550.

Bishnoi, N.R. and Garima, G. (2005). Fungus: An alternative for bioremediation of heavy metal containing wastewater: A review. J. Sci. Ind. Res., 64 (2): 93–100.

Boll, M., Estelmann, S. and Heider, J. (2020). Anaerobic degradation of hydrocarbons: Mechanisms of hydrocarbon activation in the absence of oxygen. pp. 3–29. *In*: Boll, M. (ed.). Anaerobic Utilization of Hydrocarbons, Oils, and Lipids: Handbook of Hydrocarbon and Lipid Microbiology. Springer Cham. https://doi.org/10.1007/978-3-319-50391-2_2.

Brandvik, P.J., Daling, P.S., Leirvik, F. and Krause, D.F. (2019). Interfacial tension between oil and seawater as a function of dispersant dosage. Mar. Pollut. Bull., 143: 109–114. https://doi.org/10.1016/j.marpolbul.2019.04.019.

Bucher, V.V.C., Hyde, K.D., Pointing, S.B. and Reddy, C.A. (2004). Production of wood decay enzymes, mass loss, and lignin solubilization in wood by marine ascomycetes and their anamorphs. Fungal Diver., 15: 1–14.

Camerlenghi, A. (2018). Drivers of seafloor geomorphic change. pp. 135–159. *In*: Micallef, A., Krastel, S. and Savini, A. (eds.). Submarine Geomorphology. Springer Geology. Springer, Cham. https://doi.org/10.1007/978-3-319-57852-1_9.

Cecchi, G., Cutroneo, L., di Piazza, S., Vagge, G., Capello, M. et al. (2020). From waste to resource: Mycoremediation of contaminated marine sediments in the SEDITERRA Project. J. Soils Sediments, 20(6): 2653–2663. https://doi.org/10.1007/s11368-019-02527-9.

Chen, Y.J., Leung, P.M., Cook, P.L., Wong, W.W., Hutchinson, T. et al. (2022). Hydrodynamic disturbance controls microbial community assembly and biogeochemical processes in coastal sediments. ISME J., 16(3) 750–763. https://doi.org/10.1038/s41396-021-01111-9.

D'Souza-Ticlo, D., Sharma, D. and Raghukumar, C. (2009). A thermostable metal-tolerant laccase with bioremediation potential from a marine-derived fungus. Mar. Biotechnol., 11(6): 725–737. https://doi.org/10.1007/s10126-009-9187-0.

Davis, M.W., Glaser, J.A., Evans, J.W. and Lamar, R.T. (1993). Field evaluation of the lignin-degrading fungus *Phanerochaete sordida* to treat creosote-contaminated soil. Environ. Sci. Technol., 27(12): 2572–2576. https://doi.org/10.1021/es00048a040.

Dell'Anno, F., Rastelli, E., Buschi, E., Barone, G., Beolchini, F. et al. (2022). Fungi can be more effective than bacteria for the bioremediation of marine sediments highly contaminated with heavy metals. Microorganisms, 10(5): 993. https://doi.org/10.3390/microorganisms10050993.

Dermont, G., Bergeron, M., Mercier, G. and Richer-Laflèche, M. (2008). Metal-contaminated soils: Remediation practices and treatment technologies. Pract. Period. Hazard., Toxic, Radioact. Waste Manag., 12(3): 188–209. https://doi.org/10.1061/(ASCE)1090-025X.

Drake, H. and Ivarsson, M. (2018). The role of anaerobic fungi in fundamental biogeochemical cycles in the deep biosphere. Fungal Biol. Rev., 32(1): 20–25. https://doi.org/10.1016/j.fbr.2017.10.001.

Dua, M., Singh, A., Sethunathan, N. and Johri, A.K. (2002). Biotechnology and bioremediation: Successes and limitations. Appl. Microbiol. Biotechnol., 59(2-3): 143–152. https://doi.org/10.1007/s00253-002-1024-6.

El-Sharkawy, M., Alotaibi, M.O., Li, J., Du, D. and Mahmoud, E. (2025). Heavy metal pollution in coastal environments: Ecological implications and management strategies: A review. Sustainability, 17(2): 701. https://doi.org/10.3390/su17020701.

Farrington, J.W. (2020). Need to update human health risk assessment protocols for polycyclic aromatic hydrocarbons in seafood after oil spills. Mar. Pollut. Bull., 150: 110744. https://doi.org/10.1016/j.marpolbul.2019.110744.

Fernando, L.D., Pérez-Llano, Y., Dickwella-Widanage, M.C., Jacob, A., Martínez-Ávila, L. et al. (2023). Structural adaptation of fungal cell wall in hypersaline environment. Nat. Commun., 14(1): 7082. https://doi.org/10.1038/s41467-023-42693-6.

Flegal, A.R., Schaule, B.K. and Patterson, C.C. (1984). Stable isotopic ratios of lead in surface waters of the Central Pacific. Mar. Chem., 14(3): 281–287. https://doi.org/10.1016/0304-4203.

Foght, J. (2008). Anaerobic biodegradation of aromatic hydrocarbons: Pathways and prospects. J. Mol. Microbiol. Biotechnol., 15(2-3): 93–120. https://doi.org/10.1159/000121324.

Fonti, V., Dell'Anno, A. and Beolchini, F. (2013). Influence of biogeochemical interactions on metal bioleaching performance in contaminated marine sediment. Water Res., 47(14): 5139–5152. https://doi.org/10.1016/j.watres.2013.05.052.

Gadd, G.M. (1992). Microbial control of heavy metal pollution. pp. 59–88. *In*: Fry, J.C., Gadd, G.M., Herbert, R.A., Jones, C.W. and Watson-Craik, I. (eds.). Microbial Control of Pollution. Cambridge University Press.

Gadd, G.M. (2007). Geomycology: Biogeochemical transformations of rocks, minerals, metals, and radionuclides by fungi, bioweathering, and bioremediation. Mycol. Res., 111(1): 3–49. https://doi.org/10.1016/j.mycres.2006.12.001.

Gonçalves, M.F.M., Esteves, A.C. and Alves, A. (2022). Marine Fungi: Opportunities and challenges. Encyclopedia, 2(1): 559–577. https://doi.org/10.3390/encyclopedia2010037.

Hallin, C., van Ijzendoorn, C., Homberger, J.M. and de Vries, S. (2023). Simulating surface soil moisture on sandy beaches. Coast. Eng., 185: 104376. https://doi.org/10.1016/j.coastaleng.2023.104376.

Hauton, C., Brown, A., Thatje, S., Mestre, N.C., Bebianno, M.J. et al. (2017). Identifying toxic impacts of metals potentially released during deep-sea mining: A synthesis of the challenges to quantifying risk. Front. Mar. Sci., (4): 368. https://doi.org/10.3389/fmars.2017.00368.

Hawkes, S.J. (1997). What is a "heavy metal"? J. Chem. Educ., 74(11): 1374. https://doi.org/10.1021/ed074p1374.

Heintz, R.A. (2007). Chronic exposure to polynuclear aromatic hydrocarbons in natal habitats leads to decreased equilibrium size, growth, and stability of pink salmon populations. Integr. Environ. Assess. Manag., 3: 351–363. https://doi.org/10.1002/ieam.5630030306.

Hicks, R.E. and Newell, S.Y. (1984). The growth of bacteria and the fungus Phaeosphaeriatypharum (Desm.) Holm (Eumycota: Ascomycotina) in salt-marsh microcosms in the presence and absence of mercury. J. Exp. Mar. Biol. Ecol., 78(1-2): 143–155. https://doi.org/10.1016/0022-0981(84)90075-3.

Holinsworth, B. and Martin, J.D. (2009). Siderophore production by marine-derived fungi. Biometals, 22(4): 625–632. https://doi.org/10.1007/s10534-009-9239-y.

Holland, K.T. and Elmore, P.A. (2015). A review of heterogeneous sediments in coastal environments. Earth-Sci. Rev., 89(3-4): 116–134. https://doi.org/10.1016/j.earscirev.2008.03.003.

Honda, M., Qiu, X., Koyama, J., Uno, S., Undap, S.L. et al. (2018). The wharf roach, Ligia sp., A novel indicator of polycyclic aromatic hydrocarbon contamination in coastal areas. Int. J. Environ. Res., (12): 1–11. https://doi.org/10.1007/s41742-017-0063-x.

Honda, M. and Suzuki, N. (2020). Toxicities of polycyclic aromatic hydrocarbons for aquatic animals. Int. J. Environ. Res. Public Health, 17(4): 1363. https://doi.org/10.3390/ijerph17041363.

Hong, H., Xu, L., Zhang, L., Chen, J.C., Wong, Y.S. et al. (1995). Environmental fate and chemistry of organic pollutants in the sediment of Xiamen harbor and Victoria harbor. Mar. Pollut. Bull., 31: 229–236.

Hosono, T., Su, C., Delinom, R., Umezawa, Y., Toyota, T. et al. (2011). Decline in heavy metal contamination in marine sediments in Jakarta Bay, Indonesia due to increasing environmental regulations. Estuar. Coast. Shelf Sci., (92): 297–306. https://doi.org/10.1016/j.ecss.2011.01.010.

Hussein, I. and Mona, S.M. (2016). A review on polycyclic aromatic hydrocarbons: Source, environmental impact, effect on human health, and remediation. Egypt. J. Pet., 25 (1): 107–123. https://doi.org/10.1016/j.ejpe.2015.03.011.

Imam, A., Suman, S.K., Kanaujia, P.K. and Ray, A. (2022). Biological machinery for polycyclic aromatic hydrocarbons degradation: A review. Bioresour. Technol., 343: 126121. https://doi.org/10.1016/j.biortech.2021.126121.

Johnsen, A.R., Wick, L.Y. and Harms, H. (2005). Principles of microbial PAH-degradation in soil. Environ. Pollut., 133(1): 71–84. https://doi.org/10.1016/j.envpol.2004.04.015.

Kadri, T., Rouissi, T., Kaur-Brar, S., Cledon, M., Sarma, S. et al. (2017). Biodegradation of polycyclic aromatic hydrocarbons (PAHs) by fungal enzymes: A review. J. Environ. Sci., 51: 52–74. https://doi.org/10.1016/j.jes.2016.08.023.

Khambhaty, Y., Mody, K., Basha, S. and Jha, B. (2008). Hg (II) removal from aqueous solution by dead fungal biomass of marine *Aspergillus niger*: Kinetic studies. Sep. Sci. Technol., 43(5): 1221–1238. https://doi.org/10.1080/01496390801888235.

Kostka, J.J., Overholt, S., Bubenheim, W., Hackbusch, P., Larter, S. et al. (2020). Biodegradation of petroleum hydrocarbons in the deep sea. pp. 107–124. *In*: Murawski, S.A., Ainsworth, C.H., Gilbert, S., Hollander, D.J., Paris, C.B. et al. (eds.). Deep Oil Spills: Facts, Fate, and Effects. Springer Cham. https://doi.org/10.1007/978-3-030-11605-7_7.

Kuppan, N., Padman, M., Mahadeva, M., Srinivasan, S. and Devarajan, R. (2024). A comprehensive review of sustainable bioremediation techniques: Ecofriendly solutions for waste and pollution management. Waste Manag. Bull., 2(3): 154–171. https://doi.org/10.1016/j.wmb.2024.07.005.

Lamichhane, S., Krishna, K.C.B. and Sarukkalige, R. (2016). Polycyclic aromatic hydrocarbons (PAHs) removal by sorption: A review. Chemosphere, (148): 336–353. https://doi.org/10.1016/j.chemosphere.2016.01.036.

Liang, Y., Tse, M.F., Young, L. and Wong, M.H. (2007). Distribution patterns of polycyclic aromatic hydrocarbons (PAHs) in the sediments and fish at Mai Po Marshes Nature Reserve, Hong Kong. Water Res., 41(6): 1303–1311. https://doi.org/10.1016/j.watres.2006.11.048.

Lotlikar, N.P., Damare, S.R., Meena, R.M., Linsy, P. and Mascarenhas, B. (2018). Potential of marine-derived fungi to remove hexavalent chromium pollutant from culture broth. Indian J. Microbiol., 58: 182–192. https://doi.org/10.1007/s12088-018-0719-z.

Luo, W., Vrijmoed, L.L.P. and Jones, E.B.G. (2005). Screening of marine fungi for lignocellulose-degrading enzyme activities. Bot. Mar., 48: 379–386. https://doi.org/10.1515/BOT.2005.051.

Ma, X., Ling Wu, L. and Fam, H. (2014). Heavy metal ions affecting the removal of polycyclic aromatic hydrocarbons by fungi with heavy-metal resistance. Appl. Microbiol. Biotechnol., 98(23): 9817–9827. https://doi.org/10.1007/s00253-014-5905-2.

Manjunatha, B.K., Vamshee, V.K., Poojitha, N.G., Kiran, R.J., Mythili, S.T. et al. (2025). Bioremediation of polycyclic aromatic hydrocarbons (acenaphthene, anthracene, fluoranthene, fluorene) by *Aspergillus niger* HQ170509. 1 fungus. Int. J. Environ. Sci. Technol., 22: 12567–12582. https://doi.org/10.1007/s13762-025-06370-z.

Mansur, M., Arias, M.E., Copa-Patino, J.L., Flärdh, M. and González, A.E. (2003). The white-rot fungus *Pleurotus ostreatus* secretes laccase isozymes with different substrate specificities. Mycologia, 95(6): 1013–1020. https://doi.org/10.1080/00275514.2003.12063179.

Morel, M., Meux, E., Mathieu, Y., Thuillier, A., Chibani, K. et al. (2013). Xenomic networks variability and adaptation traits in wood decaying fungi. Microb. Biotechnol., 6(3): 248–263. https://doi.org/10.1111/1751-7915.12015.

Naser, H.A. (2013). Assessment and management of heavy metal pollution in the marine environment of the Arabian Gulf: A review. Mar. Poll. Bull., 72(1): 6–13. https://doi.org/10.1016/j.marpolbul.2013.04.030.

Pande, V., Pandey, S.C., Sati, D., Bhatt, P. and Samant, M. (2022). Microbial interventions in bioremediation of heavy metal contaminants in agroecosystem. Front. Microbiol., 13: 824084. https://doi.org/10.3389/fmicb.2022.824084.

Passarini, M.R., Rodrigues, M.V., da Silva, M. and Sette, L.D. (2011). Marine-derived filamentous fungi and their potential application for polycyclic aromatic hydrocarbon bioremediation. Mar. Pollut. Bull., 62(2): 364–370. https://doi.org/10.1016/j.marpolbul.2010.10.003.

Peng, R.H., Xiong, A.S., Xue, Y., Fu, X.Y., Gao, F. et al. (2008). Microbial biodegradation of polyaromatic hydrocarbons. FEMS Microbiol. Rev., 32(6): 927–955. https://doi.org/10.1111/j.1574-6976.2008.00127.x.

Peng, J.F., Song, Y.H., Yuan, P., Cui, X.Y. and Qiu, G.L. (2009). The remediation of heavy metals contaminated sediment. J. Hazard. Mater., 161(2-3): 633–640. https://doi.org/10.1016/j.jhazmat.2008.04.061.

Philp, R.P. (2020). Composition and properties of petroleum. pp. 267–288. *In*: Wilkes, H. (ed.). Hydrocarbons, Oils, and Lipids: Diversity, Origin, Chemistry, and Fate. Handbook of Hydrocarbon and Lipid Microbiology. Springer Cham. https://doi.org/10.1007/978-3-319-54529-5_13-1.

Pisanti, F.A., Lucadamo, L., Vanzanell, F. and Totáro, E.A. (1988). Effect of heavy metal ions on microorganism aging. Mar. Pollut. Bull., 19(7): 328–333. https://doi.org/10.1016/0025-326X(88)90429-8.

Priyadarshini, E., Priyadarshini, S.S., Cousins, B.G. and Pradhan, N. (2021). Metal-Fungus interaction: Review on cellular processes underlying heavy metal detoxification and synthesis of metal nanoparticles. Chemosphere, 274: 129976. https://doi.org/10.1016/j.chemosphere.2021.129976.

Prol-Ledesma, R.M., Canet, C., Torres-Vera, M.A., Forrest, M.J. and Armienta, M.A. (2004). Vent fluid chemistry in Bahía Concepción coastal submarine hydrothermal system, Baja California Sur, Mexico. J. Volcanol. Geoth. Res., 137(4): 311–328. https://doi.org/10.1016/j.jvolgeores.2004.06.003.

Qeshmi, F.I., Homaei, A., Fernandes, P., Hemmati, R., Dijkstra, B.W. et al. (2020). Xylanases from marine microorganisms: A brief overview on scope, sources, features, and potential applications. Biochim. Biophys. Acta - Proteins Proteom., 1868(2): 140312. https://doi.org/10.1016/j.bbapap.2019.140312.

Quander, J. and Kingscott, J. (1991). Innovative-Treatment Technologies: Overview and Guide to Information Sources. Office of Solid Waste and Emergency Response, Technology Innovation Office, Washington, D.C. 20460. (https://nepis.epa.gov/Exe/ZyPDF.cgi/2000KFUQ.PDF?Dockey=2000KFUQ.PDF. (Accessed on 25 Aug 2024).

Raghukumar, C., Raghukumar, S., Chinnaraj, A., Chandramohan, D., D'Souza, T.M. et al. (1994). Laccase and other lignocellulose modifying enzymes of marine fungi isolated from the coast of India. Bot. Mar., 37(6): 515–524. https://doi.org/10.1515/botm.1994.37.6.515.

Rangel, M. and Falkenberg, M. (2015). An overview of the marine natural products in clinical trials and on the market. J. Coast. Life Med., 3(6): 421–428. https://doi:10.12980/JCLM.3.2015JCLM-2015-0018.

Roy, H., Kallmeyer, J., Adhikari, R.R., Pockalny, R., Jørgensen, B.B. et al. (2012). Aerobic microbial respiration in 86-million-year-old deep-sea red clay. Science, 336(6083): 922–925. https://doi.org/10.12980/JCLM.3.2015JCLM-2015-0018.

Say, R., Yilmaz, N. and Denizli, A. (2003). Removal of heavy metal ions using the fungus *Penicillium canescens*. Adsorp. Sci. Technol., 21(7): 643–650. https://doi.org/10.1260/026361703772776420.

Sierra-Garcia, I.N. and de Oliveira, V.M. (2013). Microbial hydrocarbon degradation: Efforts to understand biodegradation in petroleum reservoirs. Biodegradation - Eng. Technol., 10: 55920. https://dx.doi.org/10.5772/55920.

Taboski, M.A.S., Rand, T.G. and Piórko, A. (2005). Lead and cadmium uptake in the marine fungi *Corollospora lacera* and *Monodictys pelagica*. FEMS Microbiol. Ecol., 53(3): 445–453. https://doi.org/10.1016/j.femsec.2005.02.009.

Totáro, E.A., Cuomo, V. and Pisanti, F.A. (1986). Influence of environmental stress on lipofuscin production. Arch. Gerontol. Geriatr., 5(4): 343–349. https://doi.org/10.1016/0167-4943(86)90037-3.

Tucker, S.L., Thornton, C.R., Tasker, K., Jacob, C., Giles, G. et al. (2004). A fungal metallothionein is required for pathogenicity of Magnaporthe grisea. Plant Cell, 16(6): 1575–1588. https://doi.org/10.1105/tpc.021279.

USEPA (U.S. Environmental Protection Agency). (2015). Update to the 1998 U.S. EPA Supplemental Environmental Projects Policy. https://www.epa.gov/enforcement/2015-update-1998-us-epa-supplemental-environmental-projects-policy. (Accessed on 6 May 2024).

Vaksmaa, A., Guerrero-Cruz, S., Ghosh, P., Zeghal, E., Hernando-Morales, V. et al. (2023). Role of fungi in bioremediation of emerging pollutants. Front. Mar. Sci., 10. https://doi.org/10.3389/fmars.2023.1070905.

Vala, A.K., Sachaniya, B. and Dave, B.P. (2018). Marine-derived fungi: Promising candidates for enhanced bioremediation. pp. 281–300. *In*: Prasad, R. and Aranda, E. (eds.). Approaches in Bioremediation: The New Era of Environmental Microbiology and Nanobiotechnology. Springer Cham. https://doi.org/10.1007/978-3-030-02369-0_12.

Valdes, J. (2025). Evaluación del uso potencial de micromicetes provenientes de sistemas hidrotermales en la biorremediación de plomo y zinc. (Tesis de licenciatura). Universidad Nacional Autónoma de México. https://hdl.handle.net/20.500.14330/TES01000871966.

Vonk, J.W. and Sijpesteijn, A.K. (1973). Studies on the methylation of mercuric chloride by pure cultures of bacteria and fungi. Antonie van Leeuwenhoek, 39: 505–513. https://doi.org/10.1007/BF02578894.

Wahab, A.A., Awang, A.S.A.H., Azham, Z., Tay, M.G. and Adeyemi, F.M. (2017). Biosorption of lead (II) ion using Penicillium citrinum KR706304 isolated from the mangrove soil environment of southeast Borneo. Ife J. Sci., 19(2): 341–351. https://doi.org/10.4314/ijs.v19i2.14.

Wang, H., Shu, Y., Kuang, Z., Han, Z., Wu, J. et al. (2025). Bioaccumulation and potential human health risks of PAHs in marine food webs: A trophic transfer perspective. J. Hazard Mater., 485: 136946. https://doi.org/10.1016/j.jhazmat.2024.136946.

Wolska, L., Mechlińska, A., Rogowska, J. and Namieśnik. J. (2012). Sources and fate of PAHs and PCBs in the marine environment. Crit. Rev. Environ. Sci. Technol., 42: 1172–1189. https://doi.org/10.1080/10643389.2011.556546.

Zapelini de Melo, A.P., Barcellos, R.H., Molognoni, L., Oliveira, T., Daguer, H. et al. (2022). Disasters with oil spills in the oceans: Impacts on food safety and analytical control methods. Food Res. Int., 157: 111366. https://doi.org/10.1016/j.foodres.2022.111366.

Zhang, X., Wang, X., Liang, W., Liu, M., Wang, X. et al. (2024). The occurrence, sources, and health risks of substituted polycyclic aromatic hydrocarbons (SPAHs) cannot be ignored. Environ. Int., 183: 108390. https://doi.org/10.1016/j.envint.2023.108390.

12

Potential Utilization of Deep-sea Marine Fungi in the Bioremediation of Crude Oil Spills in the Ocean

Romero-Hernandez L.,[1,*] *Tezocotitla G.*,[2] *Yáñez-Ocampo G.*,[2] *Wong-Villarreal A.*[3] and *Patricia Velez*[1]

The global reliance on crude oil as the main energy source increases the urgent need to find new bioremediation agents that can reduce the environmental impact of oil spills in marine ecosystems. Deep-sea marine fungi tolerate extreme conditions such as high hydrostatic pressure, low temperatures, and variable salinity and are, therefore, a valuable source for hydrocarbon degradation. This chapter explores key ecosystems in which hydrocarbonoclastic fungi have been found in the deep-sea, including oceanic ridges, trenches, anoxic zones, hydrothermal vents, seamounts, and sediments. It also includes a review of fungi proven to degrade crude oil hydrocarbons. The classification of biosurfactants (BS) and their role in enhancing the bioconversion of petroleum are discussed. Qualitative and quantitative methods for studying BS are described throughout this review, as well as the potential and current applications of deep-sea fungi in bioremediation.

[1] Instituto de Biología, Universidad Nacional Autónoma de México, México City, México.
[2] Facultad de Ciencias, Universidad Autónoma del Estado de México, Toluca de Lerdo, Estado de México, Mexico.
[3] División Agroalimentaria, Universidad Tecnológica de la Selva, Ocosingo, Chiapas, México.
* Corresponding author: pikiuro2015@gmail.com

DOI: 10.1201/9781003795100-15

1. Introduction

The deep-sea represents the largest biome on Earth, characterized by extreme conditions including perpetual darkness, high hydrostatic pressure, cold temperatures (except for hydrothermal vents), and often chemical extremes (Jannasch and Mottl, 1985). Despite these formidable challenges, these environments are characterized by diverse microbial life uniquely adapted to survival. Recent years have seen a significant expansion in our understanding of the microbial communities inhabiting deep-sea sediments, hydrothermal vents, and cold seeps, revealing novel metabolic pathways and ecological interactions crucial for global biogeochemical cycles (Orcutt et al., 2011; Ruff et al., 2015). This extraordinary biodiversity and the specialized adaptations of deep-sea microorganisms make these ecosystems a pivotal frontier for research, particularly in the field of bioremediation.

Marine fungi have been reported across a range of deep-sea ecosystems. These include oceanic ridges (Tang et al., 2020; Yang et al., 2020; Mahé et al., 2013), oceanic trenches (Gao et al., 2020; Xu et al., 2019; Varrella et al., 2024), anoxic zones (Zhong et al., 2024; Singh et al., 2011; Barone et al., 2019), hydrothermal vents (Xu et al., 2018; Le Calvez et al., 2009; Burgaud et al., 2009), seamounts (Yang et al., 2020; Luo et al., 2020; Connell et al., 2009), and deep-sea sediments (Nagano et al., 2017; Ogaki et al., 2020; Wei et al., 2018) (Fig. 1). This chapter explores the domain of deep-sea fungi and their prospective utilisation in the remediation of crude oil hydrocarbons through biosurfactant production. The practice of bioprospection is regarded as both eco-friendly and efficacious. Deep-sea ecosystems provide a number of crucial ecological services, including the recycling of nutrients, the storage of carbon on a long-term basis, and the preservation of unique genetic resources. Furthermore, it is evident that these ecosystems make a substantial contribution to global biogeochemical cycles and biodiversity (La Bianca et al., 2023; Fisher et al., 2016; Jobstvogt et al., 2014).

Fig. 1. Deep-sea ecosystems where marine fungi have been reported: (1) Oceanic ridge; (2) Ocean trench; (3) Anoxic zone; (4) Hydrothermal vents; (5) Seamount; (6) Surface/superficial and subsurface sediments.

2. Impacts of Crude Oil Spill

Although the green energies are replacing the crude oil in some developed regions as an energy source, the Energy Return on Investment (EROI), the easy storage, transport, and the energy produced by petroleum, position the fossil fuel as an energy source that will be relevant and necessary in the next decades in most countries (Hall and Klitgaard, 2012; Holechek et al., 2022; Nekahi, 2025). In addition, the thousands of products derived from crude oil have not been replaced by any other petroleum-independent material, such as medical plastics, raw material for construction, among many others (White, 2014; Mori, 2023).

Despite concerted efforts to transition to renewable energy, these technologies require time to develop in terms of storage, utilisation, and energy conservation of energy. Consequently, it would be a mistake to dismiss petroleum as the current and future main energy source for humanity (Epstein, 2022; Burchette, 2025). Nonetheless, the consequences of accidental crude oil spills in the ocean, resulting from production and transportation activities (e.g., tank ruptures, offshore exploration, and leaks from underwater pipelines) are significant, as is the negative impact on ecosystem and human health (Saadoun, 2015; Rodrigues et al., 2025).

Petroleum input into the ocean can be attributed to natural seeps resulting from fractures in the sea floor. However, the predominant percentage is attributable to accidents associated with maritime transportation, production, discharge, and crude oil tanker accidents (Council, 2003; Wang et al., 2022). According to data from oil spill reports, significant spill hotspots have been identified in the northeastern United States, the Gulf of Mexico, and the Mediterranean Sea (Chughtai et al., 2021). There is a concern to take actions to avoid crude oil spills in marine ecosystems, especially because the oil trade is recently relying more on maritime transportation, and although global oil trade volumes continue to increase, reported incidents of oil spillage have shown a declining trend (Chen et al., 2019).

In the initial phase of a crude oil spill, the physical processes of flotation and the dissemination of petroleum along the ocean surface are observed. These phenomena are influenced by the characteristics of the crude oil, including its viscosity, density, and water surface tension (Wang et al., 2017). The process of spread involves a series of physicochemical reactions, including evaporation, disruption, dissolution, dispersion, emulsification, and photooxidation. Subsequently, the crude oil undergoes a series of physical alterations, which is known as weathering (Prince et al., 2004; Fingas, 2025).

It has been determined that the light crude oil could undergo evaporation to a degree of 70% within a 2-day period subsequent to the occurrence of the spill, whereas the heavy crude oil has demonstrated a maximum evaporation capacity of only 10% (Lee et al., 2003; Govindarajan et al., 2021). This demonstrates that the weathering of crude oil is strongly influenced by its viscosity, with the degree of weathering varying depending on the nature of the crude oil. The precipitation of crude oil can occur in various forms, including reaching the seabed, the shoreline, or being retained in the water column as small droplets (Kingston, 2002; Radovic and Silva, 2024).

The environmental fate of crude oil in marine ecosystems is determined by the size of the oil droplets. Smaller droplets have been shown to be absorbed by a diverse array of marine organisms, where they either remain suspended in the water column in association with organic matter, or eventually settle on the seabed (Mukherjee and Wrenn, 2009; Hansen et al., 2025). While the immediate visible impacts of oil spills, such as oil-coated beaches and affected surface-dwelling fauna like birds, turtles, and marine mammals, are readily apparent, the ecological consequences are frequently more intricate. The aforementioned factors consist of both acute and chronic biological exposure due to the persistence of oil components in the environment, sublethal toxicity, and disruptions to the marine food web, particularly affecting microzooplankton (Murawski et al., 2021; de Santana Campelo et al., 2021; Peterson et al., 2003).

The extent of marine crude oil pollution has been significantly underestimated, with the contamination reaching even the most remote regions of the ocean, including the abyssal zone. The oil and gas industry has been exploiting deep-sea resources for many decades, extracting crude oil from depths ranging from 3,000 metres. It is only in recent years that studies have begun to reveal the extent of human impact in these deep waters (Douglas-Westwood, 2005; Stel et al., 2021). In the event of crude oil precipitating to the deep-sea seabed sediments as a consequence of abiotic factors such as temperature and salinity, the total petroleum hydrocarbons and polycyclic aromatic hydrocarbons can persist in the environment for decades, which has consequences for benthic habitat. This phenomenon has been demonstrated by the Deep-Water Horizon blowout in several studies (Montagna et al., 2013; Fisher et al., 2016).

The presence of crude oil in deep-sea environments has been shown to result in a decrease in the rate of hydrocarbon degradation in sediments when compared to rates observed in shallower waters or surface environments. This phenomenon can be attributed to the low temperatures and limited organic carbon and nutrient input to deepwater environments (Montagna et al., 2013). Consequently, the recovery of deep-sea communities affected by crude oil spills requires several decades or longer. This is due to the slow degradation of hydrocarbons and the slow metabolic rates of deep-sea organisms (Bacosa et al., 2022; Fisher et al., 2016).

3. Bioremediation by Fungi in the Deep-Sea

The limited number of studies conducted on deep-sea fungi have primarily provided evidence of their presence and activity. This evidence has been obtained through culture-based methods or the detection of their molecular signatures (Damare et al., 2006; Velez et al., 2020; Raghukumar et al., 2010; Vargas-Gastélum et al., 2019). It has been well documented that marine fungi possess unique physiological adaptations that enable them to thrive under extreme conditions, such as high hydrostatic pressure, low temperatures and variable oxygen disposition and salinity (Ramesh et al., 2021). Deep-sea fungi, which are capable of inhabiting sediments in depths greater than 1,000 meters below the ocean surface, exhibit these adaptation mechanisms to an even greater extent, including high hydrostatic pressure. These mechanisms enable the fungi to endure such conditions with remarkable resilience, due to their physiological and metabolic adaptations (Zhong et al., 2024).

All these adaptations confer deep-sea fungi with advantages in the field of mycoremediation. In addition to the capacity of fungi to penetrate contaminants, they also have the potential to utilise enzymes capable of degrading toxic compounds (Dell´Anno et al., 2021; Deshmukh et al., 2016). In this regard, the extremophilic hydrocarbonoclastic fungi, can be defined as "*a broad group of cultivable fungi which live optimally under extreme conditions and are characterized by having a high ability to grow using hydrocarbons as sole carbon source and energy*" (Naranjo-Briceño et al., 2019). A number of studies have been conducted that focused on the isolation of extremophilic hydrocarbonoclastic fungi, particularly those from deep-sea environments, and the assessment of their capacity to bioconvert crude oil or specific hydrocarbons (Table 1). Nevertheless, the technical challenge of obtaining the necessary substrates from this ecosystem appears to be the principal factor leading to the scarcity of reports on this subject.

Deep-sea fungi have a remarkable ability to bioremediate crude oil due to the enzymatic synthetic capacities, capable of de-aromatizing hydrocarbons, reducing the density and increasing the bioavailability of crude oil. The bioprospection of hydrocarbonoclastic fungi, from extreme environment, as the deep-sea, should be explored as long as the petroleum remains as energy source for humanity, either for upgrading, biorecovery, or bioremediation of crude oil (Naranjo-Briceño at al., 2019).

4. Microbial Biosurfactants

The biosurfactant-producing microorganisms live in water (sea, fresh-water, groundwater), soil, as well as extreme environments under a wide range of temperatures, pH, and salinity, even growing in polluted sites (Pitocchi et al., 2020). The microbial biosurfactants (BS), are amphiphilic molecules with tensioactive, emulsifying, dispersing, waterproofing, softening, lubricating, and humectant properties (Bhardwaj et al., 2013). In nature, if carbon source is not bioavailable, microorganisms as fungi, bacteria, and actinobacteria mainly produce these kinds of molecules in order to reduce surface and interfacial tension (Fig. 2), and as consequence make the carbon source bioavailable for their growth (Konishi et al., 2015; Banat et al., 2014).

Fungi produce BS under extreme biotic or abiotic conditions. Specifically marine fungal production of BS, increases bioavailability of non-soluble pollutants by establishing a desorption-solubilization equilibrium, preferentially shifting the equilibrium towards the aqueous phase and allowing the different transport processes (diffusive and convective) and pollutant biodegradation (Sajna et al., 2013). The application of marine fungal BS is gaining interest in bioremediation processes of polluted environments with persistent organic compounds (Singh et al., 2019). Table 2 summarizes most known microbial BS, their main applications in different

Table 1. Hydrocarbonoclastic fungi isolated from deep-sea.

Taxa	Isolation Depth (m) and Substrata Type	Location of Substrata	Crude Oil Type	Hydrocarbon Degradation Efficiency	References
Ascomycota, and Basidiomycota	2570 m below the surface	Amazonian margin deep-sea water	Brazilian crude oil 1%	Decrease of oil droplets	Campeão et al. (2017)
Aspergillus flavus NIOSN-SK56S22	Deep-sea (Not specified)	Arabian Sea (68°E 21°N)	Crude oil 13.35% w/v	100% degradation of Carbon chain C30	Barnes et al. (2018)
Aspergillus sydowii BOBA1	3000 m	Bay of Bengal, India	Spent engine oil	71.2 and 82.5% at atmospheric and high-pressure conditions	Ganesh et al. (2021)
Aureobasidium sp., *P. brevicompactum*, *Phialocephala* sp., *Penicillium* sp., *Cladosporium* sp. 1, 2, and *C. gracilis*	Between 517 and 3240 m deep	Gulf of Mexico	Hexadecane and 1-hexadecene 1.4% each	Increase of dry weight fungal biomass with Hexadecane and 1-hexadecene as sole carbon source	Velez et al. (2020)
Penicillium ilerdanum strain NPDF1239-K3-F21 and *Aspergillus versicolor* strain NPDF190-C1-26	Deep sediment cores	Gulf of Kutch and Gulf of Cambay/ Khambhat	Phenanthrene, Anthracene, Pyrene and Fluoranthene (100 ppm each)	degradation of > 75% for all the PAHs	Mahajan et al. (2021)
Schizophyllum commune 20R-7-F01, *Irpex lacteus* 16R-2-F03 and *Bjerkandera fumosa* 26R-5-F09	Anaerobic coal associated sediments, ~1.3–2.5 km deep below the sea-floor (1118 m water depth)	Shimokita Peninsula, Japan	50 ppm phenanthrene (Phe), pyrene (Pyr) and benzo[a]pyrene (BaP)	*Schizophyllum commune* 20R-7-F01 Phe 24.8 ± 1.9 Pyr 18.1 ± 2.0 Bap 13.1 ± 2.0	ul Arifeen et al. (2022)
*Cladosporium*_GB1	1739 m water depth and 202-207 sediment depth, cold seep sediments	Sonora Margin/ Guaymas Basin transition	Naphthalene	Concept sketch for potential interactions of hydrocarbon-degrading fungi	Edgcomb et al. (2022)
Alternaria sp.	975 m, sediments	Perdido region, Gulf of Mexico	Extra-heavy crude oil	25.6% (±7.6) of the total-extra-heavy crude oil	Romero-Hernández et al. (2021)
Cadophora sp. TS2 *Pseudogymnoascus* sp. TS12	Isolated from the deep-sea sponge *Stelleta normani*	Not specified	Phenanthrene and Benzo-a-pyrene (BAP), PAHs	80%, 99.5%, 87.6, 86.7, 93.4%, 87.9	Batista-García et al. (2017)

Table 2. Applications of microbial biosurfactants and their general characteristics (Jin et al., 2022; Fenibo et al., 2019; Luna et al., 2015).

Biosurfactant and Applications	**Mechanism/Function/Property**
Environmental Applications	
Glycolipids and lipopeptides	Enhance formation of stable water-oil emulsions, breakdown oil films in the rock and reduce interfacial tension, thereby reducing the capillary forces that impede oil movement through the rock pores.
Emulsan, Alasan, Biodispersant	High molecular weight biosurfactants form a stable water-in-oil emulsion which aids soil mobility, viscosity reduction and prevents drop coalescence.
Glycolipids and trehalose lipids	Solubilization, oil bioavailable to hydrocarbon-degraders and longer shelf life, biodegradability.
Rhamnolipids, sophorolipids, surfactins, and lipopeptides	Reduction of surface and interfacial tensions lead to mobilization and consequent removal of oil from soil.
Rhamnolipids	Metal-removal by micelle entrapment.
Lipopeptides	Detergency property of biosurfactants exhibits toxicity against nematodes and insects.
Medicinal Applications	
MEL anionic surfactin isoform, rhamnolipids	Cationic liposomes bearing MEL-A effectively increase transfection of genes into mammalian cells. The antimicrobial effect of biosurfactants is demonstrated through detergent-like activities.
Sophorolipids	An antiviral agent, halt cell replication in favor of cell differentiation.
Surfactin	Immuno-modulating biosurfactants stimulate the immune system by increasing the ratio of lymphocyte transformation and migration of polymorph nuclear cells.
Sophorolipid diacetate ethyl ester, surfactin	Inactivation of viral lipid envelopes and capsids.
Sophorolipids	Biosurfactant adsorption to a substrate modifies surface hydrophobicity thereby interfering with microbial adhesion and desorption processes.
Biodispersan	Used as skin detergent, emulsifier; for tanning and wetting.
Industrial Applications	
Trehalose tetraester unspecified	Removal of lipophilic components from fiber surface as a pre-treatment, removal of oil from fibres and enhanced dispersion of dyes for uniform and better penetration into fiber.
Biodispersan	Used for washing and deresinification of pulp by defoaming, dispersion, and color levelling. Used in calendaring through wetting, leveling, coating, and coloring.
Biodispersan	Employed as a dispersant and a wetting agent during grinding and stabilization for improved mixing properties.
Rhamnolipids and polymeric biosurfactants	Modification of the rheological characteristics of the food to a desired consistency and texture.
Sophorolipids, rhamnolipids, mannosylerythritol lipids (MELs)	Application of biosurfactants in cosmetics is due to their low irritancy, cytoprotective effect, anti-ageing, acts like antioxidant, wettability, moisturizing properties, healing, and skin toning features.
Sophorolipids, MEL	Properties such as foaming, surface tension reduction, and solubilization make it suitable for detergent making.

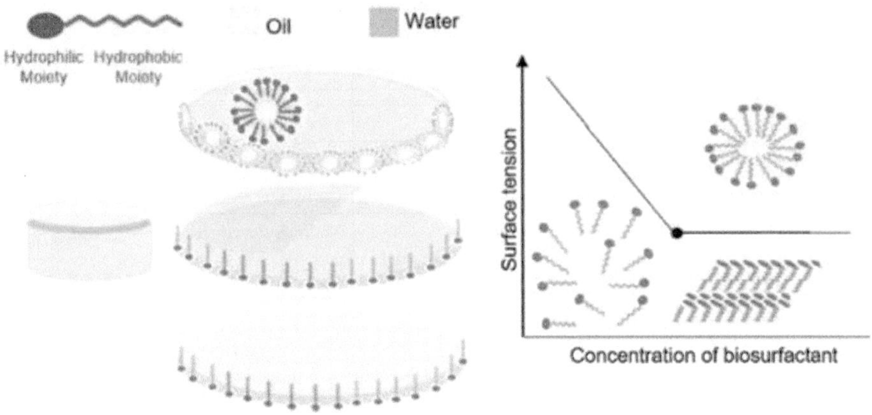

Fig. 2. Scheme of mode of action and main activity of BS (modified from da Silva et al., 2021).

areas and a brief description of their characteristics. A high quality and purity grade of BS is not required. On the other side, for medical, cosmetic and alimentary applications, microbial BS must have high quality and purity grade is obligatory, in order to ensure innocuousness and potentiate tensioactive property.

5. Marine Fungal Biosurfactants (MFBs)

The MFBs are amphipathic molecules composed of two regions, a motile hydrophilic or polar region made up of carbohydrates (rhamnose, trehalose, mannose, sophorose), amino acids or peptides, and a motile hydrophobic or non-polar region made up of a hydrocarbon chain of variable length (α-branched- β-hydroxy fatty acids) (Satpute et al., 2010; Banat et al., 2010; Ali et al., 2024). According to their biochemical composition, they are named as glycolipids, lipopolysaccharides, lipopeptides, and lipoproteins (da Silva et al., 2021; Satpute et al., 2025) (Table 3).

Hydrophobins and cerato-platanins are other kinds of surface-active extracellular proteins produced by filamentous fungi under low soluble carbon source stress like crude oil (Landeta et al., 2024). Both are amphiphilic proteins, like glycolipids and lipopolysaccharides, can lower surface tension and stabilize emulsions, helping hyphae, fruiting bodies and spores adhere to surfaces and even host-pathogen interactions (Pitocchi et al., 2020).

Cerato-platanin is a small protein cysteine-rich and hydrophobins are high-molecular weight BS fungal proteins (Pennacchio et al., 2021). These hydrophobic proteins have the ability to self-assemble forming amyloid-like aggregates and play an essential role in the growth and development of the filamentous fungi (Landeta-Salgado et al., 2021).

Table 3. Marine Fungal Biosurfactant producers and their type of molecules (Shekhar et al., 2015; Bhardwaj et al., 2013; Pitocchi et al., 2020; Landeta-Salgado, 2021; da Silva et al., 2021).

Fungi	**Type of Biosurfactants**
Aspergillus flavus	Carbohydrate-lipid-protein complexes
Aspergillus mulundensis	Lipopeptides
Aspergillus ustus	Glycolipoprotein
Aspergillus ustus	Carbohydrate-lipid-protein complexes
Candida antarctica T-34, *Pseudozyma rugulosa, Pseudozyma parantarctica*, and *Pseudozyma tsukubaensis*	Mannosylerythritol lipid-A (MEL-A)
Candida antarctica T-35, *Pseudozyma aphidis, Pseudozyma siamensis, Pseudozyma fusifornata*	Mannosylerythritol lipid-B (MEL-B)
Candida batistae	Sophorolipid
Candida bombicola ATCC22214	Sophorolipids
Candida bombicola ATCC22214	Cetylalcoholsophorolipid
Candida bombicola ATCC22214	Sophorolipids
Candida ishiwadae	Glycolipid
Candida lipolytica	Protein-lipid-polysaccharide complex
Candida tropicalis and *Geotrichum candidum*	Carbohydrate-lipid-protein complexes
Fusarium sp.	Lipopeptides
Kluyveromyces marxianus	Polymeric-glycoproteins
Penicillium chrysogenum SNP5	Lipopeptides
Penicillium citrinum	Carbohydrate-lipid-protein complexes
Pichia anamola	Sophorolipid
Saccharomyces cerevisiae	Polymeric-glycoproteins
Starmerella bombicola	Sophorolipids
Torulopsis bombicola	Sophorose lipid
Torulopsis species	Sophorolipids
Torulopsis petrophilum	Glycolipid
Trichosporon ashii	Sophorolipid
Ustilago maydis	Glycolipid
Ustilago zeae, Ustilago maydis	Cellobiolipids
Wickerhamiella domercqiae	Di-acetylatedlactonic
Wickerhamiella domercqiae	Sophorolipids
Yarrowia lipolytica	Polymeric-glycoproteins
Yarrowia lipolytica MTCC9520	Lipopeptides
Trichoderma harzianum	Cerato-platanin ThCP
Paradendryphiela salina and *Talaromyces pinophilus*	Cerato-platanins and hydrophobins
Aspergillus terreus MUT271	AtCP cerato-platanin
Trichoderma harzianum MUT290	ThCP cerato-platanin

6. Methods for Studying Biosurfactants

According to Saranraj et al. (2022) and Youssef et al. (2004) there are qualitative and quantitative methods for BS study (Fig. 3), considered as a first set of assays for screening and selecting microbial BS production (Table 4). These methods are applicable to screen MBF as well, representing a cost-effective and efficient alternative.

Purification, characterization, and chemical analysis of biosurfactants can be performed using techniques such as ammonium sulphate fractionation, chilled acetone and hexane treatments, acid precipitation, and silica-gel column chromatography. Through spectroscopic techniques, reverse phase high performance liquid chromatography (RP-HPLC), ultra performance liquid chromatography-tandem mass spectrometry (UPLC-MS), and liquid chromatography-mass spectrometry (LC-MS), it has been possible to elucidate novel biosurfactant chemical structures and physicochemical properties (Abbot et al., 2022; Ribeiro et al., 2012).

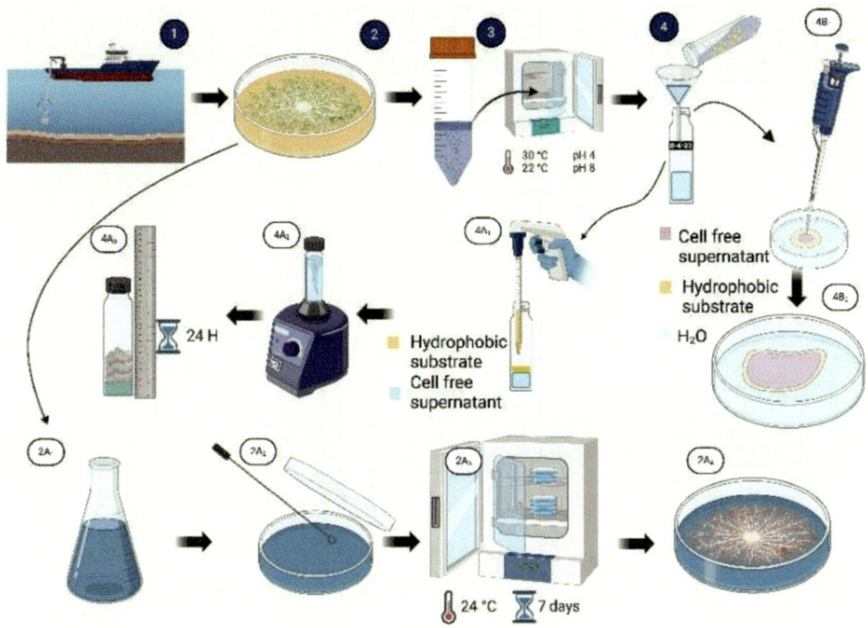

Fig. 3. Experimental diagram. (1) Sampling, (2) Isolation of cultivable fungi, ($2A_1$) Methylene blue agar, ($2A_2$) Inoculation of isolates in Petri dishes ($2A_3$), Incubation of Petri dishes at 24°C for 7 days. ($2A_4$), Growth of the isolate after incubation (3), *In vitro* biosurfactant production bioassays at two temperatures (22°C and 30°C) and two pH (4 and 8), (4) Cell-free supernatant filtration, ($4A_1$) Test tubes with a mixture (1:1 v/v) of hydrophobic substrate, and cell free supernatant, ($4A_2$) Homogenization, ($4A_3$) Reading results, ($4B_1$) Petri dishes with a hydrophobic layer of distilled water and a hydrophobic substrate and cell free supernatant on top, ($4B_2$) Dispersed oil.

Table 4. Methods for evidencing microbial biosurfactants production (Wong-Villarreal and Yañez-Ocampo, 2021; García-Reyes and Yañez-Ocampo, 2016).

Method	**Objective**	**Fundamentals**
Agar hexadecyltrimethylammonium bromide (CTAB) assay or agar Siegmund and Wagner, 1991	Detecting qualitative production of anionic biosurfactant, rhamnolipids, specifically	The CTAB is a cationic salt that reacts with the rhamnolipid, this complex (CTAB-rhamnolipid-blue methylene) can be visualized by the presence of translucent halo around the colony growth on the agar plates
Hemolytic activity assay	Visualize indirectly microbial BS´s production by inoculating of bacterial strain on a blood agar plate (5% v/v)	Translucent halos around the microbial growth on the agar plates indicate BS presence with hemolytic activity
Drop collapse test	Detection of BS presence in a cell free supernatant (FCS) or crude BS extracted and concentrated from a bacterial liquid medium culture	Surficial action on a set of hydrophobic phases (v. gr. mineral oil, soja oil, motor oil) it is possible to see the collapse of the oil drop when BS are added
Oil displacement test	Measuring diameter of oil displaced after addition of crude BS or FCS	Increasing diameter is directly related to high tensoactive activity. This test provides indirect information about biological compounds with tensoactive activity
Emulsion index (IE_{24})	Assess stable emulsion formation by crude BS or FCS with a set of hydrophobic phases	IE_{24} is calculated by dividing the height of the emulsion layer by the total height of the mixture and multiplying by 100
Surface tension	Quantify surface tension from FCS or determination of critical micelle concentration from known concentrations of crude BS	Based on the strength required to separate the ring du Noüy or platinum Wilhelmy plate from the liquid surface. Is reported in mN/m

7. Conclusion

Deep-sea marine fungi have been identified as a sustainable and efficient source for the bioremediation of crude oil pollution in marine environments. The capacity of these microorganisms to flourish in extreme conditions, including low temperatures, high pressures, and oligotrophic environments, underscores their potential for degradation of persistent hydrocarbons. This capacity serves to establish mycoremediation as a highly effective method for the degradation of hydrocarbons. In order to explore the potential of deep-sea fungi and their BS in mitigating the ecological consequences of oil spills, there is a necessity for interdisciplinary collaboration.

References

Abbot, V., Paliwal, D., Sharma, A. and Sharma, P. (2022). A review on the physicochemical and biological applications of biosurfactants in biotechnology and pharmaceuticals. Heliyon, 8(8): e10149. https://doi.org/10.1016/j.heliyon.2022.e10149.

Ali, A., Aziz, A.R.A., Ismael, M.A. and Alqaed, S. (2024). Biosurfactants as an alternative eco-friendly solution for water-in-diesel emulsions: A review paper. Heliyon, 10(17): e037485. https://doi.org/10.1016/j.heliyon.2024.e37485.

Bacosa, H.P., Ancla, S.M.B., Arcadio, C.G.L.A., Dalogdog, J.R.A., Ellos, D.M.C. et al. (2022). From surface water to the deep-sea: A review on factors affecting the biodegradation of spilled oil in marine environment. J. Mar. Sci. Eng., 10(3): 426. https://doi.org/10.3390/jmse10030426.

Barnes, N.M., Khodse, V.B., Lotlikar, N.P., Meena, R.M. and Damare, S.R. (2018). Bioremediation potential of hydrocarbon-utilizing fungi from select marine niches of India. 3 Biotech, 8(1): 21. https://doi.org/10.1007/s13205-017-1043-8.

Barone, G., Varrella, S., Tangherlini, M., Rastelli, E., Dell'Anno, A. et al. (2019). Marine fungi: Biotechnological perspectives from deep-hypersaline anoxic basins. Diversity, 11(7): 113. https://doi.org/10.3390/d11070113.

Batista-García, R.A., Kumar, V.V., Ariste, A., Tovar-Herrera, O.E., Savary, O. et al. (2017). Simple screening protocol for identification of potential mycoremediation tools for the elimination of polycyclic aromatic hydrocarbons and phenols from hyperalkalophile industrial effluents. J. Environ. Manage., 198: 1–11. https://doi.org/10.1016/j.jenvman.2017.05.010.

Banat, I.M., Satpute, S.K., Cameotra, S.S., Patil, R. and Nyayanit, N.V. (2014). Cost effective technologies and renewable substrates for biosurfactants production. Front. Microbiol., 5: 697. https://doi.org/10.3389/fmicb.2014.00697.

Bhardwaj, G., Cameotra, S.S. and Chopra, H.K. (2013). Biosurfactants from fungi: A review. J. Pet. Environ. Biotechnol., 4(06): 1–6. http://dx.doi.org/10.4172/2157-7463.1000160.

Burchette, T. (2025). What has the petroleum industry ever done for us? On the positive legacies from the petroleum industry and the role of carbonate and evaporite studies in the energy transition. Geol. Soc. Spec. Publ., 548(1): 13–64. https://doi.org/10.1144/SP548-2024-103.

Burgaud, G., Le Calvez, T., Arzur, D., Vandenkoornhuyse, P. and Barbier, G. (2009). Diversity of culturable marine filamentous fungi from deep-sea hydrothermal vents. Environ. Microbiol., 11(6): 1588–1600. https://doi.org/10.1111/j.1462-2920.2009.01886.x.

Campeão, M.E., Reis, L., Leomil, L., Vianna, P.C.F., Lima, A.O. et al. (2017). The deep-sea microbial community from the amazonian basin associated with oil degradation. Front. Microbiol., 8: 1019. https://doi.org/10.3389/fmicb.2017.01019.

Chen, J., Zhang, W., Wan, Z., Li, S., Huang, T. et al. (2019). Oil spills from global tankers: Status review and future governance. J. Clean. Prod., 227: 20–32. https://doi.org/10.1016/j.jclepro.2019.04.020.

Chughtai, R. and Asif, Z. (2021). Study fate of pollutants due to oil spill in sea water through multimedia environmental modeling. Int. J. Environ. Sci. Technol., 18(2): 761–770. https://doi.org/10.1007/s13762-020-02849-z.

Connell, L., Barrett, A., Templeton, A. and Staudigel, H. (2009). Fungal diversity associated with an active deep-sea volcano: Vailulu'u Seamount, Samoa. Geomicrobiol. J., 26(8): 597–605. https://doi.org/10.1080/01490450903316174.

Council National Research Transportation. Research Board, Division on Earth, Life Studies, Marine Board, Ocean Studies Board, Committee on Oil in the Sea, Fates (2003). Oil in the Sea III: Inputs, Fates, and Effects. National Academies Press.

Damare, S., Raghukumar, C. and Raghukumar, S. (2006). Fungi in deep-sea sediments of the Central Indian Basin. Deep-Sea Res. I: Oceanogr. Res. Pap., 53(1): 14–27. https://doi.org/10.1016/j.dsr.2005.09.005.

da Silva, A.F., Banat, I.M., Giachini, A.J. and Robl, D. (2021). Fungal biosurfactants, from nature to biotechnological product: Bioprospection, production, and potential applications. Bioprocess Biosyst Eng., 44(10): 2003–2034. https://doi.org/10.1007/s00449-021-02597-5.

de Santana Campelo, R.P., de Lima, C.D.M., de Santana, C.S., Reis, G.M.D.C., Lima, A.O.S. et al. (2021). Oil spills: The invisible impact on the base of tropical marine food webs. Mar. Pollut. Bull., 167: 112281. https://doi.org/10.1016/j.marpolbul.2021.112281.

Dell'Anno, F., Rastelli, E., Sansone, C., Brunet, C., Ianora, A. et al. (2021). Bacteria, fungi and microalgae for the bioremediation of marine sediments contaminated by petroleum hydrocarbons in the omics era. Microorganisms, 9(8):1695. https://doi.org/10.3390/microorganisms9081695.

Deshmukh, R., Khardenavis, A.A. and Purohit, H.J. (2016). Diverse metabolic capabilities of fungi for bioremediation. Indian J. Microbiol., 56(3): 247–264. https://doi.org/10.1007/s12088-016-0584-6.

Douglas-Westwood, L. (2005). Marine Industries Global Market Analysis. Marine Institute.

Edgcomb, V.P., Teske, A. and Mara, P. (2022). Microbial hydrocarbon degradation in Guaymas Basin: Exploring the roles and potential interactions of fungi and sulfate-reducing bacteria. Front. Microbiol., 13: 831828. https://doi.org/10.3389/fmicb.2022.831828.

Epstein, A. (2022). Fossil future: Why global human flourishing requires more oil, coal, and natural gas: Not less. Penguin, 39(II): 58–68. https://doi.org/10.51710/jias.v39iII.260.

Fenibo, E.O., Ijoma, G.N., Selvarajan, R. and Chikere, C.B. (2019). Microbial surfactants: The next generation multifunctional biomolecules for applications in the petroleum industry and its associated environmental remediation. Microorganisms, 7(11): 581. https://doi.org/10.3390/microorganisms7110581.

Fingas, M. (2025). Introduction to spill modeling. pp. 273–307. *In*: Fingas, M. (ed.). Oil Spill Science and Technology. Elsevier. https://doi.org/10.1016/B978-0-443-21703-6.00006-0.

Fisher, C.R., Montagna, P.A. and Sutton, T.T. (2016). How did the Deepwater Horizon oil spill impact deep-sea ecosystems? Oceanography, 29(3): 182–195. https://doi.org/10.5670/oceanog.2016.82.

Ganesh Kumar, A., Manisha, D., Sujitha, K., Magesh Peter, D., Kirubagaran, R. et al. (2021). Genome sequence analysis of deep-sea *Aspergillus sydowii* BOBA1 and effect of high pressure on biodegradation of spent engine oil. Sci. Rep., 11(1): 9347. https://doi.org/10.1038/s41598-021-88525-9.

Gao, Y., Du, X., Xu, W., Fan, R., Zhang, X. et al. (2020). Fungal diversity in deep-sea sediments from East Yap Trench and their denitrification potential. Geomicrobiol. J., 37(9): 848–858. http://dx.doi.org/10.1080/01490451.2020.1789778.

García-Reyes, S. and Yañez-Ocampo, G. (2016). Microbial biosurfactants: Methods for their isolation and characterization. J. Microbiol., Biotechnol. Food Sci., 6(1): 641. https://doi.org/10.15414/jmbfs.2016.6.1.641-648.

Govindarajan, S.K., Mishra, A. and Kumar, A. (2021). Oil spill in a marine environment: Requirements following an offshore oil spill. Rudarsko-geološko-naftni zbornik, 36(4). https://doi.org/10.17794/rgn.2021.4.1.

Hall, C.A. and Klitgaard, K.A. (2012). Energy and the wealth of Nations. Energy Policy, 42(2): 720–721.

Hansen, B.H., Altin, D. and Nordtug, T. (2025). Do oil droplets and chemical dispersants contribute to uptake of oil compounds and toxicity of crude oil dispersions in cold-water copepods? J. Toxicol. Environ. Health A., 88(2): 67–84. https://doi.org/10.1080/15287394.2023.2271003.

Holechek, J.L., Geli, H.M.E., Sawalhah, M.N. and Valdez, R. (2022). A global assessment: Can renewable energy replace fossil fuels by 2050? Sustainability, 14(8): 4792. https://doi.org/10.3390/su14084792.

Jannasch, H.W. and Mottl, M.J. (1985). Geomicrobiology of deep-sea hydrothermal vents. Science, 229(4715): 717–725. https://doi.org/10.1126/science.229.4715.717.

Jin, C., Sun, T., Ampah, J.D., Liu, X., Geng, Z. et al. (2022). Comparative study on synthetic and biological surfactants' role in phase behavior and fuel properties of marine heavy fuel oil-low carbon alcohol blends under different temperatures. Renew. Energy, 195: 841–852. https://doi.org/10.1016/j.renene.2022.06.088.

Jobstvogt, N., Townsend, M., Witte, U. and Hanley, N. (2014). How can we identify and communicate the ecological value of deep-sea ecosystem services? PLoS One, 9(7): e100646. https://doi.org/10.1371/journal.pone.0100646.

Kingston, P.F. (2002). Long-term environmental impact of oil spills. Spill Science Technology Bulletin, 7(1-2): 53–61. https://doi.org/10.1016/S1353-2561(02)00051-8.

Konishi, M., Yoshida, Y. and Horiuchi, J.I. (2015). Efficient production of sophorolipids by *Starmerella bombicola* using a corncob hydrolysate medium. J. Biosci. Bioeng., 119(3): 295–300. https://doi.org/10.1016/j.jbiosc.2014.08.007.

La Bianca, G., Rees, S., Attrill, M.J., Lombard, A.T., McQuaid, K.A. et al. (2023). A standardised ecosystem services framework for the deep-sea. Front. Mar. Sci., 10: 1176230. https://doi.org/10.3389/fmars.2023.1176230.

Landeta, C., Medina-Ortiz, D., Escobar, N., Valdez, I., González-Troncoso, M.P. et al. (2024). Integrative workflows for the characterization of hydrophobin and cerato-platanin in the marine fungus *Paradendryphiella salina*. Arch. Microbiol., 206(9): 385. https://doi.org/10.1007/s00203-024-04087-0.

Landeta-Salgado, C., Cicatiello, P., Stanzione, I., Medina, D., Mora, I.B. et al. (2021). The growth of marine fungi on seaweed polysaccharides produces cerato-platanin and hydrophobin self-assembling proteins. Microbiol. Res., 251: 126835. https://doi.org/10.1016/j.micres.2021.126835.

Le Calvez, T., Burgaud, G., Mahé, S., Barbier, G. and Vandenkoornhuyse, P. (2009). Fungal diversity in deep-sea hydrothermal ecosystems. Appl. Environ. Microbiol., 75(20): 6415–6421. https://doi.org/10.1128/AEM.00653-09.

Lee, K., Stoffyn-Egli, P., Tremblay, G.H., Owens, E.H., Sergy, G.A. et al. (2003). Oil–mineral aggregate formation on oiled beaches: Natural attenuation and sediment relocation. Spill Sci. Technol. Bull., 8(3): 285–296. https://doi.org/10.1016/S1353-2561(03)00042-2.

Luna, J.M., Rufino, R.D., Sarubbo, L.A. and Campos-Takaki, G.M. (2015). Characterization, surface properties, and biological activity of a biosurfactant produced from industrial waste by *Candida sphaerica* UCP0995 for application in the petroleum industry. Colloids Surf. B Biointerfaces, 102: 202–209. https://doi.org/10.1016/j.colsurfb.2012.08.008.

Luo, Y., Wei, X., Yang, S., Gao, Y.H. and Luo, Z.H. (2020). Fungal diversity in deep-sea sediments from the Magellan seamounts as revealed by a metabarcoding approach targeting the ITS2 regions. Mycology, 11(3): 214–229. https://doi.org/10.1080/21501203.2020.1799878.

Mahajan, M., Manek, D., Vora, N., Kothari, R.K., Mootapally, C. et al. (2021). Fungi with high ability to crunch multiple Polycyclic Aromatic Hydrocarbons (PAHs) from the pelagic sediments of Gulfs of Gujarat. Mar. Pollut. Bull., 167: 112293. https://doi.org/10.1016/j.marpolbul.2021.112293.

Mahé, S., Rédou, V., Calvez, T.L., Vandenkoornhuyse, P. and Burgaud, G. (2013). Fungi in deep-sea environments and metagenomics. pp. 325–354. *In*: Martin, F. (ed.). The Ecological Genomics of Fungi. Wiley, pp., 325–354. http://dx.doi.org/10.1002/9781118735893.ch15.

Montagna, P.A., Baguley, J.G., Cooksey, C., Hartwell, I., Hyde, L.J. et al. (2013). Deep-sea benthic footprint of the Deepwater Horizon blowout. PLoS One, 8(10): e70540. https://doi.org/10.1371/journal.pone.0070540.

Mori, R. (2023). Replacing all petroleum-based chemical products with natural biomass-based chemical products: A tutorial review. RSC Sustainability, 1: 179–212. https://doi.org/10.1039/D2SU00014H.

Mukherjee, B. and Wrenn, B.A. (2009). Influence of dynamic mixing energy on dispersant performance: Role of mixing systems. *In*: Nazaroff, W.W. and Alvarez-Cohen, L. (eds.). Environmental Engineering Science. John Wiley & Sons, 26(12): 1725–1737. https://doi.org/doi:10.1089/ees.2009.0159.

Murawski, S.A., Grosell, M., Smith, C., Sutton, T., Halanych, K.M. et al. (2021). Impacts of petroleum, petroleum components, and dispersants on organisms and populations. Oceanography, 34(1): 136–151. https://www.jstor.org/stable/27020066.

Nagano, Y., Miura, T., Nishi, S., Lima, A.O., Nakayama, C., Pellizari, V.H. and Fujikura, K. (2017). Fungal diversity in deep-sea sediments associated with asphalt seeps at the Sao Paulo Plateau. Deep-Sea Res Part II: Top Stud. Oceanogr., 146: 59–67. http://dx.doi.org/10.1016/j.dsr2.2017.05.012.

Naranjo-Briceño, L., Pernía, B., Perdomo, T., González, M., Inojosa, Y. et al. (2019). Potential role of extremophilic hydrocarbonoclastic fungi for extra-heavy crude oil bioconversion and the sustainable development of the petroleum industry. pp. 559–586. *In*: Tiquia-Arashiro, S. and Grube, M. (eds.). Fungi in Extreme Environments: Ecological Role and Biotechnological Significance. Springer. http://dx.doi.org/10.1007/978-3-030-19030-9_28.

Nekahi, A., Anil Kumar, M.R., Deng, S., Li, X., Petropoulos, A. et al. (2025). Toward green renewable energies and energy storage for the sustainable decarbonization and electrification of society. Electrochem. Energy Rev., 8(1): 12. https://doi.org/10.1007/s41918-025-00247-y.

Ogaki, M.B., Coelho, L.C., Vieira, R., Neto, A.A., Zani, C.L. et al. (2020). Cultivable fungi present in deep-sea sediments of Antarctica: Taxonomy, diversity, and bioprospecting of bioactive compounds. Extremophiles, 24(2): 227–238. https://doi.org/10.1007/s00792-019-01148-x.

Orcutt, B.N., Bach, W., Becker, N., Brazelton, W.J., Edgcomb, V. et al. (2011). Microbial activity in the subseafloor basaltic crust: A synthesis of results from the Juan de Fuca Ridge Flanks. Geochem. Geophys. Geosyst., 12(9). https://doi.org/10.3389/fmicb.2013.00189.

Pennacchio, A., Pitocchi, R., Varese, G.C., Giardina, P. and Piscitelli, A. (2021). *Trichoderma harzianum* cerato-platanin enhances hydrolysis of lignocellulosic materials. Microb. Biotechnol., 14(4): 1699–1706. https://doi.org/10.1111/1751-7915.13836.

Peterson, C.H., Rice, S.D., Short, J.W., Esler, D., Bodkin, J.L. et al. (2003). Long-term ecosystem response to the Exxon Valdez oil spill. Science, 302(5653): 2082–2086. https://doi.org/10.1126/science.1084282.

Pitocchi, R., Cicatiello, P., Birolo, L., Piscitelli, A., Bovio, E. et al. (2020). Cerato-platanins from marine fungi as effective protein biosurfactants and bioemulsifiers. Int. J. Mol. Sci., 21(8): 2913. https://doi.org/10.3390/ijms21082913.

Prince, R.C. and Lessard, R.R. (2004). Crude oil releases to the environment: Natural fate and remediation options. pp. 727–736. *In*: Encyclopedia of Energy (Vol. 1). Academic Press. http://dx.doi.org/10.1016/B978-0-12-409548-9.09112-0.

Radovic, J.R. and Silva, R.C. (2024). Ultrahigh-resolution mass spectrometry advances for biogeochemical analysis: From seafloor sediments to petroleum and marine oil spills. J. Am. Soc. Mass Spectrom., 36(1): 7–33. https://doi.org/10.1021/jasms.4c00266.

Raghukumar, C., Damare, S.R. and Singh, P. (2010). A review on deep-sea fungi: Occurrence, diversity, and adaptations. Bot. Mar., 53(6): 479–549. http://dx.doi.org/10.1515/bot.2010.076.

Ramesh, T., Yamunadevi, R., Sundaramanickam, A., Thangaraj, M., Kumaran, R. et al. (2021). Biodiversity of the fungi in extreme marine environments. pp. 75–100. *In*: Vijay Kumar Sharma, V.K., Shah, M.P., Parmar, S. and Kumar, A. (eds.). Fungi Bio-prospects in Sustainable Agriculture, Environment and Nano-technology. Academic Press.

Ribeiro, I.A., Bronze, M.R., Castro, M.F. and Ribeiro, M.H.L. (2012). Optimization and correlation of HPLC-ELSD and HPLC-MS/MS methods for identification and characterization of sophorolipids. J. Chromatogr. B, 899: 72–80. https://doi.org/10.1016/j.jchromb.2012.04.037.

Rodrigues, B.R., de Oliveira, D.H., Santos, P.G. and das Chagas, M.M.J. (2025). Ecosystem model for quantifying oil spill effects on humans: Case study in an industrial region in Brazil. Hum. Ecol. Risk Assess., 31(7-8): 1158–1185. http://dx.doi.org/10.1080/10807039.2025.2538551.

Romero-Hernández, L., Velez, P., Betanzo-Gutiérrez, I., Camacho-López, M.D., Vázquez-Duhalt, R. et al. (2021). Extra-heavy crude oil degradation by *Alternaria* sp. isolated from deep-sea sediments of the Gulf of Mexico. Appl. Sci., 11(13): 6090. https://doi.org/10.3390/app11136090.

Ruff, S.E., Aronson, J.L. and Teske, A. (2015). Microbial community structure and activity in marine sediments of the Guaymas Basin: Insights from 40Ar/39Ar dating and stable isotope geochemistry. Front. Microbiol., 6: 1246. https://doi.org/10.3389/fmicb.2021.633649.

Saadoun, I.M.K. (2015). Impact of oil spills on marine life. *In*: Emerging pollutants in the Environment: Current and Further Implications. IntechOpen, pp. 10–25. https://doi.org/10.5772/60455.

Sajna, K.V., Sukumaran, R.K., Jayamurthy, H., Reddy, K.K., Kanjilal, S. et al. (2013). Studies on biosurfactants from *Pseudozyma* sp. NII 08165 and their potential application as laundry detergent additives. Biochem. Eng. J., 78: 85–93. https://doi.org/10.1016/j.bej.2012.12.014.

Saranraj, P., Sayyed, R., Sivasakthivelan, P., Hasan, M.S., Rahman, A. et al. (2022). Microbial biosurfactants: Methods of investigation, characterization, current market value, and applications. pp. 19–34. *In*: Sayyed, R.Z. (ed.). Microbial Surfactants (Volume 3)., CRC Press.

Satpute, S.K., Banat, I.M., Dhakephalkar, P.K., Banpurkar, A.G. and Chopade, B.A. (2010). Biosurfactants, bioemulsifiers, and exopolysaccharides from marine microorganisms. Biotechnol. Adv., 28(4): 436–450. https://doi.org/10.1016/j.biotechadv.2010.02.006.

Satpute, S.K., Banat, I.M., Waghmode, S.R., Hulkane, S., Bagayatkar, M. et al. (2025). Screening strategies and production of biosurfactants (BSs)/bioemulsifiers (BEs) from marine yeasts and fungi. AIMS Microbiol., 11(3): 542–573. https://doi.org/10.3934/microbiol.2025023.

Shekhar, S., Sundaramanickam, A. and Balasubramanian, T. (2015). Biosurfactant producing microbes and their potential applications: A review. Crit. Rev. Environ. Sci. Technol., 45(14): 1522–1554. https://doi.org/10.1080/10643389.2014.955631.

Singh, P., Patil, Y. and Rale, V. (2019). Biosurfactant production: Emerging trends and promising strategies. J. Appl. Microbiol., 126(2): 2–13. https://doi.org/10.1111/jam.14057.

Singh, P., Raghukumar, C., Verma, P. and Shouche, Y. (2011). Fungal community analysis in the deep-sea sediments of the Central Indian Basin by culture-independent approach. Microb. Ecol., 61(3): 507–517. https://doi.org/10.1007/s00248-010-9765-8.

Stel, J.H. (2021). Exploring and exploiting deep ocean space. pp. 65–92. *In*: Koutsopoulos, K.C. and Stel, J.H. (eds.). Ocean Literacy: Understanding the Ocean. Key Challenges in Geography. Springer, Cham. https://doi.org/10.1007/978-3-030-70155-0_4.

Tang, X., Yu, L., Xu, W., Zhang, X., Xu, X. et al. (2020). Fungal diversity of deep-sea sediments in Mid-Oceanic Ridge area of the East Pacific and the South Indian Oceans. Bot. Mar., 63(2): 183–196. http://dx.doi.org/10.1515/bot-2018-0112.

ul Arifeen, M.Z., Ma, Y., Wu, T., Chu, C., Liu, X. et al. (2022). Anaerobic biodegradation of polycyclic aromatic hydrocarbons (PAHs) by fungi isolated from anaerobic coal-associated sediments at 2.5 km below the seafloor. Chemosphere, 303: 135062. https://doi.org/10.1016/j.chemosphere.2022.135062.

Vargas-Gastélum, L., Chong-Robles, J., Lago-Lestón, A., Darcy, J.L., Amend, A.S. et al. (2019). Targeted ITS1 sequencing unravels the mycodiversity of deep-sea sediments from the Gulf of Mexico. Environ. Microbiol., 21(11): 4046–4061. https://doi.org/10.1111/1462-2920.14754.

Varrella, S., Barone, G., Corinaldesi, C., Giorgetti, A., Nomaki, H. et al. (2024). Fungal abundance and diversity in the Mariana Trench, the deepest ecosystem on earth. J. Fungi, 10(1): 73. https://doi.org/10.3390/jof10010073.

Velez, P., Gasca-Pineda, J. and Riquelme, M. (2020). Cultivable fungi from deep-sea oil reserves in the Gulf of Mexico: Genetic signatures in response to hydrocarbons. Mar. Environ. Res., 153: 104816. http://dx.doi.org/10.1016/j.marenvres.2019.104816.

Wang, Y., Zheng, X., Yu, X. and Liu, X. (2017). Temperature and salinity effects in modeling the trajectory of the 2011 Penglai 19-3 oil spill. Mar. Georesour. Geotec., 35(7): 946–953. https://doi.org/10.1080/1064119X.2016.1261971.

Wang, J., Zhou, Y., Zhuang, L., Shi, L. and Zhang, S. (2022). Study on the critical factors and hot spots of crude oil tanker accidents. Ocean Coast. Manag., 217: 106010. https://doi.org/10.1016/j.ocecoaman.2021.106010.

Wei, X., Guo, S., Gong, L.F., He, G., Pang, K.L. et al. (2018). Cultivable fungal diversity in deep-sea sediment of the East Pacific Ocean. Geomicrobiol. J., 35(9): 790–797. http://dx.doi.org/10.1080/01490451.2018.1473531.

White, K.H. (2014). Fossil Fuels: The Moral Case. Texas Public Policy Foundation, pp. 1–35.

Wong-Villarreal, A. and Yañez-Ocampo, G. (2021). Biosurfactants: Fermentation, recovery and formulation process, and constraints. pp. 108–121. *In*: Sayyed, R.Z. (ed.). Microbial Surfactants. CRC Press.

Xu, W., Gao, Y.H., Gong, L.F., Li, M., Pang, K.L. et al. (2019). Fungal diversity in the deep-sea hadal sediments of the Yap Trench by cultivation and high throughput sequencing methods based on ITS rRNA gene. Deep-Sea Res. I: Oceanogr. Res. Pap., 145: 125–136. https://doi.org/10.1016/j.dsr.2019.02.001.

Xu, W., Gong, L.F., Pang, K.L. and Luo, Z.H. (2018). Fungal diversity in deep-sea sediments of a hydrothermal vent system in the Southwest Indian Ridge. Deep-Sea Res. I: Oceanogr. Res. Pap., 131: 16–26. http://dx.doi.org/10.1016/j.dsr.2017.11.001.

Yang, S., Xu, W., Gao, Y., Chen, X. and Luo, Z.H. (2020). Fungal diversity in deep-sea sediments from Magellan seamounts environment of the western Pacific revealed by high-throughput Illumina sequencing. J. Microbiol., 58(10): 841–852. https://doi.org/10.1007/s12275-020-0198-x.

Youssef, N.H., Duncan, K.E., Nagle, D.P., Savage, K.N., Knapp, R.M. and McInerney, M.J. (2004). Comparison of methods to detect biosurfactant production by diverse microorganisms. J. Microbiol. Methods, 56(3): 339–347. https://doi.org/10.1016/j.mimet.2003.11.001.

Zhong, M., Li, Y., Deng, L., Fang, J. and Yu, X. (2024). Insight into the adaptation mechanisms of high hydrostatic pressure in physiology and metabolism of hadal fungi from the deepest ocean sediment. mSystems, 9(1): e01085-23. https://doi.org/10.1128/msystems.01085-23.

13

Bioremediation of Textile Dyes by Marine-Derived Fungi

Neha Redkar[1] and *Cathrine Sumathi Manohar*[1,2,*]

The untreated wastewater released from the garment industries release a significantly high quantity of residual colourants, that are recalcitrant, toxic chromophoric chemical compounds. This can cause various environmental hazards, especially in the aquatic habitats, as it causes hindrance to sunlight penetration and thus interferes in the primary productivity process. These pollutants, which exhibit low biodegradability, making them resistant to the conventional removal techniques and the physicochemical dye removal techniques generally adapted in the industries, are expensive, require high energy, and generate secondary pollutants. However, the bioremediation approach, which utilises microorganisms as a whole or the extracellular enzymes, offers a viable alternative. Among microorganisms, the robust and versatile fungal enzymes are known to efficiently degrade complex chemical compounds into simpler forms through the bioremediation process. Fungi, especially those derived from the marine environment, are potential candidates, as they can adapt to conditions like high salt concentrations and varying pH range as compared to terrestrial fungi. The oxidoreductive extracellular enzymes synthesised by them play an important role in mineralising not only dyes but also other pollutants, including polycyclic aromatic hydrocarbons and heavy metals. This chapter emphasises the significance and importance of marine-derived fungi for the detoxification and breakdown of the various complex textile dyes. The marine fungal remediation approach, which can

[1] Biological Oceanography Division, CSIR-National Institute of Oceanography, Dona Paula, Goa, India.
[2] Academy of Scientific and Innovative Research (AcSIR), Ghaziabad, 201002, India.
* Corresponding author: cathrine.nio@csir.res.in/csmanohar.nio@gmail.com

DOI: 10.1201/9781003795100-16

interact and degrade recalcitrant dyes by enzymatic and non-enzymatic methods, is discussed. The key factors that help in enhancing the enzyme activity and thus influencing the decolorisation and degradation process are also discussed.

1. Introduction

The textile industry is globally distributed and contributes approximately 7% of the global exports. It employs around 35 million workers and generates a trillion-dollar economy (AbuQamar et al., 2023). The primary environmental hazard caused by the textile industry is the release of untreated coloured wastewater into aquatic bodies, which typically accounts for 80% of the total discharge. The untreated wastewater released from the textile industries contains a significantly high amount of toxic and recalcitrant compounds, particularly dyes (Dutta et al., 2024). Textile dyes are chemical substances with chromophoric groups present in their molecular structure that impart colour to the given surface (Hameed and Ismail, 2018). These dyes exhibit low biodegradability, making them resistant to the conventional removal techniques. This causes hindrance to sunlight penetration and raises the chemical oxygen demand (COD), thus hampering the marine ecosystem (Zahmatkesh et al., 2023). The physicochemical dye removal techniques generally adapted in the industries are costly, less effective, require high energy, and generate secondary pollutants. However, the bioremediation approach, which utilises microorganisms as a whole or the extracellular enzymes, offers a viable alternative (Deb et al., 2023).

Among microorganisms, the robust and versatile fungal enzymes are known to efficiently degrade complex chemical compounds into simpler forms through the mycoremediation process (D'Souza-Ticlo et al., 2006; 2009). Fungi are known as "potential degraders", especially those derived from the marine environment, as they can adapt to extreme conditions like high salt concentrations, varying oxygen concentrations, and broader pH range as compared to terrestrial fungi (Manohar and Raghukumar, 2013; Khandavilli, 2016; Barathikannan et al., 2017). The extracellular oxidoreductive enzymes produced by fungal isolates from different marine habitats have an inherent ability to breakdown pollutants such as colourants, heavy metals, and oil pollutants including polycyclic aromatic hydrocarbons (PAH) (Raghukumar et al., 2008; Varjani et al., 2020; Chen et al., 2022). The ecological role played by the fungal communities and the mechanisms for the detoxification and breakdown of the various complex textile dyes are described in detail in this chapter. The marine fungal remediation approach involves enzymatic and non-enzymatic methods, which are explained in detail. We also discuss the various key factors that influence the decolourisation and detoxification process of dyes.

2. Classification of Textile Dyes

The primary constituents of the colouring agents used in the garment and related industries are chromophores, with various chemical groups and several functional

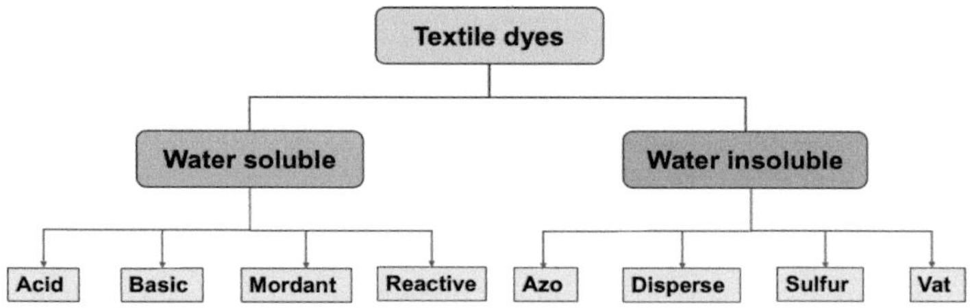

Fig. 1. Classification of colouring agents used as textile dyes based on its solubility in water.

groups including -C = O, O = (C_6H_4) = O, -N = N-, and $-NO_2$, that contributes for imparting colour to the textiles and other products manufactured. Chemical compounds classified as auxochromes are the second group of dyes, with different functional groups like -OH, -COOH, $-NH_3$, $-SO_3H$, and that increases the hydrophobicity of the fibre by either donating or removing electrons (Srinivasan and Viraraghavan, 2010). Dyes are classified into two classes, as water soluble and insoluble dyes which largely determines their ecotoxicological impact (Fig. 1). The former includes the acid, basic, reactive, direct, and mordant dyes and the latter include azo, disperse, sulfur, vat, and solvent dyes. Depending on their particle charge and ability to dissolve in aqueous solution, anionic, cationic, and non-ionic dyes are typically utilised extensively in the textile industry (Rauf and Salman, 2012).

2.1 *Water-Soluble Dyes*

2.1.1 *Acid Dyes*

The anionic dyes that undergo the dyeing process in an acidic pH range of 2–6 in an aqueous solution are referred to as acid dyes. Because they include sulphonic acid groups, they are the most soluble dye class in the colour index (Pisoni et al., 2013). They attach to the cationic NH_4^+ ions of fabrics, which contain nitrogen, such as wool, silk, polyamide, and modified acryl, and hence are often employed for colouring these materials. The majority of acid dyes, including azo, anthraquinone, triarylmethane, xanthenes, and diphenylamine, are categorised into this, based on their chromophoric groups (Wen et al., 2011).

2.1.2 *Basic Dyes*

Basic or cationic dyes are obtained from organic compounds and ionize in water, forming cations that are coloured. They are generally used to colour fibres with acid groups, which are typically synthetic fibres such as modified polyacryl. Because of the strong ionic contact between the negative charge of the co-polymer, the functional groups of these dyes, such as $-NR^{3+}$ or $=NR^{2+}$, are effective on acrylics. The most prevalent chemical classes include azine, diphenylmethane, oxazine, triarylmethane, and xanthenes (Clark, 2011).

2.1.3 Reactive Dyes

The reactive groups present in the anionic dyes bind covalently to various substrates such as cellulosic (cotton and rayon) and protein (silk and wool) fibres. Any reactive group that may react with fibre sites like OH-, NH_2^-, or SH- in wool can be included in a reactive dye (Zhang et al., 2016). For efficient dye transfer to the fibre , an electrolyte like NaCl or Na_2SO_4 is often added during the industrial process. This has lasting effect, maximum pigmentation, ease of handling across a broad temperature range, and excellent wash fastness on the fibre which are some of the well-known properties. The specific properties of these colours can be applied in the garment industry to a diverse array of fibres, including silk, most cellulosic materials, and to a lesser extent, wool. The most commonly known dyes such as aminochloro-s-triazine, aminofluoro-s-triazine, dichlorotriazin, dichloroquinoxaline, difluorochloropyrimide, trichloropyrimidine, and sulphatoethylsulphone belong to these reactive groups (Khatri et al., 2015).

2.1.4 Mordant Dyes

There are about 600 various mordant dyes mentioned in the colour index, of which 23% are utilised in the industry. These substances that can adhere to the fibre and subsequently mix with the dye to form an insoluble complex are hence known as mordants. The most common mordant dyes are Orange 10 and Solochrome Red B, which are widely used to colour modified cellulose fibres, paper, wool, leather, and silk (Shahmoradi et al., 2021).

2.2 Water-Insoluble Dyes

2.2.1 Azo Dyes

The most versatile, economical, and easy to use colourants, that have high stability and pronounced intensity of colour, hence accounting for up to 60–70% of all textile dyes are the azo groups. The functional groups of these dyes with -N = N-, provide the chemical bonding of two symmetrical or asymmetrical, non-azo aryl or alkyl radicals. They also have heterocyclic rings, which provide a bright and intense shade. Based on the number of azo groups in their structure, azo dyes could be grouped as mono, di, and poly. This range of dyes which include the Direct Red 2 and Congo Red, have an aromatic unsaturated heterocyclic compound and an unsaturated heterocycle, carboxyl, sulphonyl, or aliphatic group on either side (Liu et al., 2020).

2.2.2 Disperse Dyes

Water-insoluble dyes known as disperse dyes are associated with multiple hydrophobic properties of the synthetic fibres, such as acryl, cellulose acetate, polyamide, polyester, etc. Dye diffusion leads to swelling of the fibre, which can be achieved through high temperatures or with the help of chemical fabric softeners. These disperse dyes according to the colour index, form the third-largest group of dyes. This dye category includes Disperse Blue 56, Disperse Blue 183, Disperse Violet 1, and Disperse Yellow 13 among the myriads of dyes in this group (Jamil et al., 2020).

2.2.3 *Sulphur Dyes*

This class of dyes have excellent colour-imparting properties, are economical and have the highest ease for application. However, its complex structure with a disulfide (S-S) bridge limits its utilisation to a smaller proportion. The colouration process uses sodium sulphide as the reducing agent to convert the insoluble dye to a soluble form during the dyeing process, which is then oxidised back to its insoluble, coloured form within the fibre. Hence, they are referred to as vat dyes. The most widely used sulphur dyes are Sulphur Blue 15, Leuco Sulphur Black 1, and Sulphur Green 3 (Khattab et al., 2019).

2.2.4 *Vat Dyes*

The colourants from this group are known to dissolve well only in hot water and in the presence of chemical agents such as sodium carbonate. Because of their affinity and application on nanofibres, these dyes are mostly applied to cellulosic fibres. Widely used dyes in this category include indigo or indigotin, Vat Yellow 1, Vat Blue 1, and Vat Black 25. Vat dyes are widely recognised for their superior brightness properties and improved colour fastness (Sirianuntapiboon et al., 2006).

3. Ecotoxicological Impact of Textile Dyes

A vast majority of the textile dyes are made from petrochemical compounds and have properties like stability, a broad spectrum of colour pigments, and fast colouration. The presence of polluting colourants from garment and textile industries in water bodies causes pH fluctuations and increases the organic lead. This leads to deterioration of the major water quality due to changes in the biochemical oxygen demand (BOD), COD, and higher total suspended solids (Croce et al., 2017). The uncontrolled release of coloured contaminants causes an imbalance in the ocean that impacts aquatic fauna and flora and alters the quality of drinking water (Cardoso et al., 2012). The colour associated with the textile dyes blocks the light penetrating the water, leads to oxygen depletion below the required levels and impacts the rate of photosynthesis and largely deteriorates the aesthetics of water bodies (Hassan and Carr, 2018; Khatri et al., 2018). In addition, wastewater contaminated with dyes can clog the pores of the soils and lead to an increase in soil productivity when permitted to flow through agricultural fields (Khattab et al., 2019). Following prolonged exposure to reactive dyes, workers in the textile sector and associated industries may develop contact dermatitis, allergic reactions, rhinitis, or asthma. Studies have also reported cases of skin irritation due to contact and exposure to textile dyes. All the textile dyes at higher concentrations are known to cause various ecotoxicological impacts on humans, aquatic flora, and fauna (Table 1).

Table 1. Ecotoxicological impact of textile dyes.

Name of the Dye	Class of the Dye	Species	Toxicity	References
Acid Blue 92	Azo	*Hydrocotyle vulgari*	Reduction in growth and photosynthetic rates, disruption of the lipid peroxidation and antioxidant activity	Vafaei et al. (2013)
Azure B	Azo	Human	Causes mutations on the DNA helix and double-stranded RNA	Haq and Raj (2018)
Congo red	Azo	*Pseudokirchneriella subcapitata*	Exhibits sub-inhibitory actions that impact the rate of population growth	Hernández-Zamora and Martínez-Jerónimo (2019)
Direct Blue 86	Direct	Human	Reports on skin irritation and respiratory issues were observed	Shoaib et al. (2023)
Drimarene blue	Reactive	*Spirulina platensis*	Metabolic disturbance in algal growth due to changes in the photosynthetic efficiency	Gita et al. (2021)
Indigo dye	Vat	*Scenedesmus quadricauda*	Growth rate, biomass production, chlorophyll a content, and size of coenobium were negatively affected	Chia and Musa (2014)
Lanasyn brown	Acid	*Spirulina platensis*	Reduced growth rate, protein synthesis, and concentrations of C, N, H, and S decreased	Dutta et al. (2024); Gita et al. (2021)
Malachite green	Basic	*Chlorella pyrenoidosa* and *Daphnia magna*	Produces cellular and morphological alterations and generates genetic damage by modifying protein patterns and increasing ROS levels	Kanhere et al. (2014)
Methylene blue	Acid	*Spirulina platensis* and *Chlorella vulgaris*	Reduction in the rate of photosynthesis by inhibiting the growth rate, protein, and chlorophyll content was widely observed	Moorthy et al. (2021)
Metomega Chrome Orange GL	Acid mordant	*Nostoc muscorum*	A decrease in the protein, phycocyanin, chlorophyll a, and carotenoid content leading to a reduction in photosynthetic activity was reported	Shetty and Krishnakumar (2020)
Reactive 22	Azo	*Anabaena* species	Reduction in the biomass accumulation and growth of *Anabaena* sp.	Hu and Wu (2001)
Sudan I	Azo	Mammals Human	Impact on the urinary bladder, liver, and changes in DNA framework, micronuclei frequencies reported from human hepatoma HepG2 cells	An et al. (2007); Dalmaz and Özak (2023)
Sudan III	Azo	*Salmonella typhimurium*	Impacts metabolic activity by generating frameshift mutations	Zanoni et al. (2013)
Vat Blue 20	Vat	*Lemna gibba*	Causes detrimental effects such as necrosis, frond detachment, and decrease in photosynthetic efficiency	Hocini et al. (2019)

4. Remediation of Dyes in the Effluents from Textile Industries

Conventional wastewater treatment includes several methods for the treatment of textile effluents to degrade the coloured pollutants using either chemical, physical, or biological methods (Fig. 2). The chemical methods involve electrolysis, oxidation, ozonation, and ion exchange for the elimination of textile dyes from wastewater. Although these methods mineralise dyes completely, the main shortcomings are that they are highly susceptible to environmental factors such as pH, temperature, salts, etc., release toxic compounds, and consume high energy (Li et al., 2015). The physical treatment of textile dye-based effluents includes filtration, adsorption, osmosis, and flocculation. These methods are simple, fast, and have high efficiency; however, the major drawback is the formation of sludge, fouling, high energy consumption, and saturation issues (Zuorro and Lavecchia, 2014).

Intense research on the efficiency of microorganisms over the past few decades has reported the bioremediation potential of microbial communities of the bacterial, fungal and algal groups. But the effectiveness of these microbe-based degradation processes largely is influenced by their ability to adapt to the harsh conditions and on the activity of its extracellular enzymes for the degradation of dyes. Based on these inherent properties of microbes, the removal of pollutants and xenobiotics is carried out using biological systems. This environmentally friendly approach, for the neutralisation of the coloured pollutants through decolourisation and eventually detoxification is a challenge, as dyes are designed to be stable and function as long-lasting colourants. Hence, they are highly resistant and not easily degraded or broken down into simpler or less toxic compounds (Gupta et al., 2008; Zhao et al., 2020).

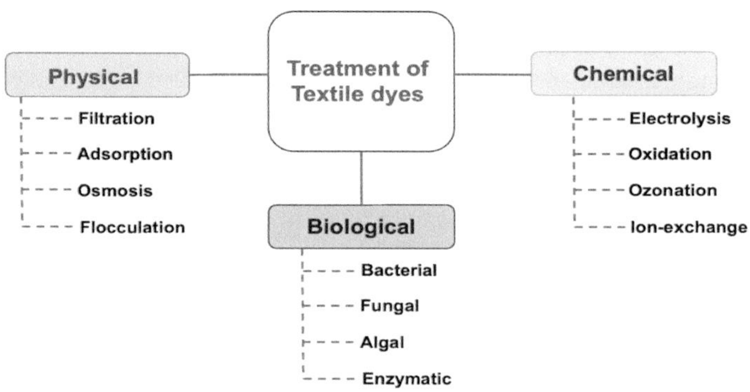

Fig. 2. Various treatment methods used for textile dye treatment.

4.1 *Fungi*

Fungal bioremediation is a well-studied field, due to the diverse extracellular enzymes produced by various isolates. The major fungal group of Basidiomycetes, largely comprising the white rot fungi, are particularly effective because they possess the inherent ability to break down complex lignin containing plant material. These ligninolytic fungi have been studied the most for their dye degradation potential

(D'Souza-Ticlo et al., 2006, 2009; Barathikannan et al., 2017). White rot fungi are the most efficient in extensive lignin degradation. To degrade this complex molecule, fungi possess an unspecific enzyme system, such as oxidoreductases, referred to as lignin-modifying enzymes (LMEs) that can act on a wide range of pollutants (Vaksmaa et al., 2023).

4.2 *Yeasts*

The biosorption of textile dyes by yeast strains is exhibited through the major functional groups, such as protein and nitrogen-containing ligands present on the cell wall. They are equipped with specialised molecules like peptidomannan, peptidoglycan, and other active groups like amino acids, acidic polysaccharides, and lipids (Danouche et al., 2021). The yeast cells act as absorbents through two mechanisms, which are chemisorption, through the formation of a strong bond between dye and yeast cells. Physio-sorption, through the weak interactions like Van der Waals, hydrogen bonds, polarity, and dipole-dipole interactions are also reported in the marine microbial community. Among many yeast strains studied, *Saccharomyces cerevisiae* is commonly studied for biosorption of synthetic dyes, heavy metals, and other aromatics (Dilarri et al., 2016).

4.3 *Microalgae and Bacteria*

The bioremediation potential of microalgae for the degradation of azo dyes is observed in many species like *Chlorella*, *Oscillatoria*, and *Spirogyra* sp. (Premaratne et al., 2021). These microalgal cells are known to adsorb dyes through their porous cell wall. The functional groups like carboxyl, hydroxyl, amino, and phosphate are shown to bind with pollutants like heavy metals and dyes efficiently. This facilitates potential adsorption for the enzymatic degradation and breakdown of complex dyes (Pathak et al., 2015).

The bioremediation potential of bacteria to metabolise and decolourise azo dyes is attributed to the cleavage reduction of the azo bonds under low oxygen, anaerobic, or microaerophilic conditions, which leads to the build-up of toxic aromatic amines. Though many report on the potential of bacteria using extracellular enzymes to break down azo dyes under aerobic conditions, they are not readily metabolised (Jamee and Siddique, 2019). However, studies have have demonstrated several taxa, such as *Aeromonas hydrophila, Bacillus cereus, B. subtilis, Proteus mirabilis, Pseudomonas luteola*, and *Pseudomonas* sp., have the potential for the biodegradation of azo dyes under anoxic conditions (Bera et al., 2022).

5. Advantages of using Marine Fungi

While the ecological functions of fungi in terrestrial ecosystems are well-documented and extensively studied, understanding their roles within marine environments has posed greater challenges for researchers. In coastal marine environments, mangrove plants and seagrasses are the primary source of lignocellulose substrate, with 50% of their structural polymers being complex polysaccharides. Epibiotic bacteria and fungi colonise mangrove leaves, wood, and seagrasses that fall into the intertidal zones. The marine ecosystems largely include facultative marine fungi that are adapted to the marine environment (Raghukumar et al., 2008). Marine fungi have the inherent capacity to break down cell walls as they thrive on woody substrates by secreting cell wall-degrading enzymes. This leads to the remineralisation of the plant biomass into dissolved organic carbon, which is utilised by bacteria to generate biomass and particulate organic carbon in the coastal and aquatic systems, which is a common food source for detritus feeders such as larvae, crabs, and shrimp (Kawaida et al., 2021; Ghose et al., 2025; Nishamol et al., 2025; Oliveira et al., 2025).

Marine-derived fungi are least explored for treating industrial effluents as a large majority of the studies have focused largely on bacterial and microalgal communities. Halo-tolerant marine fungi in the environment have osmo-regulatory systems that signal the synthesis of amino acids and polyols while simultaneously raising the cytoplasmic ion concentration. Marine-derived fungi thrive and produce degradative enzymes in seawater, making them potentially beneficial in the treatment of high-salt wastewaters (Balabanova et al., 2018; Bonugli-Santos et al., 2015). Fungi from marine habitats can grow in saline and at alkaline conditions (10–34 ppt salinity, pH 7.5–8.2) and produce enzymes that help break down pollutants from textile, distillery, and paper industries. Their ability to thrive in salty, alkaline conditions makes them ideal for treating these effluents (Luo et al., 2005; Raghukumar, 2000). For instance, four marine-derived isolates from coastal waters of Goa—*Aspergillus terreus*, *Alternaria* sp., *Cladosporium cladosporioides*, and *Penicillium chrysogenum*—exhibited 50–80% decolorisation of textile effluent that contained a complex mixture of dyes, mainly azo and anthraquinone dyes, in a 21-day incubation period (Verma et al., 2010). Similarly, marine fungi isolate NIOCC #2a, isolated from decaying mangrove wood, showed enhanced production of laccase and decolorisation of effluent under high salinity conditions (D'Souza-Ticlo et al., 2006). A study identified fungi belonging to *Alternaria* sp., *Cadophora* sp. and *Phoma* sp. from different marine habitats showed high laccase activity with decolorisation potential up to 98% for methyl orange and 93% for acid green (Toker et al., 2021).

6. Mechanism of Textile-Dye Degradation

Degradation of textile dyes by marine fungi occurs through both enzymatic and non-enzymatic methods. The enzymatic approach includes a mixture of extracellular, nonspecific, and nonselective enzymes to carry out biodegradation (Sen et al., 2016). Though, Ascomyecetous fungi are ubiquitous, Basidiomycetes especially the white rot fungi belonging to this group are the most extensively studied marine fungi. They are capable of producing ligninolytic enzymes, which decompose the complex chemical structures of textile dyes into less harmful compounds that can be easily assimilated by the environment. These enzymes are the major peroxidases such as lignin peroxidase (LiP), manganese peroxidase (MnP), versatile peroxidase (VPs), and laccase (Lac) (Robinson et al., 2001).

6.1 Enzymatic Mechanisms

Lignin-degrading enzymes (LDE) are oxidoreductive enzymes that partially degrade or oxidise altered polymeric compounds taken up by microbial cells before completely mineralising them. LDEs are divided into two classes: heme-containing peroxidases and copper-containing laccases. LDEs aid in the oxidation of aromatic compounds until they are cleaved, after which they are further degraded by other enzymes (Singh et al., 2015). For instance, LiP catalyses the oxidation of non-phenolic aromatic compounds such as veratryl alcohol (VA), whereas MnP converts Mn^{2+} to Mn^{3+}, which can then oxidise a wide range of phenolic compounds. Laccase, on the other hand, can directly oxidise phenolic substrates with O_2 as an electron acceptor (Janusz et al., 2013).

6.1.1 Laccase

One class of lignin-modifying enzymes that can oxidize a variety of compounds is laccases (benzenediol: oxygen oxidoreductase EC 1.10.3.2). Phenolic substances, including ortho- and para-diphenols, aminophenols, polyphenols, polyamines, and aromatic diamines, are substrates of laccase. The nature and position of the substituents on the phenolic ring influence the efficiency of oxidation by a specific laccase, where the substrate is oxidized with simultaneous radical formation, which can then spontaneously rearrange to cleave the aromatic rings or promote their polymerisation (Jones and Solomon, 2015). Laccases are copper-containing multimeric glycoproteins that give these enzymes their blue colour. Laccase contains copper, which oxidises the substrate and employs molecular oxygen to change various aromatic and nonaromatic molecules. The resulting radicals then contribute to the further degradation of the dyes. Four copper atoms are present in each monomer in the catalytic site of fungal laccases, which can be either monomeric, dimeric, or tetrameric glycoproteins. Laccases are multi-copper oxidases that are categorised according to their copper centre into varieties with low and high redox potentials. They use molecular oxygen as an electron acceptor rather than heme or hydrogen peroxide, as peroxidases do. A wide variety of substrates, including phenols, polyphenols, arylamines, and even certain inorganic ions, can be oxidised by laccases. They are perfect for industrial and bioremediation applications because of their excellent stability, particularly in extracellular forms due to glycosylation, and their catalytic applications (Janusz et al., 2020). Phenolic substrates can be converted by these enzymes into aryloxy radicals, which then produce quinones and undergo further polymerisation to produce colourful oligomers. Additionally, they are capable of decarboxylation, demethylation, and dehalogenation. Laccases use a non-specific radical process to break down azo dyes without producing harmful amines. Several laccase isozymes are produced by some species, and their expression patterns vary according to their growing environment (Sun et al., 2013).

6.1.2 Lignin Peroxidases

First discovered by Tien and Kirk, the extracellular oxygenase, LiP (EC 1.11.1.14) was reported from the white-rot fungus *Phanerochaete chrysosporium*. Later, it was discovered that other white-rot fungal taxa also generate lignin peroxidase isoenzymes in the extracellular environment (Tien and Kirk, 1988). The oxidoreductase family includes the extracellular, monomeric heme-containing enzymes known as lignin peroxidases (LiPs). They perform oxidative functions by producing radical cations through oxidation, which is followed by ligand modifications such as side-chain cleavage, demethylation, followed by intramolecular additions, and molecular rearrangements (Kumar and Chandra, 2020). Lignin peroxidase (LiP) may oxidise both phenolic and non-phenolic substrates, including azo dyes, since it has a larger redox potential than laccase (Lac) and manganese peroxidase (MnP). Research has indicated that *Phanerochaete chrysosporium*-produced LiPs are essential for decolorising a variety of azo dyes. The breakdown of structurally varied azo dyes is greatly aided by the production of LiPs by fungi other than *P. chrysosporium*, including *Phanerochaete ostreatus*, *Polyporus species*, and *Trametes species* (Liu et al., 2004).

The molecular weight of LiP, an extracellular N-glycosylated heme protein, ranges from 38 to 47 kDa. Its catalytic mechanism is similar to that of classical peroxidase, and its active site comprises heme (Tien and Kirk, 1988). A broad range of substrates, such as phenolic chemicals, PAHs, and other pollutants, are oxidised by LiP in a hydrogen peroxide (H_2O_2)-dependent manner. This is accomplished through a one-electron oxidation process, which initiates non-enzymatic secondary reactions that result in a wide range of degradation products. The oxidised enzyme is subsequently reduced back to its original condition by converting veratryl alcohol into veratryl aldehyde and subsequently reduced back to veratryl alcohol. These processes are carried out by accepting an electron from lignin or structurally comparable xenobiotic chemicals, resulting in the oxidation of lignin or pollutants (Chan et al., 2020).

6.1.3 Manganese Peroxidase

Manganese peroxidases (MnPs) (EC 1.11.1.13) are the heme-containing glycosylated oxidoreductases that are mostly produced by basidiomycetes from the orders *Agaricales, Corticiales, Polyporales*, and *Hymenochaetales*. Enzymes oxidize Mn^{2+} to Mn^{3+}, which breaks down phenolic structures in lignin and other aromatic substances. Chelating compounds, such as organic acids, are necessary for the production and stability of Mn^{3+}. These molecules improve MnP catalytic efficiency and facilitate the release of Mn^{3+} from the enzyme's active site. Numerous white-rot fungi have been shown to secrete MnPs extracellularly. These enzymes have molecular weights ranging from 38 to 62.5 kDa and around 43% sequence similarity with lignin peroxidases (Zhang et al., 2021). MnPs are more adaptable than many other peroxidases because they function well in the 4.0–7.0 broad pH range, with maximum activity at pH 5.5, and they are thermally stable between 50–70°C. Zhang et al. (2020) discovered a very stable MnP from *Cerrena unicolor* BBP6 that maintained activity at different metal ion concentrations, functioning best at pH 4.5 and 60°C (Zhang et al., 2020). In comparison to horseradish peroxidase and MnPs from *Bjerkandera adusta* and *Phanerochaete chrysosporium*, another MnP, TP55, isolated from *Trametes pubescens* strain i8, revealed remarkable catalytic effectiveness for textile dye degradation and showed superior tolerance to solvents and detergents (Thampraphaphon et al., 2022).

6.1.4 Versatile Peroxidase

The enzyme class II fungal peroxidases are heme-containing glycoproteins categorized as versatile peroxidases (EC 1.11.1.16) (VPs), initially reported in *Pleurotus* species in 1999 (Sridhar, 2016). These enzymes have a wide variety of substrates and can oxidise substances with low, medium, and high redox potentials using both independent and Mn^{2+}-dependent processes. By oxidising Mn^{2+} like MnPs and veratryl alcohol via a catalytic tryptophan residue like LiPs, the VPs possess the properties of both LiPs and MnPs for catalysis as reported by Moreira et al. (2007). Azo dyes are among the phenolic and non-phenolic aromatic pollutants that can break down because of their high redox potential. VPs accomplish this by connecting the substrate to the heme center via three long-range electron transfer channels, which are particularly active at acidic pH (Sáez-Jiménez et al., 2015). VPs from fungi such as *I. lacteus* and *P. ostreatus* are reported to be efficiently applied in dye decolourisation, demonstrating their unique ability to oxidise resistant chemicals that other peroxidases cannot (Fernández-Fueyo et al., 2014).

6.2 Non-Enzymatic Mechanisms

The physicochemical process of biosorption involves several mechanisms, including surface complexation, ion exchange, adsorption, absorption, and precipitation. It describes how a sorbate, such as an atom, molecule, or ion, interacts with a biosorbent, usually a biological substance, causing the sorbate to accumulate on the surface of the biosorbent. The concentration of the sorbate in the solution is lowered by this buildup. By attaching to functional groups on either living or non-living biomass, the degradation and biotransformation of organic or inorganic ions including dyes and heavy metals from a variety of material from aqueous systems are established through biosorption. Carboxyl, amino, phosphate, hydroxyl, sulphydryl, imidazole, phenol, thioether, carbonyl, sulfate, and amide moieties are examples of these functional groups.

Biosorption has been used to remove dyes using both living and inactive microbial organisms. Using biomass from WRF and other microbial cultures, decolourisation of wastewater containing dyes has been the subject of numerous investigations. Studies have also reported the application of Zygomycetes and Ascomycetes in mycelial form for decolourisation and degradation of various textile dyes using enzymatic and non-enzymatic pathways (Table 2).

Table 2. Fungal species reported for decolourisation of synthetic textile dyes.

Fungal Strain	Source	Dye	Decolorisation Reported (%)	Mechanism of Action	References
Enzymatic Method					
Achaetomium strumarium	Dye-contaminated soil	Acid Red 88	99	Laccase	Bankole et al. (2018)
Alternaria (D11 and D21), *Cadophora* sp. D43 *Cadophora luteo-olivacea* D51	Çakalburnu Lagoon in Izmir Bay, Aegean Sea, Turkey	Methyl Orange, Acid Green 3, Methylene Blue	98 93 78	Laccase	Toker et al. (2021)
Daedalea dickinsii	Brown-rot fungi	Methylene blue	54	Ligninolytic enzymes	Rizqi and Purnomo (2017)
Flavodon flavus isolate 312	Marine environments	Azure B, Brilliant Green, Congo Red, Crystal Violet, Poly-R	90	LiP, MnP, Lac	Raghukumar (2000)
Ganoderma sp. # En3 (white-rot fungus)	Tzu-chin Mountain Forest, China	Remazol Brilliant Blue R (RBBR), Indigo Carmine, Methyl Green, and Reactive Orange 16		Laccase gene Lac-En3-1	Zhuo et al. (2011)
Penicillium citrinum CBMAI 853	Marine filamentous, North coast of São Paulo State, Brazil	Remazol Brilliant Blue R	100	Extracellular oxidative enzymes	Da Silva et al. (2008)
Pestalotiopsis sp.	Marine environment	Azure Blue, Reactive Black 5, Acid Yellow 17, Nitrosulfonazo III	100	Laccase genes PsLac1 and PsLac2	Wikee et al. (2019)
Peyronellaea prosopidis	Environmental source	Scarlet RR	90	Laccase and peroxidases	Kapoor et al. (2021)
Rhodonia vinctus TBRC 6770	Wood-decaying fungi	Consortia of textile dyes	60	Lacasse and MnP	Laothanachareon et al. (2023)
Non-enzymatic Methods					
Aspergillus niger	Deposit of the André Tosello Foundation for Research & Technology Culture collection	Procion Red MX-5B	30	Biosorption	Almeida and Corso (2014)
Myrothecium verrucaria, Colletotrichum dematium, Corynespora cassiicola, Dictyosporium zhejiangensis	Freshwater streams in Zhejiang Province, China	Acid Blue 40, Acid Blue 62, Reactive Red 11, Acid Blue 193	Varying capacities for adsorption and degradation	Biosorption and degradation	Yang et al. (2016)

7. Factors Affecting Bioremediation Efficiency

Several physical and chemical factors significantly influence the ability to degrade textile dyes; some of the major factors are pH, salinity, growth media composition, and redox modifiers.

7.1 *Effect of pH*

The efficiency of textile dye degradation is impacted by changes in pH. Kunjadia et al. (2016) studied the degradation of three azo dyes, Coralene Dark Red (CDR), Coralene Navy Blue (CNB), and Coralene Golden Yellow (CGY), by *Pleurotus* spp. which showed up to 88% degradation efficiency at pH 6–6.5 (Kunjadia et al., 2016). The variation in pH also is shown to affect the enzymatic activity, dye solubility, and ionisation state of the dye molecule and the fungal biomass. Study using the fungal strains, *Aspergillus fumigatus* (SN8c) and *Aspergillus terreus* (SN40b), was carried out to understand the decolourisation of Remazol Brilliant Blue (RBB). The decolourisation efficiencies of *A. fumigatus* and *A. terreus* were 99% and 93%, respectively, at acidic pH values (3–5). Nevertheless, these values decreased to 71% and 73% at an alkaline pH of 11. The decrease in efficiency in alkaline circumstances was ascribed to the inhibition of enzyme activity, which had an adverse effect on the interaction between dye molecules and fungal biomass (Nabeela et al., 2023).

7.2 *Effect of Salinity*

The adaptation of the ligninolytic enzymes to marine environments, particularly in the presence of high salt concentrations, is poorly understood. According to Venkatachalam et al. (2019), the fungus *Talaromyces albobiverticillius*, which is obtained from marine sources, demonstrated an increase in biomass yield to 9.22 ± 0.09 g/L at 9% salt content (Venkatachalam et al., 2019). The impact of hypersaline conditions on MnP expression and activity in marine-derived fungi, *Phlebia* sp., a white rot fungus, was investigated to understand the molecular mechanisms. They discovered that three MnP-encoding genes, MGmnp1, MGmnp2, and MGmnp3, were upregulated with increased salinity, indicating a specific response to hypersaline stress (Kamei et al., 2008). Likewise, the expression of laccase genes, including PsLac1 and PsLac2, was investigated in the presence of salt concentrations ranging from 0% to 5%. PsLac2 gene expression significantly increased up to 360% at 5% sea salt concentration in comparison to PsLac1 gene expression (Wikee et al., 2019).

7.3 *Effect of the Growth Media and its Nutrient Composition*

Growth and enzyme activity of the fungal isolates is predominantly controlled by the components of the growth media which is one of the most important factors in the degradation process. Carbon sources such as sucrose, fructose, maltose, xylose, glycerol, starch, and xylan have been identified as having an impact on the growth of marine fungi, particularly white rot fungi (Pavko, 2011). The addition of glycerol increased the decolorisation efficiency of Reactive Black 5 (RB5) utilising *Trametes versicolor* to about 100%. Furthermore, glycerol significantly increased laccase activity compared to sucrose, implying that glycerol is more effective at stimulating the enzymes involved in dye degradation (Ottoni et al., 2014). Ben et al. (2020) investigated the influence of carbon sources like sucrose, glucose, and starch on laccase enzyme activity. In comparison to glucose and starch, sucrose produced 290 U L^{-1} laccase-like activity (Ben et al., 2020).

7.4 *Effect of Redox Mediators*

White-rot fungi that degrade lignin exclusively using laccase enzymes resulted in the discovery of enzyme mediators, which are low molecular weight molecules. These laccase-mediator systems work by oxidising tiny molecules into stable radicals, which subsequently act as redox mediators, allowing laccase to oxidise a broader spectrum of chemicals (Ghattavi and Homaei, 2023). The influence of redox mediators such as 1-hydroxybenzotriazole, 1-naphthol, 2-naphthol, 4-nitrophenol, ABTS, vanillin, vanillic acid and veratyl alcohol, are essential for the activity of laccase enzyme for the decolorisation of synthetic dyes. The impact of the mediators such as ABTS and 1-hydroxybenzotriazole studied on the decolorisation of crystal violet dye showed a pronounced influence (Shankar and Nill, 2015). They acted as electron shuttles, facilitating the oxidation of dye molecules that are otherwise recalcitrant to direct enzymatic action (Shankar and Nill, 2015). Another study indicated that the addition of redox mediators enhanced the decolorisation of Congo red and Poly-R. These redox mediators directly enhanced the laccase activity of the white rot fungus *Trametes hirsuta* (Moilanen et al., 2010). Similarly, ABTS decolorised Rhodamine B by over 80% in 48 hours, compared to only 20% without the mediator (Kumar et al., 2024).

8. Challenges and Future Perspectives

Over the past two decades, bioremediation employing fungi derived from marine sources has drawn a lot of interest as a potential strategy for mitigation of detrimental effects of dye effluents. The majority of research is carried out at the batch or small lab scale. Moreover, research has mostly concentrated on treating wastewater contaminated by a single dye using a single fungal species; however, actual wastewater or industrial effluents contain a variety of dyes that make it challenging for these microorganisms to handle. As a result, it is critical to expand research by selecting effective fungal species or consortia of fungi and their enzymes that exhibit synergistic effects and can assist in dye degradation in a real wastewater system. Furthermore, the screening for potential indigenous strains like marine-derived fungi from different locations can be utilised to completely mineralise the dyes from the system. This can assist in resolving the scale-up issue and facilitating the transfer of laboratory trials to an industrial setting for actual wastewater treatment.

In addition, there is a shortage of information on marine-derived fungi that are employed in bioremediation applications, and further study is needed to specify the mechanism by which different dye classes are degraded. Therefore, omics studies, including genomics, transcriptomics, proteomics, and metabolomics, should be employed to elucidate the metabolic pathways involved in degradation studies. Genetic engineering can also be used to modify the fungal strains, which can exhibit relatively superior cleanup abilities under extreme and unfavourable conditions. Lastly, synergistic approaches such as combining physical-chemical co-remediation techniques such as chemical oxidation and reduction to improve the efficacy of dyestuff treatment. The enzyme production can be increased by nanoparticles, or the enzymes can be transported by the nanoparticles, acting as a catalyst to accelerate dye degradation.

9. Conclusion

Marine-derived fungi are a promising candidate for the bioremediation of synthetic textile dyes. They have unique traits, adapt to extreme environments, and possess enzymatic activity, which puts them at an advantage over the terrestrial fungi. Marine fungi can interact and degrade recalcitrant dyes by enzymatic and non-enzymatic methods. Enzymatic methods involve extracellular enzymes, majorly peroxidases such as laccase, manganese peroxidase, lignin peroxidase, and versatile peroxidase, whereas non-enzymatic methods include biosorption, precipitation, ion exchange, and adsorption. The bioremediation and degradation process of coloured effluents are shown to be influenced by key factors such as pH, temperature, salinity, and various redox mediators, which help in enhancing the enzyme activity.

Acknowledgements

The first author is thankful to the Director, Council of Scientific and Industrial Research (CSIR), National Institute of Oceanography, Goa for the permissions and support during her Masters Dissertation work. This is CSIR-NIO contribution number 12546.

References

AbuQamar, S.F., Abd El-Fattah, H.I., Nader, M.M., Zaghloul, R.A., Abd El-Mageed, T.A. et al. (2023). Exploiting fungi in bioremediation for cleaning-up emerging pollutants in aquatic ecosystems. Mar. Environ. Res., 190: 106068. https://doi.org/10.1016/j.marenvres.2023.106068.

Almeida, E.J.R. and Corso, C.R. (2014). Comparative study of toxicity of azo dye procion red MX-5B following biosorption and biodegradation treatments with the fungi *Aspergillus niger* and *Aspergillus terreus*., 112: 317–322. https://doi.org/10.1016/j.chemosphere.2014.04.060.

An, Y., Jiang, L., Cao, J., Geng, C. and Zhong, L. (2007). Sudan I induces genotoxic effects and oxidative DNA damage in hepg2 cells. Mut. Res., 627: 164–170. https://doi.org/10.1016/j.mrgentox.2006.11.004.

Balabanova, L., Slepchenko, L., Son, O. and Tekutyeva, L. (2018). Biotechnology potential of marine fungi degrading plant and algae polymeric substrates. Front. Microbiol., 9: 1527. https://doi.org/10.3389/fmicb.2018.01527.

Bankole, P.O., Adekunle, A.A. and Govindwar, S.P. (2018). Enhanced decolorization and biodegradation of acid red 88 dye by newly isolated fungus, *Achaetomium strumarium*. J. Environ. Chem. Eng., 6: 1589–1600. https://doi.org/10.1016/j.jece.2018.01.069.

Barathikannan, K., Ramaswamy, K.P., Manohar, C.S. and Meena, R.M. (2017). Diversity of fungi and decoularisation potential of a fungus, F-38 isolated from the coral reef regions of Kavaratti. Indian J Geo-Mar, Sci., 46(03): 497–503.

Ben Ali, W., Chaduli, D., Navarro, D., Lechat, C., Turbé-Doan, A. et al. (2020). Screening of five marine-derived fungal strains for their potential to produce oxidases with laccase activities suitable for biotechnological applications. BMC Biotechnol., 20(1): 27. https://doi.org/10.1186/s12896-020-00617-y.

Bera, S.P., Shah, M.P. and Godhaniya, M. (2022). Microbial remediation of textile dye acid orange by a novel Bacterial Consortium SPB92. Front. Environ. Sci., 10: 930616. https://doi.org/10.3389/fenvs.2022.930616.

Bonugli-Santos, R.C., dos Santos Vasconcelos, M.R., Passarini, M.R., Vieira, G.A., Lopes, V.C. et al. (2015). Marine-derived fungi: Diversity of enzymes and biotechnological applications. Front. Microbiol., 6: 269. https://doi.org/10.3389/fmicb.2015.00269.

Cardoso, N.F., Lima, E.C., Royer, B., Bach, M.V., Dotto, G.L. et al. (2012). Comparison of *Spirulina platensis* microalgae and commercial activated carbon as adsorbents for the removal of reactive red 120 dye from aqueous effluents. J. Hazard. Mater., 241-242: 146–153. https://doi.org/10.1016/j.jhazmat.2012.09.026.

Chan, J.C., Paice, M. and Zhang, X. (2020). Enzymatic oxidation of lignin: Challenges and barriers toward practical applications. ChemCatChem., 12: 401–425. https://doi.org/10.1002/cctc.201901480.

Chen, G., Bai, R., Zhang, Y., Zhao, B. and Xiao, Y. (2022). Application of metagenomics to biological wastewater treatment. Sci. Total. Environ., 807: 150737. https://doi.org/10.1016/j.scitotenv.2021.150737.

Chia, M.A. and Musa, R.I. (2014). Effect of indigo dye effluent on the growth, biomass production, and phenotypic plasticity of *Scenedesmus quadricauda* (chlorococcales). An. Acad. Bras. Ciênc., 86: 419–428. https://doi.org/10.1590/0001-3765201420130225.

Clark, M. (2011). Handbook of Textile and Industrial Dyeing. https://doi.org/10.1533/9780857093974.

Croce, R., Cinà, F., Lombardo, A., Crispeyn, G., Cappelli, C.I. et al. (2017). Aquatic toxicity of several textile dye formulations: Acute and chronic assays with *Daphnia magna* and *Raphidocelis subcapitata*. Ecotoxicol. Environ. Saf., 144: 79–87. https://doi.org/10.1016/j.ecoenv.2017.05.046.

D'Souza-Ticlo, D., Sharma, D. and Raghukumar, C. (2009). A thermostable metal-tolerant laccase with bioremediation potential from a marine-derived fungus. Mar. Biotechnol., 11(6): 725–737. https://doi.org/10.1007/s10126-009-9187-0.

D'Souza-Ticlo, D., Tiwari, R., Sah, A.K. and Raghukumar, C. (2006). Enhanced production of laccase by a marine fungus during treatment of colored effluents and synthetic dyes. Enzyme Microb. Technol., 38(3-4): 504–511. https://doi.org/10.1016/j.enzmictec.2005.07.005.

Da Silva, M., Passarini, M.R., Bonugli, R.C. and Sette, L.D. (2008). Cnidarian-derived filamentous fungi from Brazil: Isolation, characterisation, and RBBR decolourisation screening. Environ. Technol., 29: 1331–1339. https://doi.org/10.1080/09593330802379466.

Dalmaz, A. and Özak, S.S. (2023). Environmentally-friendly supramolecular solvent microextraction method for rapid identification of Sudan I–IV from food and beverages. Food Chem., 414: 135713. https://doi.org/10.1016/j.foodchem.2023.135713.

Danouche, M., El Arroussi, H., Bahafid, W. and El Ghachtouli, N. (2021). An overview of the biosorption mechanism for the bioremediation of synthetic dyes using yeast cells. Environ. Technol. Rev., 10: 58–76. https://doi.org/10.1080/21622515.2020.1869839.

Deb, M., Redkar, N., Manohar, C.S., Jagtap, A.S., Saxena, S. et al. (2023). *Bacillus* sp. based on nano-bio hybrids for efficient water remediation. Environ. Pollut., 326: 121490. https://doi.org/10.1016/j.envpol.2023.121490.

Dilarri, G., de Almeida, É.J., Pecora, H.B. and Corso, C.R. (2016). Removal of dye toxicity from an aqueous solution using an industrial strain of *Saccharomyces cerevisiae* (Meyen). Water Air Soil Pollut., 227(8): 269. https://doi.org/10.1007/s11270-016-2973-1.

Dutta, S., Adhikary, S., Bhattacharya, S., Roy, D., Chatterjee, S. et al. (2024). Contamination of textile dyes in aquatic environment: Adverse impacts on aquatic ecosystem and human health, and its management using bioremediation. J. Environ. Manag., 353: 120103. https://doi.org/10.1016/j.jenvman.2024.120103.

Fernández-Fueyo, E., Acebes, S., Ruiz-Duenas, F.J., Martínez, M.J., Romero, A. et al. (2014). Structural implications of the C-terminal tail in the catalytic and stability properties of manganese peroxidases from ligninolytic fungi. Acta Crystallogr. D Biol. Crystallogr., 70: 3253–3265. https://doi.org/10.1107/s1399004714022755.

Ghattavi, S. and Homaei, A. (2023). Marine enzymes: Classification and application in various industries. Int. J. Biol. Macromol., 230: 123136. https://doi.org/10.1016/j.ijbiomac.2023.123136.

Ghose, M., Toraskar, A., Nishamol, M., Parab, A.S., Ramteke, D. et al. (2025). Bacterial communities and their functional profiles from mangroves, a blue carbon ecosystem and terrestrial forest habitats. Biologia, 80: 1023–1035. https://doi.org/10.1007/s11756-025-01884-w.

Gita, S., Shukla, S.P., Deshmukhe, G., Choudhury, T.G., Saharan, N. et al. (2021). Toxicity evaluation of six textile dyes on growth, metabolism, and elemental composition (C, H, N, S) of Microalgae *Spirulina platensis*: The environmental consequences. Bull. Environ. Contam. Toxicol., 106: 302–309. https://doi.org/10.1007/s00128-020-03074-7.

Gupta, V.K., Mittal, A., Gajbe, V. and Mittal, J. (2008). Adsorption of basic fuchsin using waste materials—bottom ash and deoiled soya—as adsorbents. J. Colloid. Interface Sci., 319: 30–39. https://doi.org/10.1016/j.jcis.2007.09.091.

Hameed, B.B. and Ismail, Z.Z. (2018). Decolorization, biodegradation, and detoxification of reactive red azo dye using non-adapted immobilized mixed cells. Biochem. Eng. J., 137: 71–77. https://doi.org/10.1016/j.bej.2018.05.018.

Haq, I. and Raj, A. (2018). Biodegradation of azure-B dye by *Serratia liquefaciens* and its validation by phytotoxicity, genotoxicity, and cytotoxicity studies. Chemosphere, 196: 58–68. https://doi.org/10.1016/j.chemosphere.2017.12.153.

Hassan, M.M. and Carr, C.M. (2018). A critical review on recent advancements of the removal of reactive dyes from Dyehouse effluent by ion-exchange adsorbents. Chemosphere, 209: 201–219. https://doi.org/10.1016/j.chemosphere.2018.06.043.

Hernández-Zamora, M. and Martínez-Jerónimo, F. (2019). Congo red dye diversely affects organisms of different trophic levels: A comparative study with microalgae, Cladocerans, and zebrafish embryos. Environ. Sci. Pollut. Res. Int., 26: 11743–11755. https://doi.org/10.1007/s11356-019-04589-1.

Hocini, I., Benabbas, K., Khellaf, N., Djelal, H. and Amrane, A. (2019). Can duckweed be used for the biomonitoring of textile effluents? Euro-Mediterr. J. Environ. Integr., 4(34). https://doi.org/10.1007/s41207-019-0126-9.

Hu, T.L. and Wu, S.C. (2001). Assessment of the effect of azo dye RP2B on the growth of a nitrogen fixing cyanobacterium—*Anabaena* sp. Bioresour. Techno., 77: 93–95. https://doi.org/10.1016/s0960-8524(00)00124-3.

Jamee, R. and Siddique, R. (2019). Biodegradation of synthetic dyes of textile effluent by microorganisms: An environmentally and economically sustainable approach. Eur. J. Microbiol. Immunol., 9: 114–118. https://doi.org/10.1556/1886.2019.00018.

Jamil, A., Bokhari, T.H., Javed, T., Mustafa, R. and Sajid, M. (2020). Photocatalytic degradation of disperse dye violet-26 using Tio_2 and Zno nanomaterials and process variable optimization. J. Mater. Res. Technol., 9: 1119–1128. https://doi.org/10.1016/j.jmrt.2019.11.035.

Janusz, G., Kucharzyk, K.H., Pawlik, A., Staszczak, M. and Paszczynski, A.J. (2013). Fungal laccase, manganese peroxidase and lignin peroxidase: Gene expression and regulation. Enzyme Microb. Technol., 52(1): 1–12. https://doi.org/10.1016/j.enzmictec.2012.10.003.

Janusz, G., Pawlik, A., Świderska-Burek, U., Polak, J., Sulej, J. et al. (2020). Laccase properties, physiological functions, and evolution. Int. J. Mol. Sci., 21: 966. https://doi.org/10.3390/ijms21030966.

Jones, S.M. and Solomon, E.I. (2015). Electron transfer and reaction mechanism of laccases. Cell Mol. Life Sci., 72: 869–883. https://doi.org/10.1007/s00018-014-1826-6.

Kamei, I., Daikoku, C., Tsutsumi, Y. and Kondo, R. (2008). Saline-dependent regulation of manganese peroxidase genes in the hypersaline-tolerant white rot fungus *Phlebia* sp. strain MG-60. Appl. Environ. Microbiol., 74: 2709–2716. https://doi.org/10.1128/aem.02257-07.

Kanhere, J., Gopinathan, R. and Banerjee, J. (2014). Cytotoxicity and genotoxicity of malachite green on non-target aquatic organisms: *Chlorella pyrenoidosa* and *Daphnia magna*. Water Air Soil Pollut., 225: 2123. https://doi.org/10.1007/s11270-014-2134-3.

Kapoor, R.T., Danish, M., Singh, R.S., Rafatullah, M. and HPS, A.K. (2021). Exploiting microbial biomass in treating azo dyes contaminated wastewater: Mechanism of degradation and factors affecting microbial efficiency. J. Water Process Eng., 43: 102255. https://doi.org/10.1016/j.jwpe.2021.102255.

Kawaida, S., Nanjo, K., Ohtsuchi, N., Kohno, H. and Sano, M. (2021). Crabs assimilating cellulose materials drive the detritus food chain in a mangrove estuary. Food Webs, 26: e00180. https://doi.org/10.1016/j.fooweb.2020.e00180.

Khandavilli, R. (2016). Fungal phylogenetic diversity in estuarine sediments of Gautami Godavari River, Andhra Pradesh, India. Curr. Res. Environm. Appl. Mycol., 6: 268–276. https://doi.org/10.5943/cream/6/4/4.

Khatri, A., Peerzada, M.H., Mohsin, M. and White, M. (2015). A review on developments in dyeing cotton fabrics with reactive dyes for reducing effluent pollution. J. Clean Prod., 87: 50–57. https://doi.org/10.1016/j.jclepro.2014.09.017.

Khatri, J., Nidheesh, P.V., Anantha Singh, T.S. and Suresh Kumar, M. (2018). Advanced oxidation processes based on zero-valent aluminium for treating textile wastewater. Chem. Eng. J., 348: 67–73. https://doi.org/10.1016/j.cej.2018.04.074.

Khattab, T.A., Abdelrahman, M.S. and Rehan, M. (2019). Textile dyeing industry: Environmental impacts and remediation. Environ. Sci. Pollut. Res. Int., 27: 3803–3818. https://doi.org/10.1007/s11356-019-07137-z.

Kumar, A. and Chandra, R. (2020). Ligninolytic enzymes and its mechanisms for degradation of lignocellulosic waste in environment. Heliyon, 6(2): e03170. https://doi.org/10.1016/j.heliyon.2020.e03170.

Kumar, V., Pallavi, P., Sen, S.K. and Raut, S. (2024). Harnessing the potential of white rot fungi and ligninolytic enzymes for efficient textile dye degradation: A comprehensive review. Water Environ. Res., 96(1): e10959. https://doi.org/10.1002/wer.10959

Kunjadia, P.D., Sanghvi, G.V., Kunjadia, A.P., Mukhopadhyay, P.N. and Dave, G.S. (2016). Role of ligninolytic enzymes of white rot fungi (*Pleurotus* spp.) grown with azo dyes. SpringerPlus, 5: 1487. https://doi.org/10.1186/s40064-016-3156-7.

Laothanachareon, T., Kongtong, K., Saeng-Kla, K., Kanokratana, P., Leetanasaksakul, K. et al. (2023). Evaluating the efficacy of wood decay fungi and synthetic fungal consortia for simultaneous decolorization of multiple textile dyes. World J. Microbiol. Biotechnol., 39: 226. https://doi.org/10.1007/s11274-023-03672-7.

Li, H., Zhang, R., Tang, L., Zhang, J. and Mao, Z. (2015). Manganese peroxidase production from cassava residue by *Phanerochaete chrysosporium* in solid state fermentation and its decolorization of Indigo Carmine. Chin. J. Chem. Eng., 23: 227–233. https://doi.org/10.1016/j.cjche.2014.11.001.

Liu, S.H., Tsai, S.L., Guo, P.Y. and Lin, C.W. (2020). Inducing laccase activity in white rot fungi using copper ions and improving the efficiency of azo dye treatment with electricity generation using microbial fuel cells. Chemosphere, 243: 125304. https://doi.org/10.1016/j.chemosphere.2019.125304.

Liu, W., Chao, Y., Yang, X., Bao, H. and Qian, S. (2004). Biodecolorization of azo, anthraquinonic, and triphenylmethane dyes by white-rot fungi and a laccase-secreting engineered strain. J. Ind. Microbiol. Biotechnol., 31: 127–132. https://doi.org/10.1007/s10295-004-0123-z.

Luo, W., Vrijmoed, L.L.P. and Jones, E.B.G. (2005). Screening of marine fungi for lignocellulose-degrading enzyme activities. Bot. Mar., 48: 379–386. https://doi.org/10.1515/bot.2005.051.

Manohar, C.S. and Raghukumar, C. (2013). Fungal diversity from various marine habitats deduced through culture-independent studies. FEMS Microbiol. Lett., 341: 69–78. https://doi.org/10.1111/1574-6968.12087.

Moilanen, U., Osma, J.F., Winquist, E., Leisola, M. and Couto, S.R. (2010). Decolorization of simulated textile dye baths by crude laccases from *Trametes hirsuta* and *Cerrena unicolor*. Eng. Life Sci., 10: 242–247. https://doi.org/10.1002/elsc.200900095.

Moorthy, A.K., Rathi, B.G., Shukla, S.P., Kumar, K. and Bharti, V.S. (2021). Acute toxicity of textile dye methylene blue on growth and metabolism of selected freshwater microalgae. Environ. Toxicol. Pharmacol., 82: 103552. https://doi.org/10.1016/j.etap.2020.103552.

Moreira, P.R., Almeida-Vara, E., Malcata, F.X. and Duarte, J.C. (2007). Lignin transformation by a versatile peroxidase from a novel *Bjerkandera* sp. strain. Int. Biodeterior. Biodegradation, 59: 234–238. https://doi.org/10.1016/j.ibiod.2006.11.002.
Nabeela, Khan, S.A., Mehmood, S., Shabbir, S.B. and Ali, S. (2023). Efficacy of fungi in the decolorization and detoxification of remazol brilliant blue dye in aquatic environments. Microorganisms, 11: 703. https://doi.org/10.3390/microorganisms11030703.
Nishamol, M., Ghose, M., Kumar, A. and Manohar, C.S. (2025). Rhizosphere microbiomes and carbon dynamics in mangrove ecosystems are shaped by host species, patch age, and environmental drivers. Rhizosphere, 16: 101153. https://doi.org/10.1016/j.rhisph.2025.101153.
Oliveira, B.M.J., Dario Corrêa, J., Parente, C.E.T. and Frases, S. (2025). Fungi in mangrove: Ecological importance, climate change impacts, and the role in environmental remediation. Microorganisms, 3: 878. https://doi.org/10.3390/microorganisms13040878.
Ottoni, C., Lima, L., Santos, C. and Lima, N. (2014). Effect of different carbon sources on decolourisation of an industrial textile dye under alkaline–saline conditions. Curr. Microbiol., 68: 53–58. https://doi.org/10.1007/s00284-013-0441-3.
Pathak, V.V., Kothari, R., Chopra, A.K. and Singh, D.P. (2015). Experimental and kinetic studies for phycoremediation and dye removal by chlorella pyrenoidosa from textile wastewater. J. Environ. Manag., 163: 270–277. https://doi.org/10.1016/j.jenvman.2015.08.041.
Pavko, A. (2011). Fungal decolourization and degradation of synthetic dyes, some chemical engineering aspects. pp. 65–88. *In*: Einschlag, F.S.G. (ed.). Waste Water—Treatment and Reutilization. InTech. https://doi.org/10.5772/16120.
Pisoni, D.D.S., de Abreu, M.P., Petzhold, C.L., Rodembusch, F.S. and Campo, L.F. (2013). Synthesis, photophysical study and BSA Association of water-insoluble Squaraine dyes. J. Photochem. Photobiol. A Chem., 252: 77–83. https://doi.org/10.1016/j.jphotochem.2012.11.013.
Premaratne, M., Nishshanka, G.K.S.H., Liyanaarachchi, V.C., Nimarshana, P.H.V. and Ariyadasa, T.U. (2021). Bioremediation of textile dye wastewater using microalgae: Current trends and future perspectives. J. Chem. Technol. Biotechnol., 96: 3249–3258. https://doi.org/10.1002/jctb.6845.
Raghukumar, C. (2000). Fungi from marine habitats: An application in bioremediation. Mycol. Res., 104: 1222–1226. https://doi.org/10.1017/s095375620000294x.
Raghukumar, C., D'Souza-Ticlo, D. and Verma, A. (2008). Treatment of colored effluents with lignin-degrading enzymes: An emerging role of marine-derived fungi. Crit. Rev. Microbiol., 34: 189–206. https://doi.org/10.1080/10408410802526044.
Rauf, M.A. and Salman, A.S. (2012). Survey of recent trends in biochemically assisted degradation of Dyes. Chem. Eng. J., 209: 520–530. https://doi.org/10.1016/j.cej.2012.08.015.
Rizqi, H.D. and Purnomo, A.S. (2017). The ability of brown-rot fungus *Daedalea dickinsii* to decolorize and transform methylene blue dye. World J. Microbiol. Biotechnol., 33: 92. https://doi.org/10.1007/s11274-017-2256-z.
Robinson, T., Chandran, B. and Nigam, P. (2001). Studies on the production of enzymes by white-rot fungi for the decolourisation of textile dyes. Enzyme Microb. Technol., 29: 575–579. https://doi.org/10.1016/s0141-0229(01)00430-6.
Sáez-Jiménez, V., Fernández-Fueyo, E., Medrano, F.J., Romero, A., Martínez, A.T. et al. (2015). Improving the pH-stability of versatile peroxidase by comparative structural analysis with a naturally-stable manganese peroxidase. PLoS One, 10(11): e0143267. https://doi.org/10.1371/journal.pone.0143267.
Sen, S.K., Raut, S., Bandyopadhyay, P. and Raut, S. (2016). Fungal decolouration and degradation of azo dyes: A review. Fungal Biol. Rev., 30: 112–133. https://doi.org/10.1016/j.fbr.2016.06.003.
Shahmoradi, G.F., Moghaddam, M.K. and Tehrani, M. (2021). Comparison of the effect of metal Mordants and bio-mordants on the colorimetric and antibacterial properties of natural dyes on cotton fabric. Color Technol., 137: 689–698. https://doi.org/10.1111/cote.12569.
Shankar, S. and Nill, S. (2015). Effect of metal ions and redox mediators on decolorization of synthetic dyes by crude laccase from a novel white rot fungus *Peniophora* sp. (NFCCI-2131). Appl. Biochem. Biotechnol., 175: 635–647. https://doi.org/10.1007/s12010-014-1279-2.
Shetty, K. and Krishnakumar, G. (2020). Algal and cyanobacterial biomass as potential Dye biodecolorizing material: A review. Biotechnol. Lett., 42: 2467–2488. https://doi.org/10.1007/s10529-020-03005-w.

Shoaib, A.G., Ragab, S., El Sikaily, A., Yılmaz, M. and El Nemr, A. (2023). Thermodynamic, kinetic, and isotherm studies of direct blue 86 dye absorption by cellulose hydrogel. Sci. Rep., 13: 5910. https://doi.org/10.1038/s41598-023-33078-2.

Singh, R.L., Singh, P.K. and Singh, R.P. (2015). Enzymatic decolorization and degradation of azo dyes: A review. Int. Biodeterior. Biodegradation, 104: 21–31. https://doi.org/10.1016/j.ibiod.2015.04.027.

Sirianuntapiboon, S., Chairattanawan, K. and Jungphungsukpanich, S. (2006). Some properties of a sequencing batch reactor system for removal of VAT dyes. Bioresour. Technol., 97: 1243–1252. https://doi.org/10.1016/j.biortech.2005.02.052.

Sridhar, M. (2016). Versatile peroxidases: Super peroxidases with potential biotechnological applications—A mini review. J. Dairy Vet. Anim. Res., 4: 277–280. https://doi.org/10.15406/jdvar.2016.04.00116.

Srinivasan, A. and Viraraghavan, T. (2010). Decolorization of dye wastewaters by biosorbents: A review. J. Environ. Manag., 91: 1915–1929. https://doi.org/10.1016/j.jenvman.2010.05.003.

Sun, X., Bai, R., Zhang, Y., Wang, Q., Fan, X. et al. (2013). Laccase-catalyzed oxidative polymerization of phenolic compounds. Appl. Biochem. Biotechnol., 171: 1673–1680. https://doi.org/10.1007/s12010-013-0463-0.

Thampraphaphon, B., Phosri, C., Pisutpaisal, N., Thamvithayakorn, P., Chotelersak, K. et al. (2022). High potential decolourisation of textile dyes from wastewater by manganese peroxidase production of newly immobilised *Trametes hirsuta* PW17-41 and FTIR analysis. Microorganisms, 10(5): 992. https://doi.org/10.3390/microorganisms10050992.

Tien, M. and Kirk, T.K. (1988). Lignin peroxidase of Phanerochaete *Chrysosporium*. Methods Enzymol., 161: 238–249. https://doi.org/10.1016/0076-6879(88)61025-1.

Toker, S.K., Evlat, H. and Koçyiğit, A. (2021). Screening of newly isolated marine-derived fungi for their laccase production and decolorization of different dye types. Reg. Stud. Mar. Sci., 45: 101837. https://doi.org/10.1016/j.rsma.2021.101837.

Vafaei, F., Movafeghi, A. and Khataee, A. (2013). Evaluation of antioxidant enzymes activities and identification of intermediate products during phytoremediation of an anionic dye (c.i. acid blue 92) by Pennywort (*Hydrocotyle vulgaris*). J. Environ. Sci., 25: 2214–2222. https://doi.org/10.1016/s1001-0742(12)60306-4.

Vaksmaa, A., Guerrero-Cruz, S., Ghosh, P., Zeghal, E., Hernando-Morales, V. et al. (2023). Role of fungi in bioremediation of emerging pollutants. Front. Mar. Sci., 10: 1070905. https://doi.org/10.3389/fmars.2023.1070905.

Varjani, S., Rakholiya, P., Ng, H.Y., You, S. and Teixeira, J.A. (2020). Microbial degradation of Dyes: An overview. Bioresour. Technol., 314: 123728. https://doi.org/10.1016/j.biortech.2020.123728.

Venkatachalam, M., Gérard, L., Milhau, C., Vinale, F., Dufossé, L. et al. (2019). Salinity and temperature influence growth and pigment production in the marine-derived fungal strain *Talaromyces albobiverticillius* 30548. Microorganisms, 7: 10. https://doi.org/10.3390/microorganisms7010010.

Verma, A.K., Raghukumar, C., Verma, P., Shouche, Y.S. and Naik, C.G. (2010). Four marine-derived fungi for bioremediation of raw textile mill effluents. Biodegradation, 21: 217–233. https://doi.org/10.1007/s10532-009-9295-6.

Wen, Y., Chen, H., Shen, C., Zhao, M. and Liu, W. (2011). Enantioselectivity tuning of chiral herbicide dichlorprop by copper: Roles of reactive oxygen species. Environ. Sci. Technol., 45: 4778–4784. https://doi.org/10.1021/es2003793.

Wikee, S., Hatton, J., Turbé-Doan, A., Mathieu, Y., Daou, M. et al. (2019). Characterization and dye decolorization potential of two laccases from the marine-derived fungus *Pestalotiopsis* sp. Int. J. Mol. Sci., 20: 1864. https://doi.org/10.3390/ijms20081864.

Yang, P., Shi, W., Wang, H. and Liu, H. (2016). Screening of freshwater fungi for decolorizing multiple synthetic dyes. Braz. J. Microbiol., 47: 828–834. https://doi.org/10.1016/j.bjm.2016.06.010.

Zahmatkesh, S., Karimian, M., Pourhanasa, R., Ghodrati, I., Hajiaghaei-Keshteli, M. et al. (2023). Wastewater treatment with algal based membrane bioreactor for the future: Removing emerging contaminants. Chemosphere, 335: 139134. https://doi.org/10.1016/j.chemosphere.2023.139134.

Zanoni, T.B., Lizier, T.M., das Dores Assis, M., Zanoni, M.V.B. and De Oliveira, D.P. (2013). CYP-450 isoenzymes catalyze the generation of hazardous aromatic amines after reaction with the azo dye sudan III. Food Chem. Toxicol., 57: 217–226. https://doi.org/10.1016/j.fct.2013.03.035.

Zhang, H., Zhang, X. and Geng, A. (2020). Expression of a novel manganese peroxidase from *Cerrena unicolor* BBP6 in *Pichia pastoris* and its application in dye decolorization and PAH degradation. Biochem. Eng. J., 153: 107402. https://doi.org/10.1016/j.bej.2019.107402.

Zhang, Y., Dong, Z., Luo, Y., Yang, E., Xu, H. et al. (2021). The manganese peroxidase gene family of *Trametes trogii*: Gene identification and expression patterns using various metal ions under different culture conditions. Microorganisms, 9: 2595. https://doi.org/10.3390/microorganisms9122595.

Zhang, Y.Q., Wei, X.C. and Long, J.J. (2016). Ecofriendly synthesis and application of special disperse reactive dyes in waterless coloration of wool with supercritical carbon dioxide. J. Clean. Prod., 133: 746–756. https://doi.org/10.1016/j.jclepro.2016.05.187.

Zhao, Y.H., Geng, J.T., Cai, J.C., Cai, Y.F. and Cao, C.Y. (2020). Adsorption performance of Basic Fuchsin on alkali-activated diatomite. Adsorpt. Sci. Technol., 38: 151–167. https://doi.org/10.1177/0263617420922084.

Zhuo, R., Ma, L., Fan, F., Gong, Y., Wan, X. et al. (2011). Decolorization of different dyes by a newly isolated white-rot fungi strain *Ganoderma* sp.en3 and cloning and functional analysis of its laccase gene. J. Hazard Mater., 192: 855–873. https://doi.org/10.1016/j.jhazmat.2011.05.106.

Zuorro, A. and Lavecchia, R. (2014). Evaluation of UV/H2O2 advanced oxidation process (AOP) for the degradation of diazo dye reactive green 19 in aqueous solution. Desalination Water Treat., 52: 1571–1577. https://doi.org/10.1080/19443994.2013.787553.

Future Directions

14

Applications of Deepsea Mycobiota

An Untapped Resource of Blue Biotechnology

*Ivonne Rodríguez-Ramírez** and *Mónica Vallejo-Arróliga*

The deepsea is an underexplored reservoir of bioactive compounds with remarkable biotechnological potential. In the field of medicine, up to 40% of deepsea natural products are drug-like and two-thirds fall among the known drugs, suggesting promising therapeutic leads. Marine-derived *Penicillium chrysogenum* strains are capable of producing more than 10 distinct compounds with potential applications in industries. Deepsea fungal ability to produce multiple biomolecules makes them highly lucrative metabolites of nutritional, industrial, and medicinal significance, including long-duration space missions. Advancing in deepsea fungal research needs an interdisciplinary approach to bridge marine biotechnology, synthetic biology, and drug discovery arenas. Cultivation of fungi under simulated deepsea conditions, scaling up production of deepsea fungi and optimization of bioactivity tasks to cater to the applications. Deepsea fungi are a promising source of antibiotics, anticancer agents, biosurfactants, enzymes, biocontrol agents, bioremediation, and biodegradation of recalcitrants.

Escuela de Biología, Universidad de Costa Rica, San Pedro de Montes de Oca, San José, Costa Rica.
* Corresponding author: ivonne.rodriguezramirez@ucr.ac.cr

DOI: 10.1201/9781003795100-18

1. Introduction

The deepsea, which accounts for approximately 60% of the ocean's area, comprises regions covered by seawater at depths greater than 2,000 meters (Zhao et al., 2022). It is a complex and extreme ecosystem characterized by high hydrostatic pressure, low or elevated temperatures (hydrothermal vents), absence of light, oligotrophic conditions, and low oxygen concentrations. These extreme conditions have driven the evolution of metabolic plasticity in deepsea organisms, enabling them to produce unique secondary metabolites with diverse biological activities. Among these organisms, deepsea fungi have emerged as particularly promising sources of bioactive compounds (Cong et al., 2022). Fungi are well recognized as major producers of secondary metabolites with diverse biological activities; however, most research on marine fungi has focused on species from nearshore or shallow waters, while deepsea fungi remain relatively underexplored. Despite this research bias, which limits the amount of available information on fungi from deep environments, several strains are already known to harbor great potential for future solutions, as summarized 47 spp. in Table 1. Their adaptation to extreme environments suggests that these understudied fungi may harbor unique metabolites not found in terrestrial or shallow-water species, offering a largely untapped reservoir for drug discovery and biotechnological applications. In this chapter, we document metabolites obtained from deepsea fungi that show potential for medicinal and biotechnological applications.

2. Sources of Medicinally Relevant Metabolites

Research on deepsea-derived fungi has revealed their value as producers of secondary metabolites with remarkable biomedical potential. Because of the extreme conditions where they live (high pressure, low temperatures, and scarce nutrients), these fungi have developed a flexible metabolism that allows them to produce unique compounds, many of which have shown promising effects against important human diseases. Recent studies have found deepsea fungal metabolites with inhibitory effects on key enzymes involved in Type 2 diabetes, cytotoxic properties against a variety of cancer cell lines, and activities targeting hypertension, thrombosis, immune modulation, antimicrobial resistance, osteoporosis, and food allergies. By combining advanced isolation techniques with molecular docking, pharmacokinetic analyses, and bioassays, researchers continue to uncover new molecules and mechanisms of action, underscoring the deep ocean as a promising frontier for drug discovery.

2.1 *Regulators of Glucose Metabolism*

Among the key enzymes involved in glucose metabolism, α-amylase, α-glucosidase, and hexokinase-II (HK-II) are responsible for breaking down ingested carbohydrates; thus, their inhibition can delay sugar absorption, which is beneficial for managing Type 2 diabetes (Patel et al., 2017). Some metabolites from deepsea fungi have shown potential as inhibitors of these enzymes based on *in silico* analyses. Virtual screening,

Table 1. Deepsea fungal strains with the potential to be used in future applications.

Fungal Strain	Source of Strain	Metabolite/Activity	Industrial Use	References
Acremonium sp. GA31	South China deepsea sediments	Nitrous oxide (N_2O) under nitrite conditions	Nitrogen cycling	Gao et al. (2020)
Acremonium fusidioides	Points in the South Shetlands Islands, Antartica	Phytotoxicity	Agriculture	Ogaki et al. (2020)
Alternaria sp. CIGOM4	Deepsea sediments, Gulf of Mexico	Antimicrobial activity	Antimicrobial pharmaceuticals	Villanueva-Silva et al. (2021)
Antarctomyces psychrotrophicus	Antarctic deepsea sediments	Biosurfactant activity	Remineralization processes, food, dairy, detergent, pharmaceutical and leather industries	da Silva et al. (2024)
Aspergillus chevalieri MCCC M23426	-	Diketopiperazines (DKPs) known as neoechinulin D	Oncology, Antimicrobial	Lv et al. (2022)
Aspergillus flavipes DS720	Mariana Trench	Flavonoid (C14H15NO)	Oncology	Xu et al. (2022)
Aspergillus sydowii SYX6	Mariana Trench	Enhanced the production of dorsinol	Oncology, Antimicrobial	Deng et al. (2023)
Aspergillus sydowii LF660	Sediment of the Mediterranean Sea	Inhibitor of the human protein tyrosine phosphatase 1B (PTP1B), a target in the signaling pathway of insulin	Endocrinology	Wiese et al. (2017)
Aspergillus ustus	deepsea sediments of the Central Indian Basin	protease	Food, dairy, detergent, pharmaceutical and leather	Damare et al. (2006)
Aspergillus versicolor GA8	South China deepsea sediments	Nitrous oxide (N_2O) under nitrite conditions	Nitrogen cycling	Gao et al. (2020)
Aspergillus versicolor SCAU214	Mariana Trench	Exopolysaccharide, AVP-214-1	Immunology	Li et al. (2025)
Aspergillus westerdijkiae (JX156359)	Deepsea sediment samples collected in the South China Sea	Antimicrobial molecules	Antimicrobial pharmaceuticals	Zhang et al. (2013)

Table 1 contd. ...

...Table 1 contd.

Fungal Strain	Source of Strain	Metabolite/Activity	Industrial Use	References
Aspergillus sp. SCSIO 41029	Deepsea sediments in South China	Lumazine peptide, penulimamide K ($C_{19}H_{18}N_6O_7$), butyroscavin, butyrolactone I, butyrolactone V, aspernolide D, and aspulvinone E	Endocrinology/Clinical pharmacology	Chen et al. (2021)
Cadophora malorum Mo12	Rainbow hydrothermal site on the Mid-Atlantic Ridge	6 type I polyketide synthases, 5 nonribosomal peptide synthetases, 2 hybrid polyketide synthase/nonribosomal peptide synthesis (PKS/NRPS), and 2 terpene synthase genes, 101 glycoside hydrolases, 74 glycosyltransferases, 1 polysaccharide lyase, 10 carbohydrate esterases, 29 carbohydrate-binding modules	More applied science is needed to determine its true potential	Rédou et al. (2016)
Chaetomium sp CS1	Deepsea sediment samples of the West Pacific	Lignin degradation	Biofuels and biochemicals	Ling et al. (2025)
Cladosporium cladosporioides HDN14-342	Indian Ocean	Cladosporol C	Endocrinology/Clinical pharmacology	Alanzi et al. (2023)
Cladosporium oxysporum 170103	-	Antimicrobial molecules, ergosterol	Antimicrobial pharmaceuticals	Xie et al. (2022)
Cladosporium ramotenellum CONTIG7	Deepsea sediments, Gulf of Mexico	Antimicrobial activity	Antimicrobial pharmaceuticals	Villanueva-Silva et al. (2021)
Cladosporium sp.	Antarctic deepsea sediments	Protease, Cellulase, Esterase, Lipase	Remineralization processes, food, dairy, detergent, pharmaceutical and leather industries	da Silva et al. (2024)
Cystobasidium laryngis IV17-028	Indian Ocean Ridge	Phenazostatin J	Oncology	Lee et al. (2022)
Dichotomomyces cejpii	sediments from the South China Sea	Gliotoxin	Biomedical industry	Ye et al. (2018)

Emericellopsis maritima SFI-F16 and SFI-D6	northwest Irish coast deepsea sediments	Xylanase, α-L-arabinofuranosidase, β-Penicillium antarcticumD-xylosidase, β-D-glucosidase, cellobiohydrolase (against β-D-cellobioside and β-D-lactoside), β-D-mannosidase, α-L-fucosidase, α-L-rhamnosidase and β-D-galactosidase	Biofuels, paper pulp	Dowd and Tuohy et al. (2023)
Geosmithia pallida FS140	South China Sea	Gliotoxin-like compounds	Cardiology	Sun et al. (2018)
Hortaea werneckii	Deepsea Sediments from East Yap Trench and from the Costarican Pacific	Lipase, urease, protease, laccase, chitinase, and cellobiase	polyurethane degradation	Han et al. (2024); Rodriguez-Ramirez et al. (2025)
Meyerozyma guilliermondii	Antarctic deepsea sediments	Protease and lipase	Remineralization processes	da Silva et al. (2024)
Nigrospora sp. NIOT	Depth of 800 m at the Arabian Sea	Cytotoxicity and Antimicrobial activity	Oncology	Arumugam et al. (2015)
Penicillium allii-sativi	Points in the South Shetlands Islands, Antartica	Phytotoxicity	Agriculture	Ogaki et al. (2020)
Penicillium antarcticum SFI-F25	Northwest Irish coast deepsea sediments	Xylanase,α-L-arabinofuranosidase, β-D-xylosidase, β-D-glucosidase, cellobiohydrolase (against β-D-cellobioside and β-D-lactoside), β-D-mannosidase, α-L-fucosidase and β-D-galactosidase	Biofuels, paper pulp	Dowd and Tuohy, et al. (2023)
Penicillium chrysogenum SFI-D13 and SCSIO 41001	Antarctic deepsea sediments, points in the South Shetlands Islands, northwest Irish coast deepsea sediments and sediment from the Indian Ocean	Protease, Agarase, Carrageenan, Cellulase, Esterase, Invertase, Lipase, Pectinase peroxidase/polyhenol oxidase, α-L-arabinofuranosidase, β-D-xylosidase, cellobiohydrolase (against β-D-cellobioside and β-D-lactoside), β-D-mannosidase, α-L-rhamnosidase and β-D-galactosidase, Phytotoxicity, cytotoxic activities against K562, A549, and Huh-7 cell lines	Remineralization processes, food, dairy, detergent, pharmaceutical and leather, agriculture, oncology	da Silva et al. (2024); Dowd and Tuohy et al. (2023); Ogaki et al. (2020); Chen et al. (2017); Wang et al. (2018)
Penicillium citreonigrum XT20-134	Indian Ocean	Circumdatin G	Endocrinology/ Clinical pharmacology	Alanzi et al. (2023)

Table 1 contd. ...

...Table 1 contd.

Fungal Strain	**Source of Strain**	**Metabolite/Activity**	**Industrial Use**	**References**
Penicillium citrinum W23	-	Neotricitrinol B	Orthopedic	He et al. (2023)
Penicillium echinulatum CONTIG4	Deepsea sediments, Gulf of Mexico	Antimicrobial activity	Antimicrobial pharmaceuticals	Villanueva-Silva et al. (2021)
Penicillium griseofulvum MCCC 3A00225	Indian Ocean sediment	(−)-cyclopenol	Immunology	Xing et al. (2021)
Penicillium cf. *palitans* SFI-F25	Antarctic deepsea sediments, points in the South Shetlands Islands	Biosurfactant activity Protease, Agarase, Carrageenan, Cellulase, Esterase, Invertase, Lipase, Pectinase phytotoxicity	Remineralization Processes,food, dairy, detergent, pharmaceutical and leather industries, agriculture	da Silva et al. (2024); Ogaki et al. (2020)
Penicillium solitum	Points in the South Shetlands Islands, Antartica	Phytotoxicity, Viridicatol (VDC)	Agriculture Orthopedics	Ogaki et al. (2020); Xie et al. (2025)
Penicillium viridicatum	Atlantic Ocean	Activity of the p53 response element (p21 promoter)	Oncology	Zhang et al. (2025)
Penicillium sp. F11	-	Penicillactam	Endocrinology/Clinical pharmacology	Alanzi et al. (2023)
Penicillium sp. JMF034	Sediments of Suruga Bay, Japan	bis(dethio)-10a-methylthio-3a-deoxy-3,3a-didehydrogliotoxin (1) and 6-deoxy-5a,6-didehydrogliotoxincytotoxic activity	Oncology	Sun et al. (2011)
Penicillium sp. SCSIO06871	Indian Ocean	Monomeric sorbicillin (C13H16O4)	Endocrinology/Clinical pharmacology, Antimicrobial pharmaceuticals	Pang et al. (2021)
Phomopsis lithocarpus FS508	Indian Ocean	Tenellone F and ozazino-cyclo-(2,3-dihydroxy-Trp-Tyr)	Endocrinology/ Clinical pharmacology	Alanzi et al. (2023)

Pseudeurotium cf. *Bakeri*	Antarctic deepsea sediments	Agarase, carrageenase and cellulase	Remineralization processes, food, dairy, detergent, pharmaceutical and leather industries	da Silva et al. (2024)
Pseudogymnoascus verrucosus	Points in the South Shetlands Islands, Antartica	Phytotoxicity	Agriculture	Ogaki et al. (2020)
Purpureocillium lilacinum FDZ8Y1	Mariana Trench	Cytotoxic activity against multiple cancer cell lines	Oncology	Li et al. (2024)
Simplicillium obclavatum EIODSF 020	Deepsea derived	Peptaibiotics, β-alanine, tyrosine, or tyrosine O-sulfate,	Food and agriculture	Liang et al. (2024)
Stachybotrys longispora	East China Sea	Isoindole alkaloid, FGFC1 ($C_{51}H_{68}N_2O_{10}$)	Cardiology	Zhang et al. (2024)
Talaromyces stolii SFI-F17	northwest Irish coast deepsea sediments	Xylanase,β-D-galactosidase, β-D-glucosidase, β-D-xylosidase, α-L-arabinofuranosidase and β-D-lactopyranosidase	Biofuels, paper pulp	Dowd and Tuohy et al. (2023)
Thelebolus balaustiformis	Antarctic deepsea sediments	Lipase, biosurfactant activity	Remineralization processes, food, dairy, detergent, pharmaceutical and leather industries	da Silva et al. (2024)

molecular docking, and pharmacokinetic predictions identified cladosporol C, from *Cladosporium cladosporioides* HDN14-342, tenellone F and ozazino-cyclo-(2,3-dihydroxy-Trp-Tyr) from *Phomopsis lithocarpus* FS508, penicillactam from *Penicillium* sp. F11, and circumdatin G from *Penicillium citreonigrum* XT20-134 as potential inhibitors of α-amylase, α-glucosidase, pancreatic lipoprotein lipase, HK-II, and protein tyrosine phosphatase-1B, respectively (Alanzi et al., 2023). Experimental validation is still required to confirm these predicted activities. For other deepsea fungal metabolites, experimental evidence is already available. For example, a novel lumazine peptide, penulimamide K ($C_{19}H_{18}N_6O_7$), together with five known metabolites (butyroscavin, butyrolactone I, butyrolactone V, aspernolide D, and aspulvinone E) isolated from *Aspergillus* sp. SCSIO 41029, a strain obtained from deepsea sediments in South China, exhibited significant α-glucosidase inhibitory activity, with IC_{50} values ranging from 18.61 to 109.06 μM, much lower than that of the positive control, acarbose (IC_{50} = 357 μM) (Chen et al., 2021). Moreover, a monomeric sorbicillin ($C_{13}H_{16}O_4$) isolated from *Penicillium* sp. SCSIO06871 displayed stronger α-glucosidase inhibition than acarbose, with an IC_{50} value of 36.0 μM (Pang et al., 2021).

2.2 *Anticancer Agents*

Cancer is primarily driven by oncogenic somatic and/or germline alterations in oncogenes and tumor suppressor genes. Among these, approximately 50% of human tumors harbor mutations in the TP53 tumor suppressor gene, resulting in the loss of the p53 protein's critical tumor-suppressive functions, such as DNA repair, cell apoptosis, and autophagy (Wang et al., 2023a). Consequently, activating the p53 signaling pathway in tumor cells represents a promising strategy for cancer treatment. Statins have been identified as inducers of degradation of misfolded mutant p53 proteins and as activators of the p53 signaling pathway. Notably, a statin isolated from the deepsea fungus *Penicillium viridicatum* significantly enhanced the activity of the p53 response element (p21 promoter) in p53-deficient H1299 cells at 10 μM (Zhang et al., 2025), exhibiting greater activity than PRIMA-1, a well-known mutant p53 reactivator. Another species from this genus with anticancer potential is *P. chrysogenum* SCSIO 41001, some of its metabolites showed cytotoxic activities against human chronic myeloid leukemia (K562), lung adenocarcinoma (A549), and hepatocellular carcinoma (Huh-7) cell lines with IC_{50} values of 6.78, 6.94, and 2.59 μM, respectively (Chen et al., 2017).

Some studies have also suggested that deepsea fungi might serve as natural reservoirs of bioactive secondary metabolites with anticancer potential. For example, it has been reported that *Nigrospora sp.* NIOT has shown GI_{50} (growth inhibition by 50%) of 1.35, 3.2, 0.13, and 0.35 μg/ml against cell lines histiocytic lymphoma (U937), breast adenocarcinoma (MCF-7), Ewing's sarcoma (A673) and T-cell leukemia (Jurkat), respectively (Arumugam et al., 2015). Also, the crude extract of *Purpureocillium lilacinum* FDZ8Y1, isolated from sediments of the Mariana Trench, exhibited cytotoxic activity against multiple cancer cell lines using the MTT assay. At 500 μg/mL, the extract induced cell mortality rates of 91.4% in lung cancer

(NCI-H460), 64.0% in breast carcinoma (MDA-MB-231), and 55.8% in breast carcinoma (MCF-7) after 24 hours of treatment (Li et al., 2024).

Similarly, *Aspergillus flavipes* DS720, another fungus from the Mariana Trench, contains a flavonoid ($C_{14}H_{15}NO$) that exhibited broad-spectrum cytotoxicity using the CCK-8 assay. At a concentration of 20 μM, after 48 h it inhibited over 90% of HPV-positive (HeLa), bladder cancer (5637), thyroid cancer (CAL-62), pancreatic adenocarcinoma (PATU8988T), malignant melanoma (A-375), and Ewing's sarcoma (A-673) cell lines (Xu et al., 2022). Another potential anticancer metabolite isolated from deepsea fungi is indole-containing diketopiperazines (DKPs), which are well recognized for their diverse biological activities, including anticancer and immunomodulatory effects (Wang et al., 2008). From the deepsea-derived fungus *Aspergillus chevalieri* MCCC M23426, a DKP known as neoechinulin D was obtained from the fermentation culture and determined as a potential anticancer agent by using clonogenic assay, cell apoptosis tests, and CCK-8 assay. The compound reduced the viability of gastric cancer (MKN1) cells, showing an IC_{50} value of 4.6 μM (Lv et al., 2022). In addition, a new phenazostatin was isolated from the culture broth of *Cystobasidium laryngis* IV17-028, derived from a deepsea sediment sample of the Indian Ocean Ridge. Evaluated by the sulforhodamine B (SRB), it exhibited cytotoxicity with an IC_{50} value of 7.7 nM against NUGC-3 (stomach) cells, displaying 19-fold stronger activity than the positive control, adriamycin (Lee et al., 2022).

Researchers have also discovered that cultivating some deepsea fungi under extreme conditions resembling their natural environments can trigger changes in gene expression and lead to the production of distinct metabolites. For example, exposure to high hydrostatic pressure in *Aspergillus sydowii* SYX6, isolated from the Mariana Trench, significantly enhanced the production of the known bioactive compound dorsinol. In MTT assays, dorsinol exhibited strong cytotoxic effects against several human cancer cell lines, including lung cancer (NCI-H460, 90% inhibition), liver cancer (HepG2, 55% inhibition), and breast carcinoma (MCF-7, 57% inhibition; MDA-MB-231, 78% inhibition), at a concentration of 200 μM (Deng et al., 2023). Another example of triggering the anticancer activity is the research by Dong et al. (2014), where the authors activated dormant pathways on *A. versicolor* ZBY-3 by introducing neomycin resistance to spores, the mutant's metabolites showed antitumoral activities tested by the MTT assay against chronic myeloid leukemia with inhibition rates between 23–73%.

2.3 *Cardiovascular Regulators*

Despite advances in treatment and preventive strategies, cardiovascular diseases continue to rise and contribute significantly to global mortality and healthcare costs (Chong et al., 2025). Among these, hypertension is one of the most common cardiovascular conditions, affecting over one billion people worldwide (Kario et al., 2024). Angiotensin-converting enzyme (ACE) plays a major role in hypertension by catalyzing the conversion of angiotensin I into angiotensin II, a peptide that increases blood pressure. Interestingly, gliotoxin-like compounds identified in the deepsea-derived fungus *Geosmithia pallida* FS140 have demonstrated bioactivity by inhibiting ACE, highlighting their potential in hypertension treatment (Sun et al., 2018).

Thrombosis is another common complication of cardiovascular diseases and a major contributor to mortality worldwide. After blood clots form due to vascular injury, thrombin-thrombomodulin (TM)-mediated activation of Thrombin Activatable Fibrinolysis Inhibitor (TAFI) prevents premature fibrinolysis, ensuring effective coagulation and protection against persistent bleeding. Consequently, direct inhibition of TAFI has been indicated as a promising pharmacological strategy for the treatment of thrombosis and other cardiovascular diseases. An isoindole alkaloid, FGFC1 ($C_{51}H_{68}N_2O_{10}$), a secondary metabolite produced by *Stachybotrys longispora* strain FG216, has been shown to exhibit remarkable fibrinolytic activity, with fibrin completely dissolved by 10 mg/kg FGFC1 in rats and proved low toxicity (EC_{50} = 5 μM). Moreover, FGFC1 solutions ranging from 0.023 to 0.736 mM effectively inhibited TAFI activation, which was further validated through molecular docking of FGFC1 to TAFI, demonstrating the compound's potential to bind the protein and modulate its activity (Zhang et al., 2024).

2.4 *Osteogenic Compounds*

With aging, osteoblasts undergo senescence and functional decline, driven by changes in hormone and growth factor levels, as well as intrinsic mechanisms associated with cellular aging. Age-related progressive bone loss is further exacerbated in patients with osteoporosis, a disease characterized by reduced bone mass, increased bone fragility, and a higher risk of fractures (Marie and Kassem, 2011). In these patients, bone marrow exhibits an increase in adipocytes accompanied by a decrease in osteoblasts. Some osteogenic compounds have been isolated from deepsea fungi and may hold potential as treatments for this disease. For example, from the deepsea-derived fungus *Penicillium citrinum* W23, three new citrinin trimers, neotricitrinols A–C, were identified. Among them, neotricitrinol B promoted osteoblastogenesis while inhibiting adipogenic differentiation, without showing cytotoxicity at concentrations up to 10 μM (He et al., 2023). Another marine compound called Viridicatol (VDC), isolated from the deepsea-derived fungus *Penicillium solitum* MCCC 3A00215, proved stimulation of osteoblasts through the Wnt/SHN3 signaling pathway, which enhanced bone formation and promoted fracture healing in mice. Furthermore, bone-targeting nanovesicles (BT-NVs) engineered using the short SDSSD (Ser-Asp-Ser-Ser-Asp) sequence efficiently delivered VDC directly to osteoblasts, effectively reducing bone loss (Xie et al., 2025).

2.5 *Immune-Stimulating and Anti-Allergic Compounds*

Macrophages are phagocytic cells of the innate immune system that reside in various tissues. Their primary functions include engulfing foreign agents, clearing apoptotic cells, recycling nutrients by digesting tissue waste products, and producing cytokines, which coordinate immune responses (Arango-Duque and Descoteaux, 2014).

Because of their central role in immune regulation, macrophages are frequently used in experimental models to evaluate the immunomodulatory effects of bioactive polysaccharides. A novel exopolysaccharide, AVP-214-1, was isolated and purified from the metabolites of the Mariana Trench fungus *Aspergillus versicolor* SCAU214. Functional assays revealed that AVP-214-1 significantly enhanced the proliferation of RAW 264.7 macrophage cells and stimulated the secretion of key cytokines, including IL-6, TNF-α, and IL-1β. Metabolomic analysis further indicated that AVP-214-1 primarily modulates pyrimidine metabolism and amino acid-related pathways, which are closely linked to immune regulation (Li et al., 2025). Collectively, these findings identify AVP-214-1 as a novel deepsea- derived immune-stimulating polysaccharide with potential immunopharmacological applications.

Allergies are systemic diseases caused by an impaired immune system (Wang et al., 2023b). From *Penicillium griseofulvum* MCCC 3A00225, a deepsea-derived fungus isolated from the Indian Ocean sediment, a compound identified as a (–)-cyclopenol, was tested for *in vitro* anti-food allergic activity using immunoglobulin (Ig) E-mediated rat basophilic leukemia (RBL)-2H3 cells. These cells serve as a model for human mast cells because they release inflammatory mediators, such as histamine, when an IgE-mediated allergic reaction is triggered. In this assay, degranulation (the process by which these mediators are released and that is responsible for allergy symptoms) was measured as an indicator of allergic response. The isolated (–)-cyclopenol significantly reduced degranulation, showing an IC_{50} of 60.3 μM, which indicates greater anti-allergic potency than the positive control loratadine, whose IC_{50} was 91.6 μM (Xing et al., 2021).

2.6 Antimicrobial Metabolites

The emergence of antimicrobial resistance (AMR) in bacterial pathogens poses a serious global health threat, creating an urgent need to discover new antimicrobial metabolites (Yadav and Kapley, 2021). Deepsea fungi represent a promising source of such novel bioactive compounds. For instance, exposure to high hydrostatic pressure in *Aspergillus sydowii* SYX6, a strain isolated from the Mariana Trench, significantly enhanced the production of the bioactive compound dorsinol. Antimicrobial assays demonstrated that dorsinol displayed notable inhibitory activity, with minimum inhibitory concentration (MIC) values of 12.50 mg/L against *Chromobacterium violaceum* and *Staphylococcus aureus*, 6.25 mg/L against *Candida albicans* and *Enterococcus faecalis*, and 25.00 mg/L against *Salmonella choleraesuis* (Deng et al., 2023).

Biogenetically related diketopiperazine (DKP) compounds obtained from the fermentation culture of the deepsea-derived fungus *Aspergillus chevalieri* MCCC M23426 exhibited selective antibacterial activity against *S. aureus* CICC 10384, achieving inhibition rates above 90% at 250 μM (Lv et al., 2022). Similarly, Sohirnone A and dihydrodemethylsorbicillin, isolated from *Penicillium* sp. SCSIO06871, displayed significant antibacterial activity against *S. aureus*, with MIC values of 10.0 and 5.0 μg/mL, respectively (Pang et al., 2021). In the same way, *Penicillium* sp. F23-2 produced penicyclones, exhibited antimicrobial activity against the Gram-positive bacterium *S. aureus* with MIC values ranging from 0.3 to 1.0 μg/mL (Guo et al., 2015).

Similarly, fungal strains isolated from deepsea sediments of the Gulf of Mexico: *Alternaria* sp. CIGOM4, *Cladosporium ramotenellum* CONTIG7, and *Penicillium echinulatum* CONTIG4, exhibited notable antimicrobial activity. *Alternaria* sp. CIGOM4, cultivated at 4°C in darkness for 60 days, completely inhibited the growth of *Bacillus subtilis* at 20 μg/mL and methicillin-sensitive *Staphylococcus aureus* (MSSA) at 200 μg/mL. The extract of *P. echinulatum* CONTIG4, grown at 20°C in darkness for 30 days, showed antimicrobial activity against *Escherichia coli*, *Salmonella typhi*, and MSSA at 200 μg/mL. Scaled-up extracts from these strains were subjected to bioactivity-guided chemical analysis. From *Alternaria* sp. CIGOM4, alternariol (1) was isolated, which completely inhibited a methicillin-resistant *S. aureus* (MRSA) strain at 387.3 μM. Alternariol also inhibited *Mycobacterium tuberculosis* growth by 87% at 50 μg/mL. From *P. echinulatum* CONTIG4, viridicatin (6) was isolated, which inhibited the growth of *M. tuberculosis* with a minimum inhibitory concentration (MIC) of 43.8 μM (Villanueva-Silva et al., 2021).

Another species that could be a potential source of antimicrobial metabolites is *Cladosporium oxysporum* 170103, which yielded 16 compounds, including ergosterol, which showed moderate antibacterial activity against *Vibrio parahaemolyticus* (Xie et al., 2022). *Aspergillus westerdijkiae* exhibited broad-spectrum antimicrobial activity (Zhang et al., 2013) and antibacterial and antifungal activities against five bacterial (*A. baumanii* ATCC 19606, *A. baumanii* ATCC 15122, *S. aureus* ATCC 16339, *S. aureus* ATCC 29213, and *K. pneumoniae* ATCC 14578) and two fungal strains (*Fusarium oxysporum* f. sp. *cucumerinu* and *Fusarium oxysporum* f. sp. *momordicae*). Also *Aspergillus* sp. SCSIO06786, with 50 μg/disc showed inhibition zones against *S. aureus*, MRSA, E. *faecalis*, and *A. baumannii* (Pang et al., 2020).

3. Other Industrial Applications

The literature mentioned in this section highlights the diverse biodegradative and bioactive potential of marine fungi, ranging from lignocellulose breakdown and protease production to plastic biodegradation. Such versatility underscores their promise of deepsea fungi as sustainable biocatalysts and sources of novel bioactive molecules for industrial and environmental applications.

3.1 *Biocontrol Agents*

Some deepsea fungal species have been shown to produce secondary metabolites with the potential to be biocontrol agents. The crude extract of *Purpureocillium lilacinum* FDZ8Y1, isolated from sediments of the Mariana Trench, exhibited potential anti-nematode activity. After 24 h of incubation, it caused 94.7% juvenile mortality in *Caenorhabditis elegans* and induced degradation of the eggshells of *C. elegans* eggs. UPLC-MS/MS analysis revealed that this fungus produces leucinostatins A and B and exhibits chitinase activity (Li et al., 2024), compounds and enzymatic functions that may contribute to its nematicidal effects. Likewise, *Simplicillium obclavatum* produced peptaibiotics with activity against plant and fish pathogens, expanding the potential of marine fungi to biotechnological and agricultural domains (Liang et al., 2024).

3.2 *Biosurfactant Activity*

The worldwide production of surfactants amounted to 17 million metric tones in 2000 (including soaps) with expected future growth rates of 3% to 4% per year globally (Rahman and Gakpe, 2008). Biosurfactants can be used in bioremediation, agriculture, commercial laundry detergents, as biopesticides, in medicine, food processing, the cosmetic industry, and oil recovery (Fakruddin, 2012). This is why the discovery of new and/or more effective biosurfactants represents an emerging field in bioprospecting, given their potential in environmental and biotechnological applications (Perfumo et al., 2018). Several deepsea fungal species have demonstrated remarkable hydrocarbon emulsifying capacity, as evidenced in a study that evaluated kerosene emulsification activity and obtained positive results for *T. balaustiformis*, *A. psychrotrophicus*, and *P. cf. palitans*, which had comparable results to commercial emulsifiers (da Silva et al., 2024).

These findings are particularly relevant considering that oil spills in marine environments represent a recurrent environmental issue, for which fungal biosurfactants offer eco-friendly bioremediation alternatives (Kumar et al., 2021). In the bioremediation sector, piezotolerant marine fungi have emerged as promising candidates due to their metabolic diversity and ability to degrade recalcitrant compounds under extreme conditions (Kumar et al., 2021). A notable example is *Aspergillus sydowii*, capable of degrading crude oil under high-pressure conditions, with degradation efficiency further enhanced by supplementation with nitrogen sources such as $CO(NH_2)_2$ and $(NH_4)H_2PO_4$, leading to significant increases in both biomass production and hydrocarbon degradation (Kumar et al., 2021). These investigations highlight the potential of fungi as biosurfactant producers and key agents in new industry mycocentric strategies to find new solutions.

3.3 *Phytotoxic Activity*

Fungal metabolites have increasingly been recognized for their phytotoxic and herbicidal potential, positioning them as promising candidates for the development of novel agrochemicals (Pantović et al., 2023). Herbicidal activity has been reported among various deepsea fungal isolates, with species such as *A. fusidioides, P. allii-sativi, P. chrysogenum, P. palitans*, and *P. solitum* producing compounds capable of completely inhibiting the germination of *A. schoenoprasum and L. sativa* (Ogaki et al., 2020). Notably, *A. fusidioides* and *P. verrucosus* were also shown to produce metabolites with selective herbicidal activity, underscoring the potential of fungi as a sustainable source of bioherbicides (Ogaki et al., 2020).

Specifically for *P. cf. Palitans*, there is a report that the species displayed varying levels of phytotoxicity but in some cases with a germination inhibition of 100% (da Silva et al., 2024). Despite these promising results, these *P. cf. Palitans* strains isolated from Antarctic substrates remain insufficiently studied, and further investigations are required to evaluate their potential in agrochemical applications (da Silva et al., 2024). The development of fungi-based bioherbicides has shown increasing success for the control of various weeds, multiple mycoherbicides based on terrestrial strains have been established and made commercially available (Roberts et al., 2022). Even though deepsea fungi have the potential to be formulated as a mycoherbicide, these have yet to be analyzed and commercialized for application.

3.4 Biodegradation Activity

One of the most promising industrial applications of marine fungal metabolites lies in their ability to degrade complex molecules and digest biomass. Genomic studies of *Emericellopsis maritima* and *Cadophora malorum* Mo12 have revealed a large arsenal of carbohydrate-active enzymes (CAZymes), underscoring their importance in biomass degradation (Dowd and Tuohy et al., 2023; Rédou et al., 2016). Among the enzymatic activities with industrial importance, xylanases play a key role in hemicellulose breakdown and have been reported in *E. maritima, P. chrysogenum, P. antarcticum*, and *T. stollii*, although no lignin-degrading activity has yet been identified (Dowd and Tuohy et al., 2023).

Moreover, *E. maritima* produces a broad spectrum of enzymes, including exo-acting glucoside hydrolases and feruloyl esterases, both essential for lignocellulosic biomass deconstruction (Dowd and Tuohy et al., 2023). The marine fungi lignin degradation could contribute to biomass pretreatment. *Chaetomium* sp., CS1 degraded alkali lignin, significantly improving the digestibility of sawdust by black soldier fly larvae and thereby enhancing lignocellulose conversion into biofuels and bioproducts (Ling et al., 2025). Microbial lignin degradation mechanisms have been recognized as valuable alternatives to reduce the reliance on harmful chemical processes for lignin biowaste treatment across various industries (Atiwesh et al., 2022). Deepsea fungi hold the potential to further expand these possibilities, offering novel and sustainable strategies for lignin degradation.

Proteases from marine fungi are another class of enzymes with extensive industrial relevance, accounting for more than 25% of the worldwide enzyme market, particularly in food, dairy, detergent, pharmaceutical, and leather industries (Damare et al., 2006). For instance, a culture of *Aspergillus ustus* NIOCC #20 from deepsea sediments exhibited significant protease production (30–40 ACU mL^{-1}) under standard conditions, while showing tolerance to elevated hydrostatic pressure and low temperature, albeit with reduced enzyme output under such conditions. Interestingly, proteases from cold-adapted fungi can be classified into three groups, with group II enzymes being heat-sensitive but displaying enhanced activity at low temperatures, a property that broadens their applicability in industrial processes requiring energy-efficient biocatalysts (Damare et al., 2006).

Beyond polysaccharide and protein degradation, marine fungi also play a role in the biotransformation of environmentally relevant molecules. Denitrification studies revealed that *Acremonium* sp. GA31 and *Aspergillus versicolor* GA8 can reduce nitrite to N_2O at substantial rates (8,556.7 and 44,631.8 ppm/g-dry mass/day, respectively), highlighting their involvement in nitrogen cycling and potential impacts on greenhouse gas emissions (Gao et al., 2020). Additionally, plastic biodegradation has emerged as a critical research frontier. *Hortaea werneckii* 13, isolated from Pacific deepsea sediments, was shown to degrade waterborne polyurethane Impranil DLN by cleaving ester and urea bonds, while also utilizing solid polyurethane as the sole carbon source, leading to significant polymer erosion (Han et al., 2024). This finding demonstrates the potential of marine fungi for eco-friendly solutions to environmental problems.

4. Conclusion and Outlook

The actual research points to the deepsea as an underexplored reservoir of bioactive compounds with remarkable biotechnological potential. In the medical field for example, the fact that nearly 40% of deepsea natural products are drug-like and two-thirds fall within the known drug space suggests that systematic bioprospecting of organisms may accelerate the identification of promising therapeutic leads (Pilkington, 2019). Considering that species such as *Penicillium chrysogenum* are capable of producing more than 10 distinct compounds with potential applications across up to eight different industry sectors, it becomes evident that a vast reservoir of bioactive metabolites remains to be discovered. Deepsea fungi represent a promising alternative to the historical reliance on terrestrial sources, with potential applications in sectors such as agriculture, bioremediation, and beyond (Monaster and Lush, 2011). Their ability to produce multiple biomolecules makes them especially attractive for futuristic contexts like long-duration space missions, where organisms adapted to extreme conditions could serve as valuable reservoirs of compounds for medicine, nutrition, and other applications (Simões et al., 2023).

Looking ahead in the short term, advancing in deepsea fungal research requires an interdisciplinary framework that bridges marine biotechnology, synthetic biology, and drug discovery methodologies. Cultivation under simulated deepsea conditions with controlled stimuli, combined with innovative bioprocessing strategies, may unlock the full biotechnological capacity of these fungi. At the same time, scaling up production of deepsea fungi and bioactivity optimization is essential to move from laboratory discoveries to real-world applications. Deepsea fungi could play a transformative role in addressing global challenges, from antimicrobial resistance and cancer to developing sustainable biocontrol agents and eco-friendly biodegradation technologies. Therefore, to keep advancing in the exploration of deepsea fungal applications, protecting the deepsea ecosystems from threats like deepsea mining will be crucial to unlock the potential hidden within the deep ocean's fungal world.

References

Alanzi, A.R., Parvez, M.K., Alqahtani, M.J. and Al-Dosari, M.S. (2023). Deepsea fungal metabolites as potential inhibitors of glucose-regulatory enzymes: *In silico* structure–activity analysis. Saudi Pharm. J., 31(11): 101776. https://doi.org/10.1016/j.jsps.2023.101776.

Arango Duque, G. and Descoteaux, A. (2014). Macrophage cytokines: Involvement in immunity and infectious diseases. Front. Immunol., 5: 491. https://doi.org/10.3389/fimmu.2014.00491.

Arumugam, G.K., Srinivasan, S.K., Joshi, G., Gopal, D. and Ramalingam, K. (2015). Production and characterization of bioactive metabolites from piezotolerant deepsea fungus *Nigrospora* sp. in submerged fermentation. J. Appl. Microbiol., 118(1): 99–111. https://doi.org/10.1111/jam.12693.

Atiwesh, G., Parrish, C.C., Banoub, J. and Le, T.A.T. (2022). Lignin degradation by microorganisms: A review. Biotechnol. Prog., 38(2): e3226. https://doi.org/10.1002/btpr.3226.

Chen, W., Chen, C., Long, J., Lan, S., Lin, X. et al. (2021). Bioactive secondary metabolites from the deepsea derived fungus *Aspergillus* sp. SCSIO 41029. J. Antibiot., 74(2): 156–159. https://doi.org/10.1038/s41429-020-00378-y.

Chen, S., Wang, J., Wang, Z., Lin, X., Zhao, B. et al. (2017). Structurally diverse secondary metabolites from a deepsea-derived fungus *Penicillium chrysogenum* SCSIO 41001 and their biological evaluation. Fitoterapia, 117: 71–78. https://doi.org/10.1016/j.fitote.2017.01.005.

Chong, B., Jayabaskaran, J., Jauhari, S.M., Chan, S.P., Goh, R. et al. (2025). Global burden of cardiovascular diseases: Projections from 2025 to 2050. Eur. J. Prev. Cardiol., 32(11): 1001–1015. https://doi.org/10.1093/eurjpc/zwae281.

Cong, M., Pang, X., Zhao, K., Song, Y., Liu, Y. et al. (2022). Deepsea natural products from extreme environments: Cold seeps and hydrothermal vents. Mar. Drugs, 20(6): 404. https://doi.org/10.3390/md20060404.

Damare, S., Raghukumar, C., Muraleedharan, U.D. and Raghukumar, S. (2006). Deepsea fungi as a source of alkaline and cold-tolerant proteases. Enzyme Microb. Technol., 39(2): 172–181. https://doi.org/10.1016/j.enzmictec.2006.03.032.

da Silva, M.K., Barreto, D.L.C., Vieira, R., Neto, A.A., de Oliveira, F.S. et al. (2024). Diversity and enzymatic, biosurfactant, and phytotoxic activities of culturable Ascomycota fungi present in marine sediments obtained near the South Shetland Islands, maritime Antarctica. Extremophiles, 28(2): 20. https://doi.org/10.1007/s00792-024-01336-4.

Deng, L., Zhong, M., Li, Y., Hu, G., Zhang, C. et al. (2023). High hydrostatic pressure harnesses the biosynthesis of secondary metabolites via the regulation of polyketide synthesis genes of hadal sediment-derived fungi. Front. Microbiol., 14: 1207252. https://doi.org/10.3389/fmicb.2023.1207252.

Dong, Y., Cui, C.-B., Li, C.-W., Hua, W., Wu, C.-J. et al. (2014). Activation of dormant secondary metabolite production by introducing neomycin resistance into the deepsea fungus *Aspergillus versicolor* ZBY-3. Mar. Drugs, 12(8): 4326–4352. https://doi.org/10.3390/md12084326.

Dowd, B. and Tuohy, M.G. (2023). Induction and characterisation of lignocellulolytic activities from novel deepsea fungal secretomes. Fermentation, 9(9): 780. https://doi.org/10.3390/fermentation9090780.

Fakruddin, M.D. (2012). Biosurfactant: Production and application. J. Pet. Environ. Biotechnol., 3(4): 124. https://doi.org/10.4172/2157-7463.1000124.

Gao, Y., Du, X., Xu, W., Fan, R., Zhang, X. et al. (2020). Fungal diversity in deep-sea sediments from East Yap Trench and their denitrification potential. Geomicrobiol. J., 37(9): 848–858. https://doi.org/10.1080/01490451.2020.1789778.

Guo, W., Zhang, Z., Zhu, T., Gu, Q. and Li, D. (2015). Penicyclones A–E, antibacterial polyketides from the deepsea-derived fungus *Penicillium* sp. SCSIO 41058. J. Nat. Prod., 78(11): 2691–2695. https://doi.org/10.1021/acs.jnatprod.5b00655.

Han, Z., Zhang, K., Ling, M., Zhuang, P., Li, J. et al. (2024). Degradation characteristics of a deepsea polyurethane-degrading fungus *Hortaea werneckii* 13. Mycosystema, 43(5): 230219. https://doi.org/10.13346/j.mycosystema.230219.

He, Z.H., Xie, C.L., Wu, T., Zhang, Y., Zou, Z.B. et al. (2023). Neotricitrinols A–C, unprecedented citrinin trimers with anti-osteoporosis activity from the deepsea-derived *Penicillium citrinum* W23. Bioorg. Chem., 139: 106756. https://doi.org/10.1016/j.bioorg.2023.106756.

Kario, K., Okura, A., Hoshide, S. and Mogi, M. (2024). The WHO Global report 2023 on hypertension warning the emerging hypertension burden in globe and its treatment strategy. Hypertens. Res., 47(5): 1099–1102. https://doi.org/10.1038/s41440-024-01622-w.

Kumar, A.G., Manisha, D., Sujitha, K., Peter, D.M., Kirubagaran, R. et al. (2021). Genome sequence analysis of deepsea *Aspergillus sydowii* BOBA1 and effect of high pressure on biodegradation of spent engine oil. Sci. Rep., 11: 9347. https://doi.org/10.1038/s41598-021-88525-9.

Lee, H.S., Kang, J.S., Cho, D.Y., Choi, D.K. and Shin, H.J. (2022). Isolation, structure determination, and semisynthesis of diphenazine compounds from a deepsea-derived strain of the fungus *Cystobasidium laryngis* and their biological activities. J. Nat. Prod., 85(4): 857–865. https://doi.org/10.1021/acs.jnatprod.1c00985.

Li, Y., Zhang, C., Zhong, M., Hu, S., Cui, Y. et al. (2024). Revealing the metabolic potential and environmental adaptation of nematophagous fungus, Purpureocillium lilacinum, derived from hadal sediment. Front. Microbiol., 15: 1474180. https://doi.org/10.3389/fmicb.2024.1474180.

Li, Y., Zhong, C., Kraithong, S., Gong, H., Han, Z. et al. (2025). Insights into a novel exopolysaccharide from Mariana Trench-derived *Aspergillus versicolor* SCAU214: Structure and immune activity. Int. J. Biol. Macromol., 308: 142660. https://doi.org/10.1016/j.ijbiomac.2025.142660.

Liang, X., Yang, J.-F., Huang, Z.-H., Ma, X., Yan, Y. et al. (2024). New antibacterial peptaibiotics against plant and fish pathogens from the deepsea-derived fungus *Simplicillium obclavatum* EIODSF 020. J. Agric. Food Chem., 72(12): 6402–6413. https://doi.org/10.1021/acs.jafc.4c00493.

Ling, M., Zheng, P., Huang, X., Fan, G., Zhang, H. et al. (2025). Improving bioavailability of lignocellulosic biomass by pretreatment with the marine fungus *Chaetomium* sp. CS1. Int. Biodeterior. Biodegradation, 199: 106031. https://doi.org/10.1016/j.ibiod.2025.106031.

Lv, D., Xia, J., Guan, X., Lai, Q., Zhang, B. et al. (2022). Indole diketopiperazine alkaloids isolated from the marine-derived fungus *Aspergillus chevalieri* MCCC M23426. Front. Microbiol., 13: 950857. https://doi.org/10.3389/fmicb.2022.950857.

Marie, P.J. and Kassem, M. (2011). Osteoblasts in osteoporosis: Past, emerging, and future anabolic targets. Eur. J. Endocrinol., 165(1): 1–10. https://doi.org/10.1530/EJE-11-0132.

Montaser, R. and Luesch, H. (2011). Marine natural products: A new wave of drugs? Future Med. Chem., 3(12): 1475–1489. https://doi.org/10.4155/fmc.11.118.

Ogaki, M.B., Coelho, L.C., Vieira, R., Neto, A.A., Zani, C.L. et al. (2020). Cultivable fungi present in deepsea sediments of Antarctica: Taxonomy, diversity, and bioprospecting of bioactive compounds. Extremophiles, 24(2): 227–238. https://doi.org/10.1007/s00792-019-01148-x.

Pang, X., Lin, X., Zhou, X., Yang, B., Tian, X. et al. (2020). New quinoline alkaloid and bisabolane-type sesquiterpenoid derivatives from the deepsea-derived fungus *Aspergillus* sp. SCSIO06786. Fitoterapia, 140: 104406. https://doi.org/10.1016/j.fitote.2019.104406.

Pang, X., Zhou, X., Lin, X., Yang, B., Tian, X. et al. (2021). Structurally various sorbicillinoids from the deepsea sediment derived fungus *Penicillium* sp. SCSIO06871. Bioorg. Chem., 107: 104600. https://doi.org/10.1016/j.bioorg.2020.104600.

Pantović, J.G., Sečanski, M., Gordanić, S. and Šarčević-Todosijević, L.Š. (2023). Weed biological control with fungi-based bioherbicides. Acta Agric. Serb., 28(55): 23–37. https://doi.org/10.5937/AASer2355023G.

Patel, H., Royall, P.G., Gaisford, S., Williams, G.R., Edwards, C.H. et al. (2017). Structural and enzyme kinetic studies of retrograded starch: Inhibition of α-amylase and consequences for intestinal digestion of starch. Carbohydr. Polym., 164: 154–161. https://doi.org/10.1016/j.carbpol.2017.01.040.

Perfumo, A., Banat, I.M. and Marchant, R. (2018). Going green and cold: Biosurfactants from low-temperature environments to biotechnology applications. Trends Biotechnol., 36(3): 277–289. https://doi.org/10.1016/j.tibtech.2017.10.016.

Pilkington, L.I. (2019). A chemometric analysis of deepsea natural products. Molecules, 24(21): 3942. https://doi.org/10.3390/molecules24213942.

Rahman, P.K.S.M. and Gakpe, E. (2008). Production, characterization, and application of biosurfactants: A review. Biotechnology, 7: 360–370. https://doi.org/10.3923/biotech.2008.360.370.

Rédou, V., Kumar, A., Hainaut, M., Henrissat, B., Record, E. et al. (2016). Draft genome sequence of the deepsea ascomycetous filamentous fungus *Cadophora malorum* Mo12 from the Mid-Atlantic Ridge reveals its biotechnological potential. Genome Announc., 4(4): e00467-16. https://doi.org/10.1128/genomeA.00467-16.

Roberts, J., Florentine, S., Fernando, W.G.D. and Tennakoon, K.U. (2022). Achievements, developments and future challenges in the field of bioherbicides for weed control: A global review. Plants, 11(17): 2242. https://doi.org/10.3390/plants11172242.

Rodríguez Ramírez, I., Solano-González, S., Cortés, J. and Rojas-Jiménez, K. (2025). Deepsea fungi of the eastern tropical Pacific of Costa Rica: Morphological, genetic, and enzymatic characterization. Front. Mar. Sci., 12. https://doi.org/10.3389/fmars.2025.1514874.

Simões, M.F., Cortesão, M., Azua-Bustos, A., Bai, F.Y., Canini, F. et al. (2023). The relevance of fungi in astrobiology research: Astromycology. Mycosphere, 14(1): 1190–1253. https://doi.org/10.5943/mycosphere/14/1/13.

Sun, Z.H., Gu, J., Ye, W., Wen, L.X., Lin, Q.B. et al. (2018). Geospallins A–C: New thiodiketopiperazines with inhibitory activity against angiotensin-converting enzyme from a deepsea-derived fungus *Geosmithia pallida* FS140. Mar. Drugs, 16(12): 464. https://doi.org/10.3390/md16120464.

Sun, Y., Takada, K., Takemoto, Y., Yoshida, M., Nogi, Y., Okada, S. and Matsunaga, S. (2011). Gliotoxin analogues from a marine-derived fungus *Penicillium* sp. strain JMF034. J. Nat. Prod., 75(12): 2201–2205. https://doi.org/10.1021/np200740e.

Villanueva-Silva, R., Velez, P., Riquelme, M., Fajardo-Hernández, C.A., Martínez-Cárdenas, A. et al. (2021). Chemical diversity and antimicrobial potential of cultivable fungi from deepsea sediments of the Gulf of Mexico. Molecules, 26(23): 7328. https://doi.org/10.3390/molecules26237328.

Wang, F., Fang, Y., Zhu, T., Zhang, M., Lin, A. et al. (2008). Seven new prenylated indole diketopiperazine alkaloids from holothurian-derived fungus *Aspergillus fumigatus*. Tetrahedron, 64(34): 7986–7991. https://doi.org/10.1016/j.tet.2008.06.013.

Wang, H., Guo, M., Wei, H. and Chen, Y. (2023a). Targeting p53 pathways: Mechanisms, structures, and advances in therapy. Signal Transduct. Target Ther., 8(1): 92. https://doi.org/10.1038/s41392-023-01347-1.

Wang, J.-F., Zhou, L.-M., Chen, S.-T., Yang, B., Liao, S.-R. et al. (2018). New chlorinated diphenyl ethers and xanthones from a deepsea-derived fungus *Penicillium chrysogenum* SCSIO 41001. Fitoterapia, 125: 49–54. https://doi.org/10.1016/j.fitote.2017.12.012.

Wang, J., Zhou, Y., Zhang, H., Hu, L., Liu, J. et al. (2023b). Pathogenesis of allergic diseases and implications for therapeutic interventions. Signal Transduct. Target Ther., 8(1): 138. https://doi.org/10.1038/s41392-023-01344-4.

Wiese, J., Aldemir, H., Schmaljohann, R., Gulder, T.A.M. and Imhoff, J.F. (2017). Asperentin B, a new inhibitor of the protein tyrosine phosphatase 1B. Marine Drugs, 15(6): 191. https://doi.org/10.3390/md15060191.

Xie, C.L., Ye, S.H., Yue, Y.T., Shi, B.H., Xu, J.P. et al. (2025). Viridicatol from the deepsea-derived fungus alleviates bone loss by targeting the Wnt/SHN3 pathway. Adv. Sci., 12(21): 2416140. https://doi.org/10.1002/advs.202416140.

Xie, M.-M., Jiang, J.-Y., Zou, Z.-B., Xu, L., Zhang, Y. et al. (2022). Chemical constituents of the deepsea-derived fungus *Cladosporium oxysporum* 170103 and their antibacterial effects. Chem. Biodivers., 19(12): e202200963. https://doi.org/10.1002/cbdv.202200963.

Xing, C.P., Chen, D., Xie, C.L., Liu, Q., Zhong, T.H. et al. (2021). Anti-food allergic compounds from *Penicillium griseofulvum* MCCC 3A00225, a deepsea-derived fungus. Mar. Drugs, 19(4): 224. https://doi.org/10.3390/md19040224.

Xu, A., Xu, X.N., Zhang, M., Li, C.L., Liu, L. et al. (2022). Cytotoxic indole alkaloids and polyketides produced by a marine-derived fungus *Aspergillus flavipes* DS720. Front. Microbiol., 13: 959754. https://doi.org/10.3389/fmicb.2022.959754.

Yadav, S. and Kapley, A. (2021). Antibiotic resistance: Global health crisis and metagenomics. Biotechnol. Rep., 29: e00604. https://doi.org/10.1016/j.btre.2021.e00604.

Ye, W., Zhang, W., Liu, T., Huang, Z., Zhu, M. et al. (2018). *De novo* transcriptome sequencing of the deepsea-derived fungus *Dichotomomyces cejpii* and analysis of gliotoxin biosynthesis genes. Int. J. Mol. Sci., 19(7): 1910. https://doi.org/10.3390/ijms19071910.

Zhang, H., Diao, X., Jiang, T., Wei, M., Su, Y. et al. (2024). Novel Deepsea Isoindole Alkaloid FGFC1 exhibits its fibrinolytic effects by inhibiting thrombin-activatable fibrinolysis inhibitor. Pharmaceuticals, 17(10): 1401. https://doi.org/10.3390/ph17101401.

Zhang, M., Chao, R., Wang, J.J., Xu, Z.H., Zhang, J.H. et al. (2025). Statins diversity revealed by the Deepsea-derived fungus *Penicillium viridicatum*. Mar. Drugs, 23(2): 87. https://doi.org/10.3390/md23020087.

Zhang, X.-Y., Zhang, Y., Xu, X.-Y. and Qi, S.-H. (2013). Diverse deepsea fungi from the South China Sea and their antimicrobial activity. Curr. Microbiol., 67(5): 525–530. https://doi.org/10.1007/S00284-013-0394-6.

Zhao, G., Tang, W., Zhang, J., Shi, P., Li, Y. et al. (2022). Deepsea-derived fungi as valuable producers of cytotoxic secondary metabolites and their leads potential. Front. Mar. Sci, 9: 929561. https://doi.org/10.3389/fmars.2022.929561.

15

Geomycology of Marine Environments

Recent Advances and Future Perspectives

Sujata Dabolkar

The study of geomycology, which focuses on the interactions between fungi and geological or geochemical processes, has been extensively explored in terrestrial environments. However, the roles of fungi in marine ecosystems remain significantly under-investigated, despite their precise contributions to biogeochemical cycles, biomineralization, and pollutant degradation. Marine fungi influence the cycling of key elements such as carbon, nitrogen, sulfur, and metals through their metabolic processes and enzymatic activities. Additionally, they play essential roles in the degradation of complex organic compounds, mineral precipitation, and the bioremediation of environmental contaminants. With advancements in metagenomics, high-throughput sequencing, and deep-sea exploration, recent studies have begun to reveal the hidden diversity and functionality of marine fungi. These organisms have been found in various habitats, including coastal estuaries, coral reefs, and the extreme environments of hydrothermal vents and deep-sea sediments. Their unique physiological adaptations enable them to thrive under high pressure, elevated salinity, and low oxygen levels, making them crucial players in marine sediment stabilization, ocean chemistry regulation, and climate change mitigation. This chapter provides a comprehensive overview of the diversity and ecological roles of marine fungi, highlighting their contributions to

Department of Botany, Government College of Arts, Science and Commerce, Quepem, Goa, India.
Email: sujatadabolkar@gmail.com

DOI: 10.1201/9781003795100-19

biogeochemical cycling, biomineralization, and environmental applications. Furthermore, we explore the industrial potential of marine fungi, particularly in bioremediation, biomining, and sustainable biotechnology.

1. Introduction

Fungi are among the most versatile and ecologically significant groups of microorganisms on Earth. Traditionally, studies on fungal interactions with geological and chemical processes—collectively termed geomycology—have primarily focused on terrestrial fungi (Gadd, 2016). These fungi play pivotal roles in mineral weathering, metal transformations, and nutrient cycling, which contribute to soil formation and the stabilization of terrestrial ecosystems. However, fungi's role in marine environments has remained relatively understudied until recently. Unlike their terrestrial counterparts, marine fungi thrive in highly dynamic and extreme conditions, ranging from sunlit coastal waters to high-pressure abyssal plains and toxic, mineral-rich hydrothermal vents (Le Calvez et al., 2009. Geomycology is indeed a fascinating field, emphasizing the significant role fungi play in geological processes and their biogeochemical influence (Gadd et al., 2011). The way fungi interact with minerals and metals, transforming them, and how they contribute to processes like weathering and biomineral formation have implications for both natural ecosystems and human industries (Podia et al., 2023). The mutualistic relationships fungi form with other organisms, like in lichens and mycorrhizas, are crucial for many ecosystems, enhancing nutrient exchange and supporting the growth of plants and phototrophic organisms. The environmental and industrial applications of geomycology, like in detoxification, leaching, and recovery of metals, show just how important this field is for tackling current challenges in environmental management and sustainability. Geomycology is defined as the study of the roles and significance of fungi in geologically relevant processes, forming a vital subset of the wider field of geomicrobiology (Gadd, 2010).

In marine environments, fungi play a role in organic matter decomposition, sediment stabilization, biomineralization, and pollutant degradation (Raghukumar, 2017; Behera and Das, 2023). Despite these crucial ecological functions, marine fungi have historically been overshadowed by studies focusing on bacteria and archaea. This is mainly due to technical challenges associated with cultivating marine fungi under laboratory conditions and the difficulty of extracting fungal DNA from complex marine substrates. However, a more complete picture of marine fungal diversity is emerging with recent advancements in next-generation sequencing, metagenomics, and environmental DNA (eDNA) analysis (Garlapati et al., 2019). Beyond their ecological significance, marine fungi also have important industrial and environmental applications. Their ability to degrade complex hydrocarbons, metals, and even plastic waste makes them promising agents for bioremediation. Additionally, the study of fungal-mediated biomineralization has significant implications for biomining, climate change mitigation, and sustainable material development (Vaksmaa et al., 2023).

As research in marine geomycology expands, new opportunities for harnessing the potential of marine fungi in biotechnology, environmental restoration, and sustainable resource management continue to arise (Table 1). Notably, recent fossil records from igneous oceanic crust have revealed extensive fungal presence, showing the presence of the largest fungal habitat on Earth (Ivarsson et al., 2016).

Looking forward, future research must focus on fungal-prokaryotic interactions across subsurface environments, soils, plant-root zones, and aquatic systems. The filamentous growth of fungi allows them to colonize and penetrate complex substrates like soil, translocating nutrients and forming mycelial highways that support bacterial migration and colonization (Warmink et al., 2011). Such fungal networks (Martin et al., 2012) can enhance organic pollutant degradation, soil fertility, and long-term carbon storage—notably through the oxalate-carbonate pathway. In principle, we must embrace geomycology as a fundamental concept in Earth and environmental sciences due to the ubiquity and activity of fungi. Integration of molecular biology, metagenomics, microscopy, and classical culturing will be essential to characterize biodiversity and understand functional roles. A renewed focus on fungal-prokaryotic interactions, phototrophic symbioses, and fungal contributions to mineral transformations can profoundly expand our understanding of Earth's biosphere. As boundaries blur between disciplines, interdisciplinary collaboration will be key to unlocking the full potential of fungi as geoengineers in both natural and engineered environments. The chapter outlines future research directions and emphasizes the need for interdisciplinary studies, deep-sea bioprospecting, and genomic exploration of marine fungi to unlock their potential in environmental and industrial applications.

2. Diversity and Adaptations

Fungi from marine environments represent a diverse array of taxonomic groups, including Ascomycota, Basidiomycota, Chytridiomycota, and Mucoromycota. They inhabit various ecological niches, from intertidal zones and coral reefs to deep-sea hydrothermal vents (Pachiadaki and Edgcomb, 2022; El-Regal and Satheesh, 2023). Some of these fungi exist as free-living saprotrophs, decomposing organic material found in the water column and sediments, while others form symbiotic relationships with marine plants, corals, and even animals. The adaptations that enable marine fungi to survive and thrive in these diverse environments offer valuable insights into their evolutionary history and ecological significance.

One of the key adaptations of marine fungi is their ability to tolerate high salinity. Unlike terrestrial fungi, which are often inhibited by elevated salt concentrations, marine fungi have developed specialized mechanisms for osmotic regulation. These mechanisms involve the production of compatible solutes, such as glycerol and mannitol, which help maintain cellular homeostasis in saline environments. Additionally, marine fungi secrete extracellular enzymes that enable them to break down complex organic materials, including cellulose, lignin, and hydrocarbons, allowing them to play a crucial role in organic matter recycling. The marine environment is remarkably complex and harbors a vast spectrum of fungal diversity. The unique physicochemical conditions of this ecosystem have likely endowed marine fungi with distinctive physiological adaptations, making them promising candidates for biotechnological applications. This chapter focuses on marine fungi

Table 1. Selected recent developments in marine fungal research.

Fungal Taxon	Marine Habitat	Ecological Role	Notable Species	Recent Advances/Findings: Conditions and Marine Stress Adaptation	References
Aspergillus spp.	Coastal waters, mangroves, marine sediments	Decomposition, biosorption of metals	*Aspergillus sydowii*, *Aspergillus niger*	Metal tolerance, bioremediation of heavy metals like lead and cadmium; production of bioactive compounds	Gazaem (2012); Bicholkar (2016)
Alternaria spp.	Marine algae, sand, sea foam	Endophyte, pathogen	*Alternaria alternata*	Produces bioactive compounds under saline	Peng et al. (2023)
Aureobasidium pullulans	Marine biofilms, ship hulls	Biofilm former, decomposer	*Aureobasidium pullulans*	Produces melanin and pullulan; adapts to metal toxicity	Nakkeeran et al. (2018); Rensink et al. (2024)
Basidiomycetes (marine mushrooms)	Coral debris, marine wood	Lignin decomposers	*Schizophyllum commune* (marine isolates)	Very rare, but DNA found in marine metagenomes	Raghukumar (2017)
Cladosporium spp.	Surface of marine organisms, marine aerosols	Endophyte, pathogen, decomposer	*Cladosporium cladosporioides*	Tolerates hypersaline conditions, involved in degradation of hydrocarbons	Yovchevska et al. (2025)
Candida spp.	Marine animals and sediments	Symbiont, decomposer	*Candida albicans* (marine isolate)	Studied for salt tolerance, metabolic plasticity	Kandasamy et al. (2012)
Debaryomyces hansenii	Seawater, marine sediments	Halotolerant yeast	*Debaryomyces hansenii*	Genetic studies reveal genes responsible for halotolerance; biotechnological potential	Andreu et al. (2022)
Emericellopsis spp.	Deep-sea sediments	Antibiotic production	*Emericellopsis maritima*	Produces antimicrobial peptides under marine pressure	Hagestad (2021)
Exophiala spp.	Coral reefs, marine sediments	Black yeast, pathogen	*Exophiala dermatitidis*	Forms biofilms in saline waters; survives extreme marine stress	Raghukumar (2017)
Fusarium spp.	Estuarine and coastal sediments	Pathogen, mycotoxin producer	*Fusarium solani*	Involved in biocorrosion; adaptation to marine stress; pathogenic to marine plants	Wekesa (2024)

Table 1 contd. ...

...Table 1 contd.

Fungal Taxon	**Marine Habitat**	**Ecological Role**	**Notable Species**	**Recent Advances/Findings: Conditions and Marine Stress Adaptation**	**References**
Lulworthia spp.	Marine wood, seaweed	Saprobe, wood decay	*Lulworthia fucicola*	Efficient lignin degraders in saline conditions	Venkateswara (2018)
Myceliophthora spp.	Hot marine sediments	Thermophilic decomposer	*Myceliophthora thermophila*	Enzyme production for industrial marine biotech	Singh (2016); Raghukumar (2008)
Nigrospora spp.	Marine macroalgae	Endophyte, decomposer	*Nigrospora oryzae*	Produces novel mycotoxins; salt and pH stress tolerance	Flewelling (2015)
Penicillium spp.	Marine sediments, decaying seaweed	Decomposer, secondary metabolite production	*Penicillium citrinum*, *Penicillium chrysogenum*	Marine isolates produce novel antimicrobials and show biosorption of gold, arsenic	Behera and Das (2023)
Trichoderma spp.	Coastal mangroves, marine driftwood	Decomposer, mycoparasite	*Trichoderma harzianum*	Biocontrol potential in aquaculture; enzyme production	Wekesa (2024); Trianto et al. (2021)
Uncultured marine fungi	Deep-sea hydrothermal vents	Unknown	-	Detected via metagenomics; involved in extreme nutrient cycling	Burgaud et al. (2014); Mahe et al. (2011); Nagahama and Nagano (2011); Zhou et al. (2022)
Zygomycetes (e.g., *Mucor* spp.)	Intertidal zones	Saprotroph	*Mucor circinelloides*	Studies on lipid production for marine biodiesel	Richardson (2009)

inhabiting extreme environments and their specialized ecological niches. These fungi often exist as endophytes or are associated with marine flora and fauna. Particularly intriguing are the deep-sea habitats—characterized by high hydrostatic pressure, low temperatures, hydrothermal vents with extreme heat, elevated metal concentrations, and anoxic marine sediments—which remain largely unexplored sources of fungal diversity. Understanding the adaptations of extremotolerant fungi in such habitats may offer profound insights into the evolution of molecular mechanisms in eukaryotes and open new avenues for the discovery of novel genes.

2.1 *Extreme Environments*

Marine environments are not uniform; they range from shallow coastal zones to some of the most hostile and extreme habitats on Earth, including deep-sea trenches, hydrothermal vents, and hypersaline basins. Fungi that inhabit these extreme marine ecosystems are known as extremophiles, and they exhibit a remarkable array of physiological and molecular adaptations that allow them to survive and thrive under such challenging conditions. The deep sea, for instance, is characterized by high hydrostatic pressure, low temperatures, absence of light, and limited nutrient availability. In contrast, hydrothermal vents present the opposite spectrum of extremes, with temperatures soaring above 100°C, high concentrations of heavy metals, and fluctuating pH. Hypersaline basins, such as Deep Hypersaline Anoxic Basins (DHABs), present yet another unique habitat where salinity may exceed five times that of seawater, coupled with anoxic conditions and high pressure (Stoeck et al., 2014; Merlino et al., 2018; Varrella et al., 2020). These environments, once considered uninhabitable, are now known to support complex microbial communities, including extremotolerant and extremophilic fungi.

Marine fungi living under such conditions have evolved unique biochemical and structural adaptations that confer resistance to environmental stressors. One key adaptation is the modification of cell membrane composition. To maintain fluidity under extreme pressures and low temperatures, these fungi alter the saturation levels of fatty acids in their membrane lipids. This ensures proper membrane function and nutrient transport even under high hydrostatic pressures found at depths exceeding 6,000 meters. Another major adaptive mechanism is the production of compatible solutes or osmolytes, such as trehalose, mannitol, and glycine betaine. These compounds help stabilize proteins and enzymes, prevent dehydration in hypersaline conditions, and maintain osmotic balance without interfering with normal cellular functions. Additionally, many extremophilic marine fungi produce specialized stress proteins and chaperones that assist in protein folding and protect cellular components from thermal or oxidative damage.

A particularly fascinating adaptation is the production of extremozymes. Unlike their mesophilic counterparts, extremozymes can operate at very high or low temperatures, in high salinity, extreme pH, or high pressure. These enzymes play a significant role in biotechnology due to their potential application in industrial processes requiring robust catalytic functions under harsh operational environments. Genomic and transcriptomic studies have revealed that extremophilic marine fungi possess a wide range of genes associated with DNA repair, oxidative stress response, and metal detoxification. These genetic features enable them to survive mutagenic conditions such as radiation and high concentrations of toxic elements like arsenic, lead, and mercury, often found near hydrothermal vents or in polluted marine sediments (Giddings and Newman, 2022; Das et al., 2022). The exploration of these unique adaptations not only deepens our understanding of fungal resilience and evolution but also opens avenues for bioprospecting, where marine fungi are increasingly being recognized as a source of novel biomolecules. These may include thermostable enzymes, antifreeze proteins, metal-binding peptides, and other extremotolerant biomolecules that could be harnessed in industrial, pharmaceutical, or environmental applications. Moreover, the study of these adaptive traits contributes to evolutionary biology by offering insights into how eukaryotic life can evolve under intense selective pressures. As sequencing technologies and *in situ* sampling improve, we are likely to uncover an even broader repertoire of fungal adaptations, potentially revolutionizing our understanding of life in the ocean's most extreme environments.

2.2 *Deep Sea*

The most extreme and inhospitable environments on Earth, characterized by hypersalinity, anoxia, and high hydrostatic pressure, are found in deep-sea hypersaline anoxic basins. Despite these harsh conditions, DHABs support remarkable microbial biodiversity, with recent studies highlighting fungi as a significant and quantitatively relevant group among the resident eukaryotes (Merlino et al., 2018; Fisher et al., 2021). These fungi thrive in DHABs partly due to the accumulation of substantial organic material, which serves as a nutrient source. Marine fungi, particularly halophilic and halotolerant species, are known producers of bioactive molecules, including enzymes and secondary metabolites with significant industrial, pharmaceutical, and environmental applications. In deep-sea environments, fungi must endure extreme hydrostatic pressure and low temperatures. Certain deep-sea fungal species, referred to as barophiles, have evolved pressure-resistant cellular structures and membrane modifications that enable them to function under these harsh conditions. Some marine fungi are also psychrophilic, meaning they are adapted to cold environments and produce specialized enzymes that remain active at low temperatures. These unique adaptations make deep-sea fungi particularly intriguing for biotechnological applications, such as the development of cold-active industrial enzymes and pressure-resistant biomaterials.

3. Role in Biogeochemical Cycles

Fungi have an important role in biogeochemical cycles, affecting the movement and transformation of essential elements in marine ecosystems. In the carbon cycle, marine fungi play a crucial role in decomposing organic matter (Hyde et al., 1998; Raghukumar, 2017; Grossart et al., 2019), by breaking down complex polymers like chitin, cellulose, and lignin, which contribute to the biological carbon pump, which regulates the sequestration of carbon from surface waters into deep-sea sediments. This process has important implications for global climate regulation, as it helps manage levels of atmospheric CO_2. In addition to their role in the carbon cycle, marine fungi also contribute to nitrogen and sulfur cycles. Some fungi are involved in denitrification, a process that converts nitrate into nitrogen gas, thus affecting nitrogen availability in the ocean. Others interact with sulfur-reducing bacteria, participating in the transformation of sulfur compounds within deep-sea sediments. This is particularly significant in anoxic environments, such as hydrothermal vents, where sulfur cycling is a vital component of the ecosystem. They play a critical role in metal cycling and biomineralization (Hyde et al., 1998) and can mediate the oxidation and reduction of metals like iron, manganese, and copper, influencing their solubility and availability in marine systems. Some fungi secrete organic acids that help dissolve minerals, while others aid in forming metal sulfides and oxides. These processes are essential for the formation of marine sediments, hydrothermal vent deposits, and even deep-sea nodules, which are important sources of rare earth elements and transition metals.

The deep biosphere is a fascinating and largely unexplored ecosystem beneath the Earth's surface, comprising microorganisms that thrive in extreme conditions, such as several kilometers below the ground and the ocean floor. Microorganisms are found in environments such as deep-sea hydrothermal vents (Jorgensen et al., 1992), marine sediments (Parkes et al., 2005; Le Calvez et al., 2009; Orsi et al., 2013), terrestrial sedimentary (Fredrickson et al., 1995), igneous rocks, and subseafloor igneous rocks (Schrenk et al., 2010), which host active ecosystems despite their isolation and nutrient limitations. Traditionally, microbiological studies in the deep biosphere have focused heavily on prokaryotes, particularly bacteria, due to their survival ability in low-energy environments (Wu et al., 2016). However, recent discoveries have suggested that eukaryotes, such as fungi, also inhabit these deep ecosystems. The findings that fungi have been identified in a wide range of deep environments such as marine sediments (Connell et al., 2009; Nagano and Nagahama, 2012), deep continental crust (Drake et al., 2017a), and subseafloor basalt (Ivarsson et al., 2012, 2016a) show their potential role in energy cycling and capacity to survive under extreme, often anoxic, conditions (Jebaraj et al., 2010; Raghukumar et al., 2010), indicating that anaerobic conditions are not necessarily a limiting factor for fungal growth. Instead, fungi appear to have adapted to anaerobic metabolism, challenging previous assumptions about the limitations of life in such conditions (Ivarsson et al., 2016b). Recently, fungi have also been described in consortia with archaea in marine sediments (Picard, 2017), and such interactions may extend to other deep biosphere environments (Ivarsson et al., 2016b; Drake et al., 2017a). There are reports on an experimental study on the ecological and biogeochemical role of calcimycocavites from the sand samples of Goa in understanding carbon fluxes in the age of global warming and climate change (Dabolkar and Kamat, 2024).

The latest research on fungi-prokaryote consortia in the igneous crust (Drake et al., 2017a) further supports the idea that anaerobic fungi could serve as a previously neglected biogeochemical force in subsurface ecosystems. By producing H_2, these fungi may fuel H_2-dependent chemoautotrophic microorganisms in the deep biosphere, expanding our understanding of microbial interactions and energy cycling in such environments. This research challenges the conventional view that anoxic conditions in deep biosphere habitats are too extreme to support significant eukaryotic life. Their presence in a variety of deep environments opens exciting possibilities for understanding how life can adapt and thrive in extreme conditions, highlighting the resilience and adaptability of life on Earth.

4. Industrial and Environmental Applications

Marine fungi have garnered attention in biotechnology and environmental remediation due to their unique abilities to interact with metals, degrade pollutants, and influence mineral formation. Many species of marine fungi possess biosorption properties. Additionally, certain marine fungi are being studied for their capacity to degrade hydrocarbons, making them potential candidates for cleaning up oil spills and addressing marine pollution. Researchers are exploring the role of aquatic fungi in biomining and metal recovery, by utilizing fungal-mediated bioleaching, scientists are investigating environmentally sustainable methods for extracting valuable metals from deep-sea mineral deposits. Recent literature highlights that extremophilic organisms, such as fungi inhabiting DHABs, have the potential to produce native proteins as well as homologous or heterologous recombinant enzymes with diverse industrial applications, particularly in the realm of "White Biotechnology" (Barone et al., 2019). Also referred to as industrial biotechnology, white biotechnology harnesses living cells and enzymes to synthesize bio-based products that are biodegradable, energy-efficient, and environmentally friendly, generating less waste. Extremozymes, either as isolated molecules or produced directly by extremophilic fungi—including those from DHABs—often display polyextremophilicity. This refers to their stability and catalytic activity across multiple extreme conditions, such as high salinity, extreme pH (acidic or basic), and elevated temperatures ranging from 55 to 113°C. Many marine extremozymes have been investigated for applications in biotechnology, pharmaceutical industries, biofuel, and textile industries (Raddadi et al., 2015; Jin et al., 2019; Kakkar et al., 2022).

Marine fungi produce enzymes such as peroxidases, laccases, which can be used for industrial and environmental applications, including lignin-degrading enzymes. Enzymes derived from marine environments often possess unique characteristics such as high salt tolerance, thermal stability, and adaptations to extreme conditions, making them suitable for harsh industrial environments. Since they have such interesting properties, they are used in breaking down pollutants and waste, chitinases for processing agricultural waste, amylases and lipases in food, biofuel, and pharmaceutical production, and cellulases for various biotechnological uses. There are reports on the potential of marine-derived fungi that can be used in the bioremediation of industrial pollutants (Verma, 2011; Raghukumar et al., 2008).

4.1 *Biotechnological Potential*

The potential of marine fungi is vast and largely untapped (Yovchevska et al., 2025; Zhou et al., 2022). These organisms, shaped by the selective pressures of extreme marine environments, produce a rich arsenal of bioactive compounds, along with the enzymes and very useful secondary metabolites that offer promising applications across diverse sectors, including pharmaceuticals, environmental management, industrial bioprocessing, and aquaculture. As our understanding of marine mycobiota expands, so too does the scope for discovering novel biotechnological tools with high commercial and societal value (Luna, 2015; Kempken, 2023; Damare et al., 2011; Nicoletti and Andolfi, 2018; Patil and Selvaraj, 2025; Selvakumar, 2025).

Marine fungi can produce extremozymes (Bonugli et al., 2015). These enzymes (Dalmaso et al., 2015; Jin et al., 2019), often derived from fungi living in deep-sea habitats, hypersaline basins, or hydrothermal vents, function efficiently at elevated temperatures, extreme pH levels, high salinities, and under pressure. Enzymes like cellulases, proteases, lipases, and laccases derived from marine fungi have already shown promise and consideration in pilot-scale applications (Muthusamy and Murugaiyan, 2023).

Marine fungi are producers of secondary metabolites. Many of these molecules have been isolated for their antimicrobial, antiviral, anticancer, anti-inflammatory, and antioxidant properties. Notable examples include alkaloids, polyketides, peptides, and terpenoids. These compounds are particularly attractive for pharmaceutical research, as they often exhibit novel structures and mechanisms of action not found in terrestrial analogs. Given the rising global concern over antibiotic resistance, marine fungal metabolites offer a reservoir of next-generation antimicrobial agents. In the realm of environmental biotechnology, marine fungi show potential as bioremediation agents. Their ability to tolerate and transform heavy metals, degrade hydrocarbons, and decompose synthetic compounds makes them ideal candidates for cleaning up marine pollutants. Fungi such as *Aspergillus sydowii* and *Penicillium* spp. have demonstrated the capacity to degrade petroleum-based contaminants, an attribute that could be harnessed for oil spill remediation in coastal and open ocean environments. Furthermore, their bioaccumulative and biosorptive capabilities position them as natural filters for detoxifying sediments and effluents contaminated with heavy metals. Marine-derived fungi can be used to synthesize biodegradable plastics, biosurfactants, bioethanol, and organic acids. Their enzymatic systems, tailored to function in saline and variable-temperature conditions, are particularly suited for industrial processes that prioritize efficiency, eco-friendliness, and reduced energy consumption. Over the past decades, marine fungi have been a significant source of a wide range of secondary metabolites with uses in industrial, pharmacological, and environmental fields (Ameen et al., 2021).

New marine fungal species are identified, such as *Simplicillium lamellicola, Leptosphaerulina* sp., *Penicillium citrinum, Penicillium chrysogenum,* and *Aspergillus sydowii* (Agrawal et al., 2018). These isolates have high antimicrobial activity against the acne-inducing bacteria *Cutibacterium acnes* and *Staphylococcus epidermidis*; thus, the extracts of these fungi show an interesting option for obtaining antibiotics for *acne vulgaris*. Three new and four known natural derivatives are obtained from the extracts of the marine-derived fungus *Aspergillus alabamensis* (Yang et al., 2018).

Marine fungi are also gaining attention in aquaculture and agriculture through "Blue" and "Green" biotechnology. In aquaculture, fungal bioactive compounds can be used as natural antifouling agents, probiotics, or disease control agents.

There are studies on the use of marine microorganisms in designing anti-infective strategies for sustainable aquaculture production (Hamza and Zinjarde, 2023). In agriculture, fungal enzymes and metabolites can serve as biofertilizers, biopesticides, or growth promoters supporting sustainable farming practices without the adverse environmental effects associated with chemical inputs.

Aquaculture, a significant food production sector, faces challenges from disease outbreaks. Antibiotic treatments often fail against pathogens due to biofilm formation and resistant strains. Marine ecosystems host unique microorganisms that produce novel bioactive compounds, some of which could serve as antibiotic alternatives. Additionally, biomass and biomolecules from these microorganisms may function as feed additives to boost the health of aquaculture species and improve water quality. Bioactive substances from marine bacteria, such as *Bacillus*, *Vibrio*, *Photobacterium*, and *Pseudoalteromonas*, can inhibit biofilm-related infections through bactericidal activity, surfactant production, anti-adhesive effects, and quorum-sensing inhibition. Marine fungi also produce antibacterial agents effective against aquaculture pathogens. Another approach is using bacterial, yeast, and microalgae biomass as feed supplements, probiotics, or immunostimulants to reduce infection severity. Marine microalgae, as sustainable alternatives to fish oil and meal, can maintain nutritional quality while enhancing growth, survival, and water quality. Marine microorganisms, via bioactive compounds and feed supplementation, could make aquaculture more sustainable in the future.

4.2 *Role in Bioremediation*

Fungi have emerged as potent agents in the biodegradation of persistent environmental pollutants, including complex industrial compounds such as textile dyes. Notably, numerous fungal species possess the enzymatic machinery necessary to break down and decolorize synthetic dyes that resist conventional treatment methods (Haritash and Kaushik, 2009). Marine-derived fungi hold significant promise due to their inherent adaptations to extreme saline and alkaline conditions prevalent in marine habitats. These adaptations render them particularly effective in treating saline textile effluents, which often contain high salt concentrations and elevated pH levels. A number of studies have highlighted the potential of the degradation of dyes such as Congo Red and Brilliant Green by marine fungi. The ability of marine fungi to facilitate efficient decolorization of these pollutants has been well-documented, underscoring their relevance in the treatment of textile wastewater (DeSouza-Ticlo, 2008). Their robustness in saline environments positions them ideal candidates for the remediation of polluted marine and estuarine ecosystems, especially those contaminated with polycyclic aromatic hydrocarbons (PAHs) and industrial effluents. Moreover, salt-tolerant fungal species, along with their extremotolerant enzymes—particularly lignin-degrading enzymes—have been effectively utilized in the bioremediation of various environmental contaminants (Passarini et al., 2011). These enzymes, which include laccases and peroxidases, retain their functionality under high salinity and are integral to the breakdown of recalcitrant organic pollutants. Further investigations have expanded the scope of marine fungi in biodegradation processes, demonstrating their enzymatic capability to degrade lignocellulosic materials. A range of hydrolytic enzymes—such as cellulases, α-amylases, alginates, laminarinases, pectinases, and xylanases—produced by marine fungi have been reported in *in vitro* studies (Balabanova et al., 2018). These enzymes are critical for the deconstruction of plant biomass and organic matter, thereby facilitating nutrient cycling and pollutant breakdown in marine ecosystems.

In addition, specific marine fungal taxa such as ascomycetes and basidiomycetes—including *Nia vibrissa*—have shown the ability to solubilize lignin from wood substrates under laboratory conditions (Bucher et al., 2004). The concerted action of hydrolytic enzymes results in the degradation of lignin (Jones and Choeyklin, 2008). Overall, the intrinsic physiological adaptations and enzymatic proficiencies of marine fungi underscore their immense potential as natural bioremediators in both marine and terrestrial contexts. Their exploitation for environmental cleanup strategies not only aligns with ecological sustainability but also opens new avenues in industrial and environmental biotechnology.

5. Conclusion and Outlook

Understanding marine geomycology is vital for both ecological and industrial applications (Fig. 1). As research continues to grow, marine fungi have the potential to transform sustainable biotechnology, biomining, and climate change mitigation. Future studies should concentrate on deep-sea exploration, fungal genomics, and interdisciplinary collaboration to fully realize the potential of these remarkable organisms.

Despite this tremendous potential, several challenges remain in bringing marine fungal biotechnology to full fruition. Cultivating marine fungi under laboratory conditions that mimic their natural environments is often difficult. Many marine fungal species remain unculturable or dormant in standard media, limiting our access to their metabolic repertoire. Advances in metagenomics, synthetic biology, and heterologous gene expression are beginning to address these obstacles by enabling the expression of marine fungal genes in more tractable microbial hosts, such as bacteria like *Escherichia coli* or yeast like *Saccharomyces cerevisiae*. In summary, marine fungi, with their unique adaptations and rich metabolic capabilities, represent a powerful and largely underexploited frontier in biotechnology. As interdisciplinary research continues to unravel its biology, the transition from potential to practical application is expected to accelerate, unlocking innovations for medicine, industry, and environmental sustainability.

The exploration of marine fungi and their multifaceted roles in marine ecosystems, biotechnology, and geochemistry is rapidly advancing, thanks to the integration of cutting-edge technologies and interdisciplinary research approaches. As the field of geomycology matures, several emerging trends and methodologies are reshaping how researchers discover, identify, and utilize marine fungal diversity. In recent years, there has been the rise of molecular and omics-based techniques, which have revolutionized the way scientists investigate microbial communities. Traditional culture-dependent methods, though valuable, fail to capture the full extent of fungal diversity in marine ecosystems, as a large proportion of marine fungi are either non-culturable or require highly specific conditions for growth.

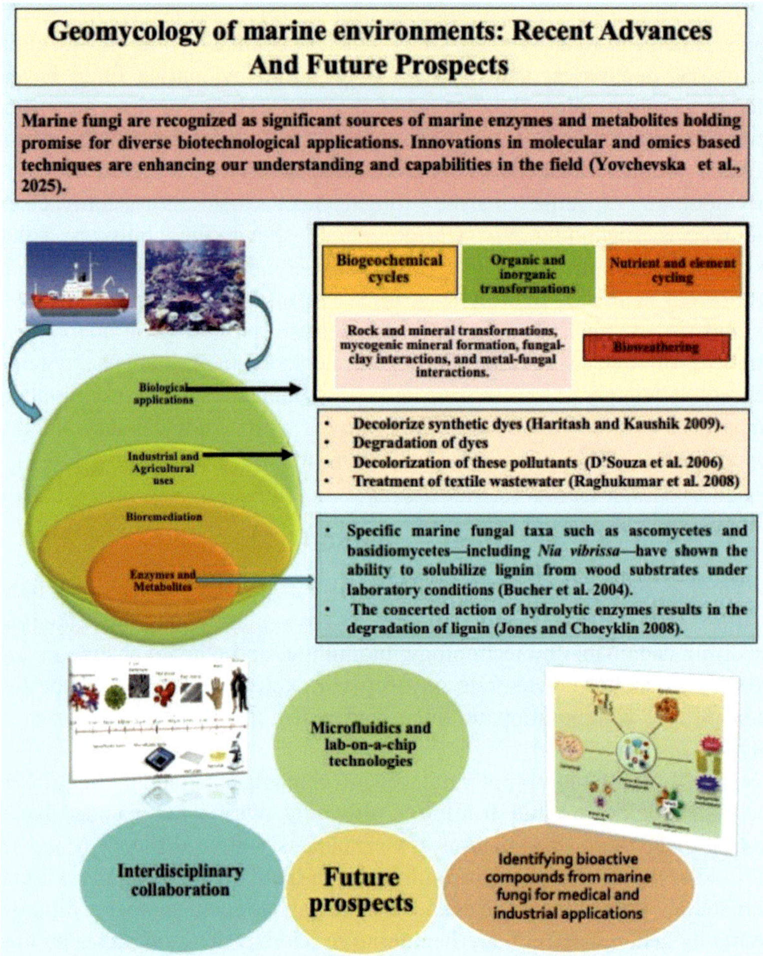

Fig. 1. Recent advances and prospects of geomycology in the marine environment (compiled from Bucher et al., 2004; Jones and Choeyklin, 2008; Raghukumar et al., 2008; D'Souza et al., 2006; Jaritash and Kaushik, 2009).

Metagenomic studies are now complemented by metatranscriptomics, metaproteomics, and metabolomics, each providing different layers of functional insights. Together, these multi-omics approaches help build a holistic understanding of fungal function and interaction in the marine environment (Yadav et al., 2023; Kumar and Yadav, 2024). Advancements in bioinformatics and machine learning are further enhancing our ability to process and interpret the vast datasets generated by omics analyses. Algorithms can now predict enzyme activity, protein structure, and biosynthetic gene clusters from raw sequence data, enabling the targeted search for novel compounds or extremozymes with industrial relevance. Furthermore, machine learning models can be trained to identify patterns or ecological signatures that might otherwise be overlooked, allowing for more strategic bioprospecting. Another notable trend is the development of synthetic biology platforms to harness the biosynthetic machinery of marine fungi. Many marine fungal metabolites are encoded by complex gene clusters that are silent or poorly expressed under laboratory conditions. Synthetic biology allows researchers to reconstruct and express these gene clusters in model organisms, such as *Saccharomyces cerevisiae* or *Aspergillus nidulans*, thus bypassing the culturing limitations of the native marine species. This approach not only facilitates the production of valuable metabolites but also enables genetic engineering to enhance yields or modify properties (Parages et al., 2016; Wagner et al., 2016).

The integration of remote sensing, underwater robotics, and autonomous sampling devices has also expanded our capacity to collect samples from previously inaccessible marine habitats. Precise sampling from deep-sea trenches, hydrothermal vents, and polar seas—regions where extreme marine fungi are most likely to occur, can now be allowed due to the use of Remotely Operated Vehicles (ROVs) and Autonomous Underwater Vehicles (AUVs) (Wynn et al., 2014; Liang et al., 2021). These tools, combined with real-time geospatial data and environmental sensors, offer a robust framework for linking fungal diversity to environmental gradients and geochemical conditions. Another emerging area of research involves fungal interactions with other marine organisms, including bacteria, archaea, microalgae, sponges, and corals. Understanding these interactions is crucial not only for ecological modeling but also for identifying co-metabolic pathways and novel compound production.

Microfluidics and lab-on-a-chip technologies are providing new opportunities to study marine fungi under controlled microenvironmental conditions that mimic their natural habitats. These miniaturized systems allow researchers to observe fungal growth, behavior, and metabolite production at single-cell resolution, significantly reducing the amount of biomass or reagents required for experimentation. Such technologies are invaluable in characterizing rare or slow-growing species. Finally, there is a growing emphasis on interdisciplinary collaboration, where marine mycologists, geologists, chemists, bioengineers, and data scientists work together to decode the complexities of marine fungi. This collaborative spirit is essential for advancing fundamental knowledge and translating discoveries into applied solutions in sectors ranging from medicine to environmental sustainability. As research tools continue to evolve and our conceptual understanding deepens, marine fungi are likely to emerge from the peripheries of marine microbiology to become central players in future ocean science and biotechnological innovation.

The geomycology of marine fungi is an emerging field with immense promise for both fundamental science and applied innovation. Future research directions emphasize the need for interdisciplinary approaches that integrate marine biology, geology, microbiology, chemistry, and artificial intelligence (AI) to unravel the complex roles fungi play in ocean ecosystems. The significance of geomycology in marine and marine-influenced ecosystems is increasingly recognized, as fungi are vital for biogeochemical processes, biomineralization, and overall resilience. They also significantly impact greenhouse gas fluxes, which are crucial for refining climate models. Through the biomineralization of various compounds, including carbonates, oxalates, and metal oxides, marine fungi are pivotal in developing sustainable bioremediation techniques for addressing heavy metal, hydrocarbon, and radionuclide contamination in coastal and deep-sea sediments. Deep-sea bioprospecting stands out as a particularly rich avenue, as extreme habitats such as hydrothermal vents, abyssal plains, and hypersaline lagoons harbor unique fungal taxa capable of producing novel enzymes, metabolites, and biomineralization processes with direct relevance to biotechnology, pharmaceuticals, and environmental sustainability. Furthermore, mineral-associated marine fungi are emerging as valuable sources of new secondary metabolites for pharmaceutical applications, particularly from extreme environments like hydrothermal vents and hypersaline lagoons. Advances in genomic and metagenomic exploration will be crucial for unlocking the metabolic potential of marine fungi, particularly in understanding genes involved in biogeochemical cycling, pollutant degradation, and mineral transformations. Such insights may pave the way for the development of fungal-based bioremediation technologies for heavy metals, hydrocarbons, and microplastics in marine sediments, as well as biomineralization-driven carbon capture strategies to mitigate climate change. From an industrial perspective, mineral-associated fungi represent a valuable but underexplored reservoir of bioactive compounds with antimicrobial, antiviral, and anticancer properties.

Research in geomycology could offer effective methods for conserving submerged cultural heritage and safeguarding coral reef ecosystems from fungal biodeterioration. Investigations into fungal relationships with sponges, corals, seagrasses, and algae may enhance ecosystem resilience against challenges such as ocean acidification and rising temperatures. Furthermore, geomycological studies may contribute to the blue economy, supporting sustainable aquaculture, seaweed farming, and marine infrastructure through eco-friendly biomaterials and fungal probiotics. Recent advancements in omics technologies, isotope probing, and nanoscale imaging are expected to uncover hidden fungal diversity and deepen our understanding of mineral interactions at unprecedented levels. This research aims to connect fundamental ecological knowledge with innovative practical applications. Ultimately, marine geomycology plays a critical role in global change science, providing insights into ocean acidification, deoxygenation, and shifts in microbial communities, making substantial contributions to sustainability efforts and climate change mitigation. AI is set to significantly advance geomycology in marine and marine-influenced environments by integrating large datasets, advanced modeling, and predictive analytics. Utilizing data from metagenomics, metatranscriptomics, metabolomics, and imaging, AI and machine learning can unveil hidden fungal diversity, identify new gene clusters, and predict metabolic pathways involved in biomineralization, pollutant breakdown, and secondary metabolite production. Deep learning technologies will assist in classifying fungi from high-resolution microscopy and environmental DNA sequences, reducing reliance on traditional cultivation methods. AI-based simulations and predictive models can also forecast how fungal-mineral interactions will adapt to environmental changes, providing valuable insights for global change science. Merging AI with experimental geomycology will expedite discoveries, enhance predictive accuracy, and transform marine fungal research into sustainable strategies for climate resilience, biotechnology, and the blue economy. To fully harness this potential, future efforts must rely on interdisciplinary collaboration, cutting-edge tools such as omics technologies, nanoscale imaging and AI-driven data integration, ensuring that the geomycology of marine fungi evolves into a key discipline for addressing global environmental challenges and advancing industrial innovation. Figure 1 summarizes recent advances and future prospects of geomycology in the marine and marine-influenced environments.

Acknowledgement

The authors acknowledge the use of OpenAI's ChatGPT for assistance with language editing and improving the manuscript's readability.

References

Agrawal, S., Adholeya, A., Barrow, C.J. and Deshmukh, S.K. (2018). *In vitro* evaluation of marine derived fungi against *Cutibacterium acnes*. Anaerobe, 49: 5–13. https://doi.org/10.1016/j.anaerobe.2017.10.010.

Ameen, F., Al-Nadhari, S. and Al-Homaidan, A.A. (2021). Marine microorganisms as an untapped source of bioactive compounds. Saudi J. Biol. Sci., 28(1): 224–231. https://doi.org/10.1016/j.sjbs.2020.09.052.

Andreu, C., Zarnowski, R. and del Olmo, M.L. (2022). Recent developments in the biology and biotechnological applications of halotolerant yeasts. World J. Microbiol. Biotechnol., 38(2): 27. https://doi.org/10.1007/s11274-021-03213-0.

Balabanova, L., Slepchenko, L., Son, O. and Tekutyeva, L. (2018). Biotechnology potential of marine fungi degrading plant and algae polymeric substrates. Front. Microbiol., 9: 1527. https://doi.org/10.3389/fmicb.2018.01527.

Barone, G., Varrella, S., Tangherlini, M., Rastelli, E., Dell'Anno, A. et al. (2019). Marine fungi: Biotechnological perspectives from deep-hypersaline anoxic basins. Diversity, 11(7): 113. https://doi.org/10.3390/d11070113.

Behera, A.D. and Das, S. (2023). Ecological insights and potential application of marine filamentous fungi in environmental restoration. Rev. Environ. Sci. Biotechnol., 22(2): 281–318. https://doi.org/10.1007/s11157-023-09655-2.

Bicholkar, A.A. (2016). Fungi from Estuarine Habitats and their Potential for Heavy Metal Bioremediation. Doctoral dissertation, Goa University, Goa, India. http://irgu.unigoa.ac.in/drs/handle/unigoa/4983.

Bonugli-Santos, R.C., dos Santos Vasconcelos, M.R., Passarini, M.R., Vieira, G.A., Lopes, V.C. et al. (2015). Marine-derived fungi: Diversity of enzymes and biotechnological applications. Front. Microbiol., 6: 269. https://doi.org/10.3389/fmicb.2015.00269.

Bucher, V.V., Pointing, S.B., Hyde, K.D. and Reddy, C.A. (2004). Production of wood decay enzymes, loss of mass, and lignin solubilization in wood by diverse tropical freshwater fungi. Microb Ecol., 48(3): 331–337. https://doi.org/10.1007/s00248-003-0132-x.

Burgaud, G., Meslet-Cladière, L., Barbier, G. and Edgcomb, V.P. (2014). Astonishing fungal diversity in deep-sea hydrothermal ecosystems: An untapped resource of biotechnological potential? pp. 85–98. *In*: La Barre, S. and Kornprobst, J.M. (eds.). Outstanding Marine Molecules: Chemistry, Biology, Analysis. John Wiley & Sons. https://doi.org/10.1002/9783527681501.ch04.

Connell, L., Barrett, A., Templeton, A. and Staudigel, H. (2009). Fungal diversity associated with an active deep-sea volcano: Vailulu'u Seamount, Samoa. Geomicrobiol. J., 26(8): 597–605. https://doi.org/10.1080/01490450903316174.

Dabolkar, S. and Kamat, N.M. (2024). Experimental study on ecological and biogeochemical role of calcimycocavites in understanding carbon fluxes in the age of global warming and climate change. Geomicrobiol. J., 41(5): 577–583. https://doi.org/10.1080/01490451.2024.2356520.

Dalmaso, G.Z., Ferreira, D. and Vermelho, A.B. (2015). Marine extremophiles: A source of hydrolases for biotechnological applications. Mar. Drugs, 13(4): 1925–1965. https://doi.org/10.3390/md13041925.

Damare, S., Singh, P. and Raghukumar, S. (2012). Biotechnology of Marine Fungi. pp. 277–297. *In*: Raghukumar, C. (ed.). Biology of Marine Fungi. Progress in Molecular and Subcellular Biology (Vol. 53). Springer, Berlin, Heidelberg. https://doi.org/10.1007/978-3-642-23342-5_14.

Das, T., Al-Tawaha, A.R., Pandey, D.K., Nongdam, P., Shekhawat, M.S. et al. (2022). Halophilic, acidophilic, alkaliphilic, metallophilic, and radioresistant fungi: Habitats and their living strategies. pp. 171–193. *In*: Sahay, S. (ed.). Extremophilic Fungi: Ecology, Physiology, and Applications. Springer Nature, Singapore. https://doi.org/10.1007/978-981-16-4907-3_9.

DeSouza-Ticlo, D. (2008). The Lignin-Degrading Enzyme, Laccase from Marine Fungi: Biochemical and Molecular Approaches (Doctoral dissertation). http://drs.nio.org/drs/handle/2264/3469.

Drake, H. and Ivarsson, M. (2018). The role of anaerobic fungi in fundamental biogeochemical cycles in the deep biosphere. Fungal Biol. Rev., 32(1): 20–25. https://doi.org/10.1016/j.fbr.2017.10.001.

Drake, H., Åström, M.E., Heim, C., Broman, C., Åström, J., Whitehouse, M. and Sjövall, P. (2015). Extreme 13C depletion of carbonates formed during oxidation of biogenic methane in fractured granite. Nat. Commun., 6(1): 7020. https://doi.org/10.1038/ncomms8020.

El-Regal, M.A. and Satheesh, S. (2023). Biodiversity of marine ecosystems. pp. 1–42. *In*: Ibrahim, H.A. and El-Sheekh, M.M. (eds.). Marine Ecosystems: A Unique Source of Valuable Bioactive Compounds (Vol. 3). https://doi.org/10.2174/9789815051995123030l.

Fisher, L.A., Pontefract, A., M Som, S., Carr, C.E., Klempay, B. et al. (2021). The current state of athalassohaline deep-sea hypersaline anoxic basin research: Recommendations for future work and relevance to astrobiology. Environ. Microbiol., 23(7): 3360–3369. https://doi.org/10.1111/1462-2920.15414.

Flewelling, A.J., Currie, J., Gray, C.A. and Johnson, J.A. (2015). Endophytes from marine macroalgae: Promising sources of novel natural products. Curr. Sci., 109(1): 88–111.

Fredrickson, J.K., McKinley, J.P., Nierzwicki-Bauer, S.A., White, D.C., Ringelberg, D.B. et al. (1995). Microbial community structure and biogeochemistry of Miocene subsurface sediments: Implications for long-term microbial survival. Mol. Ecol., 4(5): 619–626. https://doi.org/10.1111/j.1365-294X.1995.tb00262.x.

Gadd, G.M. (ed.). (2006). Fungi in Biogeochemical Cycles (Vol. 24). Cambridge University Press.

Gadd, G.M. (2013). Praeger review: Geomycology: Fungi as agents of biogeochemical change. Biol. Environ. Proc. R. Ir. Acad., 113(2): 139–153. Royal Irish Academy. https://doi.org/10.1353/bae.2013.0006.

Gadd, G.M. (2016). New horizons in geomycology. Environ. Microbiol. Rep., 9(1): 4–7. https://doi.org/10.1111/1758-2229.12480.

Gadd, G.M. (2017). The geomycology of elemental cycling and transformations in the environment. Microbiol. Spectr., 5(1): 10–1128. https://doi.org/10.1128/microbiolspec.funk-0010-2016.

Gadd, G.M., Rhee, Y.J., Stephenson, K. and Wei, Z. (2012). Geomycology: Metals, actinides, and biominerals. Environ. Microbiol. Rep., 4(3): 270–296. https://doi.org/10.1111/j.1758-2229.2011.00283.x.

Garlapati, D., Charankumar, B., Ramu, K., Madeswaran, P. and Ramana Murthy, M.V. (2019). A review of the applications and recent advances in environmental DNA (eDNA) metagenomics. Rev. Environ. Sci. Biotechnol., 18: 389–411. https://doi.org/10.1007/s11157-019-09501-4.

Gazaem, M.A. (2012). Sorption of Heavy Metal by Marine Derived Aspergillus Species. Doctoral dissertation, Goa University, Goa, India. http://hdl.handle.net/10603/226310.

Giddings, L.A. and Newman, D.J. (2022). Extremophilic fungi from marine environments: Underexplored sources of antitumor, anti-infective, and other biologically active agents. Mar. Drugs, 20(1): 62. https://doi.org/10.3390/md20010062.

Grossart, H.P., Van den Wyngaert, S., Kagami, M., Wurzbacher, C., Cunliffe, M. et al. (2019). Fungi in aquatic ecosystems. Nat. Rev. Microbiol., 17(6): 339–354. https://doi.org/10.1038/s41579-019-0175-8.

Hagestad, O.C. (2021). Bioprospecting of Marine Fungi from the High Arctic: A Study of High Latitude Marine Fungi from Understudied Taxa: Bioactivity Potential, Taxonomy and Genomics. Thesis for Doctor of Philosophiae, The Arctic University of Norway. https://doi.org/10.13140/RG.2.2.33484.54401.

Hamza, F. and Zinjarde, S. (2023). Use of marine microorganisms in designing anti-infective strategies for sustainable aquaculture production. J. Appl. Microbiol., 134(7): lxad128. https://doi.org/10.1093/jambio/lxad128.

Haritash, A.K. and Kaushik, C.P. (2009). Biodegradation aspects of polycyclic aromatic hydrocarbons (PAHs): A review. J. Hazard. Mater., 169(1-3): 1–15. https://doi.org/10.1002/9783527681501.ch04.

Hyde, K.D., Jones, E.G., Leaño, E., Pointing, S.B., Poonyth, A.D. et al. (1998). Role of fungi in marine ecosystems. Biodivers. Conserv., 7: 1147–1161. https://doi.org/10.1023/A:1008823515157.

Ivarsson, M., Bengtson, S. and Neubeck, A. (2016). The igneous oceanic crust: Earth's largest fungal habitat? Fungal Ecol., 20: 249–255. https://doi.org/10.1016/j.funeco.2016.01.009.

Ivarsson, M., Bengtson, S., Belivanova, V., Stampanoni, M., Marone, F. et al. (2012). Fossilized fungi in subseafloor Eocene basalts. Geology, 40(2): 163–166. https://doi.org/10.1130/G32590.1.

Ivarsson, M., Bengtson, S., Skogby, H., Belivanova, V. and Marone, F. (2013). Fungal colonies in open fractures of subseafloor basalt. Geo-Mar. Lett., 33: 233–243. https://doi.org/10.1007/s00367-013-0321-7.

Ivarsson, M., Schnürer, A., Bengtson, S. and Neubeck, A. (2016). Anaerobic fungi: A potential source of biological H_2 in the oceanic crust. Front. Microbiol., 7: 674. https://doi.org/10.3389/fmicb.2016.00674.

Jørgensen, B.B., Isaksen, M.F. and Jannasch, H.W. (1992). Bacterial sulfate reduction above 100 C in deep-sea hydrothermal vent sediments. Science, 258(5089): 1756–1757. DOI: 10.1126/science.258.5089.1756.

Jebaraj, C.S., Raghukumar, C., Behnke, A. and Stoeck, T. (2010). Fungal diversity in oxygen-depleted regions of the Arabian Sea was revealed by targeted environmental sequencing combined with cultivation. FEMS Microbiol. Ecol., 71(3): 399–412. https://doi.org/10.1111/j.1574-6941.2009.00804.x.

Jin, M., Gai, Y., Guo, X., Hou, Y. and Zeng, R. (2019). Properties and applications of extremozymes from deep-sea extremophilic microorganisms: A mini-review. Mar. Drugs, 17(12): 656. https://doi.org/10.3390/md17120656.

Jones, E.B.G. and Choeyklin, R. (2008). Ecology of marine and freshwater basidiomycetes. In British Mycological Society Symposia Series, 28: 301–324. Academic Press. https://doi.org/10.1016/S0275-0287(08)80018-5.

Kakkar, P. and Wadhwa, N. (2022). Extremozymes are used in the textile industry. J. Text. Inst., 113(9): 2007–2015. https://doi.org/10.1080/00405000.2021.1948251.

Kandasamy, K., Alikunhi, N.M. and Subramanian, M. (2012). Yeasts in marine and estuarine environments. J. Yeast Fungal Res., 3(6): 74–82. https://doi.org/10.5897/JYFR12.003.

Kaushik, B. D. (2009). Developments in cyanobacterial biofertilizer. Algal Biology and Biotechnology. New Delhi: IK International Publishing House Pvt. Ltd, 97–108.

Kempken, F. (2023). Marine fungi: A treasure trove of novel natural products and for biological discovery. PLoS Pathog., 19(9): e1011624. https://doi.org/10.1371/journal.ppat.1011624.

Kumar, A. and Yadav, A. (2024). Next-generation sequencing in metagenomics and metatranscriptomics. pp. 49–75. *In*: Mani, I. and Singh, V. (eds.). Multi-omics Analysis of the human Microbiome: From Technology to Clinical Applications. Springer Nature Singapore. https://doi.org/10.1007/978-981-97-1844-3_3.

Le Calvez, T., Burgaud, G., Mahé, S., Barbier, G. and Vandenkoornhuyse, P. (2009). Fungal diversity in deep-sea hydrothermal ecosystems. Appl. Environ. Microbiol., 75(20): 6415–6421. https://doi.org/10.1128/AEM.00653-09.

Liang, J., Feng, J.C., Zhang, S., Cai, Y., Yang, Z. et al. (2021). Role of deep-sea equipment in promoting the forefront of studies on life in extreme environments. iScience, 24(11): 103299. https://doi.org/10.1016/j.isci.2021.103299.

Luna, G.M. (2015). Biotechnological potential of marine microbes. pp. 651–661. *In*: Kim, S.K. (ed.). Springer Handbook of Marine Biotechnology. Springer Handbooks. Springer, Berlin, Heidelberg. https://doi.org/10.1007/978-3-642-53971-8_26.

Mahé, S., Rédou, V., Calvez, T.L., Vandenkoornhuyse, P. and Burgaud, G. (2013). Fungi in deep-sea environments and metagenomics. pp. 325–354. *In*: Martin, F. (ed.). The Ecological Genomics of Fungi. John Wiley & Sons, Inc. https://doi.org/10.1002/9781118735893.ch15.

Martin, G., Guggiari, M., Bravo, D., Zopfi, J., Cailleau, G. et al. (2012). Fungi, bacteria, and soil pH: The oxalate–carbonate pathway as a model for metabolic interaction. Environ. Microbiol., 14(11): 2960–2970. https://doi.org/10.1111/j.1462-2920.2012.02862.x.

Merlino, G., Barozzi, A., Michoud, G., Ngugi, D.K. and Daffonchio, D. (2018). Microbial ecology of deep-sea hypersaline anoxic basins. FEMS Microbiol. Ecol., 94(7): fiy085. https://doi.org/10.1093/femsec/fiy085.

Muthusamy, C. and Murugaiyan, K. (2023). Bioprospecting marine fungal enzymes-scope and challenges. Kavaka, 59(1): 33–47. https://doi.org/10.36460/Kavaka/59/1/2023/33-47.

Nagahama, T. and Nagano, Y. (2011). Cultured and uncultured fungal diversity in deep-sea environments. pp. 173–187. *In*: Raghukumar, C. (ed.). Biology of Marine Fungi: Progress in Molecular and Subcellular Biology (PSMB) (Vol. 53). Springer, Berlin, Heidelberg. https://doi.org/10.1007/978-3-642-23342-5_9.

Nakkeeran, E., Rathna, R. and Viveka, R. (2018). Mechanism and action of Aureobasidium pullulans on biosorption of metals. pp. 215–231. *In*: Varjani, S., Gnansounou, E., Gurunathan, B., Pant, D. and Zakaria, Z. (eds.). Waste Bioremediation: Energy, Environment, and Sustainability. Springer Singapore. https://doi.org/10.1007/978-981-10-7413-4_11.

Nicoletti, R. and Andolfi, A. (2018). The marine-derived filamentous fungi in biotechnology. pp. 157–189. *In*: Rampelotto, P. and Trincone, A. (eds.). Grand Challenges in Marine Biotechnology. Springer Cham. https://doi.org/10.1007/978-3-319-69075-9_4.

Orsi, W., Biddle, J.F. and Edgcomb, V. (2013). Deep sequencing of subseafloor eukaryotic rRNA reveals active fungi across marine subsurface provinces. PLOS One, 8(2): e56335. https://doi.org/10.1371/journal.pone.0056335.

Pachiadaki, M. and Edgcomb, V. (2022). Marine extreme habitats. pp. 579–619. *In*: Stal, L.J. and Cretoiu, M.S. (eds.). The Marine Microbiome: The Microbiomes of Humans, Animals, Plants, and the Environment (Vol. 3). Springer, Cham. https://doi.org/10.1007/978-3-030-90383-1_14.

Parages, M.L., Gutiérrez-Barranquero, J.A., Reen, F.J., Dobson, A.D. and O'Gara, F. (2016). Integrated (meta) genomic and synthetic biology approaches to develop new biocatalysts. Mar. Drugs, 14(3): 62. https://doi.org/10.3390/md14030062.

Parkes, R.J., Webster, G., Cragg, B.A., Weightman, A.J., Newberry, C.J. et al. (2005). Deep sub-seafloor prokaryotes stimulated at interfaces over geological time. Nature, 436(7049): 390–394. https://doi.org/10.1038/nature03796.

Patil, A. and Selvaraj, S. (2025). Bioprospecting of bioactive compounds from Marine Fungi. pp. 1–27. *In*: Uppuluri, K.B. and Selvasembian, R. (eds.). Bioprospecting of Multi-tasking Fungi for Therapeutic Applications (Vol. II) Springer, Singapore. https://doi.org/10.1007/978-981-96-2975-6_1.

Peng, Q., Li, Y., Fang, J. and Yu, X. (2023). Effects of epigenetic modification and high hydrostatic pressure on polyketide synthase genes and secondary metabolites of *Alternaria alternata* derived from the Mariana trench sediments. Mar. Drugs, 21(11): 585. https://doi.org/10.3390/md21110585.

Picard, K.T. (2017). Next-Generation Approaches to Understanding the Diversity and Evolution of Marine Fungi. Doctoral dissertation, Duke University, USA. https://hdl.handle.net/10161/16339.

Podia, M., Yadav, P., Hooda, S., Diwan, P. and Gupta, R.K. (2023). Microbial weathering of rocks in natural habitat. pp. 265–302. *In*: Singh, V.B., Madhav, S., Pant, N.C. and Shekhar, R. (eds.). Genetic Basis and Omics-Based Exploration. Weathering and Erosion Processes in the Natural Environment. John Wiley & Sons Ltd. https://doi.org/10.1002/9781394157365.ch11.

Raddadi, N., Cherif, A., Daffonchio, D., Neifar, M. and Fava, F. (2015). Biotechnological applications of extremophiles, extremozymes, and extremolytes. Appl. Microbiol., Biotechnol., 99: 7907–7913. https://doi.org/10.1007/s00253-015-6874-9.

Raghukumar, C. (2008). Marine fungal biotechnology: An ecological perspective. Fungal Divers., 31: 19–35.

Raghukumar, C., D'Souza-Ticlo, D. and Verma, A. (2008). Treatment of colored effluents with lignin-degrading enzymes: An emerging role of marine-derived fungi. Crit. Rev. Microbiol., 34(3-4): 189–206 https://doi.org/10.1080/10408410802526044.

Raghukumar, S. (2017). Fungi in coastal and oceanic marine ecosystems. *In*: Marine Fungi. Springer Cham, 387 pp. https://doi.org/10.1007/978-3-319-54304-8.

Rensink, S., van Nieuwenhuijzen, E.J., Sailer, M.F., Struck, C. and Wösten, H.A. (2024). Use of *Aureobasidium* in a sustainable economy. Appl. Microbiol. Biotechnol., 108(1): 202 https://doi.org/10.1007/s00253-024-13025-5.

Richardson, M. (2009). The ecology of the Zygomycetes and its impact on environmental exposure. Clinical Microbiology and Infection, 15: 2–9. https://doi.org/10.1111/j.1469-0691.2009.02972.x.

Schrenk, M.O., Huber, J.A. and Edwards, K.J. (2010). Microbial provinces in the subseafloor. Ann. Rev. Mar. Sci., 2: 279–304. https://doi.org/10.1146/annurev-marine-120308-08100.

Selvakumar, P. (2025). Fungal bioprospecting: Discovering novel bioactive compounds. pp. 327–348. *In*: Kashyap, A.K. and Jogawat, A. (eds.). Mycological Inventions for Sustainable Agriculture and Food Production. IGI Global Scientific Publishing.

Singh, B. (2016). *Myceliophthora thermophila* syn. Sporotrichum thermophile: A thermophilic mold of biotechnological potential. Crit. Rev. Biotechnol., 36(1): 59–69. https://doi.org/10.3109/07388551.2014.923985.

Stoeck, T., Filker, S., Edgcomb, V., Orsi, W., Yakimov, M.M. et al. (2014). Living at the limits: Evidence for microbial eukaryotes thriving under pressure in deep anoxic, hypersaline habitats. Adv. Ecol., 2014(1): 532687. https://doi.org/10.1155/2014/532687.

Thatoi, H., Behera, B.C. and Mishra, R.R. (2013). Ecological role and biotechnological potential of mangrove fungi: A review. Mycology, 4(1): 54–71. https://doi.org/10.1080/21501203.2013.785448.

Trianto, A., Radjasa, O.K., Subagiyo, S., Purnaweni, H., Bahry, M.S. et al. (2021). The potential of fungi isolated from a mangrove ecosystem in Northern Sulawesi, Indonesia: Protease, cellulase, and anti-microbial capabilities. Biodiversitas., 22(4): 1717–1724. https://doi.org/10.1080/21501203.2013.785448.

Vaksmaa, A., Guerrero-Cruz, S., Ghosh, P., Zeghal, E., Hernando-Morales, V. et al. (2023). Role of fungi in bioremediation of emerging pollutants. Front. Mar. Sci., 10: 1070905. https://doi.org/10.3389/fmars.2023.1070905.

Varrella, S., Barone, G., Tangherlini, M., Rastelli, E., Dell'Anno, A. et al. (2021). Diversity, ecological role and biotechnological potential of antarctic marine fungi. J. Fungi, 7(5): 391. https://doi.org/10.3390/jof7050391.

Varrella, S., Tangherlini, M. and Corinaldesi, C. (2020). Deep hypersaline anoxic basins as an untapped reservoir of polyextremophilic prokaryotes of biotechnological interest. Mar. Drugs, 18(2): 91. https://doi.org/10.3390/md18020091.

Verma, A.K. (2011). Potential of Marine-derived Fungi and Their Enzymes in Bioremediation of Industrial Pollutants. Doctoral dissertation, Goa University, Goa, India. http://drs.nio.org/drs/handle/2264/4022.

Venkateswara Sarma, V. (2018). Obligate Marine Fungi and bioremediation. pp. 307–323. *In*: Prasad, R. (ed.). Mycoremediation and Environmental Sustainability. Fungal Biology. Springer Cham. https://doi.org/10.1007/978-3-319-77386-5_12.

Wagner, J.M. and Alper, H.S. (2016). Synthetic biology and molecular genetics in non-conventional yeasts: Current tools and future advances. Fungal Genet. Biol., 89: 126–136. https://doi.org/10.1016/j.fgb.2015.12.001.

Warmink, J.A., Nazir, R., Corten, B. and Van Elsas, J.D. (2011). Hitchhikers on the fungal highway: The helper effect for bacterial migration via fungal hyphae. Soil Biology and Biochemistry, 43(4): 760–765. https://doi.org/10.1016/j.soilbio.2010.12.009.

Wekesa, T.B. (2024). Evaluation of Extremophilic Bacteria from Lakes Bogoria and Magadi for Biocontrol of Fusarium Solani and Rhizoctonia Solani in Phaseolus Vulgaris L. Doctoral dissertation, JKUAT-IBR. http://ir.jkuat.ac.ke/handle/123456789/6231.

Wu, X., Holmfeldt, K., Hubalek, V., Lundin, D., Åström, M., Bertilsson, S. and Dopson, M. (2016). Microbial metagenomes from three aquifers in the Fennoscandian shield terrestrial deep biosphere reveal metabolic partitioning among populations. The ISME Journal, 10(5): 1192–1203. https://doi.org/10.1038/ismej.2015.185.

Wynn, R.B., Huvenne, V.A., Le Bas, T.P., Murton, B.J., Connelly, D.P. et al. (2014). Autonomous Underwater Vehicles (AUVs): Their past, present, and future contributions to the advancement of marine geoscience. Mar. Geol., 352: 451–468. https://doi.org/10.1016/j.margeo.2014.03.012.

Yadav, A.N. (2021). Recent Trends in Mycological Research (Vol. 2): Environmental and Industrial Perspective. Springer International Publishing, Berlin, 532 pp.

Yadav, B.N.S., Sharma, P., Maurya, S. and Yadav, R.K. (2023). Metagenomics and metatranscriptomics as potential driving forces for the exploration of diversity and functions of micro-eukaryotes in soil. 3 Biotech, 13(12): 423. https://doi.org/10.1007/s13205-023-03841-3.

Yang, S.Q., Li, X.M., Li, X., Chi, L.P. and Wang, B.G. (2018). Two new diketomorpholine derivatives and a new highly conjugated ergostane-type steroid from the marine algal-derived endophytic fungus *Aspergillus alabamensis* EN-547. Mar. Drugs, 16(4): 114. https://doi.org/10.3390/md16040114.

Yovchevska, L., Gocheva, Y., Stoyancheva, G., Miteva-Staleva, J., Dishliyska, V. et al. (2025) Halophilic Fungi: Feature and potential applications. Microorganisms, 13(1): 175. 10.3390/microorganisms13010175.

Zhou, Z., St. John, E., Anantharaman, K. and Reysenbach, L. (2022). Global patterns of diversity and metabolism of microbial communities in deep-sea hydrothermal vent deposits. Microbiome, 10(1): 241. https://doi.org/10.1186/s40168-022-01424-7.

16

Role of Synthetic Biology in Marine Fungi

A. Sai Ramesh,[1] *T. Naveen Kumar,*[2] *A.P. Roshan Gershom*[3] and *L. Karthik*[4,*]

Marine fungi are fungal organisms present in marine habitats. The potential of marine fungi remains less recognised due to the lack of research and exploration of marine fungal strains because of the various challenges, such as hard-to-culture strains, a lack of understanding, limited product rates in metabolic pathways, and much more. To exploit the potential resources from these unconditional sources of natural products, techniques of synthetic biology could be applied to harvest these natural products produced by marine fungal strains. Synthetic biology techniques involve genetic engineering (like CRISPR), pathway-specific engineering, transcriptional, and epigenetic regulation. All these techniques could possibly be used to increase the metabolite production of marine fungi by modifying the metabolic pathways, RNA transcription, and to overcome all the existing challenges present in harvesting metabolites from marine fungi. The natural products produced by the marine fungi have numerous applications, which include enzyme production, pharmaceutical and biomedical applications, bioremediation and

[1] Department of Bio-Engineering, Vels Institute of Science, Technology and Advanced Studies, Chennai 600117, Tamil Nadu, India.

[2] Department of Biotechnology, Vel Tech High Tech Dr. Rangarajan Dr. Sakunthala Engineering College, Avadi, Chennai 600062, Tamil Nadu, India.

[3] Department of Biotechnology, Prathyusha Engineering College, Poonamallee Tiruvallur Road, Aranvoyal Kuppam, Tiruvallur 602025, Tamil Nadu, India.

[4] Arqgene, Vellore, Tamil Nadu, India.

* Corresponding author: karthiksynbio@gmail.com

DOI: 10.1201/9781003795100-20

environmental sustainability. Since the production of metabolites by marine fungi is increased, their application in various fields could be utilised to its fullest, and thus could be a potential alternative to the existing metabolites that are produced from conventional microbes and other species. This book chapter aims to understand the role of synthetic biology in marine fungi as an unconditional reservoir of natural products whose potential remains untapped and could be exploited using synthetic biology techniques, overcoming existing challenges.

1. Introduction

Marine fungi are among the organisms that lack exploration, but have huge hidden potential and occupy different ecological niches in marine ecosystems. These fungi, in order to survive the harsh and extreme conditions (like high salinity, lack of nutrients, and high pressure), have undergone numerous evolutions, which have made them adapt to numerous environments present in the marine hydrosphere. Along with these adaptations, evolution also resulted in the development of unique metabolic pathways that produce highly productive secondary metabolites and have potential applications in various fields such as pharmacology, bioremediation, bioprocessing, biocontrol, agriculture, and industrial biotechnology. Despite all its advantages, the harnessing of these compounds remains a challenge due to the complexities in culturing and studying the organisms under laboratory conditions. Recent advancements in synthetic biology are becoming a solution to face all the existing challenges in using marine fungi. Marine fungi represent a fairly untapped bioresource for new natural product discovery, although over the past decade marine fungi have handed over a number of new secondary metabolites, which is discussed in Agrawal et al. (2018).

Synthetic biology has become a promising field, offering various tools and methodologies that help us to unlock the hidden potential of marine fungi. By combining synthetic biology with other disciplines of research such as molecular biology and bioinformatics, bioprocessing will help us to develop novel strategies to enhance the metabolite production and to maintain optimised conditions to develop marine fungal cultures under laboratory conditions. Further advancements in this field will help us to overcome the existing limitations by transforming marine fungi into a valuable resource in the development of unique and sustainable solutions to the existing and upcoming global challenges. This review is a blend of marine fungi and synthetic biology. It aims to give an overview of marine fungi and its classifications, the types of bioactive compounds produced by marine fungi, and the effect of those compounds in various fields such as pharmacology, bioremediation, and bioprocessing.

2. Synthetic Biology in Marine Fungi

Synthetic biology is a multidisciplinary field that is a combination of principles from engineering biology, computational biology, and techniques from biotechnology,

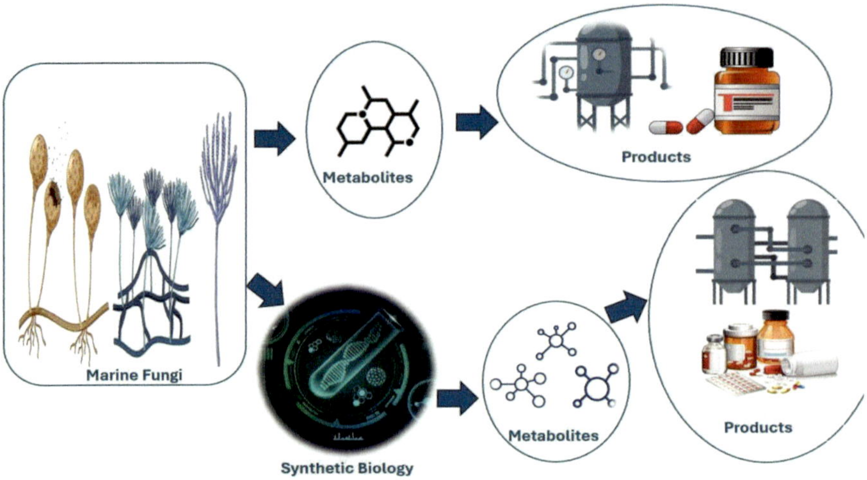

Fig. 1. Schematic view of marine fungal synthetic biology.

which has risen as a mighty tool for exploring the hidden potential of marine fungi (Fig. 1). These eukaryotic organisms are found in various marine environments and possess a reliable ability to produce a wide spectrum of secondary metabolites; many of these metabolites have potential in various applications such as therapeutics and industries (Table 1).

Recent advancements in DNA sequencing, genome editing, and computational tools have enabled researchers to dig deeper into the genetic structure of marine fungi, enabling new fields for engineering these organisms to improve their biosynthetic capabilities.

3. Genetic Engineering Tools

3.1 CRISPR Cas9 and Gene Editing Strategies

Genetic engineering tools are used to change genes, which are responsible for the various characters present in the body of the host organism. CRISPR stands for Clustered Regularly Interspaced Short Palindromic. It is a genetic engineering tool introduced recently and became popular because of its application. It is used to cut, add, or replace DNA at its various regions.

CRISPR Cas is a gene editing technology that uses a bacterial restriction enzyme to alter the DNA. Cas stands for CRISPR-associated protein. Therefore, CRISPR is used to enhance the metabolite production of marine fungi by changing the individual genes responsible for the production of specific metabolites.

Marine fungi are developing as new sources of new secondary metabolites with different bioactivities. Still, the lack of effective tools limits the discovery of new composites, and the identification of biosynthesis mechanisms (Chen et al., 2022) developed an effective cut-mediated chemical metamorphosis system for protoplasts in two marine-derived fungi, *Spiromastix* sp. SCSIO F190 and *Aspergillus* sp.

Table 1. Depicting a few metabolites synthesized by synthetic biology techniques.

Compound	Fungi	Function	Synthetic Biology Technique/Tool Used	References
2 Alkenyl tetrahydropyran	*Trichoderma applanatumin* and *Aspergillus nidulans*	Antifungal	Heterologous expression	Lyu et al. (2021)
4-Hydroxy-α-Pyrones	*Aspergillus niger*	Antibacterial	OSMAC	Ding et al. (2019)
Alkaloid	*Streptomyces* sp. HZP-2216E	Anti-MRSA	OSMAC	Pinedo-Rivilla et al. (2022)
Angucyclines	*Penicillium* sp. KMU18029	Antibacterial (weak)	Epigenetic regulation	Li et al. (2024)
anthranilic acid	*Penicillium paneum* SD-44	Anticancer	OSMAC	Li et al. (2013)
Caspofungin	*Aspergillus fumigatus*	Antifungal	Epigenetic regulation	Valero et al. (2021)
Cytochalasans	*Aspergillus flavipes* PJ03-11	Cytotoxic activities against human cancer cell lines	OSMAC	Si et al. (2018)
Diketopiperazines	*Aspergillus puulaauensis* F77	Neuroinflammatory	OSMAC	Wu et al. (2025)
Gibberellic acid	*Fusarium fujikuroi*	Plant growth regulator	CRISPR/Cas9	Shi et al. (2019)
L-Malic acid	*Aspergillus niger*	Food additive	CRISPR/Cas9	Wu et al. (2024)
Polyketides	*Aspergillus aculeatus*	Mycotoxin	CRISPR/Cas9	Wolff et al. (2020)
Polyketides	*Diaporthe goulteri*	Antioxidant	Epigenetic modification	Ma et al. (2025)
Polyketides	*Sarcopodium* sp. QM3–1	Antimicrobial	Genome sequencing	Zhang et al. (2025)
Suberoylanilide hydroxamic acid (SAHA)	*Aspergillus sydowii* DL1045	Anticancer	Epigenetic regulation	Shi et al. (2024)
Sesquisterpenes	*Eutypella* sp.	Anti-inflammatory	Epigenetic manipulation	Niu et al. (2021)
Talaronoid	*Aspergillus flavipes*	Anti-neurodegeneration	Heterologous expression	Zhang et al. (2022)
Terpene	*Penicillium ochrochloron*	Antioxidant	Co-culturing of fungi and bacteria	Oppong-Danquah et al. (2020)
Terpene	Hypomontagnella monticulosa	Anticancer	OSMAC	Li et al. (2013)

SCSIO SX7S7. Next, they developed a simple CRISPR Cas9-based gene dislocation strategy by perceiving a target fungus with a single plasmid. In addition, the histone deacetylase gene rpd3 was transported using the developed CRISPR Cas9 system, thereby bringing out new secondary metabolites that were not produced in the original wild-type strain. Taken together, a flexible CRISPR Cas9-grounded gene dislocation system was developed, which will facilitate the discovery of new natural products and further natural studies.

CRISPR technology is also used to identify the part of specific genes or gene clusters that are responsible for producing certain secondary metabolites. Gakuubi et al. (2022) worked on *Lophiotrema* sp. F6932, where the crude extracts from the fungus exhibited strong antibacterial activity, and bioassay-guided insulation and structure identification of bioactive ingredients led to the insulation of palmarumycin C8 and a new analog, palmarumycin CP30.

Whole genome sequencing analysis helped in the identification of an apparent Type 1 iterative PKS (iPKS) suspected to be involved in the biosynthesis of palmarumycins. To confirm the involvement of the palmarumycin (confidante) gene cluster in the biosynthesis of these composites, we employed RiboNucleoProtein (RNP)-mediated CRISPR Cas9 to induce targeted omission of the KetoSynthase (KS) sphere in confidante. Double Strand Breaks (DSBs) upstream and downstream of the KS sphere were followed by homology-directed form (HDR) with a hygromycin resistance mail adjoined by 50 bp of homology on both sides of the DSBs. The attendant rejection mutants displayed fully different phenotypes compared to the original wild-type strain, as they had different colony morphology and were no longer able to produce palmarumycins or melanin. This study, thus, confirms the involvement of confidante in the biosynthesis of palmarumycins and paves the way for enforcing an alternative approach in the identification of other gene clusters of interest in this undiscovered fungal strain.

In addition to the regulation of metabolite products and identification of genes or gene clusters, CRISPR technology could also be used to study species specifically for their exceptional capacities in metabolite production and survival. *H. werneckii* has a unique cell cycle division and can grow at high swab attention. Due to most isolates being intraspecific mongrels with significant phenotypic variation, *H. werneckii* can become a model for understanding how hybridisation leads to new phenotypes. Still, the lack of tools has limited the capability to study this organism, Hernandez-Rodriguez et al. (2025) developed two different metamorphosis approaches to further study this unique organism. First, they've developed a plasmid-based ectopic integration protocol. This system can be used to express cellular labels for live cell imaging. For illustration, it could be used to fit a histone GFP plasmid and cover nuclear dynamics. Second, they developed a marker-free CRISPR/Cas9 protocol to target specific genes. Finally, they established a marker-free CRISPR/Cas9 protocol for targeted gene rejection using the melanin conflation pathway as a visual journalist of successful metamorphosis. These metamorphosis strategies will allow testing suspicions related to *H. werneckii* cell biology and physiology.

Marine-derived fungi are developing as large sources of structurally new natural products (NPs) with different bioactivities. Still, the limited availability of selection labels deprives the information of cryptic NPs. Recyclable labels thus have the ability to awaken silent SM clusters. Chen et al. (2023) established both pyrG and amdS-based recyclable marker cassettes and successfully applied them in the marine-derived fungi *Aspergillus* sp. SCSIO SX7S7 and *Spiromastix* sp. SCSIO F190, independently. Using the pyrG recyclable marker, a marker-less 7S7 Δ depH strain with a simplified HPLC background was established by inactivating a polyketide synthase (PKS) gene depH and dragging out the pyrG recyclable marker after depH rejection. Meanwhile, an amdS recyclable marker system was also developed to help strains that are too delicate to use pyrG markers. By employing the amdS marker, a backbone gene spm11, responsible for one major product of *Spiromastix* sp. SCSIO F190 was inactivated, and the amdS marker was removed after using, generating a fairly clean F190 Δspm11 strain for further activation of new NPs. The collection of two different reclaim labels will guarantee flexible operation in marine-derived fungi with different inheritable backgrounds, enabling the discovery of new structures in various fungi species with different genome mining strategies.

The CRISPR-Cas-based technologies will really continue to improve in clinical and biotechnological exploration. Still, the road ahead is not free of obstacles. One similar problem is the implicit immunogenicity to CRISPR-Cas9 proteins. The most extensively used Cas9 proteins are from *S. aureus* and *S. pyogenes*. Especially since these bacteria are responsible for contagious complaints in humans at high frequency, a recent report proved that more than half of humans may formerly have had pre-existing humoral and cell-intermediated adaptive immune responses to Cas9 proteins. Thus, as the CRISPR-Cas9 system moves forward into clinical trials, this factor must be considered. Studying and understanding similar challenges will enable us to determine their limitations and ways to overcome them. To this end, one proposed result to the immunogenicity problem could be to identify and use orthogonal CRISPR-Cas9 proteins to which we as humans haven't been introduced, which is also suggested by Adli (2018).

Using AI in GED is imperative and holds the pledge to revise the healthcare sector. CRISPR-based editing technologies like CRISPR-Cas9 allow for precise and targeted editing of the genes of organisms, which is a major advancement in biotechnology, which is discussed by Tyagi et al. (2020). Still, AI's integration with CRISPR improves the overall GED channel, furnishing new perceptivity, capabilities, and openings for manipulating and understanding the genes as suggested by Dixit et al. (2023).

3.2 *Alternative Genetic Manipulation Approaches*

Heterologous expression involves the introduction of a gene reproducing a protein of interest from one species into the cell of another species, therefore allowing for the host cells to express the foreign protein. Frequently, heterologous expression can affect very low protein expression in the host cells. The production of heterologous proteins by marine fungi has faced numerous challenges that limit its production. Once changed through synthetic biology, the desired metabolite product production is often less than that of its original product production, this is mentioned by Sun et al. (2019) and Meyer et al. (2021). For optimal heterologous gene expression, multiple factors similar to the codon operation of the host species, guanine and cytosine (GC) composition, Kozak sequence, Shine-Dalgarno sequence, runner ribonucleic acid (mRNA) stability, and splicing pattern of the gene have to be examined. Codon operation is the most important factor to be considered for advanced expression, especially when the gene patron and gene host species belong to a prokaryote and eukaryote or vice versa (Watts et al., 2021).

The study done by Cochereau et al. (2023) relies on a targeted genome mining approach to express the presence of an apparent vanadium-dependent haloperoxidase-reproducing gene in the deep-ocean hydrothermal fungus *Hortaea werneckii* UBOCC A 208029. This study described only three fungal vanadium-dependent haloperoxidases that have been discovered: one from the terrestrial species *Curvularia inaequalis*, one from the fungal factory pathogen *Botrytis cinerea*, and one from a marine-derived species, *Alternaria didymospora*. This study describes a new vanadium chloroperoxidase from the black species *H. werneckii*, successfully reproduced and overexpressed in a bacterial host, which possesses a higher affinity for platitude (K_m = 26 μM) than chloride (K_m = 237 mM). The enzyme was biochemically analysed, and they have estimated its eventuality for biocatalysis by determining its stability and forbearance in organic detergents. They also describe its implicit three-dimensional structure by establishing a model using the AlphaFold 2 artificial intelligence tool. This model shows some conservation of the 3D structure of the active point compared to the vanadium chloroperoxidase from *C. inaequalis*, but it also points out some differences in the active point entrance and the volume of the active point fund, focusing on its originality.

Lovastatin has wide operations thanks to its multiple pharmacological goods. The turmoil by filamentous fungi represents the major way of lovastatin production. Still, the current lovastatin productivity by fungal turmoil is limited and needs to be better. In the study done by Na et al. (2024), the lovastatin-producing strains of *Aspergillus terreus* from marine terrain were screened, and their lovastatin products were further improved by genetic engineering.

Five strains of *A. terreus* were isolated from marine surroundings. Their secondary metabolites were identified by metabolomics analysis using Ultra Performance Liquid Chromatography-Mass Spectrometry (UPLC-MS) with Global Natural Products Social Molecular Networking (GNPS), revealing that the product of secondary metabolites was variable among different strains. Remarkably, the strain of *A. terreus* MJ106 could basically biosynthesise the target medicine lovastatin, which was verified by High-Performance Liquid Chromatography (HPLC) and gene expression analysis. In a one-factor trial, lactose was set up to be the carbon source for *A. terreus* MJ106 to produce lovastatin. To improve the lovastatin titer in *A. terreus* MJ106, genetic engineering was applied to this strain. Originally, a series of strong promoters was linked by transcriptomic and green fluorescent protein journalist analysis. Also, three named strong promoters were used to overexpress the recap factor gene lovE, reproducing the major transactivator for lov gene cluster expression. The results revealed that compared to *A. terreus* MJ106, all lovE overexpression mutants displayed significantly further production of lovastatin and advanced gene expression. One of them, lovE b19, showed the highest lovastatin productivity at a titer of 1512 mg/L, which represents the highest product position reported in *A. terreus* (Na et al., 2024).

The Marine Fungi Natural Products (MaFNaP) community visualises that in the near future innovative ways will address the issue of cultivability. On the other hand, innovative biotechnological approaches and effective heterologous expression systems will be demanded for full disquisition of cultivable and non-cultivable marine fungi in all stages of the discovery-to-development channel. This knowledge is of interest to all fungal scientists, as mentioned by Tasdemir (2017).

The metabolite product of marine fungi could be regulated using methods like transcriptional regulation and epigenetic regulation. Fungal secondary metabolites are small boats that parade different natural conditions, which are utilised in drugs, assiduity, and husbandry. Their biosynthesis is regulated by co-expressed genes that frequently co-localise in gene clusters. Another challenge for marine natural products identification is the silence of BGCs under laboratory conditions (Overy et al., 2023; Valiante, 2023). The vast majority of these secondary metabolite gene clusters are inactive under laboratory conditions, which is due to tight transcriptional regulation. Variations of chromatin, the complex of DNA, and histone proteins impacting DNA availability, play an important part in this regulation. Still, tinkering with well-characterised chemical and genetic variations that affect the chromatin and alter the expression of only many biosynthetic gene clusters, and therefore the regulation of the vast majority of biosynthetic pathways, remains enigmatic (Collemare and Seidl, 2019).

4. Transcriptional and Epigenetic Regulation

Transcriptional and epigenetic regulations are two linked mechanisms that regulate gene expression by altering how accessible DNA is to the recap machinery, with recap factors directly linking to DNA sequences to manipulate gene exertion, while epigenetic differences like DNA methylation and histone differences modify the chromatin structure to laterally regulate the access to genes, which impacts the quantity of protein product within a cell.

Trichoderma reesei pattern C 30 is a well-known robust producer of cellulolytic enzymes, which are used to degrade lignocellulosic biomass for the sustainable production of biofuels and biochemicals. However, studies of its secondary metabolism and regulation remain scarce. Ypr1 was previously described as a controller of the biosynthesis of sorbicillin (a bioactive agent with great pharmaceutical interest) in *T. reesei* and several other fungi. Still, how this controller affects global gene expression has not been explored. The study done by Yang et al. (2023) reports the effect of Ypr1 on the regulation of both the secondary and primary metabolism of *T. reesei* pattern C 30.

A global gene expression profile was attained using a relative transcriptomic analysis of the wild-type strain *T. reesei* pattern C 30 and its ypr1 deletion mutant. The results of this analysis suggest that, in addition to its part in regulating sorbicillin and the major extracellular (hemi) cellulases, Ypr1 also affects the arrangement of genes, reproducing several other secondary metabolites. Although the primary metabolism of *T. reesei* Δ ypr1 became less active compared with that of *T. reesei* pattern C 30, several gene clusters involved in its secondary metabolism were regulated, similar to the gene clusters for the biosynthesis of specific polyketides and nonribosomal peptides, together with the "sorbicillinoid-cellulase" supercluster, indicating that specific secondary metabolites and cellulases may be coregulated in *T. reesei* pattern C 30. The results presented in this study may benefit the development of genetic engineering strategies for the production of sorbicillin by *T. reesei* pattern C 30 and enhance the sorbicillin production in other filamentous fungal species.

The work done by Valero et al. (2021) gives the complete characterization of ZnfA, a transcription factor (TF) that was identified during the previous screening studies that was associated with the *A. fumigatus* responses to calcium and CSP. These transcription factors perform the important function of regulation of iron homeostasis and cell wall arrangement in response to high CSP concentrations as expressed by chromatin immunoprecipitation coupled to DNA sequencing (ChIP-seq) analysis. ZnfA also acts collaboratively with the key transcription factor CrzA in regulating the response to calcium as well as cell wall and osmotic stresses.

According to the research by Lei et al. (2020), four new polyketide derivatives, pestalotiopols A–D, along with seven other compounds, were isolated from a chemical-epigenetic culture of *Pestalotiopsis* sp. The structure and configurations of the discovered compounds were analysed by spectroscopic analysis, M_{O2} induced CD, and electronic circular dichroism (ECD). The compounds that were isolated were analysed for their cytotoxic activities. Among the discovered compounds (pestalotiopols A and B; heterocornols A and E) exhibited cytotoxicity against four human cancer lines with IC_{50} values from 16.5 to 56.5 μM. Wu et al. (2020) performed chemical epigenetic modification on a marine-derived fungus, *Aspergillus terreus* RA2905, with the help of the histone deacetylase inhibitor suberoylanilide hydroxamic acid (SAHA), which resulted in a predominant change in the metabolic profile. The structures with benzyl furanone or benzyl pyrone skeletons were isolated from natural products for the first time by Wu et al. (2020). Compounds (±)-asperpyranones A, (+)-asperpyranones A, (−)-asperpyranones A, and asperpyranones B exhibited antifungal properties against *Candida albicans* with MIC values of 32, 16, 64, and 64 μg/mL and PTP1B inhibitory activities with the IC_{50} values of 45.79, 17.32, 35.50, and 42.32 μM, respectively. Asperpyranones B displayed antibacterial property against *Pseudomonas aeruginosa* with the MIC value of 32 μg/mL.

Epigenetic manipulation of deep ocean deposits derived *Spiromastix* sp. fungus using suberoylanilide hydroxamic acid (SAHA) induction contributed to the activation of a terpene-related biosynthetic gene cluster, and nine new guaiane-type sesquiterpenes, spiromaterpenes A I (1–9) were isolated. Their structures were determined using various spectroscopic ways, in association with the modified Mosher's system, reckoned with ECD gamuts, and chemical conversion for configurational assignments. This study fills the gap in knowledge regarding the anti-neuroinflammatory activity of guaiane-type sesquiterpenoids at the cellular position and suggests that spiromaterpene E is promising for further optimisation as a multifunctional agent for anti-neuroinflammation (Guo et al., 2021).

5. Pathway-Specific Engineering for Enhanced Production

Organisms could be altered in the production of a wide variety of composites by either enhancing endogenous metabolic pathways or by introducing exogenous pathways that are, moreover, espoused from other organisms or *de novo* designed pathways unknown to nature. While overexpression of tailback enzymes and omission of contending pathways remain at the core of metabolic pathway engineering, there are numerous other crucial rudiments that need to be considered to successfully develop strains for the production of precious products (Alper and Avalos, 2018). Therefore, by engineering specific pathways of marine fungi, we would be able to recover advanced yields of natural products. The genome sequencing of *Aspergillus terreus* reveals that the vast number of prognosticated biosynthetic gene clusters have not been reflected by the metabolic profile when observed under conventional culture conditions.

In the study done by Zheng et al. (2024), a silent azaphilone biosynthetic gene cluster was regulated by overexpressing a pathway-specific recap factor gene2642 in marine-derived fungus *A. terreus* RA2905. The bioactivity assay revealed that the composites displayed significant anti-inflammatory activity. Based on the circumstances of different interceders and the apparent gene functions, a presumptive biosynthetic pathway of these composites was proposed. The results demonstrated that overexpression of the pathway-specific recap factor presents a promising approach for perfecting fungal secondary metabolites and accelerating the targeted discovery of new biosynthetic products.

Terrein, a major secondary metabolite from *Aspergillus terreus*, shows great capabilities in biomedical and agrarian operations. Still, the low turmoil yield of terrein in wild *A. terreus* strains limits its artificial operations. Yao et al. (2022) constructed a cell plant grounded on the marine-deduced *A. terreus* RA2905, allowing for overproducing terrein by using bounce as the sole carbon source. Originally, the pathway-specific recap factor TerR was overexpressed under the control of a native gpdA protagonist of *A. nidulans*, performing in 5–16 crowds upregulating in terR reiterations compared to WT. As anticipated, the titer of terrein was altered in the two tested terR OE mutants when compared to WT. Secondly, the global controller gene stuA, which was demonstrated to suppress the terrein conflation in our analysis, was deleted, leading to a greatly enhanced product of terrein.

In addition, LS MS/MS analysis showed that deletion of StuA caused dropped conflation of the major derivative butyrolactones. To achieve an optimal strain, Yao et al. (2022), further refactored the inheritable circuit by combining deletion of stuA and overexpression of terR; an advanced terrein yield was achieved with a lower background of derivations in double mutants.

In addition, it was also set up that loss of StuA (both ΔstuA and ΔstuAOEterR) contributed to a conidial morphology but a slightly faster growth rate than that of

WT. The results demonstrated that refactoring both global and pathway-specific recap factors (StuA and TerR) provides a highly effective strategy to enhance terrein product, which could be espoused for large-scale production of terrein or other secondary metabolites in marine-derived filamentous fungi.

6. Biotechnological Applications of Engineered Marine Fungi

Marine fungi serve as a fat source for the production of multiple metabolites that hold great significance in various fields similar to medicinal, diligence, and bioremediation. By enhancing the product of marine fungi, it would be suitable to meet the utmost of the current demand posed from the below-mentioned fields.

6.1 *Artificial Bioconversion and Enzyme Product*

The natural products produced by marine fungi play a significant part in various diligences. Marine produces various enzymes that serve as catalysts in several artificial processes. In the study done by Vidgren et al. (2020), two marine fungi, a *Trichoderma* sp. and a *Coniochaeta* sp., which can grow on d-galacturonic acid and pectin, were named as hosts to produce the mucic acid product, assessing the ability of marine fungi for the production of platform chemicals. The pathway for biotechnological production of mucic galactaric acid from d-galacturonic acid is simple and requires minimum revision of the genome, optimally one deletion and one insertion. D-galacturonic acid, the main element of pectin, is an implicit substrate for bioconversion, since pectin-rich waste is abundant. *Trichoderma* sp. LF328 and *Coniochaeta* sp. MF729 were altered using CRISPR Cas9 to oxidize d-galacturonic acid to mucic acid, breaking the endogenous pathway for d-galacturonic acid catabolism when fitting a gene reproducing bacterial uronate dehydrogenase. The uronate dehydrogenase was expressed under control of a synthetic expression system, which functioned in both marine strains.

The marine *Trichoderma* transformants produced 25 g/L mucic acid from d-galacturonic acid in equimolar quantities; the yield was 1.0–1.1 g of mucic acid (per g of d-galacturonic acid employed). D-xylose and lactose were the preferred co-substrates. The altered marine *Trichoderma* sp. was more productive than the traditional *Trichoderma reesei* strain (D 161646) described in the literature to date, which had been altered to produce mucic acid. With marine *Coniochaeta* transformants, d-glucose was the preferred co-substrate, but the highest yield was 0.82 g/g a portion of d-galacturonic acid was still metabolised. *Coniochaeta* sp. transformants produced acceptable pectinases to produce mucic acid from pectin, but *Trichoderma* sp. transformants did not. Both marine species were successfully altered using CRISPR Cas9, and the synthetic expression system was functional in both the species.

Ethanologenic white spoilage fungus *Phlebia* sp. MG 60 P2 produces ethanol directly from several lignocelluloses. Effective gene silencing styles are demanded for metabolic engineering of this fungus for biorefinery use. In the study done by Motoda et al. (2019), they estimated the effectiveness of RNAi-intermediated silencing of the pyruvate decarboxylase gene of *Phlebia* sp. MG 60 P2 (MGpdc1). The RNAi lines generated showed a variety of repression situations of ethanol product and MGpdc1 expression, and two named strains led to different metabolic fluxes, performing in rapid-fire accumulation of xylitol from xylose. Gene silencing using RNAi for MGpdc1 in the ethanologenic white spoilage fungus *Phlebia* sp. MG 60 P2 is an effective first step in metabolic engineering to produce other chemicals besides ethanol. This high effectiveness of metamorphosis and silencing effect will make it possible to co-transform with multiple expression vectors, which enhance the minor metabolic pathway or introduce exogenous metabolic response in *Phlebia* sp. MG 60 P2.

Numerous challenges encountered in marine fungi strain engineering stem from the deep lack of understanding about marine fungi physiology, cellular and metabolic processes, nonsupervisory pathways, and its performance of goods on metabolite and enzyme products. This is further complicated by the civilization environment, including turmoil type (e.g., solid-state or submerged turmoil), turmoil vessel, and culture media feedstock as discussed by Jo et al. (2023).

6.2 Pharmaceutical and Biomedical Applications

Marine fungi produce a large amount of new bioactive secondary metabolites, some of them representing new carbon fabrics. Similar secondary metabolites are of applicability as distinguished structures for drugs, including antimicrobial, antiviral, antiprotozoal, antidiabetic, antituberculosis, antioxidant, cytotoxic, antitumour, or anti-inflammatory conditioning (Kushveer et al., 2021). Prenylated indole diketopiperazines represent a different set of alkaloids with complex chemical structures and with a wide range of natural conditioning. Aiming to discover bioactive metabolites with structural novelty, genomic reflection in association with the MS/MS-based molecular networking demonstrated a deep-ocean-derived fungus, *Aspergillus puulaauensis* F77, containing a profile of diketopiperazines. Targeted separation of the desired fungus led to the isolation of 19 undescribed austamide-type diketopiperazines, videlicet versicoines A-S. Their structures were illustrated by the 2D NMR data, in association with Snatzke's method, ECD computations, and single demitasse X-ray diffraction data for configurational assignments. Versicoine N S represents a unique class of austamide-type alkaloids with a spirocenter at C 3. Further mechanistic information demonstrated the significant inhibition of versicoine N against p65 expression and its nuclear transposition, along with the inhibition toward phosphorylation of IKK/IκB in the NF κB signalling pathway. In addition, versicoine N also inhibited NLRP3 inflammasome activation and its affiliated proteins, including caspase 1, pro-caspase 1, IL 1β, and pro-IL-1β. This study largely extends the chemical diversity of austamide-type alkaloids and provides promising lead composites for anti-neuroinflammation (Wu et al., 2025). The research done by Ren et al. (2022) found the compound 5-chloro-6-hydroxymellein as a potent anti-inflammatory compound which is derived along with 15 other compounds from the marine endophytic fungus *Amorosia* sp. SCSIO 41026.

In the study done by Oppong-Danquah et al. (2020), organic extracts of six fungal isolates that belong to the genera *Penicillium*, *Cladosporium*, *Emericellopsis*, and *Plectosphaerella* were analysed for their antimicrobial property against plant and human pathogens and anticancer property. This analysis led to the discovery of new derivatives like new analogs of the antimicrobial tetrapeptide, fungisporin. Regarding the bioactivity, the Penicillium sp. isolate 31.68F1B exhibited a strong and wide range of activity against seven plant and human pathogens, with the phytopathogen *Magnaporthe oryzae* and the human pathogen *Candida albicans* being the most prone candidates (IC50 values 2.2 and 6.3 μg/mL, respectively).

The soft coral-derived fungus *Trichoderma harzianum* (XS-20090075) was identified to be a potential strain to produce substantial new compounds in the study done by Shi et al. (2020). In order to explore its ability to produce more metabolites, chemical epigenetic manipulation was used to wake up the inactive biosynthetic gene clusters, which led to the predominant changes of its secondary metabolites by using a histone deacetylase (HDAC) inhibitor.

The main difference was the original main products, harziane diterpenoids, were transformed into cyclonerane sesquiterpenoids. The fungal culture treated with 10 μM sodium butyrate produced three new terpenoids along with cleistanthane diterpenoid, harzianolic acid A, harziane diterpenoid, harzianone E, and cyclonerane sequiterpenoid, 3,7,11-trihydroxy-cycloneran, with the other 11 sequiterpenoids. The new compounds, the cleistanthane diterpenoid harzianolic acid A, the harziane diterpenoid harzianone E, and the cyclonerane sesquiterpenoid, 3,7,11-trihydroxy-cycloneran, were analysed as antimicrobial agents against various bacterial strains like *Staphylococcus aureus* ATCC 27154, *Escherichia coli* ATCC 25922, *Pseudomonas aeruginosa* ATCC 10145, and one pathogenic fungal strain (*Candida albicans* ATCC 76485).

Fungi are a prospective resource of bioactive composites, but conventional styles of medicine discovery are not effective enough to completely explore their metabolic reservoir. The study done by Khan et al. (2020) aimed to develop a fluently attainable system to evoke the metabolic reservoir of fungi using *Aspergillus nidulans* laeA as a feedback regulation tool. In this study, functional analysis of *A. nidulans* laeA (AnLaeA) and *Aspergillus* sp. Z5 laeA (Az5LaeA) was done in the fungus *Aspergillus* sp. Z5. Heterologous AnLaeA and native Az5LaeA overexpression displayed analogous phenotypic goods and caused an increase in the production of a bioactive emulsion, diorcinol, in *Aspergillus* sp. Z5, which proved the conserved function of this global controller. In particular, heteroexpression of AnLaeA showed a significant impact on the expression of velvet complex genes, diorcinol conflation-related genes, and different feedback factors (TFs). So also, heteroexpression of AnLaeA told the whole genome gene expression of *Aspergillus* sp. Z5 and initiated the upregulation of numerous genes. Overall, these findings suggest that heteroexpression of AnLaeA in fungi serves as a simple and easy system to explore their metabolic eventuality. In relation to this, AnLaeA was overexpressed in the fungus *Penicillium* sp. LC1 4, which contributes to increased production of quinolactacin A. All of the above-mentioned composites produced have antibiotic parcels.

6.3 Bioremediation and Environmental Operations

Marine fungi have the capability to survive in extreme conditions and in different substrates; this makes them ideal in the treatment of waste in various surroundings, consuming the waste as substrates. By enhancing the pathways responsible for the removal of waste material, marine fungi would become ideal agents in the bioremediation of contaminated sites.

The marine-derived fungus *Peniophora* sp. CBMAI 1063 is a hyperactive producer of laccase. Laccase is a multicopper oxidase enzyme that is suitable to oxidize different sweet composites while reducing molecular oxygen to water. Laccases from various environments, such as terrestrial surroundings, have been isolated, purified, and characterised successfully. But still laccases from marine-derived fungus are less known. In this study done by Otero et al. (2025), Pnh_Lac1 (Lac1) derived from the fungus *Peniophora* sp. The CBMAI 1063 was heterologously expressed in *Pichia pastoris*, purified, analysed, and used for the detoxification of synthetic dyes. Lac1 (~72 kDa) exhibited optimal activity at 60°C and pH 3, with good thermostability and high tolerance to essence ions and organic detergents. The Lac 1 decolorized different classes of dyes under low enzyme attention (0.2–0.02 U/mL), with an outstanding performance regarding the decolorization of Indigo Carmine (93 after 2 h) in the presence of syringaldehyde. Lac 1-midddleman system was used to degrade 65% of the azo colour Reactive Black 5 into simpler metabolites, which results in the significant reduction of phytotoxicity. These results conclude that the marine-derived Lac1 is a fungal laccase highly active under low attention, which eventually improves environmental adulterants in biodegradation strategies based on biocatalysis.

Most of the research on the cycling of carbon in the open oceans is concentrated on heterotrophic prokaryotes and eukaryotic phytoplankton, but the potential of pelagic fungi lacks attention. Baltar et al. (2021) performed a global ocean multi-omics analysis of all pelagic fungal carbohydrate active enzymes (CAZymes), which are crucial enzymes in the cycling of carbon. Baltar et al. (2021) studied the condition, expression, diversity, functional bracket, and taxonomic cooperation of the genes producing all pelagic fungal CAZymes from the epi- and mesopelagic environments. Pelagic fungi activate at lower carbon concentrations by an improved rate of CAZyme reiterations per gene. Dothideomycetes in epipelagic and Leotiomycetes in mesopelagic waters, both belonging to the phylum Ascomycota, are the main pelagic fungi responsible for carbohydrate declination in the ocean. Mesopelagic waters have more advanced cornucopia, expression, and diversity when compared to that of epipelagic waters with respect to the distribution of prokaryotic CAZymes. The results of Baltar et al. (2021) reveal a wide range of applications for different types of CAZymes by pelagic fungi, showing an active and therefore largely unexplored participation of fungi in the pelagic carbon cycling, where pelagic prokaryotes and fungi attract different ecological niches, and fungi get fairly more important with depth.

7. Challenges and Future Perspectives

7.1 Challenges and Limitations

There are a number of significant problems that exist with using marine fungi for biotechnology. First is the lack of genetic tools to genetically engineer marine fungi. Sadly, this is the only science that requires the removal of the only usable technology for a less innovative sister organism. This situation does not apply directly to marine fungi or marine fungi commonly studied in academia. The techniques to manipulate and modify genes are primarily available in terrestrial fungi. Using marine fungi for biotechnology is even more challenging because we not only lack efficient genetic tools, but there are also very few functional selection markers available or not known at all to be able to successfully identify modified fungal cells.

Another issue is the low protein expression that often occurs with heterologous expression. When a gene from another species is added to a marine fungus, it is usually expressed at the very low-yielding levels of biosynthesis interest, requiring the researcher to modify or even adapt their condition to express interesting proteins and metabolites. Even worse, a significant portion of our interesting biosynthetic pathways that can be found in marine fungi are stowed away quietly or "silent" at typical laboratory or maybe even marine conditions, as they are still genetically coded, just not expressing. The regulatory mechanisms of these unique metabolic pathways are cloudy or opaque at best, so researchers are unable to discover the procedures to express these pathways to begin producing new compounds or interesting proteins.

The low fermentation yields of many wild fungal strains pose a very real barrier to their commercial industrial application. Even with successful genetic engineering, there is still low production efficiency for the project (which may also require specific or costly co-substrates). Ultimately, and optimally, a project where certain key or metabolic pathways are entirely closed down, so resources can be utilised efficiently, not wasting co-substrates and not producing unwanted byproducts like energy, glutamate, or ammonia. Conversely to wild fungal performance limits in the laboratory of the research study, much of the research community seems to overlook the global potential of species and strains from marine environments with relevance to health and well-being. The focus of live laboratory studies has been on traditional easier-to-study or well-defined fungi, ignoring the inherent worth or reach of unexplored and unique fungi. The lack of focused research limits our real fundamental knowledge of marine fungi and their tangible inherent capabilities that will allow us to truly explore their purpose in overall commercial development.

7.2 Future Perspectives

To overcome the challenges faced by the marine fungi researchers could incline them to innovative approaches that improve existing synthetic biology techniques. CRISPR-based technologies have immense potential to be more flexible and efficient in editing the genomes of marine fungi. The key factor is to identify novel Cas proteins that do not trigger any kind of immune response, thus expanding the therapeutic potential of marine fungal species. The efficiency of marine fungal research could be improved multiple-fold by the integration of Artificial Intelligence (AI) with genetic engineering. AI could be used to analyse enormous amounts of genomic data and support researchers in creating more efficient genetic alterations and guessing which pathways to target for maximum compound production.

Another advancement which could be made is to research these species in their natural habitats so that their metabolism could be understood more deeply and it also helps to decipher the chemical language used by these species to communicate with their surroundings which ultimately helps in uncovering new compounds and gaining deeper knowledge regarding the metabolic processes taking place in these organisms. In the end, innovative culturing methods and better systems for inducing and expressing foreign genes could play a critical role. This will allow researchers to cultivate a wider range of marine life.

8. Conclusion

Marine fungi serve as a resource that has a high potency for biotechnological applications because of their potential to produce a unique and wide range of bioactive compounds and their adaptations to numerous extreme and harsh environmental conditions. The immense potential of marine fungi remains underexplored due to the challenges and complexities in utilising these organisms to their fullest. Synthetic biology has now paved a way to access these organisms, overcoming all the existing challenges and complexities by providing various tools and methodologies to engineer metabolic pathways, enhance metabolite production and discover unique bioactive compounds. This review points out the significant advancements made in accessing marine fungal species using synthetic biology to enhance the metabolite production yields, which could be used in various fields such as bioremediation, bioprocessing, pharmaceuticals, and other metabolites.

Further research is essential to overcome the existing challenges, which are the improvisations to be made in culturing marine fungi outside its native habitat, optimisation of required factors in order to enhance the metabolite production and discovering new isolation techniques to isolate marine fungal species from its native habitat. Further advancements in synthetic biology will help us overcome all these existing challenges and meet all the future requirements. By the integration of marine fungi and synthetic biology, we can create a huge reservoir of untapped potential, which makes these organisms the most valuable resources for sustainable development and cutting-edge tools in biotechnological applications.

References

Adli, M. (2018). The CRISPR tool kit for genome editing and beyond. Nat. Commun., 9: 1911. https://doi.org/10.1038/s41467-018-04252-2.

Agrawal, S., Adholeya, A., Barrow, C.J. and Deshmukh, S.K. (2018). Marine fungi: An untapped bioresource for future cosmeceuticals. Phytochem. Lett., 23: 15–20. https://doi.org/10.1016/J.PHYTOL.2017.11.003.

Alper, H.S. and Avalos, J.L. (2018). Metabolic pathway engineering. Synth. Syst. Biotechnol., 3(1): 1–2. https://doi.org/10.1016/J.SYNBIO.2018.01.002.

Baltar, F., Zhao, Z. and Herndl, G.J. (2021). Potential and expression of carbohydrate utilization by marine fungi in the global ocean. Microbiome, 9(1): 106. https://doi.org/10.1186/s40168-021-01063-4.

Chen, Y., Cai, C., Yang, J., Shi, J., Song, Y. et al. (2022). Development of the CRISPR-Cas9 system for the marine-derived Fungi *Spiromastix* sp. SCSIO F190 and *Aspergillus* sp. SCSIO SX7S7. J. Fungi, 8(7): 715. https://doi.org/10.3390/jof8070715.

Chen, Y., Yang, J., Cai, C., Shi, J., Song, Y. et al. (2023). Development of marker recycling systems for sequential genetic manipulation in marine-derived Fungi *Spiromastix* sp. SCSIO F190 and *Aspergillus* sp. SCSIO SX7S7. J. Fungi, 9(3): 302. https://doi.org/10.3390/jof9030302.

Cochereau, B., Le Strat, Y., Ji, Q., Pawtowski, A., Delage, L. et al. (2023). Heterologous expression and biochemical characterization of a new chloroperoxidase isolated from the deep-sea hydrothermal vent black yeast *Hortaea werneckii* UBOCC-A-208029. Mar. Biotechnol., 25(4): 519–536. https://doi.org/10.1007/s10126-023-10222-7.

Collemare, J. and Seidl, M.F. (2019). Chromatin-dependent regulation of secondary metabolite biosynthesis in fungi: Is the picture complete? FEMS Microbiol. Rev., 43(6): 591–607. https://doi.org/10.1093/femsre/fuz018.

Ding, L., Ren, L., Li, S., Song, J., Han, Z. et al. (2019). Production of new antibacterial 4-hydroxy-α-pyrones by a marine fungus *Aspergillus niger* cultivated in solid medium. Mar. Drugs, 17(6): 344. https://doi.org/10.3390/md17060344.

Dixit, S., Kumar, A., Srinivasan, K., Vincent, P.M.D.R. and Ramu Krishnan, N. (2023). Advancing genome editing with artificial intelligence: Opportunities, challenges, and future directions. Front. Bioeng. Biotechnol., 11: 1335901. https://doi.org/10.3389/FBIOE.2023.1335901/PDF.

Gakuubi, M.M., Ching, K.C., Munusamy, M., Wibowo, M., Lim, C.T. et al. (2022). CRISPR/Cas9 RNP-assisted validation of palmarumycin biosynthetic gene cluster in *Lophiotrema* sp. F6932. Front. Microbiol., 13: 1012115. https://doi.org/10.3389/fmicb.2022.1012115.

Guo, X., Meng, Q., Niu, S., Liu, J., Guo, X. et al. (2021). Epigenetic manipulation to trigger production of guaiane-type sesquiterpenes from a marine-derived *Spiromastix* sp. fungus with antineuroinflammatory effects. J. Nat. Prod., 84(7):1993–2003. https://doi.org/10.1021/acs.jnatprod.1c00293.

Hernandez-Rodriguez, Y., Bullard, A.M., Busch, R.J., Marshall, A. and Vargas-Muñiz, J.M. (2025). Strategies for genetic manipulation of the halotolerant black yeast *Hortaea werneckii* : Ectopic DNA integration and marker-free CRISPR/Cas9 transformation. Microbiol. Spectr., 13(1): 1–10. https://doi.org/10.1128/spectrum.02430-24.

Jo, C., Zhang, J., Tam, J.M., Church, G.M., Khalil, A.S. et al. (2023). Unlocking the magic in mycelium: Using synthetic biology to optimize filamentous fungi for biomanufacturing and sustainability. Mater. Today, Bio., 19: 100560. https://doi.org/10.1016/J.MTBIO.2023.100560.

Khan, I., Xie, W.L., Yu, Y.C., Sheng, H., Xu, Y. et al. (2020). Heteroexpression of *Aspergillus nidulans* laeA in marine-derived fungi triggers upregulation of secondary metabolite biosynthetic genes. Mar. Drugs, 18(12): 652. https://doi.org/10.3390/MD18120652.

Kushveer, J.S., Rashmi, M. and Sarma, V.V. (2021). Bioactive compounds from marine-derived fungi and their potential applications. pp. 91–173. *In*: Sharma, V.K., Shah, M.P., Parmar, S. and Kumar A. (eds.). Fungi Bio-Prospects in Sustainable Agriculture, Environment and Nano-Technology (Vol. 3): Fungal Metabolites, Functional Genomics, and Nano-Technology. https://doi.org/10.1016/B978-0-12-821734-4.00014-9.

Lei, H., Zhang, D., Ding, N., Chen, S., Song, C. et al. (2020). New cytotoxic natural products from the marine sponge-derived fungus *Pestalotiopsis* sp. by epigenetic modification. RSC Adv., 10(62): 37982–37988. https://doi.org/10.1039/D0RA06983C.

Li, C.S., Li, X.M., Gao, S.S., Lu, Y.H. and Wang, B.G. (2013). Cytotoxic anthranilic acid derivatives from deep sea sediment-derived fungus *Penicillium paneum* SD-44. Mar. Drugs, 11(8): 3068–3076. https://doi.org/10.3390/MD11083068.

Li, C.Y., Li, H.T., Shao, Y.T., Guo, X.Y., Li, W. et al. (2024). Regulation of secondary metabolites in the endophytic fungus *Penicillium* sp. KMU18029 by the chemical epigenetic modifier 5-azacitidine. Nat. Prod. Res., 38(4): 581–588. https://doi.org/10.1080/14786419.2023.2183199.

Lyu, H.N., Zhang, J., Zhou, S., Liu, H.W., Zhuang, W.Y. et al. (2021). Heterologous expression of a single fungal HR-PKS leads to the formation of diverse 2-alkenyl-tetrahydropyrans in model fungi. Org. Biomol. Chem., 19(38): 8377–8383. https://doi.org/10.1039/D1OB01501J.

Ma, X., Wei, X., Xing, S., Cheng, X., Lu, J. et al. (2025). Polyketide production in a mangrove-associated fungus *Diaporthe goulteri*, induced by chemical epigenetic modification. Nat. Prod. Res., 1–9. https://doi.org/10.1080/14786419.2024.2441496.

Meyer, V. (2021). Metabolic engineering of filamentous fungi. *In*: Nielsen, J., Stephanopoulos, G. and Lee, S.Y. (eds.). Metabolic Engineering: Concepts and Applications: 13: 765–801. https://doi.org/10.1002/9783527823468.CH20.

Motoda, T., Yamaguchi, M., Tsuyama, T. and Kamei, I. (2019). Down-regulation of pyruvate decarboxylase gene of white-rot fungus *Phlebia* sp. MG-60 modify the metabolism of sugars and productivity of extracellular peroxidase activity. J. Biosci. Bioeng., 127(1): 66–72. https://doi.org/10.1016/j.jbiosc.2018.06.017.

Na, H., Zheng, Y.Y., Jia, Y., Feng, J., Huang, J. et al. (2024). Screening and genetic engineering of marine-derived *Aspergillus terreus* for high-efficient production of lovastatin. Microb. Cell Fac., 23(1):134. https://doi.org/10.1186/s12934-024-02396-z.

Niu, S., Liu, D., Shao, Z., Liu, J., Fan, A. et al. (2021). Chemical epigenetic manipulation triggers the production of sesquiterpenes from the deep-sea derived *Eutypella* fungus. Phytochemistry, 192: 112978. https://doi.org/10.1016/J.PHYTOCHEM.2021.112978.

Oppong-Danquah, E., Passaretti, C., Chianese, O., Blümel, M. and Tasdemir, D. (2020). Mining the metabolome and the agricultural and pharmaceutical potential of sea foam-derived fungi. Mar. Drugs, 18(2): 128. https://doi.org/10.3390/md18020128.

Otero, I.V.R., Haslbeck, M., Santello, L.C., Ferreira, H., Sieber, V. et al. (2025). Heterologous laccase from the marine environment: Purification, characterization, and degradation of synthetic dyes. Biocatal. Agric. Biotechnol., 63: 103485. https://doi.org/10.1016/J.BCAB.2024.103485.

Overy, D.P., Subramaniam, R., Matias De Carvalho Bittencourt, D., Wang, D. et al. (2023). Advances and challenges in CRISPR/Cas-based fungal genome engineering for secondary metabolite production: A review. J. Fungi, 9(3): 362. https://doi.org/10.3390/JOF9030362.

Pinedo-Rivilla, C., Aleu, J. and Durán-Patrón, R. (2022). Cryptic metabolites from marine-derived microorganisms using OSMAC and epigenetic approaches. Mar. Drugs, 20(2): 84. https://doi.org/10.3390/MD20020084.

Ren, X., Chen, C., Ye, Y., Xu, Z., Zhao, Q. et al. (2022). Anti-inflammatory compounds from the mangrove endophytic fungus *Amorosia* sp. SCSIO 41026. Front. Microbiol., 13: 976399. https://doi.org/10.3389/fmicb.2022.976399.

Shi, T., Shao, C.L., Liu, Y., Zhao, D.L., Cao, F. et al. (2020). Terpenoids from the coral-derived fungus *Trichoderma harzianum* (XS-20090075) induced by chemical epigenetic manipulation. Front. Microbiol., 11: 506817. https://doi.org/10.3389/FMICB.2020.00572.

Shi, X., Li, X., He, X., Zhang, D., Quan, C. et al. (2024). Chemical epigenetic regulation secondary metabolites derived from *Aspergillus sydowii* DL1045 with inhibitory activities for protein tyrosine phosphatases. Molecules, 29(3): 670. https://doi.org/10.3390/molecules29030670.

Si, Y., Tang, M., Lin, S., Chen, G., Feng, Q. et al. (2018). Cytotoxic cytochalasans from *Aspergillus flavipes* PJ03-11 by OSMAC method. Tetrahedron Lett., 59(18): 1767–1771. https://doi.org/10.1016/J.TETLET.2018.03.077.

Sun, X. and Su, X. (2019). Harnessing the knowledge of protein secretion for enhanced protein production in filamentous fungi. World J. Microbiol. Biotechnol., 35(4): 54. https://doi.org/10.1007/S11274-019-2630-0.

Tasdemir, D. (2017). Marine fungi in the spotlight: Opportunities and challenges for marine fungal natural product discovery and biotechnology. Fungal Biol. Biotechnol., 4(1): 1–4. https://doi.org/10.1186/S40694-017-0034-1.

Tyagi, S., Kumar, R., Das, A., Won, S.Y. and Shukla, P. (2020). CRISPR-Cas9 system: A genome-editing tool with endless possibilities. J. Biotechnol., 319: 36–53. https://doi.org/10.1016/J.JBIOTEC.2020.05.008.

Valero, C., Colabardini, A.C., de Castro, P.A., Silva, L.P., Ries, L.N.A. et al. (2021). *Aspergillus Fumigatus* ZnfA, a novel zinc finger transcription factor involved in calcium metabolism and caspofungin tolerance. Front. Fungal Biol., 2: 689900. https://doi.org/10.3389/FFUNB.2021.689900/BIBTEX.

Valiante, V. (2023). Advances in synthetic biology of fungi and contributions to the discovery of new molecules. ChemBioChem, 24(11): e202300008. https://doi.org/10.1002/CBIC.202300008.

Vidgren, V., Halinen, S., Tamminen, A., Olenius, S. and Wiebe, M.G. (2020). Engineering marine fungi for conversion of d-galacturonic acid to mucic acid. Microb. Cell Fac., 19(1): 156. https://doi.org/10.1186/s12934-020-01411-3.

Watts, A., Sankaranarayanan, S., Watts, A. and Raipuria, R.K. (2021). Optimizing protein expression in heterologous system: Strategies and tools. Meta Gene, 29: 100899. https://doi.org/10.1016/j.mgene.2021.100899.

Wolff, P.B., Nielsen, M.L., Slot, J.C., Andersen, L.N., Petersen, L.M. et al. (2020). Acurin A, a novel hybrid compound, biosynthesized by individually translated PKS- and NRPS-encoding genes in *Aspergillus aculeatus*. Fungal Genet. Biol., 139: 103378. https://doi.org/10.1016/j.fgb.2020.103378.

Wu, J., Kang, Y., Meng, Q., Jia, H., Liu, D. et al. (2025). Targeted isolation of diketopiperazines from a deep-sea derived fungus with anti-neuroinflammatory effects. Bioorg. Chem., 154: 108099. https://doi.org/10.1016/J.BIOORG.2024.108099.

Wu, J.S., Shi, X.H., Zhang, Y.H., Shao, C.L., Fu, X.M. et al. (2020). Benzyl furanones and pyrones from the marine-derived fungus *Aspergillus terreus* induced by chemical epigenetic modification. Molecules, 25(17): 3927. https://doi.org/10.3390/molecules25173927.

Wu, X., Zhang, T., Zhang, K., Zhang, R., Shi, M. et al. (2024). The forced activation of asexual condition in *Aspergillus niger* simplifies bioproduction. Synth. Syst. Biotechnol., 9(2): 277–284. https://doi.org/10.1016/j.synbio.2024.02.007.

Yang, J., Li, J.X., Zhang, F. and Zhao, X.Q. (2023). Global regulation of fungal secondary metabolism in *Trichoderma reesei* by the transcription factor Ypr1, as revealed by transcriptome analysis. Eng. Microbiol., 3(2). 100065. https://doi.org/10.1016/j.engmic.2022.100065.

Yao, G., Bai, X., Zhang, B., Wang, L., Chen, S. et al. (2022). Enhanced production of terrein in marine-derived *Aspergillus terreus* by refactoring both global and pathway-specific transcription factors. Microb. Cell Fac., 21(1): 131. https://doi.org/10.1186/s12934-022-01859-5.

Zhang, P., Wu, G., Heard, S.C., Niu, C., Bell, S.A. et al. (2022). Identification and characterization of a cryptic bifunctional Type i diterpene synthase involved in talaronoid biosynthesis from a marine-derived fungus. Org. Lett., 24(38): 7037–7041. https://doi.org/10.1021/acs.orglett.2c02904.

Zhang, W., Ling, M., Zhang, K., Liu, R., Huang, X. et al. (2025). Complete genome sequence of the marine mangrove fungus *Sarcopodium* sp. QM3–1 confirmed its high potential for antimicrobial activity. Mar., Genom., 79: 101162. https://doi.org/10.1016/j.margen.2024.101162.

Zheng, Y.Y., Mao, J.Q., Liu, Y., Han, N., Lv, L. et al. (2024). Pleiotropically activation of azaphilone biosynthesis by overexpressing a pathway-specific transcription factor in marine-derived *Aspergillus terreus* RA2905. Bioorg Chem., 153: 107832. https://doi.org/10.1016/j.bioorg.2024.107832.

Index

About the Editors

Kandikere R. Sridhar is an Adjunct Professor (Department of Biosciences, Mangalore University, India) and a Visiting Professor at NITTE (Deemed to be University), Mangalore, India. His research areas are the Diversity and Ecology of Fungi of the Western Ghats and maritime habitats. He was an NSERC fellow and visiting professor at Mount Allison University, Canada; the Helmholtz Centre for Environmental Research at UFZ and Martin Luther University, Germany; and the Centre of Biology, University of Minho, Portugal. He was a recipient of the Dr. Shome Memorial Award (2004), Vice President (2013), President (2018) of the Mycological Society of India (IMSI). He is a Fellow of the Indian Mycological Society, Kolkata (2014), Distinguished Asian Mycologist (2015), Outstanding Leader in Education and Research, Association of Agricultural Technology, Bangkok (2016), and Fellow of the Karnataka Science and Technology Academy, India (2023). He received the Lifetime Achievement Award of MSI (2019) and the Association of Fungal Biologists (AFB), India (2025). He has mentored 25 Ph.D. theses, 5 M.Phil. theses, and 25 MSc projects (mycology, food science, and myriapodology). He was a UGC-BSR Faculty Fellow, India (2014–2017) and in the top 2% of scientists in the field of mycology (2020–2025). He has published over 600 research papers, edited 22 books, and reviewed over 200 research articles, 25 project proposals, 15 book proposals, and has evaluated 25 Ph.D. theses (Australia, India, Mauritius, and Thailand).

Sunil K. Deshmukh is a Scientific Advisor to Greenvention Biotech, Uruli Kanchan, Pune, India, and AGPharm Bioinnovations LLP, Patiala, Punjab, India. He is a veteran industrial mycologist who spent a substantial part of his career in drug discovery at Hoechst Marion Roussel Limited [now Sanofi India Ltd.], Mumbai, and Piramal Enterprises Limited, Mumbai. He has also served at the TERI-Deakin Nanobiotechnology Centre, New Delhi, and as an Adjunct Associate Professor at Deakin University, Australia. He has to his credit 8 patents, 158 publications, and 28 books on various aspects of fungi and natural products of microbial origin. He is the president of the Association of Fungal Biologists (AFB) and a past president of the Mycological Society of India (MSI). Dr. Deshmukh is Managing Editor of *Mycological Spectrum*, the official journal of the Association of Fungal Biologists. He has approximately four decades of research experience dealing with bioactives from fungi, including keratinophilic fungi. He has approximately four decades of research experience in getting bioactives from fungi and various aspects of keratinophilic fungi.

For Product Safety Concerns and Information please contact our EU representative GPSR@taylorandfrancis.com Taylor & Francis Verlag GmbH, Kaufingerstraße 24, 80331 München, Germany